D1921295

Electrocatalysis in Balancing the Natural Carbon Cycle

Electrocatalysis in Balancing the Natural Carbon Cycle

Yaobing Wang

WILEY-VCH

Author

Prof. Yaobing Wang
Fujian Institute of Research on the Structure of Matter CAS
155#, YangQiao West Road
Gulou District
350002 Fuzhou
China

Cover: Courtesy of the author

All books published by **WILEY-VCH** are carefully produced. Nevertheless, authors, editors, and publisher do not warrant the information contained in these books, including this book, to be free of errors. Readers are advised to keep in mind that statements, data, illustrations, procedural details or other items may inadvertently be inaccurate.

Library of Congress Card No.:
applied for

British Library Cataloguing-in-Publication Data
A catalogue record for this book is available from the British Library.

Bibliographic information published by the Deutsche Nationalbibliothek
The Deutsche Nationalbibliothek lists this publication in the Deutsche Nationalbibliografie; detailed bibliographic data are available on the Internet at <http://dnb.d-nb.de>.

© 2021 WILEY-VCH GmbH, Boschstr. 12, 69469 Weinheim, Germany

All rights reserved (including those of translation into other languages). No part of this book may be reproduced in any form – by photoprinting, microfilm, or any other means – nor transmitted or translated into a machine language without written permission from the publishers. Registered names, trademarks, etc. used in this book, even when not specifically marked as such, are not to be considered unprotected by law.

Print ISBN: 978-3-527-34913-5
ePDF ISBN: 978-3-527-83227-9
ePub ISBN: 978-3-527-34916-6
oBook ISBN: 978-3-527-34915-9

Typesetting Straive, Chennai, India
Printing and binding CPI Group (UK) Ltd, Croydon, CR0 4YY

Printed on acid-free paper

C113771_170621

Contents

Preface

In recent years, sustaining the carbon cycle and producing valuable fuels that may considerably reduce dependence on fossil fuels using electrocatalysis has grown. A great deal of interest has arisen, leading to an increase in young researchers entering this area. This book focuses on the aspects of efficient redox catalysis, such as water and carbon dioxide electrolysis toward hydrogen and fuel-based energy systems, respectively, and remarkable energy technologies such as fuel cells and metal–air batteries. An introduction and recent progress are provided, emphasizing reaction conversions developments over various types of electrocatalysts. In addition, the book includes chapters that attract not only experimentalists but also theoretical chemists who have interest in the electrocatalyst design.

The book comprises 10 parts. The first three parts provide a solid introduction for the electrocatalytic carbon cycle addressing water splitting, carbon dioxide reduction, and its counterpart, the oxidation of small organic molecules, followed by detailing each related reaction in separated part: (i) part IV presents water-splitting devices after describing the key fundamental research developments of water oxidation and reduction reactions, (ii) part V includes hydrogen and oxygen-related electrocatalytic aspects with well-defined fuel cell and batteries models, (iii) part VI illustrates the oxidation of C1 and C2+ molecules as up-and-coming advanced power systems, and (iv) part VII reviews the fundamentals in carbon dioxide reduction reaction including well-defined catalytic electrodes as well as the current and next generation of its devices. From the experimentation developments to the theoretical approaches, part VIII focuses on the kinetics and thermodynamics of the reactions aforementioned. For the most popular current motifs, the fundamentals of computational screening, descriptors, and modeling are outlined, as well as their application toward catalyst design direction instead of trial-and-error approach. While emerging the advanced characterizations addressing the in situ techniques become shining, part IX includes modern analytical methods to uncover the surface evolution/reconstruction of a given catalyst under the electrocatalytic conditions, which drive the electrochemical surface science research toward confirming the real active sites of a desired catalytic performance. The book ends with part X evaluating the electrocatalytic carbon cycles and its involved redox reactions. The parts are detailed as follows:

In Chapter 1, *Prof. Yaobing Wang* discusses the motivation for writing this book introducing the natural carbon cycle (NCC) and the emergence of its challenges in multiple environmental and ecological systems due to the overuse of fossil fuels and the increasingly severe energy crisis. *Prof. Yaobing Wang* highlights the electrochemical carbon cycle (ECC) as an endorsed solution for such concerns. In Chapter 2, *Mr. Wei Wang* and *Dr. Jiafang Xie* explain the various aspects in NCC from organic to inorganic cycle and vice versa, as well as anthropogenic carbon standpoint. In Chapter 3, *Miss Zhen Peng* and *Dr. Jiafang Xie* present the possible involved reaction in the ECC including oxygen reduction reaction (ORR), hydrogen evolution reaction (HER) and oxygen evolution reaction (OER), electrochemical carbon dioxide reduction reaction (ECDRR), and small molecule oxidation reaction (MOR) from mechanistic viewpoint and their evaluation parameters such as Faradic efficiency (FE), Tafel slope, current density, and onset potential. In Chapter 4, *Miss Rui Yang* and *Dr. Yiyin Huang* go through water-splitting reactions in various pH media and evaluation of their catalytic parameters such as turnover frequency (TOF), stability, and FE. In Chapter 5, *Miss Rui Yang* and *Dr. Yiyin Huang* specify the regular OER over various electrocatalytic materials and further consider the photo-assisted OER. As HER is involved in water splitting, *Miss Rui Yang* and *Dr. Yiyin Huang* explore it in Chapter 6 over noble and non-noble catalysts, in addition to the overall device. In Chapter 7, *Mr. Zipeng Zeng, Dr. Parameswaram Ganji,* and *Dr. Yiyin Huang* cover the basic parameters for hydrogen oxidation reaction (HOR) and ORR. In addition, they review the catalytic materials, the possible pathways, and the final products of HOR and ORR in Chapters 8 and 9, respectively. In Chapter 10, the focus of *Mr. Zipeng Zeng, Dr. Parameswaram Ganji,* and *Dr. Yiyin Huang* goes to H_2 fuel cell and metal–air batteries as promising devices in which these reactions undergo. Moving to small MOR, *Miss Xueyuan Wang and Dr. Yiyin Huang* outline the measurement conditions in Chapter 11. MOR could be classified into C1 and C2+ reactants; thus, *Miss Xueyuan Wang and Dr. Yiyin Huang* describe C1 (including methane, methanol, and formic acid) and C2+ (including ethanol, glucose, ethylene glycol, and glycerol) reactions and their advances in Chapter 12 and 13, respectively. *Miss Xueyuan Wang and Dr. Yiyin Huang* give a general model and the advantage of the related typical device, the direct liquid fuel cell (DLFC) in Chapter 14. Parallel with schematic content, *Mr. Rahul Anil Borse* and *Dr. Jiafang Xie* elaborate the experimentation fundamentals (Chapter 15), the electrocatalytic advances of the ECDRR on various catalytic materials (Chapters 16 and 17); in addition, the available fabricated devices and their aspects (Chapter 18). Transferring from the experimentations into theoretical approach, *Mrs. Aya Gomaa Abdelkader Mohamed* provides more fundamental insights into the catalytic process addressing the electric double layer, the thermodynamics and kinetics, and the electrode potential effects in Chapter 19. In Chapter 20, *Mrs. Aya Gomaa Abdelkader Mohamed* covers the computational theories (such as DFT) and their principles such as reactivity descriptors and scaling relationships, which are related into the electrocatalytic process. *Mrs. Aya Gomaa Abdelkader Mohamed* explores how these principles can guide the rational design toward high-performance and desired catalyst in Chapter 21. Moreover, *Mrs. Aya Gomaa Abdelkader Mohamed* reveals the computational applications to get deep understanding of the electrocatalytic mechanism for ORR, OER, HER, HOR, ECO_2RR, and various MOR, as well as the electrocatalytic environment, and analyze their kinetics in Chapter 22. Recently, the in situ characterizations have become potential to identify the real active sites. Therefore, *Mr. Mostafa Ragab Hassan* and *Mrs. Aya Gomaa Abdelkader Mohamed* cooperate to give a basic background for the most reported in situ characterizations, which are classified into optical, X-ray, mass, electron-based techniques in Chapter 23. The investigation of a given catalyst in several electrocatalytic reactions by the in situ analysis is addressed in Chapter 24 by *Mrs. Aya Gomaa Abdelkader Mohamed* and *Mr. Mostafa Ragab Hassan* in different aspects such as probing the real active sites, determining the reaction mechanism, evaluating the catalyst stability/decay, and providing interfacial-related insights. Finally, *Dr. Jiafang Xie* proposes the anthropogenic ECC solution to supplement unbalanced NCC from a global viewpoint in which the advances in critical electrocatalysts and performance for ECDRR (Chapter 25) and electrochemical fuel oxidation (Chapter 26) are presented. Then, several key indexes, external managements, and general principles are proposed to evaluate, support, guide the overall efficiency of ECC in Chapter 27.

This multidisciplinary work is not just a reference for the electrochemistry researchers, but also a handy book for advanced graduate-level students in surface science-, engineering-, and theoretical-related courses, especially those with interest in developing novel catalysts for efficient energy conversions, as well as the experienced researchers seeking to expand their scope.

19 November 2020

Yaobing Wang
Fujian Institute of Research
on the Structure of Matter
Chinese Academy of Sciences
Fuzhou, Fujian
China

Acknowledgments

I much acknowledge the dedication of all the authors who worked on this work. Without their effort, the book could not have been finished. My appreciation also goes to Fujian Institute of Research on the Structure of Matter, Chinese Academy of Sciences, my family, my wife, and two daughters, as well as Wiley Publishing, specially Lifen Yang, Katherine Wong, etc.

Part I

Introduction

Electrocatalysis is considered a core technique for sustaining the carbon cycle and producing valuable fuels as an additional benefit. In recent years, a great deal of interest has arisen in efficient redox catalysis, such as water and carbon dioxide electrolysis toward hydrogen and fuel-based energy systems, respectively, and remarkable energy technologies such as fuel cells and metal–air batteries. Understanding the fundamental aspects and the catalytic behavior of such reactions plays a considerable role in further commercializing electrocatalytic energy devices, helping to close the carbon cycle. In this part, we are going to discuss the motivation for writing this book and presenting a preface to its contents. Our hope is that this book will prove useful to researchers already familiar with electrocatalysis but interested in acquiring more insights and in-depth digestion of state of the art of their catalysis research.

Electrocatalysis in Balancing the Natural Carbon Cycle, First Edition. Yaobing Wang.
© 2021 WILEY-VCH GmbH. Published 2021 by WILEY-VCH GmbH.

1
Introduction

Carbon cycle is the basic cycle on earth to maintain all the life forms. In the earth, there are four primary carbon pools [1]. Among them, the natural carbon cycle (NCC) mainly refers to the cyclic change of carbon in the three-carbon pools of atmospheric carbon pool, marine carbon pool, and terrestrial ecosystem carbon pool [1–4]. The atmospheric carbon has a direct influence on human life; therefore, it attracts great attention from researchers. The carbon in the atmospheric carbon pool mainly exists in the form of CO_2 gas. The basic process of the NCC can be expressed as follows [1, 5]: CO_2 in the atmosphere is solidified into organic carbon through photosynthesis of plants and stored in plants. Part of the organic carbon in plants releases CO_2 into the atmosphere through the plant's respiration (i.e. autotrophic respiration), the consumption of organic carbon by animals, and the decomposition of organic matter by microorganisms (i.e. heterotrophic respiration), forming a terrestrial ecology system carbon cycle process.

Since the industrial revolution and with the rapid world population/economic expansion, people utilized more and more fossil fuels for providing raw materials and electrical power, etc. [6]. Challenges in multiple environmental and ecological are emerging due to the overuse of fossil fuels and the increasingly severe energy crisis [7]. The results are that the NCC has been increasingly broken, leading to unavoidable sustainability in energy and environment, threatening the survival of human society. Thus, various strategies and various renewable energy technologies have been developed from all aspects to solve the broken NCC and maintain the sustainability of human society and the economy [8, 9]. These techniques include fuel cells, CO_2 electrolysis, metal-air batteries, water splitting, and so on. All of these techniques consist of the kernel and/or secondary components in artificial nature carbon cycle (ACC) to supplement for the NCC with synergistic effects [5, 7]. These techniques are mainly powered by solar-derived electricity, which were also defined as the artificial electrochemical carbon cycle (ECC). The ECC mainly involves electrochemical oxidation of chemicals and fuels into CO_2 (CO_2 liberation) and electrochemical reduction of CO_2 into value-added chemicals/fuels (CO_2 fixation), and also other fuel storage and transport, and other secondary reactions for supporting the carbon-based electrochemical reactions.

Electrocatalysis in Balancing the Natural Carbon Cycle, First Edition. Yaobing Wang.
© 2021 WILEY-VCH GmbH. Published 2021 by WILEY-VCH GmbH.

For the realization of ECC, the extensive fundamental and utilitarian electrochemical processes, including oxygen reduction reaction (ORR), oxygen evolution reaction (OER), small organic molecule oxidation reaction, hydrogen evolution reaction (HER), hydrogen oxidation reaction (HOR), and electrochemistry carbon dioxide reduction reaction (ECDRR), were involved. In the water splitting processes [10], water oxidation (OER) occurs on the anode, and four-electron transfer is needed for a complete OER. At the same time, the generated electron and proton will be combined on the cathode and releasing H_2 from the cathode. In various fuel cells [11, 12], HOR via the two-electron transfer process and various small organic fuel (ethanol, formic acid, methanol, glucose oxidase, ethylene, glycerol, glycol, etc.) oxidation occur on the anode, while ORR takes place on the cathode. The cathode of metal-air batteries [13] also employs ORR processes. For the CO_2 electrolysis [14], CO_2 was electrochemically transformed into various fuels, as mentioned earlier, via different electron-transfer processes.

The generally sluggish reaction kinetics is always a bottleneck that limits the overall performance of the new energy devices, hindering their progress of commercialization [15, 16]. To drive these electrochemical processes, electrocatalysts are required. Noble metal catalysts are widely used in these electrocatalytic processes due to their high activity and stability. On the other hand, the high cost hinders their commercialization. Various non-noble metal catalysts were also developed, such as carbon materials, polymer, transition metal materials, and metal-organic materials [15]. With the development of nanotechnology and nanoscience over the past decades, the research mode for developing electrocatalysts has shifted gradually from the traditional trial-and-error methods to the accurate design and fabrication of nanocatalysts at atomic and molecular levels [15, 17, 18]. Besides, other factors in the electrochemical devices, such as the device structure, electrolyte, electrode configuration, and operation temperate, should be considered toward the high performance of the devices. Among these controllable factors, the electrocatalyst design is still among the core factor. To achieve the rational design of electrocatalyst for highly efficient electrocatalytic reaction processes, studies on the active sites' recognition, reaction mechanism, and kinetic and thermodynamic processes should be conducted. In this sense, computational methods combined with in situ characterization techniques allowed the researcher to realize an in-depth and comprehensive understanding of realistic reaction conditions into the nature of the active sites and its interaction with reactants, intermediates, and products and the final overall catalytic processes.

In this book, we will discuss the reaction mechanism and core reaction parameters (e.g. turnover frequency [TOF], onset potential or overpotential, stability, Faradaic efficiency, partial current density) of these electrochemical reactions strongly to the ECC and summarize the advances of various catalysts in terms of the categories to gain an overview on the design principles for electrocatalysts toward various electrochemical reactions. The device categories and advances will also be summarized, with respect to the electrolyte, device structure, electrode, and external environment controls. Then, theoretical calculations for these electrocatalytic reactions were introduced in terms of background, concepts, processes, and applications. Besides, an overview of the common and the most crucial in situ characterization techniques was summarized to assist the theoretical calculations study and help the electrocatalyst design. Further, we have summarized the advances on electrochemical reactions highly related to the ECC, that is, ECDRR, and fuel oxidation for the chemical conversions including CO_2/CO, CO_2/HCOOH, CO_2/CH_3OH, and CO_2/CH_3CH_2OH, along with presenting the mechanistic understanding and proposed key indexes, general principles, and external managements for evaluating and optimizing the overall ECC efficiency. Finally, current challenges and future perspectives for promoting ECC to supplement NCC were concluded. It is believed that this book will provide a comprehensive, deep-going, and cutting-edge introduction on the ECC and related electrocatalysis.

References

1 Schmitz, O.J., Wilmers, C.C., Leroux, S.J. et al. (2018). *Science* 362: eaar3213.
2 Reichstein, M., Bahn, M., Ciais, P. et al. (2013). *Nature* 500: 287–295.
3 Siedler, G., Griffies, S., and Church, J. (2013). *Ocean Circulation and Climate: a 21st Century Perspective*, 257–282. International Geophysics.
4 IPCC (Intergovernmental Panel on Climate Change) (2001). Climate change 2001: the scientific basis. *KSCE J. Civil Eng.* 19 (2): 359–365.
5 Ross, M.B. (2019). *Joule* 3: 1814–1816.
6 Owusu, P.A., Asumadu-Sarkodie, S., and Dubey, S. (2016). *Cogent Eng.* 3 (1), 1167990.
7 Fukushima, T., Kitano, S., Hata, S., and Yamauchi, M. (2018). *Sci. Technol. Adv. Mater.* 19: 142–152.
8 MacElroy, J.M. (2016). *AMBIO* 45 (Suppl 1): S5–S14.
9 Mohsin, I., Al-Attas, T.A., Sumon, K.Z. et al. (2020). *Cell Rep. Phys. Sci.* 1: 100104.
10 You, B. and Sun, Y.J. (2018). *Acc. Chem. Res.* 51: 1571–1580.
11 Hacker, V. and Mitsushima, S. (2018). *Fuel Cells and Hydrogen: From Fundamentals to Applied Research*. Elsevier.
12 Mohammed, H., Al-Othman, A., Nancarrow, P. et al. (2019). *Energy* 172: 207–219.
13 Zhang, X.-b. (2018). *Metal-Air Batteries: Fundamentals and Applications*. Wiley.
14 Qiao, J., Liu, Y., and Zhang, J. (2016). *Electrochemical Reduction of Carbon Dioxide: Fundamentals and Technologies*. Boca Raton, USA: CRC Press.
15 Tang, C., Wang, H.F., and Zhang, Q. (2018). *Acc. Chem. Res.* 51: 881–889.
16 Chen, P., Tong, Y., Wu, C., and Xie, Y. (2018). *Acc. Chem. Res.* 51: 2857–2866.
17 Li, Z., Ma, X.F., and Xin, H.L. (2017). *Catal. Today* 280: 232–238.
18 Jin, Z., Wang, L., Zuidema, E. et al. (2020). *Science* 367: 193–197.

Part II

Natural Carbon Cycle

Nature's photosynthesis uses green plants to absorb carbon dioxide (CO_2) from the atmosphere, convert it into glucose and release oxygen with the participation of water, and organisms reuse glucose to synthesize other organic compounds. Only when enough geological time is given can new fossil fuels be formed naturally. Due to the continuous massive consumption of fossil fuels by human beings, a large amount of CO_2 is produced. These anthropogenic CO_2 emissions exceed the recovery capacity of natural CO_2, causing serious damage to the environment. In order to supplement the natural carbon cycle, the researchers proposed and developed a feasible chemical cycle of anthropogenic CO_2. Carbon dioxide can be captured from the atmosphere or industrial production through absorption technology. Then it can be converted into fuel through feasible chemical conversion. For example, by electrochemical reduction method, CO_2 can be efficiently converted into reusable chemical products under normal temperature and pressure, such as carbon monoxide (CO), formic acid (HCOOH), methane (CH_4), ethylene (C_2H_4), ethanol (CH_3CH_2OH), etc. The required renewable raw materials, water, and CO_2 can be used anywhere on earth. The energy required for the synthetic carbon cycle can come from any alternative sustainable clean energy, such as solar energy, wind energy, geothermal energy, and even safe nuclear energy. When fossil fuels become scarce, the anthropogenic CO_2 cycle provides a way to ensure a sustainable future for humanity.

Electrocatalysis in Balancing the Natural Carbon Cycle, First Edition. Yaobing Wang.
© 2021 WILEY-VCH GmbH. Published 2021 by WILEY-VCH GmbH.

2

Natural Carbon Cycle and Anthropogenic Carbon Cycle

2.1 Definition and General Process

Carbon is an essential part of all life forms on Earth. There are four main carbon pools in the earth system: atmospheric carbon pool, marine carbon pool, terrestrial ecosystem carbon pool, and lithosphere carbon pool. Among them, the lithospheric carbon pool mainly exists in the earth's rocks and its cycle period is the geological age scale, which is up to millions of years. It can be considered that the lithospheric carbon pool is fixed on the scale of hundreds of years, so the natural carbon cycle mainly refers to the cyclic change of carbon in the three-carbon pools of atmospheric carbon pool, marine carbon pool, and terrestrial ecosystem carbon pool. The size of the atmospheric carbon pool is about 700 GtC, which is the smallest carbon pool among the three-carbon pools, but because the atmosphere directly affects human life, the atmospheric carbon pool was the first to attract people's attention. The carbon in the atmospheric carbon pool mainly exists in the form of CO_2 gas. The ocean, which accounts for 71% of the earth's surface, is a huge carbon pool with carbon storage of about 38 000 Gt. It is more than 50 times the atmospheric carbon pool and 20 times the terrestrial carbon pool, the largest of the three-carbon pools. The primary forms of carbon in the ocean are dissolved inorganic carbon, dissolved organic carbon, carbonate, particulate organic carbon, etc., of which more than 97% is dissolved in the form of inorganic carbon. The carbon storage of terrestrial ecosystems is about 2000 Gt, of which the carbon storage of living organisms is 600–1000 Gt and the storage of soil organic carbon such as biological residues is about 1200 Gt. The basic process of the terrestrial ecosystem carbon cycle is: CO_2 in the atmosphere is solidified into organic carbon through photosynthesis of plants and stored in plants [1, 2]. The chemical expression of photosynthesis is formula (Figure 2.1). Part of the organic carbon in plants releases CO_2 into the atmosphere through the plant's own respiration (i.e. autotrophic respiration), the consumption of organic carbon by animals, and the decomposition of organic matter by microorganisms (i.e. heterotrophic respiration), forming a terrestrial ecology system carbon cycle process. In short, the natural carbon cycle includes carbon chemical and biological processes and is a comprehensive and complex system. In this article, we focus on the chemical processes of the carbon cycle.

Electrocatalysis in Balancing the Natural Carbon Cycle, First Edition. Yaobing Wang.
© 2021 WILEY-VCH GmbH. Published 2021 by WILEY-VCH GmbH.

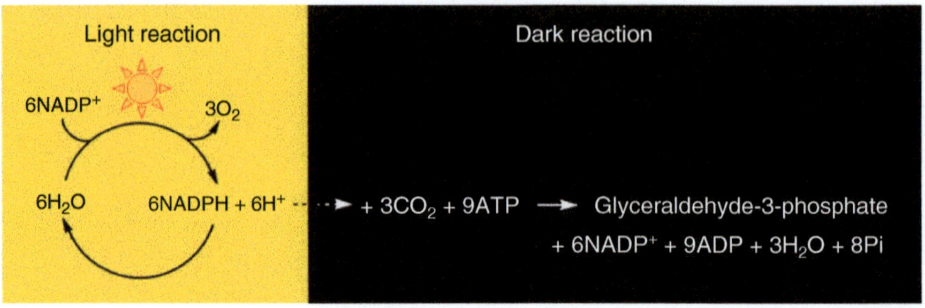

Figure 2.1 Light and dark reactions of photosynthesis. Source: Dogutan and Nocera [3]. © 2019, American Chemical Society.

2.2 From Inorganic Carbon to Organic Carbon

In the natural carbon cycle, the energy of the sun is used by vegetation, plankton, algae through photosynthesis, along with the recovery of carbon dioxide from the natural environment. In this process, water serves as a source of hydrogen and green chlorophyll as a catalyst, which ultimately creates new plant life. It will eventually be converted into fossil fuels over millions of years [4].

Photosynthesis occurs in the chloroplasts of plant cells and is divided into two stages: the light reaction stage and the dark reaction stage [3]. Under the catalysis of the photosynthetic system II (Photosystem II), water is broken down into electrons, oxygen, and protons (photoreaction, Figure 2.1). Among them, electrons are transported along the electron transfer chain to Photosystem I (Photosystem I) through a series of electron transport substances and participate in the generation of reduced coenzyme II (triphosphopyridine nucleotide, nicotinamide adenine dinucleotide phosphate [NADPH]), which is used to fix carbon dioxide in the subsequent dark cycle (dark reaction, Figure 2.1). The whole process involves not only the synthesis and decomposition of various substances but also various energy conversions: from light energy to electrical energy, to biological energy, and finally to chemical energy. This series of processes achieves energy conversion and cleverly fixes solar energy into green plants.

The hydrogen evolution reaction (HER) is a simple $2e^-/2H^+$ process, which is kinetically easier than reducing the multi-electron/multi-proton of carbon dioxide to biomass. Therefore, systems that directly reduce carbon dioxide in water face the challenge of suppressing HER with greater kinetic advantages and competitiveness. Photosynthesis does not use H_2O but uses H_2 equivalent ($NADPH/H^+$) to reduce CO_2, thereby avoiding the HER during CO_2 reduction. However, photosynthesis is relatively inefficient in converting solar energy in the form of sugar, cellulose, lignin, etc. into chemical energy. In spite of some green plants enable about 8% of solar energy to be converted into biomass, the photosynthetic efficiency of most crops is usually limited to 0.5–2%. Substances produced by photosynthesis can eventually be transformed into fossil fuels under anaerobic conditions through the decay of animals and plants. Therefore, fossil fuels are also often regarded as stored fossil solar energy. However, the natural circulation of regenerated fossil fuels requires certain special conditions to proceed, and this process is prolonged.

Carbon dioxide can be used as the only carbon source for many microorganisms found in the microbial world. There are several ways and solutions for organisms to reduce carbon dioxide to organic carbon. The most important one is Calvin reducing the pentose phosphate pathway. Throughout the evolutionary process, the immobilization, reduction, and reconstruction of CO_2 receptor molecules will be involved. Therefore, some autotrophic bacteria, almost all photosynthetic bacteria, eukaryotic algae, and some prokaryotes and green plants absorb carbon dioxide through the Calvin cycle. Reactions unique to this pathway are catalyzed by ribulose 1,5-bisphosphate (RuBP) carboxylase/oxygenase, and phosphoribulokinase (PRK). In some cases, there is a distinct sedoheptulose 1,7-bisphosphatase, separate from fructose 1,6-bisphosphatase, that may be considered unique to the Calvin cycle [5]. Although the Calvin reductive pentose phosphate pathway is the major assimilatory path used in the biosphere, many autotrophic species fix CO_2 by different routes. In particular, the acetogenic bacteria and the methanogens reduce CO_2 to acetate (and other short-chain fatty acids) or methane, respectively. The green photosynthetic bacteria appear to be unusual among photosynthetic organisms in not using the Calvin cycle to reduce CO_2.

2.3 From Organic Carbon to Inorganic Carbon

Carbon dioxide gas is part of the atmosphere (about 0.03% of the total volume of the atmosphere) and is abundant in nature. The main ways of producing carbon dioxide in nature are: the respiration of plants and animals converts part of the organic carbon taken into the body into carbon dioxide and releases it into the atmosphere, and the other part constitutes the organism's body or is stored in the body. After the death of animals and plants, the organic carbon in the debris is also converted into carbon dioxide by the decomposition of microorganisms and finally discharged into the atmosphere. The cycle of carbon dioxide in the atmosphere takes about 20 years.

2.4 Anthropogenic Carbon Cycle

By imitating the natural carbon cycle, humans have developed an artificial model to utilize CO_2 present in the atmosphere and water (Figure 2.2) [6]. The anthropogenic carbon cycle is based on the capture, and where needed, temporary storage of CO_2 followed by its chemical/physical recycling (carbon capture and recycling) into products. Compared with the natural carbon cycle, the anthropogenic carbon cycle can be industrially scaled to produce sustainable, renewable, and environmentally friendly carbon sources. At the same time, human-made carbon dioxide recycling can reduce the harmful effects of excess carbon dioxide in the atmosphere in the global environment [7]. Therefore, carbon dioxide should be regarded as a valuable industrial C1 raw material, not just a greenhouse gas that harms the global environment.

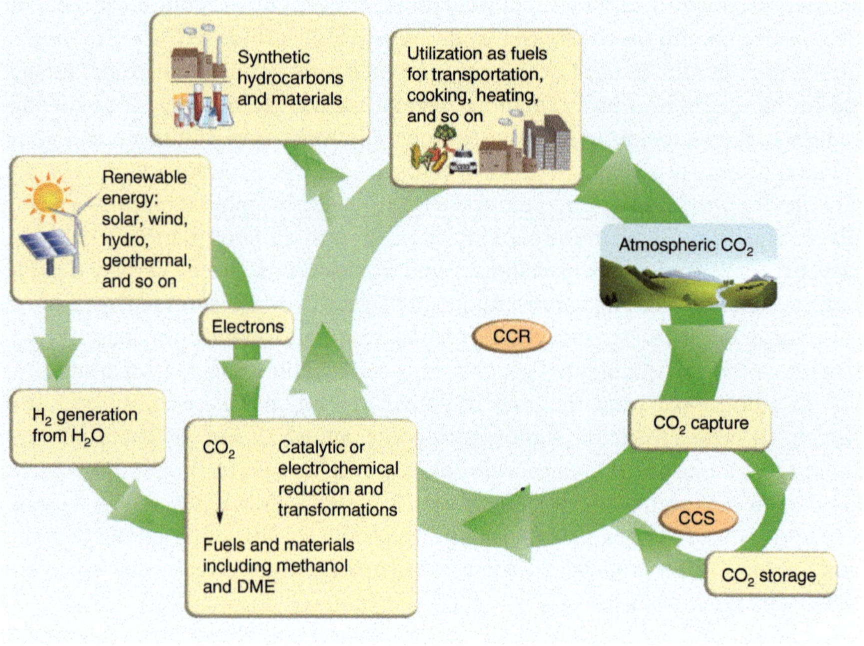

Figure 2.2 Sustainable anthropogenic recycling of atmospheric CO_2 to fuels and materials. CCR, carbon capture and recycling; CCS, carbon capture and storage; DME, dimethoxyethane. Source: Goeppert et al. [6]. © 2013, Newslands Press Ltd.

2.4.1 Anthropogenic Carbon Emissions

Anthropogenic carbon emissions refer to various forms of carbon emissions (mainly carbon dioxide) resulted from human activities. Human activities that increase net emissions include but are not limited to the burning of fossil fuels, logging of forests, changes in land-use patterns, livestock, and fertilizers. The concentration of CO_2 in the atmosphere has kept steadily increasing in the last 60 years, rising to 416 ppm from 315 ppm [8]. Global CO_2 emissions from fossil fuel combustion and industrial activities have increased every 10 years, from an average of 11.4 Gt CO_2 per year in the 1960s to an average of 34.7 ± 2 Gt CO_2 per year from 2009 to 2018. Emissions in 2018 reached a record 36.6 ± 2 Gt CO_2; the emissions from the burning of coal, oil, natural gas, cement, and natural gas account for 40%, 34%, 20%, 4%, and 1% of the total global emissions, respectively [8]. The anthropogenic CO_2 emissions mainly come from scattered small emission sources, such as heating and cooling of houses and offices, agricultural production, and the most important transportation industry. Now, it is generally believed that the excessive release of carbon dioxide along with other greenhouse gases is caused by human activities and is the main factor leading to global warming. They also caused other serious environmental damage, including ocean acidification, sea-level rise, more frequent extreme climate events, and unpredictable changes in biodiversity. Therefore, many efforts have been made worldwide to reduce carbon dioxide emissions of carbon dioxide in the atmosphere, including the famous Paris Agreement in the United Nations Framework Convention on Climate Change. Governments have promulgated laws to limit carbon dioxide emissions in industrial production. At the same time, individuals are encouraged to use clean energy and purchase clean energy vehicles that use non-gasoline fuel. In addition, scientists are actively studying the industrial application of CO_2.

2.4.2 Capture and Recycle of CO_2 from the Atmosphere

The bulk of carbon dioxide emissions comes from scattered, small-scale sources. It would be difficult and expensive to collect CO_2 from millions of or even billions of small fossil-fuel combustion plants [4]. For example, the cooling or heating of homes and offices produces a widely dispersed and limited amount of carbon dioxide. The capture and transportation of these CO_2 will require a large amount of expensive infrastructure, which is obviously impractical. Therefore, in order to deal with small-scale CO_2 emission sources and to avoid the development and construction of huge CO_2 collection infrastructure, capturing CO_2 from the air is an ideal strategy [9, 10]. The atmosphere can be considered a medium that transmits CO_2 emissions to its capture location; it is like a free "natural carbon dioxide conveyor belt." Since the concentration of CO_2 in the air in the world is in equilibrium, this will make the collection site of CO_2 not affected by the source of CO_2 emissions and can capture CO_2 from a variety of sources.

Alkaline absorbents (such as NaOH or KOH) can be used to capture CO_2 from the atmosphere. These absorbents react with CO_2 to form Na_2CO_3 and K_2CO_3, respectively. The absorption of CO_2 is an exothermic reaction and can be easily implemented by mixing CO_2 with an appropriate alkali. However, desorption, as the opposite step, is an endothermic process, so additional energy is required to regenerate the alkali and recover the CO_2. Both $CaCO_3$ and Na_2CO_3 require a lot of energy to recycle alkali, so they are not suitable for capturing and releasing CO_2. KOH is considered to be a promising absorbent. Electrolysis of K_2CO_3 in water can effectively produce CO_2, accompanied by H_2 production.

The physical absorption of carbon dioxide by solids is accomplished by the reversible adsorption of certain substances in the mixture on the solid surface. Solids with good physical adsorption properties include silica gel, zeolite, alumina, activated carbon, and other materials with large pore size distribution and specific surface area. Among them, some zeolite adsorbents show high CO_2 absorption capacity at room temperature (13X zeolite, 160 mg CO_2 g^{-1} zeolite, and 4A zeolite, 135 mg CO_2 g^{-1} zeolite) [11]. However, these adsorbents all have the problem that as the increase of temperature, the adsorption property weakens speedily. In addition, since the gas is fixed on the adsorbent by physical adsorption, the separation coefficient between the gas of different components is very low. This makes it impractical to capture CO_2 from a mixed gas source containing low CO_2 content. Therefore, in practical use, the adsorbent is often required to have high adsorption capacity and easy regeneration and high selectivity for CO_2. Currently, the metal–organic framework (MOF) has also been found to have high CO_2 storage capacity. For example, MOF-177 consists of zinc clusters and 3,5-triphenyltricarboxylic acid triphenyl ester units. Its surface area is up to 4500 m^2 g^{-1}. At 30 atm, the CO_2 storage per gram of MOF is about 1.47 g [12, 13]. However, under lower pressure and mixed gas conditions, the absorption capacity of MOF is more restricted.

2.4.3 Fixation and Conversion of CO_2

Much of the CO_2 used today is obtained not only as a byproduct of industrial processes such as ammonia production, oil and gas refineries, ethanol production, and chemical manufacturing but also natural sources such as geological formations. However, any CO_2 source, including CO_2 captured from fossil fuel-burning power plants and the atmosphere, could be used. The products for which CO_2 can be used as a feedstock are by no means limited to the ones obtained presently from CO_2. Essentially, all the fuels and materials currently derived from fossil fuels could potentially be made from CO_2 through various synthetic pathways. The utilization and recycling of CO_2 through chemical, electrochemical, photochemical, and other routes have been extensively covered and reviewed in light of the increasing concerns about climate change [14–19].

2.4.3.1 Photochemical Reduction

Due to the diffuse and intermittent nature of sunlight, the massive utilization of solar power to satisfy human demands may need the reserve of energy in serried portable media through chemical bonds. At present, there is still no low-cost and efficient technology to directly convert sunlight into useful fuel, and new technology needs to be developed. Storing solar energy as a usable fuel through photosynthesis provides a blueprint for researchers. In fact, all fossil fuels consumed by humans today come from solar energy stored by photosynthetic organisms. Inspired by natural photosynthesis, photo-generated electrons and holes can be used to drive chemical reactions without collecting electrical energy. Therefore, the collected solar energy can be converted into chemical energy stored in the form of chemical bonds. A multi-step process that mimics the function of natural photosynthesis is often called artificial photosynthesis (AP) [20].

The process of optical system and photosynthesis is extremely complicated. Its structure and components cannot be accurately modeled. However, functions and chemical processes can be replicated to enable APs to transform solar energy into chemical energy and store it as chemical fuel (Figure 2.3) [21]. Solar energy can be directly converted into chemical fuel through a photoelectric catalytic reaction, or the electrical energy generated by solar energy can be used to drive an electrocatalytic reaction, thereby indirectly converting solar energy into fuel. Hydrolysis to produce hydrogen fuel and CO_2 reduction to produce carbon-neutral fuel is widely studied artificial photosynthesis reactions.

In artificial photosynthesis, CO_2 is reduced to produce fuel or valuable compounds by simulating photosynthetic organisms [22]. At the same time, in order to make the compounds economically attractive and environmentally friendly, the raw materials for the synthesis of these compounds must be cheap and readily available (such as water and carbon dioxide) [23]. At present, in some of the best systems, the

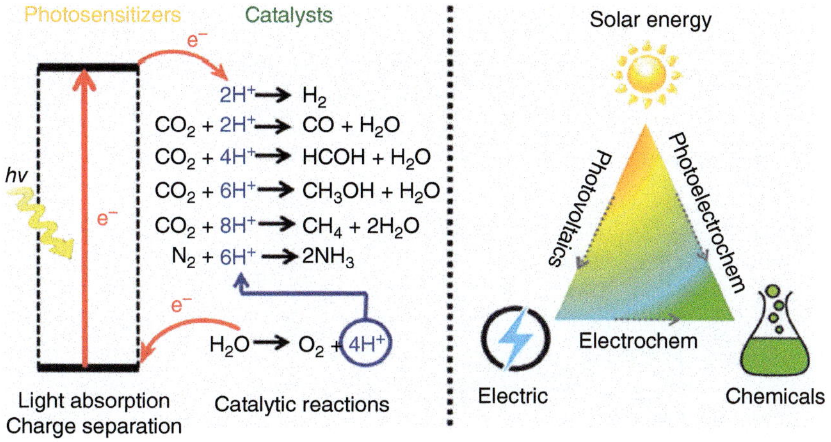

Figure 2.3 Schematic representation of the general concept of artificial photosynthesis. Source: Zhang et al. [21]. © 2019, The Royal Society of Chemistry.

total quantum yield can reach 40%. However, despite some progress in all aspects of artificial photosynthesis, there are still some unresolved problems with the selectivity of the reaction, especially the limited efficiency. Therefore, these issues must be resolved before evaluating the photoreduction potential to reduce CO_2 emissions [23, 24].

2.4.3.2 Electrochemical Reduction

The electrochemical reduction technology can be carried out at normal temperature and pressure, and the product selectivity and reaction rate can be adjusted by simply changing the electrolysis conditions. Electricity is the most important output energy among renewable energy (excluding biomass conversion), but it still accounts for only a small proportion in the chemical industry today, and there is still much room for development. As shown in Figure 2.4, natural, sustainable energy sources such as wind, solar, and tidal energy generate electricity. Electric power resources drive electrochemical cells to convert CO_2 and water into some value-added fuels and chemical products, called electric fuels. These generated electric fuels can be simply stored, distributed, or consumed. At the same time, CO_2 is released as the main waste. This released CO_2 will be captured and fed back to the reactor to make the carbon cycle form a closed loop. This technology successfully realized the recycling of CO_2 and effectively suppressed CO_2 emissions, thereby reducing the greenhouse effect. In addition, value-added chemical raw materials and fuels generated by electrochemical CO_2 reduction reaction (CRR) can effectively reduce the dependence on traditional fossil fuels. Although CO_2 electroreduction is of great significance in energy and environmental problems, difficulties and challenges still exist and deserve attention. CO_2 molecule has a very stable performance due to its linear structure and chemical inertness. The C=O bond only breaks under the excitation of high-energy electrons or photons and requires a high barrier to the activation energy, which ultimately leads to a significant overpotential. At the same time, the

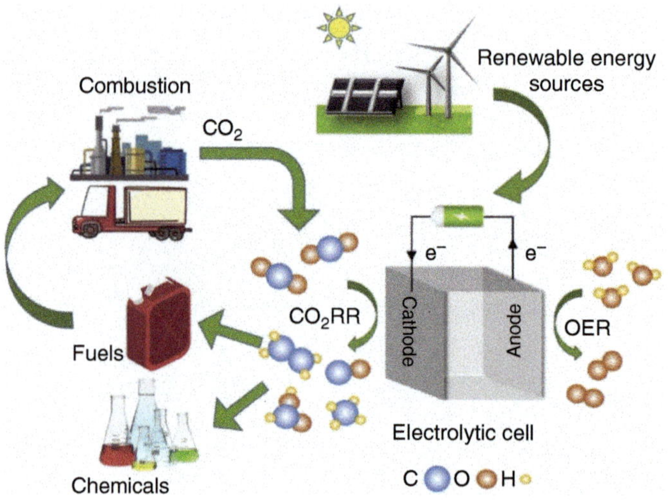

Figure 2.4 Schematic diagram of the clean energy supply of electricity to electrochemically reduction CO_2 to chemical fuel. Source: Liang et al. [25]. © 2019, Elsevier.

HER as a competitive path poses a threat to CO_2 reduction. The potential of HER is usually relatively close to that of CO_2 reduction. The lower kinetic performance and the accompanying side reactions hinder the effective reduction of CO_2. On the other hand, during the CO_2 electroreduction process, the multi-electron transfer mechanism leads to the diversity of products: carbon monoxide, methane, formic acid, methanol, ethylene, etc. However, the thermodynamic reaction potentials between different reaction paths are usually very close, and the similarity of reaction potentials directly leads to a lower selectivity for specific products. Because of this, there is a challenge in wanting to efficiently produce a specific product.

2.4.3.3 Chemical/Thermo Reforming

Carbon dioxide and methane can produce syngas through a "dry reforming" process. The reaction is strongly endothermic and is carried out at a temperature of about 800–1000 °C using a nickel-based catalyst (Ni/MgO, Ni/$MgAl_2O_4$, etc.) [15, 24, 26–28]. In the double reforming process, by combining CO_2 (dry reforming) and H_2O (steam reforming), methane can be directly converted into syngas [15]. In actual use, the primary source of methane is natural gas. In addition to methane, there are other hydrocarbons with different contents present in natural gas. These hydrocarbons can also be double-integrated to produce syngas. The main problem hindering the application of the dry reforming industry is the generation of coke, and the formation of coke will rapidly deactivate the reforming catalyst. Unless used at very high temperatures (beyond 900 °C), the generation of coke is thermodynamically advantageous.

2.4.3.4 Physical Fixation

The CO_2 capture, utilization, and storage (CCUS) technology has great potential to achieve a large-scale reduction of atmospheric CO_2 concentration in a short period of time [29]. Generally speaking, CCUS technology mainly includes the following steps: capture, transportation, utilization, or storage. How to achieve cheap and efficient CO_2 capture and a large number of safe storages is the key to the smooth implementation of this technology. After CO_2 is captured, in order to ensure that it is not discharged into the atmosphere again in a short time, it needs to be sealed up. Carbon dioxide storage refers to capturing, compressing, and transporting carbon dioxide generated by large-scale emission sources to a selected location for long-term storage, rather than releasing it into the atmosphere. There are various methods of CO_2 storage, but generally speaking, they can be divided into two categories: marine storage and geological storage. Especially geological storage is getting more and more attention and research.

To transport the captured CO_2 to the utilization or storage site, it is necessary to select an appropriate transportation method according to the specific conditions, requirements, and transportation volume of the system. The transportation state of CO_2 can be gas, supercritical, liquid, and solid, but from the perspective of the feasibility of large-scale transportation, fluid state (gaseous, supercritical, and liquid) CO_2 is convenient for large-scale transportation and pipeline transportation usually adopts supercritical state. The CO_2 transportation methods that have been practiced mainly include pipeline transportation, ship transportation, and tank truck transportation. These three modes of transportation are suitable for different transportation occasions and conditions. Pipeline transportation is suitable for directional transportation with large capacity, long-distance, and stable load. Steamers are suitable for large-capacity, ultra-long-distance, close to the ocean, or river transportation. Tanker transportation is suitable for a short-to-medium distance and small-capacity transportation, and its transportation is relatively flexible.

How to choose a suitable and safe storage address and master the law of dissolution, diffusion, and migration after CO_2 storage is the key to the smooth implementation of the actual geological storage project. In addition, the immature real-time monitoring technology is a bottleneck that significantly slows down the process of CO_2 underground storage.

2.4.3.5 Anthropogenic Carbon Conversion and Emissions Via Electrochemistry

Since electrocatalytic carbon dioxide reduction (ECDR) at metal electrodes was first reported in the 1980s, researchers have been working to mitigate the global greenhouse effect through ECDR. In the past decade, with the rapid progress of nanocatalysts, electrodes, and microfluidics, the use of ECDR for emission reduction and fuel production has the feasibility of a practical application. The most ideal way is to use renewable energy to provide electricity to drive carbon dioxide reduction, such as solar energy, wind energy, or hydropower (Figure 2.4) [25]. Theoretical calculations also show that even with non-renewable energy sources, ECDR can still reduce carbon emissions and produce high value-added chemicals or fuels.

CO_2 is linear and centrosymmetric, with two C=O [30, 31]. The energy required to dissociate the C=O bond (800 kJ mol^{-1}) of carbon dioxide is higher than other carbonyl groups because the C=O bond length (116.3 pm) is obviously shorter than the C—O single bond and other functional groups. The stability of its structure determines that activating CO_2 molecules to participate in the reaction is usually difficult. Bioreduction, photoreduction, thermal reduction, and electroreduction are commonly used to overcome the huge activation energy required for CO_2 reactions. Among these different methods, ECDR in aqueous solution is a multi-proton-coupled electron transfer process, which has some superiority for industrial applications, such as mild reaction conditions, close to neutral pH environment, and simple raw materials. In heterogeneous ECDR, the reaction is carried out under normal pressure and normal temperature. The electrolyte is usually a low-cost inorganic salt solution (such as $NaHCO_3$, $KHCO_3$, Na_2SO_4) and can provide activated protons to destroy the C=O bond. In addition, the electrolytic cell can be easily scaled up or scaled up and can also be assembled with an energy input or a gas reactant input device for mass production. Although ECDR has so far had attractive values in terms of sustainable energy use and environmental protection, it is generally inefficient and poorly selective, which still severely limits actual and large-scale deployment. Therefore, the most important challenge still lies in the rational design and implementation of ECDR electrocatalysts.

References

1 Falkowski, P., Scholes, R.J., Boyle, E. et al. (2000). *Science* 290: 291–296.
2 Hashimoto, K., Yamazaki, M., Meguro, S. et al. (2001). *J. Surf. Finish. Soc. Japan* 52: 195–200.
3 Dogutan, D.K. and Nocera, D.G. (2019). *Acc. Chem. Res.* 52: 3143–3148.
4 Olah, G.A., Prakash, G.K., and Goeppert, A. (2011). *J. Am. Chem. Soc.* 133: 12881–12898.
5 Gurevitz, M., Somerville, C.R., and McIntosh, L. (1985). *Proc. Natl. Acad. Sci.* 82: 6546–6550.
6 Goeppert, A., Czaun, M., Prakash, G.S., and Olah, G.A. (2013). *An Introduction to Green Chemistry Methods*. London: Future Science.
7 Mackay, A. (2007). *J. Environ. Qual.* 37: 2407.
8 P. Tans, NOAA/ESRL (www.esrl.noaa.gov/gmd/ccgg/trends/), R. Keeling, Scripps Institution of Oceanography (scrippsco2.ucsd.edu/), 2020.
9 Asinger, F. (1986). *Methanol – Chemie – und Energierohstoff: Die Mobilisation der Kohle*. Berlin, Heidelberg: Springer.
10 Plasseraud, L. (2010). *ChemSusChem* 3: 631–632.
11 Siriwardane, R.V., Shen, M.S., Fisher, E.P., and Poston, J.A. (2001). *Energy Fuels* 15: 279–284.
12 Banerjee, R., Phan, A., Wang, B. et al. (2008). *Science* 319: 939–943.
13 Millward, A.R. and Yaghi, O.M. (2005). *J. Am. Chem. Soc.* 127: 17998–17999.
14 Peters, M., Kohler, B., Kuckshinrichs, W. et al. (2011). *ChemSusChem* 4: 1216–1240.
15 Olah, G.A., Goeppert, A., and Prakash, G.K. (2009). *J. Org. Chem.* 74: 487–498.

16 Centi, G., Iaquaniello, G., and Perathoner, S. (2011). *ChemSusChem* 4: 1265–1273.

17 Quadrelli, E.A., Centi, G., Duplan, J.L., and Perathoner, S. (2011). *ChemSusChem* 4: 1194–1215.

18 Graves, C., Ebbesen, S.D., Mogensen, M., and Lackner, K.S. (2011). *Renewable Sustainable Energy Rev.* 15: 1–23.

19 Wang, Z.J., Song, H., Liu, H., and Ye, J. (2020). *Angew. Chem. Int. Ed.* 59: 8016–8035.

20 McEvoy, J.P. and Brudvig, G.W. (2006). *Chem. Rev.* 106: 4455–4483.

21 Zhang, B. and Sun, L. (2019). *Chem. Soc. Rev.* 48: 2216–2264.

22 Meyer, T.J. (1989). *Acc. Chem. Res.* 22: 163–170.

23 Balzani, V., Credi, A., and Venturi, M. (2008). *ChemSusChem* 1: 26–58.

24 Xu, X.D. and Moulijn, J.A. (1996). *Energy Fuels* 10: 305–325.

25 Liang, S.Y., Altaf, N., Huang, L. et al. (2020). *J. CO_2 Util.* 35: 90–105.

26 Bradford, M.C.J. and Vannice, M.A. (1999). *Catal. Rev.* 41: 1–42.

27 Ross, J.R.H. (2005). *Catal. Today* 100: 151–158.

28 Mikkelsen, M., Jorgensen, M., and Krebs, F.C. (2010). *Energy Environ. Sci.* 3: 43–81.

29 Agarwal, A.S., Rode, E., Sridhar, N., and Hill, D. (2017). *Handbook of Climate Change Mitigation and Adaptation*. Cham, Switzerland: Springer.

30 Habisreutinger, S.N., Schmidt-Mende, L., and Stolarczyk, J.K. (2013). *Angew. Chem. Int. Ed.* 52: 7372–7408.

31 Li, K., Peng, B.S., and Peng, T.Y. (2016). *ACS Catal.* 6: 7485–7527.

Part III

Electrochemical Catalysis Process

The studies of clean energy are significate in the next decades. Many advanced technologies for clean energy conversion, such as water splitting, CO_2 to fuel conversion, and fuel cells, are the subject of both extensive fundamental and utilitarian studies. The key point of these devices involves many electrochemical processes, including oxygen reduction reaction (ORR) and methanol oxidation reaction (MOR), hydrogen evolution reaction (HER) and oxygen evolution reaction (OER), or electrochemistry carbon dioxide reduction reaction (ECO_2RR) and OER. In this section, an independent foundational summary for these half-reactions and its possible reaction pathways are highlighted. In addition, parameters used to judge the activity of catalysts or a reaction condition are also summarized, including turnover frequency (TOF), onset potential or overpotential, stability, Faradic efficiency, partial current density, environmental impact and cost, common electrolytes, and electrochemical cells.

Electrocatalysis in Balancing the Natural Carbon Cycle, First Edition. Yaobing Wang.
© 2021 WILEY-VCH GmbH. Published 2021 by WILEY-VCH GmbH.

3

Electrochemical Catalysis Processes

3.1 Water Splitting

3.1.1 Reaction Mechanism

3.1.1.1 Mechanism of OER

One half of the water-splitting reaction is oxygen evolution reaction (OER) (Eq. (3.1)). The cathodic and anodic reaction parts are different in acidic environments (Eqs. (2.2) and (3.3)) and alkaline environments (Eqs. (3.9) and (3.10)). Many scientists proposed possibilities of OER mechanisms in acid (Eqs. (3.4)–(3.8)) or alkaline environments (Eqs. (3.11)–(3.15)). In addition, there are some differences and similarities between the proposed mechanisms. For example, most proposed mechanisms have the same intermediates, such as MOH and MO. However, the main difference may be around the reaction that produces oxygen. There are two different ways of producing O_2 from molybdenum intermediates. As shown in Figure 3.1, one of them is the green route, where two MOs directly combine to form O_2 (Eq. (3.6)), while the other one possesses MOOH intermediate formation (Eqs. (3.7) and (3.14)), followed by decomposing to O_2 (Eqs. (3.8) and (3.15)).

$$2H_2O \rightarrow 2H_2 + O_2 \tag{3.1}$$

Acidic conditions:

Cathode reaction:

$$4H^+ + 4e^- \rightarrow 2H_2; \quad E_c^\circ = 0\,V \tag{3.2}$$

Anode reaction:

$$2H_2O \rightarrow O_2 + 4H^+ + 4e^-; \quad E_a = 1.23\,V \tag{3.3}$$

Proposed mechanism under acidic conditions:

$$M + H_2O \rightarrow MOH + H^+ + e^- \tag{3.4}$$

$$MOH + OH^- \rightarrow MO + H_2O + e^- \tag{3.5}$$

$$2MO \rightarrow 2M + O_2 \tag{3.6}$$

Electrocatalysis in Balancing the Natural Carbon Cycle, First Edition. Yaobing Wang.
© 2021 WILEY-VCH GmbH. Published 2021 by WILEY-VCH GmbH.

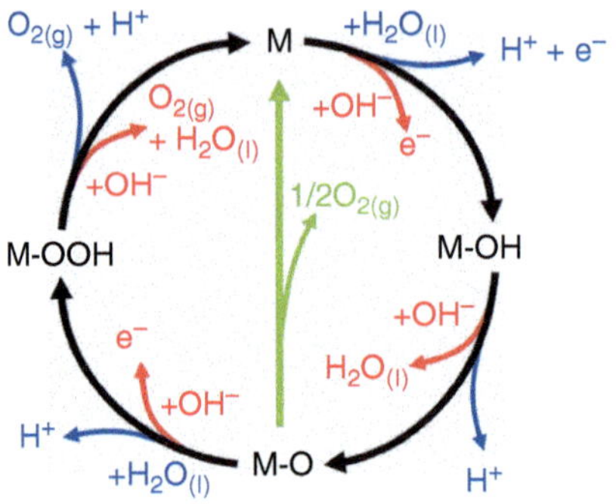

Figure 3.1 The OER mechanism for acid (blue line) and alkaline (red line) conditions. The black line indicates that the oxygen evolution involves the formation of a peroxide (M−OOH) intermediate (black line) while another route for the direct reaction of two adjacent oxo (M−O) intermediates (green) to produce oxygen is possible as well. Source: Suen et al. [1]. © 2017, Royal Society of Chemistry.

$$MO + H_2O \rightarrow MOOH + H^+ + e^- \tag{3.7}$$

$$MOOH + H_2O \rightarrow M + O_2 + H^+ + e^- \tag{3.8}$$

Alkaline conditions:

Cathode reaction:

$$4H_2O + 4e^- \rightarrow 2H_2 + 4OH^-; \quad E_c^\circ = 0.83\,V \tag{3.9}$$

Anode reaction:

$$4OH^- \rightarrow 2H_2 + 2H_2O + 4e^-; \quad E_a^\circ = 0.40\,V \tag{3.10}$$

Proposed mechanism under alkaline conditions:

$$M + OH^- \rightarrow MOH \tag{3.11}$$

$$MOH + OH^- \rightarrow MO + H_2O \tag{3.12}$$

$$2MO \rightarrow M + O_2 \tag{3.13}$$

$$MO + OH^- \rightarrow MOOH + e^- \tag{3.14}$$

$$MOOH + OH^- \rightarrow M + O_2 + H_2O \tag{3.15}$$

3.1.1.2 Mechanism of ORR

The overall equation of oxygen reduction reaction (ORR) includes 4e$^-$ pathway or two-step 2e$^-$ pathway. For fuel-cell operations, a direct 4e$^-$ process is the most desirable to achieve high efficiency; on the contrary, the reduction pathway of selective 2e$^-$ is applied in industry to produce H_2O_2 [2]. The detailed mechanisms of ORR are complex and rely on the surface properties of electrodes. For 4e$^-$ reduction process, there are two possibilities of ORR mechanism, dissociative pathway and associative pathway, proposed in alkaline solution, as follows [3, 4]:

Dissociative mechanism:

$$O_2 + 2 * \rightarrow 2O^* \tag{3.16}$$

$$2O^* + 2e^- + 2H_2O \rightarrow 2OH^* + 2OH^- \tag{3.17}$$

$$2OH^* + 2e^- \rightarrow 2OH^- + 2 * \tag{3.18}$$

Among them, * represents active sites on a specific surface. This mechanism can be expressed as the first adsorption of O_2 (Eq. (3.16)), following by the breakage of O—O bond and two adsorbed O* species are generated (Eqs. (3.17) and (3.18)). The O* species further gain two electrons and two protons to directly form the final product OH^- rather than forming OOH^-. Thus, this mechanism can be regarded as a direct $4e^-$ pathway. There is an alternative pathway for ORR, which can proceed as an associative mechanism (Eqs. (3.19)–(3.23)):

The four-electron pathway:

$$O_2 + * \rightarrow 2O_2{}^* \tag{3.19}$$

$$O_2{}^* + H_2O + e^- \rightarrow OOH^* + OH^- \tag{3.20}$$

$$OOH^* + e^- \rightarrow O^* + OH^- \tag{3.21}$$

$$O^* + H_2O + e^- \rightarrow OH^* + OH^- \tag{3.22}$$

$$OH^* + e^- \rightarrow OH^- + * \tag{3.23}$$

Or the two-electron pathway:

$$O_2 + * \rightarrow 2O_2{}^* \tag{3.24}$$

$$O_2{}^* + H_2O + e^- \rightarrow OOH^* + OH^- \tag{3.25}$$

$$OOH^* + e^- \rightarrow OOH^- + * \tag{3.26}$$

Whether the reaction is dissociated or associative relies on the initial energy barrier of O_2 dissociation on the surface [3]. For instance, a series of density functional theory (DFT) studies showed that the value of the dissociation energy barrier of the carbon surface is usually very high [5], which is not preferable for the $4e^-$ dissociative pathway. As a result, the electron transfer measurements of almost all carbon materials are experimentally <4. On the contrary, ORR usually passes through a dissociation path on most metal surfaces because O_2 adsorption is very strong. For example, the free energy of O_2 adsorption on the surface of Pt (111) is about 0.6 eV, while the dissociation energy barrier of O_2 is low to 0.3 eV [6, 7]. This theoretical result is consistent with the experimental data, in which Pt-based catalysts show an ideal $4e^-$ reduction selectivity. An exception in metals is Ag(111), where the O_2 adsorption is very weak, leading to a more than 1 eV of O_2 dissociation energy barrier [8]. Therefore, the reaction path of the Ag surface generally obeys the association mechanism with a highly selective $2e^-$ pathway.

3.1.1.3 Mechanism of HER

In this section, we reviewed the total equation (Eq. (3.27)) and stepped pathway of hydrogen evolution reaction (HER). However continuous the reaction is, the initial step of HER is to transfer an H^+ from the acid electrolyte with an e^- so that forms adsorption of H on the electrode surface (called Volmer reaction, see Eqs. (3.28) and (3.32)). In alkaline environments, it would need an extra water dissociation before, which is likely to increase an additional energy barrier of the reaction and slow the overall reaction speed. In the second step, there exist two possible pathways: the first one is the Heyrovsky reaction (see Eqs. (3.29) and (3.33)), where the adsorbed H combines with an e^- transferred from the surface of the electrode surface and an H^+ in the electrolyte to generate H_2, while another is Tafel reaction (see Eqs. (3.30) and (3.34)), where two adsorbed H atoms that are adjacent combine to generate H_2. The value of the Tafel slope from polarization curves can be used to analyze which one is the probable rate-determining step(s) [9].

Acidic solution:

Cathode reaction:

$$2H^+ + 2e^- \rightarrow H_2 \tag{3.27}$$

The proposed mechanism under acidic conditions:

$$H^+ + e^- + * \rightarrow H^+ \text{ (Volmer)} \tag{3.28}$$

$$H^+ + e^- + H^* \rightarrow H_2 \text{ (Heyrovsky)} \tag{3.29}$$

Or

$$2H^* \rightarrow H_2 \text{ (Tafel)} \tag{3.30}$$

Alkaline solution:

Cathode reaction:

$$2H_2O + 2e^- \rightarrow H_2 + 2OH^- \tag{3.31}$$

The proposed mechanism under alkaline conditions:

$$H_2O + e^- \rightarrow H^* + OH^- \text{ (Volmer)} \tag{3.32}$$

$$H_2O + e^- + H^* \rightarrow H_2 + OH^- \text{ (Heyrovsky)} \tag{3.33}$$

Or

$$2H^* \rightarrow H_2 \text{ (Tafel)} \tag{3.34}$$

3.1.2 General Parameters to Evaluate Water Splitting

3.1.2.1 Tafel Slope

The dynamics of any electron transfer process can be studied by a properly fixed polarization curve, like the Tafel curves. The slope of the linear region over the Tafel plot is related to the ability of the charge transfer, which is defined as the ratio of the change in ohmic loss compensated overpotential (η) and logarithmic current density ($\log j$), as follows [10, 11]:

$$\eta = a + b \log j \tag{3.35}$$

$$b = 2.303RT/\alpha nF \tag{3.36}$$

where j presents the current density, η presents overpotential, T presents the room temperature, R is the ideal gas constant, α is the charge transfer coefficient, n is the molar numbers of electrons generating 1 mol of O_2, and F is the Faraday constant, 96 485 C mol^{-1}. From Eq. (3.36), it can be seen that the Tafel slope and charge transfer coefficient are interdependent. It means that if the value of the Tafel slope is low, the activity of electrocatalysts will be high. In addition, the number of OER transferred electrons participating in the rate-determining step can be predicted if the charge transfer coefficient is a fixed value. When the charge transfer coefficient is 0.5 at 25 °C, if the value of transferred electrons at the rate-controlling step varies as 1, 2, 3, and 4 [11, 12], the Tafel slope can be equal to 120, 90, 60, and 30 mV dec^{-1}, respectively. Because OER is a process of four-electron transfer, an ideal catalyst should represent a Tafel slope of 30 mV dec^{-1}. However, there are few reports of OER catalysts whose Tafel slope is 30 or close to 30 mV dec^{-1}. It has been found that most of the reported OER catalysts with high activity have a Tafel slope range between 40 and 50 mV dec^{-1}. It is located between three and four electron transfer processes, which is equivalent to the release of oxygen due to the simultaneous oxidation and the formation of hydroxides [10, 11, 13]. Therefore, metals with electron preferred oxidation variables are good catalysts for catalyzing OER.

3.1.2.2 TOF

There are two parameters to judge the activity of electrocatalysts in water splitting. The first one is total electrode activity, such as current density ($mA_{cmgeometric}{}^{-2}$). The other one is the internal activity of the catalytic site, like turnover frequency (TOF), which is more beneficial to study the catalytic activity fundamentally.

$$\mathrm{TOF} = \frac{J}{nFN_{\text{active site}}} \tag{3.37}$$

Among them, n is the number of transferred electrons and $N_{\text{active site}}$ is the number of active sites.

Some electrochemical measurement parameters, such as the standardization of area J and the number of active centers, build a connection between the internal performance of catalysts and current density or overpotential. Thus, it is better to reduce the Tafel slopes and increase the value of J, particularly for high-current applications. From the experimental view, nanostructures are easy and efficient ways to improve electrocatalytic activity on the surface of some bulk electrodes. In other words, the surface area can be significantly increased due to the production of electrocatalysts with suitable nanostructures, resulting in exposing much more active sites in the unit geometric area while the intrinsic properties of electrodes will not change. As a result, it is necessary for judging its intrinsic activity to normalize the measured performance of nanostructured electrodes by unit surface area and mass loading.

3.1.2.3 Onset/Overpotential

The onset of potential plays the most crucial role in judging the performance of OER catalysts, while its exact value is difficult to determine. Therefore, the value of potential corresponding to 10 mA cm^{-2} is regarded as more reliable and widely used. What is more, overpotential is determined by the potential difference between the potentials achieving a specific current density and standard electrode potential (1.23 V vs. reversible hydrogen electrode [RHE]). Generally, if the overpotential of a catalyst is in the range of 300–400 mV, it is regarded as a suitable catalyst.

3.1.2.4 Stability

The stability of catalysts plays a vital role in a practical commercial application. The stability of catalysts is seriously affected by various aspects, such as the nature of the electrolyte, because most catalysts are very stable in an alkaline medium, while it is not good to work in an acid environment. The properties of the working electrode (WE) also play a decisive role in the stability of the catalyst. Typically, compared with the way of dropping or coating, the active materials in situ grown on WE are more stable. In addition, the practical question about the reported stability is whether it can still work well after long-term electrolysis in real devices because there is no straightforward method to evaluate the stability of catalysts. Most researchers judge stability a short period of time by exploiting chronopotentiometry with a constant potential/current density. Actually, the structure and composition of catalysts would undergo changeable during electrocatalysis, so no one knows exactly what happened, and no one knows the role of catalyst. Recently, several methods, such as X-ray diffraction (XRD), X-ray photoelectron spectroscopy (XPS), scanning electron microscopy (SEM), and transmission electron microscopy (TEM), have been used to determine the structure and composite stability of catalysts.

3.1.2.5 Electrolyte

Various electrolytes, alkaline or acid, have a great influence on the properties of electrode materials. The performance of OER is more favorable electrocatalysis in alkaline solution; however, it is very difficult in neutral and acidic solutions. At present, most researchers concentrated on exploring the OER electrocatalysts with high durability under alkaline conditions, such as carbon-based materials, oxides or hydroxyl transition metal, hybrids, etc. However, many of them are unstable in acidic conditions because of the high oxidation potential. Thus, the electrocatalysts of OER working under the full range of the whole pH scale are highly desirable.

3.2 Electrochemistry CO_2 Reduction Reaction (ECDRR)

3.2.1 Possible Reaction Pathways of ECDRR

Scientists have paid enormous efforts to explore the activation and reduction mechanism of CO_2 from both theoretical and experimental aspects [14, 15]. However, most of the mechanisms are based on the results of either DFT calculations or the Tafel slope. Thus, to fully understand the mechanical routes, more experimental work combined with DFT calculations and in situ spectroscopy is necessarily required.

3.2.1.1 Formation of $HCOO^-$ or HCOOH

$HCOO^-$ or HCOOH may be changed by intermediate compounds, which is connected to the transition metal electrode by one or two oxygen atoms (Scheme 3.1a and top route in Scheme 3.1b) [14]. This intermediate product can be produced in *H by inserting CO_2 into the M–H or protonation directly with H^+ in solution.

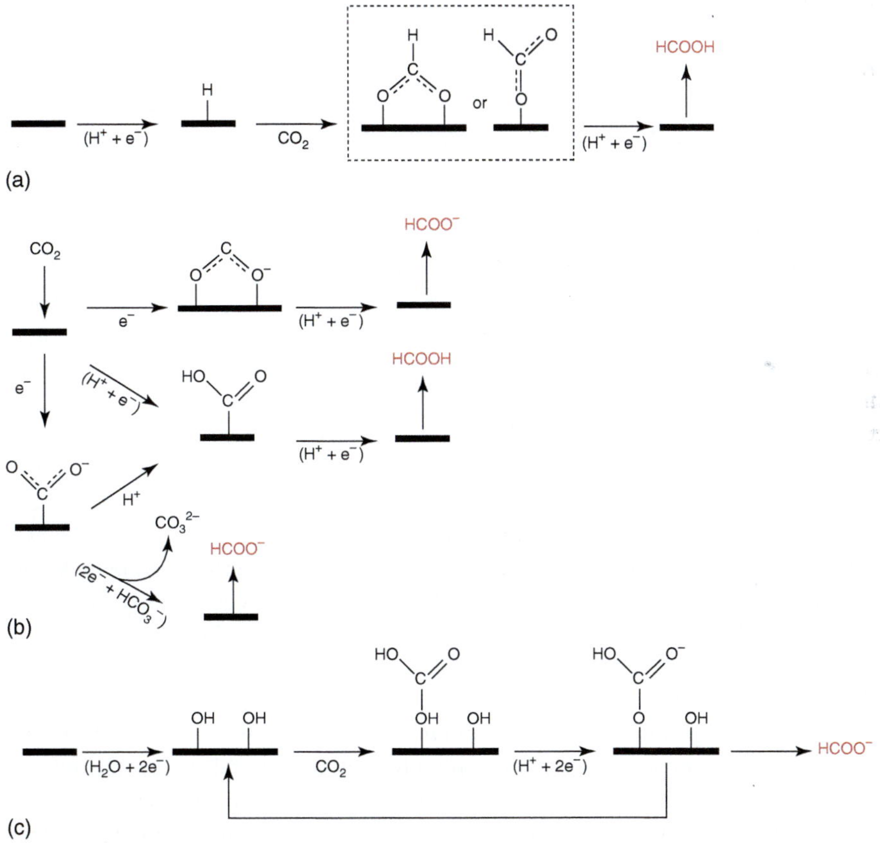

Scheme 3.1 Possible reaction pathways for the electrocatalytic reduction of CO_2 to $HCOO^-$ or HCOOH. (a) Monodentate or bidentate intermediate route. (b) $CO_2{\cdot}^-$ radical intermediate route. (c) Surface-bound carbonate intermediate route. Source: Sun et al. [16]. © 2017, Elsevier.

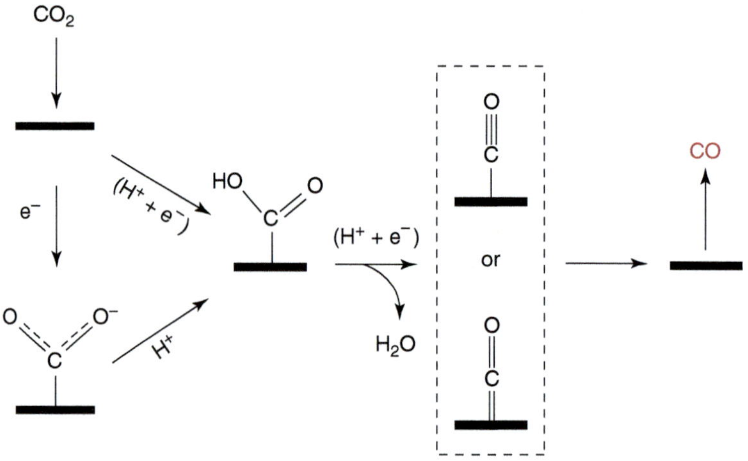

Scheme 3.2 Possible reaction pathways for the electrocatalytic reduction of CO_2 to CO. Source: Sun et al. [16]. © 2017, Elsevier.

Proceeding through a $CO_2{\cdot}^-$ radical reacting with an adjacent H^+ to form $HCOO^-$ or HCOOH on p-block metals in (Scheme 3.1b) [14, 17] is an alternative pathway. What is more, it was found that the existence of HCO_3^- ions would increase the production of $HCOO^-$ (Scheme 3.1c).

3.2.1.2 Formation of CO

Electrochemistry CO_2 reduction reaction (ECDRR) to CO is much easier with the transfer process of two-electron compared with other complicate products (Scheme 3.2). By transferring the first pair of electrons coupled with a proton (H^+/e), CO_2 is reduced to *COOH, following by the second pair of H^+/e attacking the O atom in the *COOH to generate H_2O and CO. The CO molecule would desorb from the electrode [15]. It is confirmed that the conversion process of *COOH to *CO occurs readily. However, the initial conversion step of CO_2 to *COOH is inhibited due to the poor binding force of COOH. Similarly, the last desorption step of CO from the electrode surface is difficult because of the intense CO binding. These two steps were regarded as rate controlling for the whole reaction [18]. Another possible pathway for the production of *COOH is via a transfer process of decoupled electron and proton, which involves the generation of a $CO_2{\cdot}^{2-}$ intermediate.

3.2.1.3 Formation of C_1 Products

*CO is a necessary intermediate to produce HCHO, CH_3OH, and CH_4. Peterson and his colleagues exploited DFT calculation on the surface of Cu(211). The results showed that the initial generation of *CO intermediate, following by the subsequent hydrogenation to *HCO, *H_2CO, and *H_3CO, is the thermodynamically most favorable pathway (Scheme 3.3a) [19]. Among them, the reduction of *H_3CO can form CH_3OH, as well as CH_4 and *O. However, The barrier of the formation of CH_4, through C—H bond generation with C-O dissociation, was up to 1.21 eV, which is higher than 0.15 eV for CH_3OH production [15]. It means the pathway

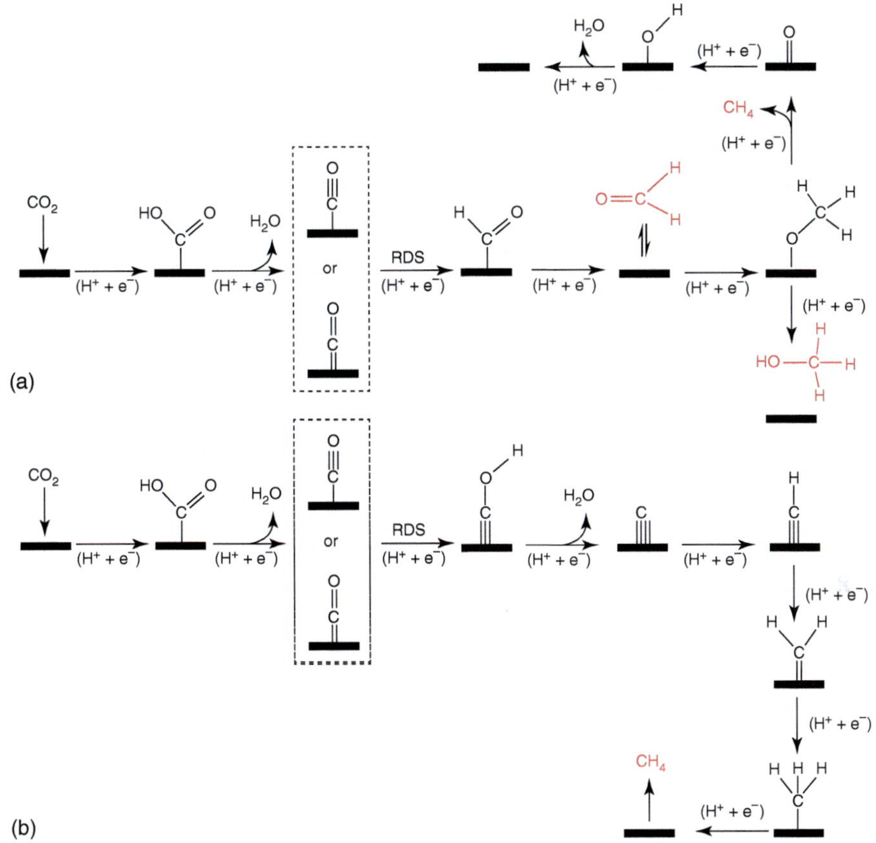

Scheme 3.3 Possible reaction paths for electrocatalytic CO_2 reduction to produce HCHO, CH_3OH, and CH_4 on Cu electrodes. (a) A thermodynamic analysis. (b) A combined thermodynamic and kinetic analysis. Source: Sun et al. [16] © 2017, Elsevier.

of CH_4 production is kinetically prohibitive. Nie et al. proposed another possible reaction pathway [15]. They regarded *COH as an intermediate to form an adsorbed *C (Scheme 3.3b), which could be observed by in situ XPS and atomic emission spectrometry (AES) on Cu electrodes [20]. The carbon species can gradually add H and reduced to form *CH, *CH_2, *CH_3, and final CH_4. The surficial oxophilicity determined by the binding energy of O_{ads} plays a significate role in the selectivity of CH_4 or CH_3OH. Only CH_3OH was identified on the poorly oxophilic surface of the Au electrode, while only CH_4 was detected on the strongly oxophilic surface of the Fe electrode [21]. From this view, controlling the binding free energy for C_{ads} and O_{ads} to change the breakage of the C—O bond allows us to change the selectivity of catalysts.

3.2.1.4 Formation of C_2 Products

The products of C_2H_4, CH_3CHO, and C_2H_5OH generated at more negative potentials than CH_4 [22, 23]. The generation of these C_2 products has many possible pathways, which involve the "carbene" mechanism and is a complex process. For example, C_2H_4 is formed via either two of $^*CH_2$ species coupling [15] or CO insertion in a Fischer–Tropsch process (Scheme 3.4a). It is also considered to be the pathway of C_2H_5OH [24]. The dimerization of two *CO with an electron transfer to form a $^*C_2O_2^-$ key intermediate, following by the protonation of *CO–COH (Scheme 3.4b), is regarded as an alternative pathway [14, 25]. The C_2H_5OH can be an important intermediate and further reduced to C_2H_4, or CH_3CHO. C_2H_4 is a dynamic, preferred product on the surface of the Cu electrode [26].

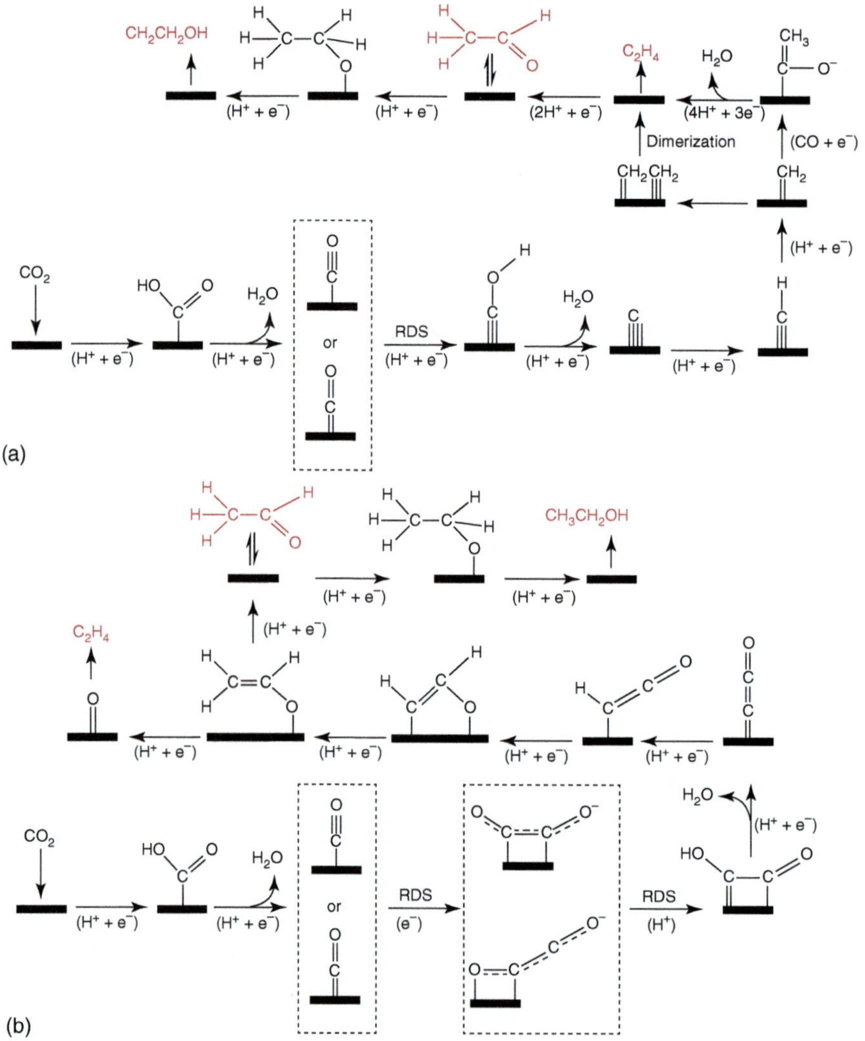

Scheme 3.4 Possible reaction paths for electrocatalytic CO_2 reduction to produce C_2H_4, CH_3CHO, and C_2H_5OH. (a) Coupling of two $^*CH_2$ species or CO insertion in a Fischer–Tropsch-like step. (b) *CO dimerization. Source: Sun et al. [16] © 2017, Elsevier.

3.2.1.5 Formation of CH_3COOH and CH_3COO^-

Recently, scientists successfully obtained other C_2 products of ECDRR on Cu-based catalysts and NDD electrode [27], like CH_3COOH or CH_3COO^-, where 91.2–91.8% Faradaic efficiency for CH_3COO^- formation at a wide potential window(0.8–1.0 V vs. RHE) on NDD was achieved [27]. The formation of CH_3COOH is related to the nuclear attack of adsorbed intermediate $CO_2\cdot^-$ (Scheme 3.5) [28, 29]. The formation of OOC–COO intermediate by two adsorbed $CO_2\cdot^-$ coupling suggested by electrokinetic data and in situ infrared (IR) is regarded as an important intermediate for the production of acetic acid.

3.2.1.6 Formation of *n*-Propanol (C_3 Product)

ECDRR to produce over C_2 products, such as C_3H_7OH, remains challenging. Up to now, the maximum Faradaic efficiency (FE) of 30% was achieved only on the graphene/ZnO/Cu_2O catalyst [30]. Based on it, a hypothesis was put forward. The adsorbed C_2 intermediate can interact with the adjacent C_1 intermediate to cause C–C coupling, followed by transferring proton-coupled electron, so that generate propionaldehyde (CH_3CH_2CHO) (Scheme 3.6) [24]. CH_3CH_2CHO is further reduced to C_3H_7OH.

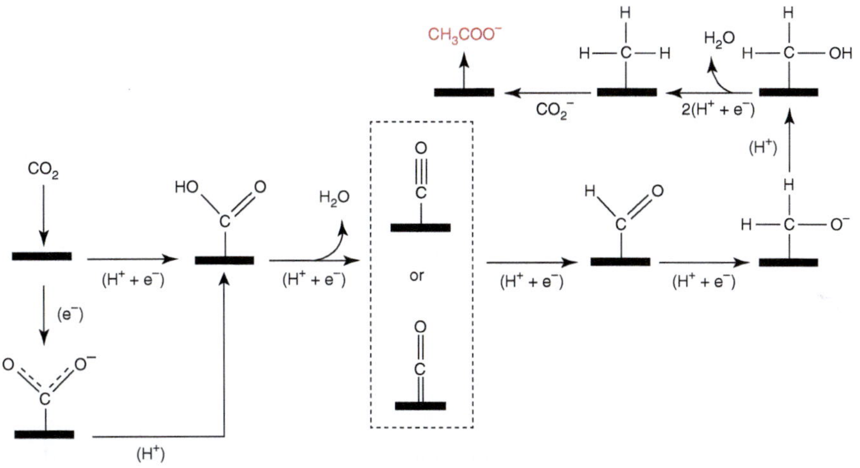

Scheme 3.5 Possible reaction paths for electrocatalytic CO_2 reduction to produce CH_3COO^- or CH_3COOH. Source: Sun et al. [16] © 2017, Elsevier.

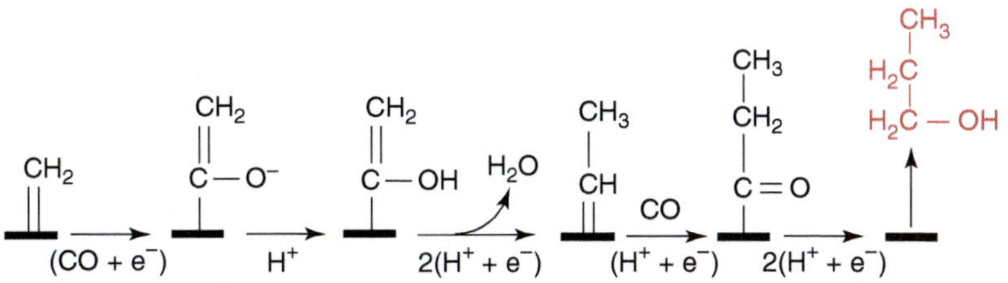

Scheme 3.6 Possible reaction paths for electrocatalytic CO_2 reduction to produce C_3H_7OH. Source: Sun et al. [16] © 2017, Elsevier.

3.2.2 General Parameters to Evaluate ECDRR

The materials of ECDRR catalysts are generally evaluated from the following aspects based on their onset potential, activity, selectivity, environmental impact, and cost, etc.

3.2.2.1 Onset Potential

The onset potential means the lowest potential that the target product can be firstly responsibly detected by the detector. For consistency, it is recommended that all potential should be converted to the reversible hydrogen electrode to avoid the effect of pH, and the possible Ohmic loss should also be compensated. The catalytic performance of a material is not only determined by the onset potential, which should be distinguished from HER and OER electrocatalysis. It is not always true that the onset potential at the polarization curve is the onset potential of ECDRR because of the competing HER. Low cathodic overpressure indicates that ECDRR can be started more easily on the catalyst, so it is very desirable. This also reduces energy consumption when ECDRR catalysts are combined with OER electrocatalysts for electrolysis with a complete splitting cell. Among all the existing electrocatalysts for the production of HCOOH, only the onset potential of Pd or its alloys is close to the standard reduction potential, although their stability is low [31].

3.2.2.2 Faradaic Efficiency

When several competing reaction processes occur at the same time, the selectivity of targeted products becomes a major problem. Because of the complicate reaction pathways and the strongly competing HER, selectivity is extremely significate to ECDRR. Usually, selectivity defined as the ratio of transferred charges to the amount of targeted product over the total charges is described by Faradaic efficiency (FE), as follows:

$$\mathrm{FE} = \alpha nF/Q \tag{3.38}$$

α presents the number of transferred electrons needed to generate a target product molecule, n is the molar number of the targeted product, F is the Faraday's constant, 96 485 C mol^{-1}, and Q is the total consumption charge, which can be obtained from electrochemistry station. The validity of Faraday reflects the dynamic benefit of the specific reaction process. High Faradaic efficiency is a target as it may decrease the additional cost of separation and purification of products. The main group of metals usually have peak efficiency for Faraday with 70–100% formic acid. Improving the selectivity of catalysts and achieving a Faradaic efficiency, close to 100% during a wide potential window, remains challenging.

3.2.2.3 Partial Current Density

Partial current density is an effective current density, which can control the generation of target products. It can be calculated by multiplying the Faradaic efficiency at given potentials with the whole current density. Partial current density depends not only on the internal activity of the catalyst but also on the experimental conditions, including the type of collector, electrochemical cell, and electrolyte, etc. Most previous studies reported that the maximum partial current density of HCOOH is less 30 mA cm^{-2} in H-type cells because of the limitation of the CO_2 solubility and diffusion. Gas diffusion electrodes (GDEs) can effectively remove this limitation and greatly increase the partial current density. They are currently actively used in commercial carbon dioxide electrolyzer.

3.2.2.4 Environmental Impact and Cost

In addition to selectivity and activity; the ideal ECDRR electrocatalyst should be environmental-friendly and rich in low-cost rare earth elements. Hori firstly reported many primary group metals that could selectively catalyze CO_2 to HCOOH, such as Hg, Pb, Cd, and Tl. However, because of their high toxicity and environmental, they are now excluded from practical applications. On the other hand, it is very expensive [32]. Thus, scientists give more attention to Sn-based and Bi-based electrocatalysts at present. In addition, these electrocatalysts should be synthetized and processed using environmentally friendly and economical methods to meet the requirements of low environmental pollution and low manufacturing costs.

3.2.2.5 Electrolytes

In the process of electroreduction of CO_2, electrolyte provides the environment for the migration of related electrons coupled with protons. The types and concentration of electrolytes play an important role in the performance of catalysts. In addition, trace metal and organic impurities in the electrolyte may cause decontamination of the catalytic active center.

The first one is aqueous solutions. Most of the researches on electrochemical reduction of CO_2 are focused on water electrolyte with low acid or alkaline saturated CO_2, which contains inorganic salts. Cationic species change the potential of the outer surface of Helmholtz and affect the hydrogen coating of the electrode by transporting H_2O from its sol shell to the electrode. At the same time, cations can change concentrations of charged particles near the electrode, so that changes the product Faraday efficiency [33]. Scientists found that the reaction rate of electrochemical reduction of CO_2 to CH_3OH would increase when the surface charge of electrolyte cations increases: $Zr^{4+} > Al^{3+} > Ba^{2+} > Ca^{2+} > Mg^{2+} > Na^{+}$ [34]. It is shown that the size of cations affects the selectivity of the product. Large positive ions are helpful to generate HCOOH on Hg electrodes or C_2H_4 on Cu electrodes [35] and CO on Ag electrodes [33]. The large size of cations was also regarded to suppress HER [33, 36].

The second is organic electrolytes, nonaqueous electrolytes, which were used in ECDRR to increase the solubility of carbon dioxide and inhibit the release of hydrogen so as to improve the efficiency of Faraday. The solubility of carbon dioxide in dimethyl sulfoxide (DMSO) and acetonitrile (AN) is about four times that of water, while the solubility in CH_3OH and propylene carbonate is five and eight times higher, respectively. Dimethylformamide (DMF) is a good CO_2 solvent whose solubility is 20 times higher than that of water in the environment. For example, it is beneficial for Sn, In, Hg, and Pd electrodes to facilitate the dimerization of $CO_2{\cdot}^{-}$

so that produces $C_2O_4^{2-}$ in nonaqueous solvents, such as DMF, AN, and DMSO [37], while Cu electrode is preferable for CH_4 production in CH_3OH [38]. It can effectively control Faraday's efficiency and reduce the selectivity of carbon dioxide emission by managing the water quantity in the organic electrolyte to control the availability of protons.

The third type is ionic liquids, including 1-butyl-3-methylimidazolium hexafluorophosphate (BMIM-PF_6), 1-butyl-3-methylimidazolium triflate (BMIM-OTF), and 1-ethyl-3-methyl-imidazolium triflate (EMIM-OTf), and 1-ethyl-3-methylimidazolium tetrafluoroborate (EMIM-BF_4), etc. There are three main advantages in using ionic liquids: (i) lowering the overpotential of CO_2 reduction by reducing the Gibbs free energy of the intermediate $CO_2{\cdot}^-$; (ii) inhibiting HER; (iii) increasing the selectivity for CO product [39, 40].

The last one is solid-oxide electrolytes, which are usually applied in high temperatures ($T > 673$ K) for the H_2O and CO_2 coelectrolysis. These electrolytes are typically zirconia, which is stabilized by yttrium oxide [41]. For example, a $BaCeO_{0.5}Zr_{0.3}Y_{0.16}Zn_{0.04}O_3$-d electrolyte could reduce CO_2 into CO and CH_4 with the help of H_2 and H_2O.

3.2.2.6 Electrochemical Cells

The design of the electrolyzer has far-reaching consequences for mass transport. There are two common cells for ECDRR. The first one is H-Type cells, which are widely exploited in a lab-scale electrochemical cell in ECDRR, consisting of two chambers (separated by a proton- or anion-exchange polymer membrane) and three electrodes, such as WE, reference electrode (RE), and counter electrode (CE) [42]. The structure of this setup not only allows H^+ to the cathodic chamber from the anodic chamber but also prevents cathodic products into the anode chamber occurring oxidation reactions. Moreover, the gas product at the cathode can be directly injected into an mass spectrometry (MS) or gas chromatography (GC) to analyze the product.

Another type is micro-flow cells, which are widely used in an industrial electrochemical cell in ECDRR, consisting of three chambers (cathodic liquid, cathodic gas chambers, and anodic chamber, separated by a proton- or anion-exchange membrane) and two electrodes, such as WE and CE [43]. Different from the H-type cell, a three-phase boundary is set up, where CO_2 gas goes through at the backside of the WE (GDE) and cathodic electrolyte flow from the forward side of GDE. This kind of continuous-flow reactor is beneficial to online detection products.

3.3 Small Organic Molecules Oxidation

3.3.1 The Mechanism of Electrochemistry HCOOH Oxidation

As with all polymer electrolyte membrane-based fuel cells, the direct formic acid fuel cell also uses an air cathode. Oxygen reduction, through a four-electron reaction at the cathode, is usually facilitated by a platinum-based catalyst. At the anode, direct oxidation of formic acid releases two electrons per molecule. The cathode, anode, and overall reactions of a direct formic acid fuel cell are described as:

$$\text{Anode: } HCOOH \rightarrow CO_2 + 2H^+ + 2e^-; \quad E^\circ = -0.25 \text{ V vs.SHE} \tag{3.39}$$

$$\text{Cathode: } \frac{1}{2}O_2 + 2H^+ + 2e^- \rightarrow H_2O; \quad E^\circ = 1.23 \text{ V vs.SHE} \tag{3.40}$$

$$\text{Overall: } HCOOH + \frac{1}{2}O_2 \rightarrow CO_2 + H_2O; \quad \text{OCV} \sim 1.48 \text{ V} \tag{3.41}$$

The fuel cell of direct formic acid has a higher electromotor force (EMF) open-circuit voltage (OCV), which is approximately 1.48 V; values vary slightly relying on the source than the fuel cells with hydrogen [44, 45] or direct methanol [46]. The theoretical energy density of HCOOH is determined as $2F \times \text{OCV} \times (\text{MW})^{-1}$. Among them, MW is the molecular weight (kg mol^{-1}), and F is the Faraday constant, 96 485 C mol^{-1}. Thus, the intrinsic energy density of HCOOH is 1725 Wh kg^{-1}. When taking the density of this liquid (1.22 kg l^{-1}) into consideration, HCOOH has a theoretical energy density of 2104 Wh l^{-1}.

The fuel cell of direct formic acid has been studied for only approximately 14 years from 1996 [47], while the history of the mechanism research of oxidation of formic acid is more extended [48–53]. At present, the "parallel or dual pathway mechanism" is the most widely accepted mechanism [53]. Direct oxidation (pathway I) occurs through a dehydrogenation reaction rather than reaction intermediate CO, as Eq. (3.42), while another reaction pathway (pathway II) forms adsorbed CO by dehydration, as Eq. (3.43):

$$\text{Pathway I: } HCOOH \rightarrow CO_2 + 2H^+ + 2e^- \tag{3.42}$$

$$\text{Pathway II: } HCOOH \rightarrow CO_{ads} + H_2O \rightarrow CO_2 + 2H^+ + 2e^- \tag{3.43}$$

For reaction pathway II, HCOOH firstly adsorbs onto the surface of catalysts so that forms an intermediate adsorbed CO species, following by the CO oxidation to the final product, the gaseous CO_2. For fuel cells of direct formic acid, dehydrogenation is an ideal reaction pathway, which not only can enhance the overall efficiency of the cell but also avoid poisoning of catalysts.

3.3.2 The Mechanism of Electro-oxidation of Alcohol

Because there is only one oxygen atom in –OH, an extra oxygen atom is required for the deep oxidation of a (primary) alcohol to CO_2, which can be provided by water, or adsorbed residue (such as adsorbed OH). Therefore, the general overall reaction of electro-oxidation of primary alcohol can be expressed as follows:

$$C_nH_{2n+1}OH + (2n - 1)H_2O \rightarrow nCO_2 + 6nH^+ + 6ne^- \tag{3.44}$$

According to the Eq. (3.44), an excellent electrocatalyst is required to activate both the chemisorption of alcohol and the water molecule. In addition, the reaction mechanisms involve several adsorbed intermediates, the formation of numerous products, and by-products.

To elucidate the reaction mechanism, electrochemical methods, including rotating disk electrodes, cyclic voltammetry, combining with other physicochemical methods, such as in situ infrared and UV–vis reflectance spectroscopy [54, 55], are needed to detect the adsorbed intermediates. Moreover, "on line" chromatographic techniques are also used to quantitatively analyze the reaction products and by-products [56].

The reaction mechanism is now well understood because enormous efforts had been paid during the past decades [57]. The overall oxidation reaction involving six electrons and one water molecule is written as follows:

$$CH_3OH + H_2O \rightarrow CO_2 + 6H^+{}_{aq} + 6e^- \tag{3.45}$$

On a Pt electrode, CO bonded to the surface with either the way of linear or bridge was firstly observed by electrochemically modulated infrared reflectance spectroscopy (EMIRS) during oxidation of methanol [58], where the coverage degree of CO bonded linearly reached 90%. As a result, most of the active sites were poisoned. In addition, other adsorbed species, such as $(^*CHO)_{ads}$ or $(^*COOH)_{ads}$, were also identified by EMIRS or Fourier transform infrared reflectance spectroscopy [59, 60].

Gas or liquid chromatography was also used to quantitatively analyze reaction products, such as CO_2, HCHO, and HCOOH [61]. The reaction mechanism involves six electrons (methanol oxidation), at least, which is relatively complex [57], where the adsorbed formyl-like species $(^*CHO)_{ads}$ play an important role. Thus, efficient electrocatalysts should favor this kind of adsorbed species and avoid the subsequent formation of adsorbed CO. Meanwhile, electrocatalysts should either prevent its formation by blocking neighboring Pt sites or oxidize it at lower potentials.

The reaction mechanisms of the oxidation of other alcohols at the anode are more complicated for us to infer because of the increasing number of exchanged electrons and the difficulty of activation of the C—C bond at moderate temperatures (50–150 °C). The complete oxidation of ethanol and *n*-propanol involving 12 and 18 electrons per molecule, respectively, can be expressed as follows:

$$C_2H_5OH + 3H_2O \rightarrow 2CO_2 + 12H^+ + 12e^- \tag{3.46}$$

$$C_3H_7OH + 5H_2O \rightarrow 3CO_2 + 18H^+ + 18e^- \tag{3.47}$$

Chromatographic techniques (high-performance liquid chromatography [HPLC], GC) [62] or by differential electrochemical mass spectrometry (DEMS) [63] provide us with a more detailed reaction mechanism of ethanol oxidation on Pt electrodes in acid medium, involving parallel and consecutive oxidation reactions, as follows:

$$CH_3CH_2OH + H_2O \rightarrow CH_3CH_2OOH + 4H^+ + 4e^- \tag{3.48}$$

$$CH_3CH_2OH \rightarrow CH_3CHO + 2H^+ + 2e^- \tag{3.49}$$

Reaction (3.48) mainly occurs at higher electrode potentials ($E > 0.8$ V vs. RHE), where the water molecule is activated to form oxygenated species at the platinum surface, whereas reaction (3.49) occurs mainly at lower I potentials ($E < 0.6$ V vs.

RHE) [62]. On the other hand, the dissociative adsorption of water would occur during intermediate potentials ($0.6 < E < 0.8$ V vs. RHE),

$$Pt + H_2O \rightarrow Pt(OH)_{ads} + H^+ + e^- \quad (3.50)$$

Thus, the adsorbed CH_3-CHO in Eq. (3.49) could be further oxidized, so that produces acetic acid as follows:

$$CH_3CHO + Pt(OH)_{ads} \rightarrow CH_3COOH + H^+ + e^- + Pt \quad (3.51)$$

It is usually difficult for further oxidation to CO on Pt electrodes at room temperature. However, poisoning species CO and CO_2 could be clearly detected by IR spectroscopy or DEMS [63], and GC, respectively, as well as traces of methane at low potential ($E < 0.4$ V vs. RHE) [63]. This may be explained as follows:

$$CH_3CHO + Pt \rightarrow Pt(COCH_3)_{ads} + H^+ + e^- \quad (3.52)$$

$$Pt + Pt\,(COCH_3)_{ads} \rightarrow Pt(CO)_{ads} + Pt(CH_3)_{ads} \quad (3.53)$$

$$2Pt + H_2O \rightarrow Pt(H)_{ads} + Pt(OH)_{ads} \quad (3.54)$$

$$Pt(CH_3)_{ads} \rightarrow Pt(H)_{ads} \rightarrow CH_4 + 2Pt \quad (3.55)$$

$$Pt(CO)_{ads} \rightarrow Pt(OH)_{ads} \rightarrow CO_2 + H^+ + e^- + Pt \quad (3.56)$$

The existence of toxic species and intermediate reaction products greatly reduces the useful energy density of fuel, requiring to development of new electrocatalysts, which can break the C—C bond at low temperatures and further oxidize adsorbed CO at lower potentials, so that reduces the oxidation overpotential.

Other electrocatalysts, such as rhodium, iridium (Ir) [64], or gold (Au) [65], were designed to electro-oxide ethanol and similar products were obtained in acid medium. The oxidation of ethanol on Rh mainly went through the formation of CH_3COOH and CO, where linearly bonded and bridge-bonded CO is formed with a similar amount at relatively low potentials and then rapidly oxidized to CO_2 when the surface of Rh began to be oxidized at 0.5–0.7 V vs. RHE [64]. In this case, acetaldehyde was the main product in the acid medium [65].

The electro-oxidation of higher alcohols usually produces amounts of adsorbed CO because of the adsorption of the molecule, which is confirmed by IR spectroscopy. It shows that it is easy for the chemisorption of primary alcohols to dissociate so that produces adsorbed CO, while the ternary alcohols do not produce any CO [57].

In conclusion, most alcohols dissociated at platinum electrodes would produce various chemisorbed species, among which adsorbed CO is usually observed by IR spectroscopy at around 2000 cm^{-1}. These species would make the active electrode sites poisoned, decreasing the stability of the fuel cell performance. Meanwhile, the amount of adsorbed CO would decrease while the number of carbon atoms increases [57].

The biggest problem of clean and sustainable energy conversion technology is to develop highly efficient, cost-effective electrocatalysts, which can catalyze the elementary reaction with long stability. The improvement of catalysts' performances greatly relies on the fundamental understanding of the electrochemical process. This section has surveyed the fundamental mechanism of ECDRR, methanol oxidation reaction (MOR), and water splitting (OER, ORR, and HER) based on representative electrodes. The basic measurement condition for these reactions was also mentioned and discussed.

References

1 Suen, N.T., Hung, S.F., Quan, Q. et al. (2017). *Chem. Soc. Rev.* 46: 337–365.
2 Kattel, S. and Wang, G. (2013). *J. Mater. Chem. A* 1: 10790–10797.
3 Yu, L., Pan, X.L., Cao, X.M. et al. (2011). *J. Catal.* 282: 183–190.
4 Jiao, Y., Zheng, Y., Jaroniec, M., and Qiao, S.Z. (2015). *Chem. Soc. Rev.* 44: 2060–2086.
5 Yan, H.J., Xu, B., Shi, S.Q., and Ouyang, C.Y. (2012). *J. Appl. Phys.* 112, 104316.
6 Bocquet, M.L., Cerda, J., and Sautet, P. (1999). *Phys. Rev. B: Condens. Matter* 59: 15437–15445.
7 Ou, L., Yang, F., Liu, Y., and Chen, S. (2009). *J. Phys. Chem. C* 113: 20657–20665.
8 Alducin, M., Busnengo, H.F., and Diez Muino, R. (2008). *J. Chem. Phys.* 129: 224702.
9 Conway, B.E. and Tilak, B.V. (2002). *Electrochim. Acta* 47: 3571–3594.
10 Fabbri, E., Habereder, A., Waltar, K. et al. (2014). *Catal. Sci. Technol.* 4: 3800–3821.
11 Anantharaj, S., Ede, S.R., Sakthikumar, K. et al. (2016). *ACS Catal.* 6: 8069–8097.
12 Anantharaj, S., Karthik, P.E., and Kundu, S. (2015). *J. Mater. Chem. A* 3: 24463–24478.
13 Lyons, M.E.G. and Burke, L.D. (1987). *J. Chem. Soc., Faraday Trans. 1* 83: 299–321.
14 Kortlever, R., Shen, J., Schouten, K.J. et al. (2015). *J. Phys. Chem. Lett.* 6: 4073–4082.
15 Nie, X., Esopi, M.R., Janik, M.J., and Asthagiri, A. (2013). *Angew. Chem. Int. Ed.* 52: 2459–2462.
16 Sun, Z., Ma, T., Tao, H. et al. (2017). *Chem* 3: 560–587.
17 Zhang, S., Kang, P., and Meyer, T.J. (2014). *J. Am. Chem. Soc.* 136: 1734–1737.
18 Hansen, H.A., Varley, J.B., Peterson, A.A., and Norskov, J.K. (2013). *J. Phys. Chem. Lett.* 4: 388–392.
19 Peterson, A.A., Abild-Pedersen, F., Studt, F. et al. (2010). *Energy Environ. Sci.* 3: 1311–1315.
20 Dewulf, D.W., Jin, T., and Bard, A.J. (1989). *J. Electrochem. Soc.* 136: 1686–1691.
21 Kuhl, K.P., Hatsukade, T., Cave, E.R. et al. (2014). *J. Am. Chem. Soc.* 136: 14107–14113.
22 Mistry, H., Varela, A.S., Bonifacio, C.S. et al. (2016). *Nat. Commun.* 7: 12123.

23 Hoang, T.T.H., Ma, S.C., Gold, J.I. et al. (2017). *ACS Catal.* 7: 3313–3321.
24 Hori, Y., Takahashi, R., Yoshinami, Y., and Murata, A. (1997). *J. Phys. Chem. B* 101: 7075–7081.
25 Yang, K.D., Lee, C.W., Jin, K. et al. (2017). *J. Phys. Chem. Lett.* 8: 538–545.
26 Bertheussen, E., Verdaguer-Casadevall, A., Ravasio, D. et al. (2016). *Angew. Chem. Int. Ed.* 55: 1450–1454.
27 Liu, Y., Chen, S., Quan, X., and Yu, H. (2015). *J. Am. Chem. Soc.* 137: 11631–11636.
28 Genovese, C., Ampelli, C., Perathoner, S., and Centi, G. (2017). *Green Chem.* 19: 2406–2415.
29 Sun, X.F., Zhu, Q.G., Kang, X.C. et al. (2017). *Green Chem.* 19: 2086–2091.
30 Geioushy, R.A., Khaled, M.M., Alhooshani, K. et al. (2017). *Electrochim. Acta* 245: 448–454.
31 Gao, D.F., Zhou, H., Cai, F. et al. (2018). *ACS Catal.* 8: 1510–1519.
32 Hori, Y., Wakebe, H., Tsukamoto, T., and Koga, O. (1994). *Electrochim. Acta* 39: 1833–1839.
33 Thorson, M.R., Siil, K.I., and Kenis, P.J.A. (2013). *J. Electrochem. Soc.* 160: F69–F74.
34 Schizodimou, A. and Kyriacou, G. (2012). *Electrochim. Acta* 78: 171–176.
35 Murata, A. and Hori, Y. (1991). *Bull. Chem. Soc. Jpn.* 64: 123–127.
36 Hori, Y. and Suzuki, S. (1982). *Bull. Chem. Soc. Jpn.* 55: 660–665.
37 Vassiliev, Y.B., Bagotzky, V.S., Khazova, O.A., and Mayorova, N.A. (1985). *J. Electroanal. Chem.* 189: 295–309.
38 Mizuno, T., Naitoh, A., and Ohta, K. (1995). *J. Electroanal. Chem.* 391: 199–201.
39 Rosen, B.A., Salehi-Khojin, A., Thorson, M.R. et al. (2011). *Science* 334: 643–644.
40 Alvarez-Guerra, M., Albo, J., Alvarez-Guerra, E., and Irabien, A. (2015). *Energy Environ. Sci.* 8: 2574–2599.
41 Ye, L., Zhang, M., Huang, P. et al. (2017). *Nat. Commun.* 8: 14785.
42 Kang, X.C., Zhu, Q.G., Sun, X.F. et al. (2016). *Chem. Sci.* 7: 266–273.
43 Bevilacqua, M., Filippi, J., Miller, H.A., and Vizza, F. (2015). *Energy Technol.* (Weinheim, Germany), 3: 197–210.
44 Demirci, U.B. (2007). *J. Power Sources* 169: 239–246.
45 Rice, C., Ha, R.I., Masel, R.I. et al. (2002). *J. Power Sources* 111: 83–89.
46 Lovic, J.D., Tripkovic, A.V., Gojkovic, S.L.J. et al. (2005). *J. Electroanal. Chem.* 581: 294–302.
47 Weber, M., Wang, J.T., Wasmus, S., and Savinell, R.F. (1996). *J. Electrochem. Soc.* 143: L158–L160.
48 Capon, A. and Parsons, R. (1973). *J. Electroanal. Chem.* 45: 205–231.
49 Capon, A. and Parsons, R. (1973). *J. Electroanal. Chem.* 44: 1–7.
50 Parsons, R. and Vandernoot, T. (1988). *J. Electroanal. Chem.* 257: 9–45.
51 Markovic, N.M., Gasteiger, H.A., Ross, P.N. et al. (1995). *Electrochim. Acta* 40: 91–98.
52 Ross, P.N. (1998). *Electroanalysis*, 43–74.
53 Macia, M.D., Herrero, E., and Feliu, J.M. (2003). *J. Electroanal. Chem.* 554: 25–34.
54 Beden, B. and Lamy, C. (1988). *Spectroelectrochemistry*. Boston, MA: Springer.

55 Kolb, D.M. (1988). *Spectroelectrochemistry*. Boston, MA: Springer.
56 Belgsir, E.M., Bouhier, E., Yei, H.E. et al. (1991). *Electrochim. Acta* 36: 1157–1164.
57 Beden, B., Leger, J., and Lamy, C. (1992). *Mod. Aspects Electrochem.* 22: 97–264.
58 Beden, B., Lamy, C., Bewick, A., and Kunimatsu, K. (1981). *J. Electroanal. Chem.* 121: 343–347.
59 Beden, B., Hahn, F., Juanto, S. et al. (1987). *J. Electroanal. Chem.* 225: 215–225.
60 Juanto, S., Beden, B., Hahn, F. et al. (1987). *J. Electroanal. Chem.* 237: 119–129.
61 Belgsir, E.M., Huser, H., Leger, J.M., and Lamy, C. (1987). *J. Electroanal. Chem.* 225: 281–286.
62 Hitmi, H., Belgsir, E.M., Leger, J.M. et al. (1994). *Electrochim. Acta* 39: 407–415.
63 Iwasita, T. and Pastor, E. (1994). *Electrochim. Acta* 39: 531–537.
64 Detacconi, N.R., Lezna, R.O., Beden, B. et al. (1994). *J. Electroanal. Chem.* 379: 329–337.
65 Tremiliosi-Filho, G., Gonzalez, E.R., Motheo, A.J. et al. (1998). *J. Electroanal. Chem.* 444: 31–39.

Part IV

Water Splitting and Devices

Considering energy and environmental issues, electrolyzed water is an efficient and clean technology that produces zero carbon emissions, high-purity products, and intermittent renewable energy (i.e. wind and solar). The anode, electrolyte, and cathode consist of the simplest archetypal cell for water electrolysis. For the water-splitting mechanism, water oxidation occurs on the anode, with the reaction process expressed as this formula: $2H_2O \rightarrow O_2 + 4e^- + 4H^+$ in an acidic electrolyte such as H_2SO_4. This reaction is also called an oxygen evolution reaction (OER). During complete OER, four electrons will be generated from the anode and transfer from the external circuit to the cathode, while four protons will also be generated from the anode and transport to the cathode across the electrolyte. Thus, the two water molecules will be decomposed into one O_2 molecule. At the same time, the generated electron and proton will be combined on the cathode and release H_2 from the cathode. This process is also called the hydrogen evolution reaction (HER) expressed as $4e^- + 4H^+ \rightarrow 2H_2$ in an alkaline medium. Typically, the minimum theoretical voltage for the occurrence of both HER and OER (named as water splitting) is 1.23 V at 25 °C. Nevertheless, these two reactions require excess energy to overcome the kinetics barrier and ensure an ideal rate for the two reactions. It is accepted that with the higher requirement in reaction rates, more energy should be imported, which represents higher reaction overpotential is needed. The overpotential is usually expressed as the symbol η. Electrocatalysts and photoelectrocatalysts are used to reduce such additional energy, thus decrease the overpotential and make the water-splitting voltage approaching 1.23 V or even lower than this value (photo-assisted electrochemical water splitting). Thus far, noble metals of Pt and IrO_2/RuO_2 are still the best electrocatalysts for water splitting. However, the high cost hinders their applications. Improvements in the noble metals or utilization of other cost-effective metal and nonmetal materials are promising. In this part, we will discuss the basic principles for water splitting and present the latest advances for the catalysts and devices.

Electrocatalysis in Balancing the Natural Carbon Cycle, First Edition. Yaobing Wang.
© 2021 WILEY-VCH GmbH. Published 2021 by WILEY-VCH GmbH.

4

Water Splitting Basic Parameter/Others

4.1 Composition and Exact Reactions in Different pH Solution

Electrochemical water splitting consists of two reactions: hydrogen escape reaction (HER) and an oxygen escape reaction (OER), at cathode and anode, respectively. As a strong uphill reaction, the water electrolyzer usually operates at relatively large voltages of 1.8–2.0 V, which is 1.23 V against the theoretical limit. Using a significant catalyst for hydrogen/oxygen releasing or evolution reactions (HER/OER), the reaction rate of delayed reaction HER/OER can be effectively improved, and overvoltage can be reduced. Figure 4.1 shows an electrolyzer used for water splitting, consisting of three parts: an anode, a cathode, and an aqueous electrolyte.

Due to the full availability of protons, HER is easy under acidic conditions and proceeds in a multistep reaction through two possible mechanisms. The mechanism of HER progression was evaluated by the Tafel slope. The first step of the multistep HER is that at the electrode surface, it produces protons, the energy called adsorbed hydrogen (H_{ads}). This is the Walmer reaction ($H^+ + e^- \rightarrow H_{ads}$). The prereaction Tafel slope is:

$$b_{1,V} = 2.303RT/\beta F \tag{4.1}$$

According to the coverage of H_{ads}, desorption of H_2 is regarded as the second step (called Heyrovsky reaction) or the chemical desorption of H_2 (called Tafel reaction). When H_{ads} showed a low value, and there are enough active sites near the position of H_{ads} on the electrode surface, the adsorbed H atoms will combine with protons and electrons at the same time, producing H_2 ($H_{ads} + H^+(aq) + e^- \rightarrow H_2(g)$). That is Heyrovsky reaction, and its Tafel slope is

$$b_{1,H} = 2.303RT/(1 + \beta)F \tag{4.2}$$

When high H_{ads} coverage, at that time, two adjacent H_{ads} chemically combine to release H_2 molecules. That is Tafel's reaction. The Tafel slope is

$$b_{1,T} = 2.303RT/2F \tag{4.3}$$

When an electrocatalyst HER showed the value of the Tafel slope is 0.029 V dec^{-1}, which means it belongs to the Tafel–Heyrovsky. Among these, the electrochemical

Electrocatalysis in Balancing the Natural Carbon Cycle, First Edition. Yaobing Wang.
© 2021 WILEY-VCH GmbH. Published 2021 by WILEY-VCH GmbH.

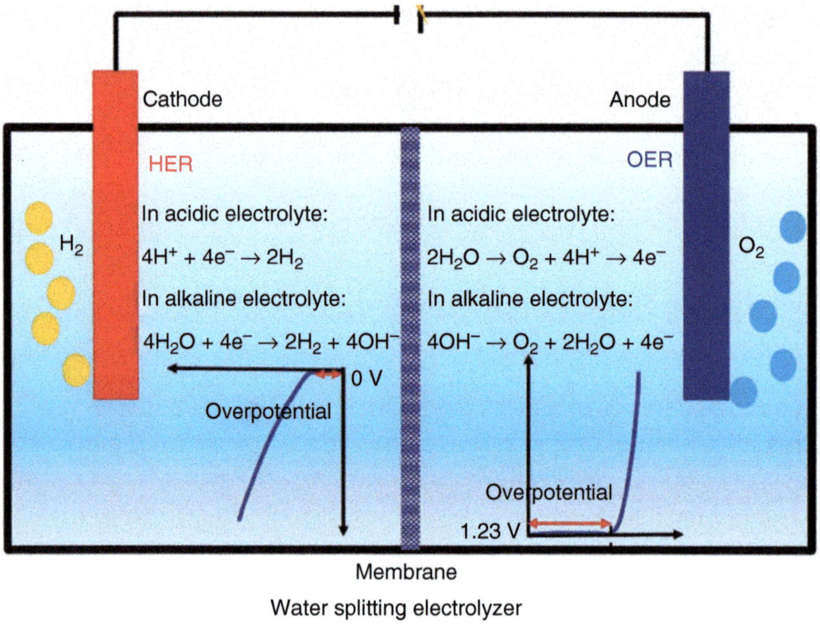

Figure 4.1 Schematic diagram of the reaction rate associated with a water-splitting electrolyzer. Source: Zhang et al. [1]. © 2019, John Wiley & Sons.

desorption process is the rate-determining step. The HER is relatively slow under alkaline conditions because it provides protons to the cathode through deprotonation of the hydroxide ion, also is directly dependent on the anodic OER, which affects the reaction rate of the HER. In this state, the following reactions occur at the cathode.

$$4e^- + 4H_2O(l) \rightarrow 4OH^-(aq) + 2H_2(g) \tag{4.4}$$

Protocols formed by the removal of hydroxide from the anode interact with a large amount of OH^- ion in the alkaline solution, making it difficult for HER to proceed.

OER is different. OER kinetics indifferent media depend on the type of material that was being absorbed. Also, noble metal catalysts precipitate OER easier than acidic media. The VIII group, 3d metals catalogs, are more sensitive to OER in alkaline media. The phenomenon is mostly related to the machines they operate. In alkaline media, all proposed methods start with the required level of coordination of the hydroxide in the active state and proceed to the various proposed steps. The kinematic constraints associated with each level of fundamentals cause the required maximum excess. In dissolution kinetics, the rate-limiting step (RDS) is the first step. The electrochemical reactions occur in different media describe in these equals:

In acid media:

$$\text{Cathode: } 2H^+ + 2e^- \rightarrow H_2 \quad E_c = 0\,V \tag{4.5}$$

$$\text{Anode: } 2H_2O \rightarrow O_2 + 4H^+ + 4e^- \quad E_a = 1.23\,V \tag{4.6}$$

In alkali:

$$\text{Cathode: } 2H_2O + 2e^- \rightarrow H_2 + 2OH^- \quad E_c = -0.83\,\text{V} \tag{4.7}$$

$$\text{Anode: } 4OH^- \rightarrow O_2 + 2H_2O + 4e^- \quad E_a = -0.40\,\text{V} \tag{4.8}$$

4.2 Evaluation of the Catalytic Activity

The main parameters for electrocatalytic characterization in electrochemical water separation include roughness, Tafel slope, current density, flow velocity, stability, Faraday efficiency (FE), and specific gravity and functions [2].

4.2.1 Overpotential

There are no electrochemical reactions that occur at potentials that can only be predicted by thermodynamic considerations, except for the movement disorders that occur in real systems. Because of this kind of interference, in order to have an impact on this kind of electrochemical reaction, further power needs to be obtained. It is called overvoltage (marked with the symbol). OER and HER have three overvoltage sources. That is, reactive overvoltage, concentration overvoltage, and overvoltage caused by the uncompensated resistance (Ru) attached to the electrochemical surface. Active overpressure is a unique characteristic of the substance that stimulates electrode reaction, and each substance is different. Therefore, an effective catalyst can be selected to minimize it. Due to the sharp decrease of concentration near the interface, the concentration of electrode reaction increases at the beginning. Stir the solution to a minimum. The resistance overvoltage can be relieved by the resistance crystal used in many electrochemical workstations. Otherwise, the result current density multiplied by Ru can be manually executed. This potential reduction, called infrared (IR) drop, should be removed from the experimental potential. As to HER, the homomechanics of HER is faster than that of real, so it can be said that the active overvoltage of the initial overvoltage is more important than other overvoltage. This could be calculated from the relationship between current density and voltage. However, in order to calculate overvoltage, OER must consider other parameters. As shown above, all mechanisms undergo water adjustment in acidic solution adjustment in an alkaline solution. After the first basic stage, they go through a series of stages. Basics go ahead. The related kinetic barriers in these stages benefit the overall activation. Mann's group using the ideal conditions of $u = 0$ and standard hydrogen electrode (SHE) to investigate the thermodynamics of the OER mechanism; in order to calculate theoretical OER overvoltage, the next equation is calculated.

$$\eta^{\text{OER}} = \left(\frac{\Delta G_{\text{max}}}{e}\right) - 1.23\,\text{V} \tag{4.9}$$

There was, however, a very large difference between the theoretical and experimental values for changing the standard free energy associated with the basic steps involved in converting oxides to peroxides. This clearly shows that the thermodynamic prediction of the overvoltage does not take movement disorders into account. Due to the different rates of these basic steps in different materials, overvoltage at constant current densities (j). 10 mA cm^{-2} are generally accepted as essential quantitative activity parameters instead of starting overvoltage. Evaluate electrocatalysts more and more. The same is used. Substances with strong redox peaks during the potential window of exhaust and well performance catalysts (current flow) with a density of more than 500 mA cm^{-2} (e.g. double hydroxide) and excess voltage of the high current density additional activity parameters such as 50 and 100 mA cm^{-2} [3–10].

4.2.2 Tafel Slope

A panel diagram for an electrocatalytic process uses plotting a linear sweep voltammogram as a logarithmic vs. η diagram.

The significant part of the Tafel diagram is in its linear part. The slope determination means the dependence between the infrared compensation overvoltage and the current density:

$$\frac{d\ \log(j)}{d\eta} = 2.303RT/\alpha nF \tag{4.10}$$

Because R, T, F, and n are constant, the Tafel slope is the charge transfer coefficient. It is inversely proportional to α. This corresponds to OER: 4 and HER: 2. This means that when there is a small gradient, the catalyst has a good charge transfer capability.

However, this approach ("converting linear sweep voltammetry [LSV] to Tafel charts") raises a number of issues that may lead to improper direction of catalyst activity.

From the experimental results, when the scanning speed of LSV is slower, the error of the value of the Tafel slope is larger. When the catalyst has high capacity, it becomes a serious problem to measure the sampling rate of the polarization curve used in the Tafel chart. Usually, the exchange current density is realized by linear matching and equilibrium potential inference. In this case, a high-pressure catalyst has a high exchange current density. It is impossible to have a high exchange current; that is, the transport of electrons through the catalyst interface is simple, and the activation energy required is very low. As a result, there was a slight increase.

In addition to the capacitive current problem, the current observed in the Faraday range is not true because it is not a normal current. Therefore, the determined table slope value is an incorrect test. This can be said that: in timing analysis ("constant potential"), the normal current is observed only after a few seconds. Even if the sampling rate of the linear scan voltage current method is 1 MW s^{-1}, the current obtained is not the real permanent current. In order to solve this problem, the conventional method can be used. In this method, the static current of catalyst is obtained by I–T chromatographic curve (such as 5 mV) obtained under different stable small interval overvoltage. The value of Tafel can be obtained, compared with the constant current, the value will be larger from cyclic voltammetry (CV) than from the constant current, and the exchange current density obtained from the timing method is more accurate.

In addition, it seems that the latter method has superiority over using the polarization curve, but it is not easy to suppress potential drop caused by uncorrected resistance, thus reflecting the unique and accurate catalytic activity of the material. This is not usually the case in the old methods of recording polarization curves using *iR* compensation [11–15].

4.2.3 Stability

The stability of electrocatalysts is obtained by using CV cycles at higher sampling rates, also called accelerated decomposition tests. As to HER, the polarization starts at 0 V compared to normal hydrogen electrode (NHE), so accelerated decomposition tests are performed for thousands of cycles. For OER, accelerated aging test cycles range from 250 to 1000. After 1000 cycles, such reports on OER catalysts with extreme stability are rarely found. Usually, when the current density is 10 mA cm^{-2}, at that time, the generated overvoltage (anode for HER, an anode for OER) is regarded as the standard of stability. At a given current density or potential, the stability is investigated over minutes to hours. Chronoamperometry now has a stable current density for 12 hours or more (e.g. 10 mA cm^{-2}), or chronopotentiometry has a negligible increase in overvoltage for 12 hours or more at 10 mA cm^{-2} current density. The generally accepted catalyst performance is the overvoltage in the catalytic OER or HER process. The strong redox peak allows long-term chronopotentiometry and chronoamperometric analysis using different standards, usually just using larger current density replace for 10 mA cm^{-2}.

4.2.4 Faradaic Efficiency

The effectiveness of Faraday is another quantitative parameter. This can be defined as the efficiency of the electric catalyst, which transfers the electrons delivered by the external catalyst to the electroactive species. The following two ways were used to determine the effectiveness of Faraday [16–20].

In the first method, a rotating ring-disk electrode was used, called RRDE. Without external interference, catalytic active substances are applied to RRDE disk. Usually, glass carbon (GC) panes and Pt rings are used as RRDEs. When using this method, one is given by the collection efficiency of RRDE. The other is experimentally determined by examining the reaction of conventional iron/Redox iron systems at different speeds. The potential is set in the same with OER, and as to oxygen reduction reaction (ORR), the potential of the Pt ring is set to a given potential. The pH value of the electrolyte influences the ORR potential in the Pt ring. In order to calculate the Faraday efficiency (FE) in OER material, the following equation is used.

$$\mathrm{FE} = \frac{I_R n_D}{I_D n_R N_{\mathrm{CL}}} \tag{4.11}$$

As for measuring the actual performance of OER catalysts, it is a good way. Other electricity that can tolerate unwanted side effects can tolerate catalysts and electric catalysts that emit heat during the electrocatalyst process. The second way to determine Faraday's effectiveness is usually with gas. The amount of gas produced is obtained by integration into a chronometric analysis. After that, you can calculate the actual gas volume (H_2/O_2) using one of the following three methods: the first way is the traditional water gas substitution way and the next way is gas chromatography. The next way is to stimulate the generated oxygen from the triple state into the singlet state, and it can only be applied to the spectral method by fluorescence relaxation. The fluorescence intensity directly measures the oxygen content. The type of catalyst will affect the determination method of actual gas quantity. In this research method, the Faraday efficiency of catalyst refers to the gas volume ratio determined in the actual and theoretical methods. The selection of specific methods is related to the selection of catalysts [21–26].

4.2.5 Turnover Frequency

Another quantitative parameter is turnover frequency (TOF) to adjust electrocatalysts on defined overvoltage. The number of moles of hydrogen and oxygen at one unit is the TOF of the catalyst. The sub equation is generally used to calculate the time rate of gas generation by electrocatalysis.

$$\mathrm{TOF} = \frac{IN_A}{AFn\Gamma} \tag{4.12}$$

I, current, N_A, Avogadro constant, A, geometric surface, Γ is the atom number. There are usually several pathways to characterize the catalytic atomic number or total concentration. The redox peak of cyclic voltammetry can be used to measure the surface concentration of catalysts after the CV cycle. According to the Avogadro method, the average size of catalyst particles can be used to calculate the total atomic concentration. Another way is a one-story house. If the catalyst converter has more than one element of the catalyst converter is not fully activated, the first way may cause an error. The second way does not reflect the precise catalytic performance. In fact, the catalyst component of the catalyst also contains atoms contained in the particle core. The third way may result in errors when the material does not show totally flat, the electrochemical condition is poor, or there is no flake. Therefore, different catalyst chooses different ways to check its performance [27–30].

References

1 Zhang, J., Zhang, Q., and Feng, X. (2019). *Adv. Mater.* 31: e1808167.
2 Shi, Q., Zhu, C., Du, D., and Lin, Y. (2019). *Chem. Soc. Rev.* 48: 3181–3192.
3 Encrenaz, T. (2008). *Annu. Rev. Astron. Astrophys.* 46: 57–87.
4 Kanan, M.W. and Nocera, D.G. (2008). *Science* 321: 1072–1075.
5 Hagiwara, H., Inoue, T., Kaneko, K., and Ishihara, T. (2009). *Chem* 15: 12862–12870.
6 Liu, Y.B., Zhou, B.X., Bai, J. et al. (2009). *Appl. Catal., B* 89: 142–148.
7 Reece, S.Y., Hamel, J.A., Sung, K. et al. (2011). *Science* 334: 645–648.
8 Brillet, J., Yum, J.H., Cornuz, M. et al. (2012). *Nat. Photonics* 6: 823–827.

9 Jacobsson, T.J., Fjallstrom, V., Sahlberg, M. et al. (2013). *Energy Environ. Sci.* 6: 3676–3683.
10 Pinaud, B.A., Benck, J.D., Seitz, L.C. et al. (2013). *Energy Environ. Sci.* 6: 1983–2002.
11 Tachan, Z., Hod, I., and Zaban, A. (2014). *Adv. Energy Mater.* 4, 1301249.
12 Gong, M., Zhou, W., Kenney, M.J. et al. (2015). *Angew. Chem. Int. Ed.* 54: 11989–11993.
13 Malara, F., Minguzzi, A., Marelli, M. et al. (2015). *ACS Catal.* 9: 5292–5300.
14 Zhou, F., McDonnell-Worth, C., Li, H. et al. (2015). *J. Mater. Chem. A* 3: 16642–16652.
15 Anantharaj, S., Ede, S.R., Sakthikumar, K. et al. (2016). *ACS Catal.* 6: 8069–8097.
16 El-Sawy, A.M., Mosa, I.M., Su, D. et al. (2016). *Adv. Energy Mater.* 6, 1501966.
17 Hernández, S., Gerardi, G., Bejtka, K. et al. (2016). *Appl. Catal., B* 190: 66–74.
18 Liu, G.Y., Karuturi, S.K., Simonov, A.N. et al. (2016). *Adv. Energy Mater.* 6, 1600697.
19 Qu, K.G., Zheng, Y., Dai, S., and Qiao, S.Z. (2016). *Nano Energy* 19: 373–381.
20 Singh, S.K., Kumar, D., Dhavale, V.M. et al. (2016). *Adv. Mater. Interfaces* 3, 1600532.
21 Wu, J.J., Liu, M.J., Chatterjee, K. et al. (2016). *Adv. Mater. Interfaces* 3, 1500669.
22 Xiao, W., Huang, X.L., Song, W.D. et al. (2016). *Nano Energy* 25: 60–67.
23 Yin, J., Fan, Q.H., Li, Y.X. et al. (2016). *J. Am. Chem. Soc.* 138: 14546–14549.
24 Zhang, C.L., Wang, B.W., Shen, X.C. et al. (2016). *Nano Energy* 30: 503–510.
25 Zhu, D.D., Liu, J.L., and Qiao, S.Z. (2016). *Adv. Mater.* 28: 3423–3452.
26 Zhao, S.L., Wang, Y., Dong, J.C. et al. (2016). *Nat. Energy* 1: 1–10.
27 da Silva, G.C., Perini, N., and Ticianelli, E.A. (2017). *Appl. Catal., B* 218: 287–297.
28 Ross, M.B., Dinh, C.T., Li, Y. et al. (2017). *J. Am. Chem. Soc.* 139: 9359–9363.
29 Hou, Y.H., Liu, Y.P., Gao, R.Q. et al. (2017). *ACS Catal.* 7: 7038–7042.
30 Huang, X.B., Zhao, G.X., and Wang, G. (2017). *J. Mater. Chem. A* 5: 24631–24635.

5

H_2O Oxidation

5.1 Regular H_2O Oxidation

5.1.1 Noble Metal Catalysts

RuO_2 and IrO_2 are generally considered to be the last oxygen evolution reaction (OER) electrocatalysts, while IrO_2 has quite a stability in alkaline acid and rough media. But it is very important to solve some public problems. Due to the complexity of the real electrochemical process, catalysts based on IrO_2 are as follows: (i) in-depth understanding of surface species and their evolution in real process; (ii) identification of actual depth of catalyst surface or volume; (iii) identification of structure–activity stability relationship dissolving metals; and (iv) clarifying the doping effect of isomers in the real process.

There are benefits of field observations using advanced technologies, for example, the ambient pressure X-ray photoelectron optical spectrometer. In the process, the former two questions were answered completely, confirming the chemical change of Ir(IV) in Ir(V) and the oxidation hydroxyl oxide channel on the catalyst surface. The surface changes of hydroxides and Ir V detection also provide an inhibition mechanism mediated by OOH.

It needs lager free energy to get OOH intermediates, which is driven by the position of an Ir V atom. X-ray photoelectron spectroscopy (XPS) results show that on the upper atomic layer, there have 14.6% of the Ir IV converted to Ir V, while the detection depth is 0.7 nm. The increase in the released kinetic energy to further increase the detection depth to 1.11 nm did not significantly change the fraction ratio between Ir(IV) and Ir V, indicating that the important impactor of the performance is the atomic layer. Recently, Chandra's group carefully fabricated intermediate hydroxyl and oxo terminal films (OH) under annealing conditions of 300–350 °C, the motivation of which is the change of surface hydroxyl. Although the very small nanoparticle (NP) (3–5 nm) provides a free channel for electrons to move across the crystal boundary rapidly, the performance of $IrO_x(OH)_y$ NPs attributed to the amorphous IrO_x and IrO_2 crystals. Thus, regulating composition in the surface is very important to provide a very effective OER catalyst.

The improvement of the catalytic performance of OER metal or IRs electrode catalyst is limited to the change of morphology in order to get better active sites or change

Electrocatalysis in Balancing the Natural Carbon Cycle, First Edition. Yaobing Wang.
© 2021 WILEY-VCH GmbH. Published 2021 by WILEY-VCH GmbH.

the metal-supported interaction. Our main focus is to optimize the alloy composition of bimetallic/trimetallic alloy, IR-based OER catalyst, including the advantages of internal catalyst varieties. In order to improve the electrochemical performance, the secondary metal was transformed and doped. Under the stimulation of the favorable influence of copper doping in the IrO_2 system, the redistribution of Ir-5d orbital electrons, and the bimetallic or trimetallic alloy based on IrCu and various nanostructures, the doping effect of Fe and Co has been explored more deeply. The alloys based on Ir are Ni, W, etc. It was agreed that in all Ir-M ("m = Cu, Fe, Co, Ni"), the metal state of Ir evolved into an "oxide state" (Ir^{3+}, Ir^{4+}, or $Ir^{>4+}$), which served as a single effective site for catalyzing the reaction. This is determined by the ratio of hydroxyl species adsorbed on the catalyst surface. Due to the occupation of part of the orbit and d-band center shift caused by the Jahn Teller effect to adjust the atomic ratio to optimize the binding force with the intermediate species adsorption, Cu addition deformed the IR electronic structure. In the Ir–Co–Ni system studied by Guo, the movement of d-band center away from the Fermi level was obtained. The binding energy of oxygen intermediates is lower than that of pure Ir, as shown in Figure 5.1a–c.

The binding energy classification of Ir M(Ir-metal) NP is obtained by calculating the state density ("back") of the d-band. The minimum of Tafel efficiency and the

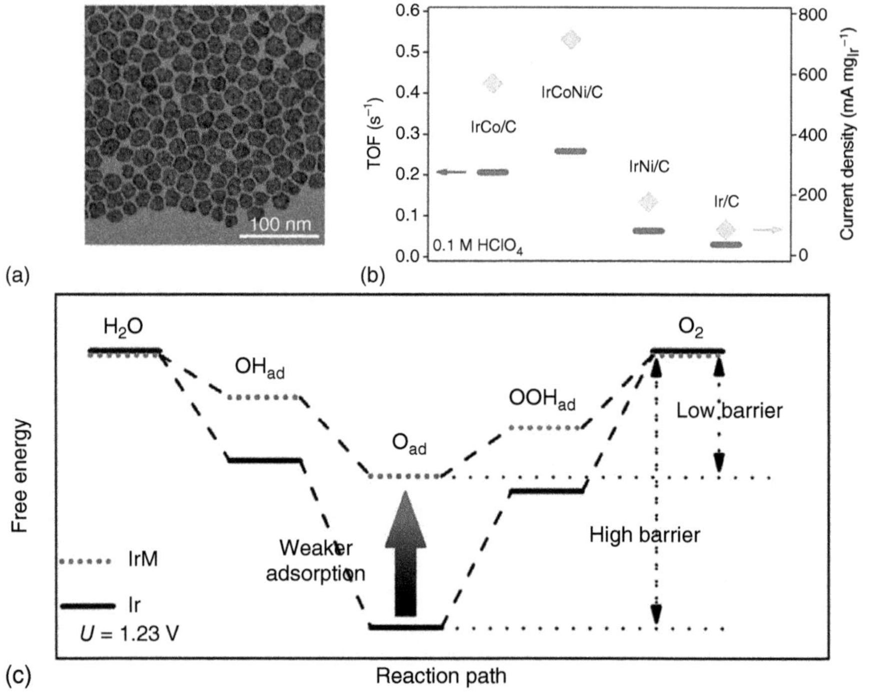

Figure 5.1 (a) TEM image and model (insert); (b) TOF value of IrM alloy; (c) OER reaction path for alloy optimization of Ir with M. Source: (a) Feng et al. [1]. Reproduced with permission of Wiley-VCH Verlag GmbH & Co. KGaA, Weinheim, (b, c) Feng et al. [1]. © 2017, John Wiley & Sons.

maximum of turnover frequency (TOF) represent the electrocatalytic activity and the reaction kinetics of the OER catalyst (Figure 5.2). W doping in IrW nanowires not only leads to the higher corrosion resistance of acid electrolyte in the OER catalyst but also the optimization effect of binding energy. Compared with IrW_2, the better corrosion resistance of IrW alloy benefits excellent stability, that is, due to an enhanced interaction between Ir^{x+} and O^{y-}, which is emphasized in the largest region of Ir_{5d} and O_{2p} orbital overlap. However, there are divergences in the interpretation of the coordination between CO doping and catalytic activity. Alias et al. hypothesized that due to the compression of IR network, the chimerism of Ir–O reduced by the increase in specific activity with Co alloy, but this conjecture shows that there is no conclusive evidence. Another hypothesis is that the Ir–OH active species produced in the rel dissolution process can improve the rel performance by dissolving Ni in the rel process. Therefore, the composition can be effectively optimized to improve the catalytic activity of rel by doping heteroatoms to adjust the binding energy or by changing the surface chemical properties to produce more active oxide species.

It is known that high OER catalytic activity can be obtained on rutile RuO_2, and the Ru surface has large redox false capacitance. Nevertheless, the relationship between the effect of pH and these properties is unclear. In this regard, Stoerzinger

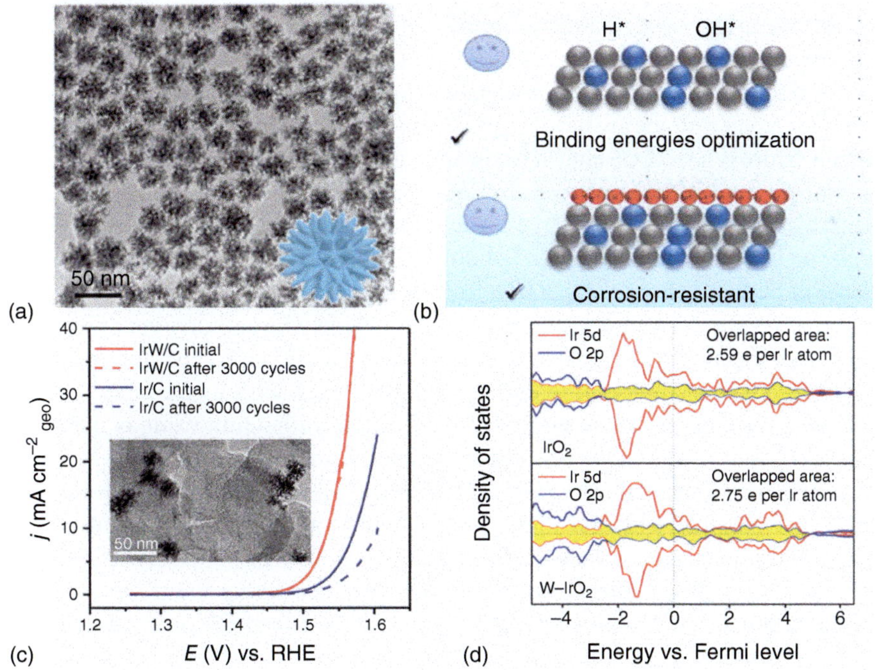

Figure 5.2 (a) TEM image of as synthesized IrW nanodendrites (NDs); (b) proposed illustration of the W-doping effect; (c) DOS; (d) poloidal curve before and after IrW/C and Ir/C periodic test; and (d) density of states (DOS). Source: (a, c) Shi et al. [2]. Reproduced with permission of The Royal society of Chemistry, (b, d) Shi et al. [2]. © 2019, Royal Society of Chemistry.

et al. studied the effect of surface termination on pH (the relationship between OER activity and pseudocapacitance of RuO_2) [3]. The OER activity is not only dependent on the active site densities but on their oxidation potentials at a given pH. The pH-dependent OER activities were found on the (001), (101), and (111) crystal planes with pH-dependent Ru redox, so they have lower activity at low pH values, in which higher potentials vs. reversible hydrogen electrode (RHE) for Ru redox occur. The relationship between the OER reaction sequence and super-Nernst transition of Ru redox shows further that the oxidation of Ru at the coordinated unsaturated "CUS" is an essential precursor for the release of O_2. In addition, the rate-determining steps on the surfaces of (001), (101), and (111) involve an intermediate. The rate of displacement and formation or binding strength of this intermediate is affected by proton activity in the electrolyte, resulting in the fact that activity is dependent on the pH values, and Ru (110) and (100) surfaces have pH-independent OER activity.

At the same time, Reshma R. Rao and coworkers used surface X-ray scattering or crystal-truncated rod (CTR) analysis to measure the relationship between the atomic structure change of in situ single-crystal RuO_2 (110) and the acid potential, which was supplemented by density functional theory (DFT) research [3]. In their work, they used DFT to calculate the energy of surface adsorbates and CTR measurements to help refine the atomic structure. Using this synergistic method, we found that although the bridge oxygen atom is protonated at 0.5 V vs. RHE, water dissociation is observed at 1.0 V RHE at the CUS position (Figure 5.3). Under these potentials, every second of water molecules dissociate to form an –OH group and every second of bridging oxygen atoms are protonated. The redox transition at ~1.1 V RHE results in a fully hydroxylated CUS site. What is more noteworthy is that we found through experiments that –OO substances are stable before oxygen is precipitated at 1.5 V RHE. This is supported by our DFT calculation, which shows that on the basis of the strong interaction between –OO and the adjacent –OH group, the unhindered transfer of hydrogen from the –OOH structure to the structure formed by the adjacent oxygen is at this potential, that is, the area is stable. The author's work provides unique experimental evidence for the OER active surface of RuO_2 (110) and proposes an OER reaction step that is different from the traditional OER mechanism.

Therefore, as demonstrated by Laha Sourav et al., ruthenium oxide is used for effective OER [5]. In their work, Ru oxide was made into 2D nanosheet form by peeling off hexagonal ruthenic acid nanosheets, which is the first report. The etched Ti plate with the dimensionally stable anode setting was used for the deposition of the exfoliated nanosheets. The OER electrocatalytic activity of the synthesized nanosheets was measured after depositing them onto the Ti plate in dimension stable anode (DSA). Three different loadings on the electrode (c. 0.5, 0.2, and 0.07 $mg\,cm^{-2}$) were initially measured. Since the nanosheets with loading at 0.2 $mg\,cm^{-2}$ showed the highest electrocatalytic activity, studies were further carried out under this optimal loading. The reference electrocatalyst had employed a commercial rutile RuO_2 with testing in the same loading. It shows the relationship between the current density (j) and the potential curve at 10 $mV\,s^{-1}$. The geometric area of the electrode was used to normalize the current. In order to

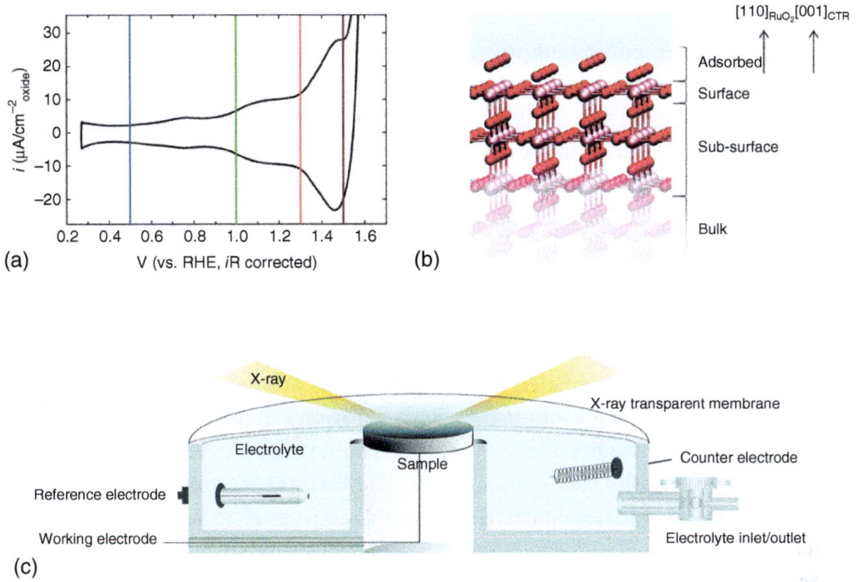

Figure 5.3 (a) Cyclic voltammogram of a RuO_2 (110) single-crystal surface in Ar-saturated 0.1 M $HClO_4$ solution (~pH 1.2) measured at a scan rate of 50 mV s^{-1}. The vertical lines at 0.5, 1.0, 1.3, and 1.5 V vs. RHE indicate the potentials at which surface diffraction measurements were performed; (b) model of the (110) surface used for fitting. Pink and red spheres represent Ru and O atoms, respectively. All atoms in the "adsorbed," "surface," and "sub-surface layer" were allowed to relax in the *z*-direction, with the constraints being tighter for the "sub-surface" layer atoms. The lattice parameter in the *z*-direction for the "bulk" and "sub-surface" layer was also allowed to relax; and (c) schematic of the three-electrode X-ray electrochemical cell and the X-ray scattering geometry. Source: Rao et al. [4]. © 2019, Royal Society of Chemistry.

obtain current at 10 and 1 mA cm^{-2}, at which value-efficient solar water splitting occurs, the overpotentials of the synthesized nanosheets are approximately 255 and 215 mV, respectively. Such performance is obviously better than the rutile RuO_2. For nanosheets, a current of approximately 9.5 mA cm^{-2} was obtained at $\eta = 250$ mV. In comparison to other precious-metal-based catalysts, the low overpotential values of these nanosheets, combined with the small Tafel slope of 38 mV dec^{-1}, show the high potential for an effective OER application. These data show that compared with rutile RuO_2 nanoparticles, when $\eta = 250$ mV, the current density of ruthenate nanosheets is increased by about 3 orders of magnitude (≈9.5 mA cm^{-2}; loading 0.2 mg cm^{-2}, etching Ti plate with DSA electrode, ≈10 μA cm^{-2}; load 0.05 mg cm^{-2} in the same electrolyte, glass–carbon rotating disk electrode). In electrochemical impedance spectroscopy (EIS), ruthenate nanosheets have low charge transfer resistance, indicating the higher catalytic charge transfer rate compared to rutile RuO_2. Further, catalytic current corrected by Ir was normalized to the normalized current density of rutile RuO_2 (1.8 A g^{-1}), rendering it to be one of the best performances reported until now. Compared with bulk materials, the authors attributed the enhanced activities to or the significant decrease in the overpotential on active sites of nanosheets or the increase in the active site number. It is known that abundant edge and surface sites contribute to the better OER performance in CoFe-layered double hydroxide (LDH) nanosheets in alkaline media. Besides, the latest research on the OER electrocatalysts of Ir–Ni oxide pointed out the key mechanism of surface OH during OER. Compared with RuO_2 nanoparticles, the number of surface hydroxyl groups in ruthenic acid nanosheets increased significantly. C_{dl} was further used to determine the electrochemical surface area (ECSA), which is strongly associated with the activity origin. The slope from the linear relationship between the scan rate and the capacitance current is equal to twice the C_{dl}. The result shows that the ECSA of ruthenate nanosheets is six times higher than that of rutile RuO_2; thus, it should be one key factor to promote nanosheet performance. It obtains the amount of deposited nanoplatelets that are mass active. In order to achieve the current of $10\,A\,g^{-1}$, an overpotential η of 225 mV is required on the nanosheets, while it is 343 mV for rutile RuO_2. The nanosheet electrode achieves a current of $\approx 42\,A\,g^{-1}$ as the η equal to 250 mV. This current is almost 4 times higher compared to rutile RuO_2 nanoparticles ($11\,A\,g^{-1}$) and more than 20 times by the mass activity. Chronopotentiometry and cycle stability were used to investigate the long-term stability of the synthesized nanosheets. The chronopotentiometer was recorded over six hours at constant currents of 1 and $10\,mA\,cm^{-2}$. The rutile RuO_2 had an overpotential of $\approx$300 mV at $1\,mA\,cm^{-2}$, while it is only $\approx$220 mV for the nanosheets. At the current of $10\,mA\,cm^{-2}$, the overpotential of the nanosheets increased from $\approx$260 to $\approx$520 mV during the test, while the overpotential of rutile RuO_2 raised from $\approx$700 to 1000 mV or more within two hours. This attenuation is due to the partial dissolution of Ru into solution, as well as the physical separation between electrode and nanosheets caused by abundant oxygen bubbles formed on the DSA, causing the nanosheets to become coarse.

In addition to pure Ru compounds, the Ru component can also be used in combination with other metals/nonmetals to enhance its activity. For example, Lin et al. reported an iridium-free, low-ruthenium oxide material ($Cr_{0.6}Ru_{0.4}O_2$), which is prepared by a metal–organic framework (MOF) and has excellent OER activity under acid conditions [6]. At the current of $10\,mA\,cm^{-2}$, it showed an extremely low overpotential at 178 mV and maintained outstanding activity during the entire 10-hour chronopotentiometric measurement at the current of $10\,mA\,cm^{-2}$ in a 0.5 M H_2SO_4 electrolyte. It was revealed from DFT calculations that the internal mechanism of the excellent OER activity highlights the effect of the Cr element for enhancing the stability and activity. On the other hand, Lei Wang et al. reported a simple solid–liquid phase chemical method which is used to produce hollow Ru–RuP_x–Co_xP polyhedrons [7]. Here, due to the unstable surface termination demonstrated by XPS, X-ray diffraction (XRD), and line scanning electrons, the in situ reconstruction of Ru occurred on the surface of 2P to form Ru–RuP_x–Co_xP polyhedrons, as revealed by electron energy loss spectroscopy (EELS) analysis. Ru–RuP_xCo_xP with abundant active sites arisen from the big specific surface area showed significantly better OER activity (the overpotential is only 291 mV at $10\,mA\,cm^{-2}$) and excellent stability (more than 10 000 cycles). It was further proven by DFT calculations that the existence of Ru reduces the energy for intermediate adsorption and also improves the electron transfer rate by increasing the density of states at the Fermi level, thereby promoting the electrocatalytic reaction kinetics.

Compared with ruthenium-based materials, iridium-based materials have been more widely used in OER. Yuan Ping et al. reported firstly the atomic mechanism of a heterogeneous electrocatalyst for OER, involving a free energy barrier under a stable potential test [8]. They employed DFT to study the OER mechanism on the surface of IrO_2 (110), involving dynamic barriers under equipotential conditions. The overpotential and Tafel slope calculated on the basis of the micro-kinetic models are consistent with the experimentally results by electrochemical measurements. Their main findings include: (i) in some cases, thermodynamically favorable reaction steps are kinetic unfavorable; (ii) rate-determined water dissociation step relied strongly on the potential test condition (though it is not the constant charge condition); and (iii) surface-dissociated O_2 changes continuously the double states when it was bound to surface to the triplet states of dissociation, and O_2 should be removed from the surface to further bind H_2O molecule on O vacancy. The top O atom, which has an unpaired spin state, was observed to be the active site, which suggests that the most unsaturated Ir atoms could be the most active on the surface orientation. It was due to the higher reactivity of exposed surfaces of unsaturated Ir atoms with H_2O. On the basis of the high binding energy of water on the surface of IrO_2, the surface OH sites will be formed, resulting in free radical characteristics on O surface on the high OER potential toward protonation. The result indicates that the IrO_2 (110) surface is less active compared to the unstable IrO_2 (100) surface. These mechanical insights related to the OER on IrO_2 (110) could offer some guidance for the design of catalysts with higher activity.

Akbashev et al. used perovskite heterostructures with ultra-thin core–shell structures (with the smallest core/shell thickness of 0.4 nm) as a model system for the design of highly active/stable catalysts [9]. All of these electrocatalysts are heterostructure with multilayer films epitaxially growing onto a substrate with a single-crystal structure. Based on the DFT calculations, the author found that new 4d electronic states in the $SrTiO_3$ (STO) band gap were introduced by the chemical modification of STO, including the Ru state, thereby increasing the electron energy level and changing the electron to hybridize and transfer easily into the intermediates adsorbed on the surface. This hypothesis was verified by employing experiments of atomically accurate heteroepitaxial deposition. They found it is sufficient for the activation of the topmost STO layer by the single unit layer of $SrRuO_3$ (SRO) during OER. The inherently unstable SRO was prevented from corrosion in OER processes by the buried SRO under the two-unit cells. In general, their hierarchical heterostructure is a model system of oxide core–shell structure, in which unstable catalysts (cores) are protected from degradation, and electrons activate the inactive shells. Through growing ultra-thin heterostructures, it can elucidate the mechanism for subsurface activation used to obtain active/stable oxide catalysts and establish a rigid platform facilitating screen of core–shell particle combinations/thickness.

Shaun M. Alia et al. have prepared Ir–Co and Ir–Ni nanowires through electric substitution and evaluated the improved durability and overall performance of electrolytic materials [10]. The performance of Ir–Co and Ir–Ni nanowires on OER was tested in half-cells equipped with a rotating disk electrode as well as a single-cell electrolyzer, and it was analyzed in comparison with the benchmarks. In the half-cell test, the mass-specific activity of the nanowire electrocatalyst was enhanced by order of magnitude compared with the commercialized Ir nanoparticles. When excess Co and Ni were removed by acid leaching, the nanowire catalyst also showed great durability. Although the two materials have some specific differences, it was found that Co and Ni templates show similar positive effects. In the single-cell electrolysis test, the performance of the nanowires is four to five times better compared to the Ir nanoparticles, which indicates that a significant reduction in the catalyst loading is possible without affecting the performance.

da Silva et al. studied the temperature effects toward the activity and stability of the IrO_x electrocatalyst layer prepared by a hydrothermal reaction for OER [11]. IrO_x particles were further processed at the different temperatures during calcination, and the OER activity/stability was tested. Several techniques are used to characterize the catalysts for physical and chemical characterization, which include transmission electron microscopy, X-ray photoelectron spectroscopy, X-ray diffraction, X-ray absorption spectroscopy, and energy-dispersive X-ray spectroscopy. It was found that the volt–ampere curve of the catalyst calcined at a temperature as high as 300 °C is almost the same as the curve of electrochemically prepared hydrated IrO_x, and the cyclic voltammetry (CV) curve is the characteristic of the thermally prepared IrO_x with calcination at high temperature. As the calcination temperature of IrO_x increases, its OER activity is reduced, but the stability of these materials shows the opposite trend. The electrocatalysts which were calcined in the temperature of 400–500 °C show a higher balance between stability and activity. Nevertheless, in spite of the high loss in performance, the unaged IrO_x catalyst could still show high mass-specific activity after aging experiment when the electrode potential relative to RHE is as high as 1.6 V. Using the same position transmission electron microscope to study the reason which causes the activity decrease of the electrode, the data indicated that the instability of the IrO_x electrode was due to the degradation of the IrO_x thin layer and the dissolution of iridium oxide.

Zhang and others have developed a simple one-step method to prepare Ir-doped OER Co-based hydroxide nanosheets with well dispersion (named Co-Ir) [12]. Through the formation of abundant Co-Ir species, the defect-rich hydroxide nanosheets were incorporated with the Ir clusters and/or single atom Ir, which is characterized by X-ray spectroscopy based on high-angle circular dark-field scanning transmission and system synchrotron radiation. Combined with the results of electron microscope measurements. The optimized CoIr with an Ir content of 9.7 wt% shows excellent electrocatalytic OER performance to reach the current density of 10 mA cm^{-2} at a low overpotential of 373 mV in a 1.0 M phosphate buffer solution, which is obviously better than the commercial IrO_2 electrocatalysts. The characterization of the electrocatalyst after OER showed that the high-valent low-coordination Ir species and unique Co oxyhydroxide were generated based on the positive electrooxidation potential, contributing indeed to the excellent OER activity.

Jongsik Park et al. prepared IrNiCu double-layered nanoframe (DNF) [13]. In the synthesis process, $IrCl_3$, $Ir(acac)_3$, $Cu(acac)_2$, $Ni(acac)_2$, 1,2-HDD (1,2-hexadecanediol), CTAC (hexadecyl trimethyl ammonium chloride), ammonium chloride, and oleylamine were mixed in a Schlenk tube under stirring. Hereinafter, the reaction compositions are taken as standard compositions. The solution was placed in a vacuum oven at 60 °C for 5 minutes, and then the Schlenk tube was directly placed in an oil bath at 260 °C for 40 minutes. The mixture was then cooled down to 25 °C under stirring action. Ethanol and toluene were added into the mixture; the reaction mixture was then treated by centrifugation at the rate of 4000 rpm for five minutes. IrNiCu DNF is synthesized by selectively removing Cu and Ni elements from the formed nanoparticles employing hydrochloric acid etchant at 60 °C. The use of single Ir precursor yielded single nanoframe structures, highlighting the importance of employing dual Ir precursors. In addition, the structure of Ir-based nanocrystals could be further controlled to DNF with octahedral morphology and CuNi@Ir core–shell structures via a simple tuning of experimental factors. The single nanoframe structure was generated by the employment of a single iridium precursor, suggesting the fact that using a dual Ir precursor is of great importance. Additionally, by adjusting experimental parameters, an octahedral structure of Ir-based nanocrystals or a CuNi@Ir core–shell structure DNF could be obtained. High OER activity over IrNiCu DNF in acidic media was obtained; this activity is greatly higher compared to the Ir/C electrocatalyst. In addition, excellent OER durability was also observed on the IrNiCu DNF. This arises from the fact that particle agglomeration and growth were hindered by the frame structure, and a strong rutile IrO_2 phase was in situ formed during long-term operation.

Emma Oakton et al. have prepared IrO_2–TiO_2 [14] and in their synthesis, titanyl sulfate hydrate ($TiOSO_4{\cdot}xH_2SO_4{\cdot}yH_2O$) and iridium acetylacetonate ($Ir(acac)_3$) are selected as chlorine-free molecular synthesis materials. With a heat treatment of this homogeneous mixture of sodium nitrate and these precursors for one hour at 350 °C and subsequent washing and drying processes, the authors successfully synthesized the IrO_2–TiO_2 catalyst at the 40 molM% Ir. According to the N_2 adsorption–desorption isotherm test, the BET surface area of the obtained black powder is 245 $m^2\ g^{-1}$ (IrO_2–TiO_2-245). The authors' previous results show that the surface area of the synthesized pure IrO_2-23 under the same conditions is 150 $m^2\ g^{-1}$. It leads to getting that and titanium oxide iridium to contribute equally to the final surfaces in IrO_2–TiO_2-245. The material has a conductivity of 0.26 $S\ cm^{-1}$, comparable with the value of IrO_2–TiO_2 (0.49 $S\ cm^{-1}$) at the same nominal content of Ir made of $IrCl_3$. IrO_2–TiO_2 is prepared step by step from molten $NaNO_3$, and it is consisted of c. 1–2. In the matrix of TiO_2 nanoparticles, there are 1–2 nm IrO_2 particles distributed, and the total BET surface area is c. 245 $m^2\ g^{-1}$. About 40 molM% of Ir is included in the material, and OER activity/stability reaches a better one compared with the commercially available benchmark electrocatalyst, as well as the prior IrO_2 with high surface area. The presence of Ir hydroxyl surface species was demonstrated by ex situ characterizations, and these species had been previously related to high performance for OER. Studies from operando X-ray absorption spectra show that with the change of applied potentials, the surface species would alter, suggesting the initial hydroxyl surface layers are transformed into an oxygen-terminated surface through anodization (OER method).

Kwon et al. investigated the IrCu alloy core–shell nanoparticles with iridium-rich shell and copper-rich core prepared by simultaneously decomposing copper and iridium precursors [15]. The rapid decomposition of the copper precursor resulted in the generation of copper-based nanoparticles, and the subsequently decomposed iridium precursor on the copper nanoparticles generated an iridium-based shell. However, based on the fact that the Ir-rich shell has impermeability, the Cu component cannot be leached under acidic conditions. By introducing an Ir-rich shell with other impurity elements which have different lattice parameters with Ir, the permeability of the Ir-rich shell may be reduced, which will cause the internally captured Cu to be leached. The authors further studied the influence of Co impurities on IrCu nanoparticle growth during the Cu and Ir precursor decomposition to verify this route to generate hollow nanocage structures of the Ir-based catalysts. For a typical synthesis of Co-doped IrCu octahedral particles, as designed with Co-IrCu octahedral nanocages (ONs), Ir(III) acetylacetonate ($Ir(acac)_3$), and Cu(II) acetate ($Cu(OAc)_2$), Co(III) was magnetically stirred to form a slurry in oleylamine, and cetyltrimethylammonium bromide (CTAB) was added in a Schlenk tube. The tube was evacuated by magnetic stirring for two minutes at 25 °C and purged with CO gas at 1 atm. Afterward, this tube was placed in an oil bath and heated to 270 °C for 0.5 hours with the presence of CO gas. In the end, the solution was cooled down to 25 °C, followed by centrifugation using methanol and toluene to obtain a dark black precipitate. Selectively removing Cu components aided with acetic acid to prepare Co-doped IrCu catalyst with octahedral hollow nanocage structure from Co-IrCu ON, which was defined as Co-IrCu octahedral hollow nanocage (ONC). The authors also compare the OER activity between Co-IrCu ONC/C catalyst and Ir/C electrocatalyst. From the observation of the OER polarization curve (Figure 5.4), it clearly shows that the Co-IrCu ONC/C catalyst has excellent OER activity compared with Ir/C. In order to obtain the current of 10 mA cm^{-2} (this current is equivalent to 10% of solar energy converted into fuel under 1 solar radiation), Co-IrCu ONC/C requires 0.293(±0.001) and 0.315(±0.005) V overpotential (η) Ir/C catalyst. More important, from comparing the reported current density, overpotential, and mass activity of the OER catalyst based on Ir nanoparticles, it shows that the Co-IrCu ONC/C electrocatalyst is one of the best electrocatalysts for OER. The Tafel curve of the catalyst shows that the Tafel slope of the two electrocatalysts is similar to the value of 50 mV dec^{-1}, which may imply that the Ir/C and Co-IrCu ONC/C electrocatalysts have similar reaction kinetics. Kinetic data and OER activity show that nanostructured alloying facilitates to enhance the activity of the Ir-based catalysts for OER, while the reaction kinetics was not altered by alloying. A long-term durability test was further used to evaluate the stability of the Ir-based catalyst. The test was performed by 200 CV cycles in the potential of 1.2–1.7 V (relative to RHE). From the OER polarization curve after cycling, it is worth noting that Co-IrCu ONC/C is highly durable, and the activity after the cycle test is only reduced by 3%. In contrast, after the durability test, Ir/C was significantly inactivated, and its activity was reduced by 50%.

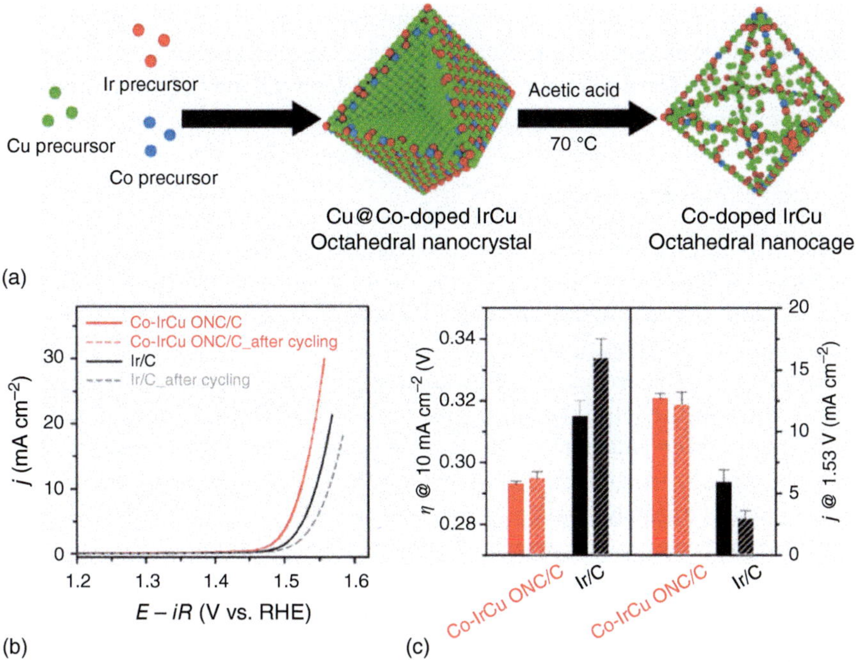

Figure 5.4 (a) Schematic illustration of the formation process of Co-doped IrCu octahedral nanocages; and electrocatalytic OER activity of Co-IrCu ONC/C and commercial Ir/C catalysts. (b) OER polarization curves of Co-IrCu ONC/C and Ir/C were measured in 0.1 M $HClO_4$. Dotted lines indicate the curves obtained after cycling tests for 2000 cycles from 1.2 to 1.7 V (vs. RHE) at 50 mV s^{-1}. (c) Comparison of overpotentials to drive 10 mA cm^{-2} and current densities at 1.53 V (vs. RHE) between the samples. Patterned bars indicate the activity parameters after the cycling tests. Error bars are the standard deviations estimated from repeated measurements of at least three times. Source: Kwon et al. [15]. © 2017, John Wiley & Sons.

In addition to Ru- and Ir-based materials, OER also reported precious metals of Au- and Ag-based materials. Alaina L. Strickler et al. reported an Au@metal oxide nanoparticle electrocatalyst with a core–shell structure has been prepared [16]. From the beneficial effects of the gold support to the high surface area particles composed of a metal oxide shell and a gold core (Au@M_xO_y, where M = Co, Fe, Ni, or CoFe). The authors established a performance trend and demonstrated the increase in general activity as using gold cores in the nanoparticles from 3d transition metal oxide via such evaluation. An overpotential of 328 ± 3 mV was observed from the most active Au@$CoFeO_x$ particle in the two-hour long-term test at 10 mA cm^{-2}, which shows that the core–shell nanoparticle morphology strategically couples the Au carrier and the mixed metal oxide, which could be a potential method to reach equipment-ready efficient electrocatalysts OER. On the other hand, Hou et al. reported a Ag@Co_xP core–shell heterogeneous catalyst prepared by the one-pot method [17]. These include the formation of silver seeds, the growth of cobalt shells, and subsequent phosphating treatment. The electrocatalyst only required a small overpotential of 310 mV to provide the current of 10 mA cm^{-2}.

It also has high electrocatalytic stability. In addition, the Ag@Co_xP catalyst had the electrocatalytic activity, which is eight times better compared to the activity on CO_2P nanocatalyst. The strong electron interaction between the Co_xP shell and the Ag core is responsible for such high performance.

5.1.2 Other Transition Metals

Catalysts without precious metals have aroused great interest and made great scientific progress. This section focuses on the recent progress in electrolytic catalyst (mainly Ni, Mn, CO, and Fe) rel precious metals classified as metal/alloy, oxide, hydroxide, sulfide, and phosphorus.

Due to its high corrosiveness, pure metal and non-noble metal catalysts cannot survive directly in the hard electrolyte, including acid and alkali medium. As a result, they are often integrated or decorated in relatively stable material (e.g. carbon materials, whether in a metal or alloy state). Xu et al. reported that an efficient OER electrocatalyst was prepared by simply coating MOF material. The Ni(-NP) nanoparticles were encapsulated in doped *n*-graphite. Metal alloys usually have synergistic effects between metals, which leads to improved catalytic activity. NiFe NP and various crystal structures of NiFe alloy are covered by doped carbon shell (*n*). The hexagonal closed phase (HCP) NiFe alloy catalyst has an ultra-low potential of 226 mV to induce OER with a current of 10 mm^{-2}. Besides the study of small NP rich in earth metals, scientists also focused on the exploration of nanocrystals and monatomic catalyst chambers to effectively improve the catalytic performance and atomic efficiency of the catalyst. Zhang's group, an effective OER electrocatalyst, has been found. Its position and atom are distributed on *n* (hollow carbon matrix [HCM]@Ni–N)-doped hollow carbon matrix.

As shown in Figure 5.5, synthesis process forms a matrix way after pyrolysis and acid etching to obtain the matrix. The characteristic of HCM@Ni–N are

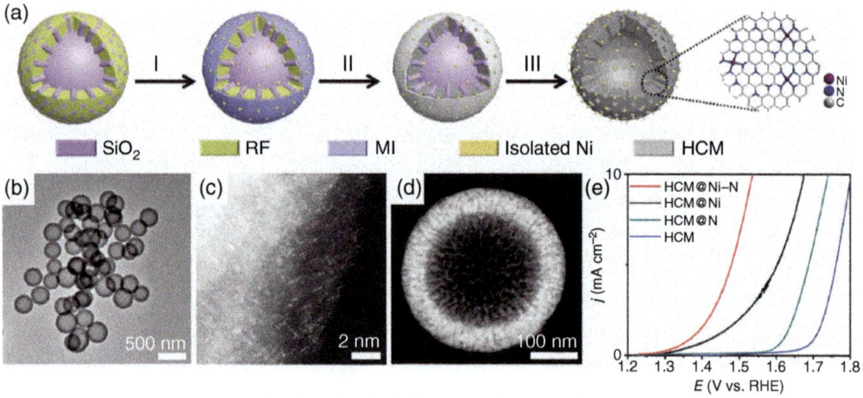

Figure 5.5 (a) Schematic diagram of HCM@Ni–N production process; (b) TEM image representing HCM@Ni–N; (c) nano layer, ADF-STEM image with atomic resolution; (d) HAADF-STEM image of HCM@Ni–N; and (e) various catalysts, OER polarization curve in 1.0 M KOH electrolyte. Source: (a, e) Wu et al. [18], (b–d) Wu et al. [18]. © 2020, John Wiley & Sons.

transmission electron microscopy (TEM) (Figure 5.5b), electron transmission ring black background microscope (Figure 5.5c), and electron transmission high-angle circular dark-field-scanning transmission electron microscope (HAADF-STEM) (Figure 5.5d). A spherical hollow structure of Ni atoms uniformly distributed in the whole particle. HCM@Ni–N showed better OER performance than others (Figure 5.5). As shown by X-ray absorption spectroscopy (XAS) and DFT calculation, the increase in catalytic activity is the result of the better electronic link between Ni and N, which mainly reduces the Fermi level and also the adsorption of L module on the main intermediate.

Oxides based on non-noble metals have attracted much attention in the field of electrocatalysis due to their low price, rich reversion, adjustable structure, and stable characteristics. The manufacturing of OER-based electrocatalyst is proved to be effective by regulating morphology, operating the components, adjusting the electronic structure using the outer metal, and integrating the hybrid structure of composite materials.

It is very effective oxide. The following part will introduce some OER catalysts on the basis of metal oxides. The real activity of single-layer non-noble metal oxide electrocatalyst depends on their type of metal, oxidation state, morphology, and support. Moreover, if the conductivity is not good, the application of OER will be greatly damaged. In order to solve the problem of poor conductivity, two strategies have been developed: (i) the structure and operation composed of two kinds of atom doping, the introduction of oxygen vacancy and the formation of multi-metal oxides and (ii) the conductive substrate (carbon material or metal substrate).

Karko geiner of non-noble metals (mainly sulfides and selenides) has done a lot of research on the application of OER. At present, the most commonly used method is to produce metal sulfides and series structures. For solvothermal method, the general supply sources are thiourea, thiourea, thiamide, etc., and the commonly used supply source of Cyrus is Cyrus, sodium selenide, sodium hydrogen selenide, etc. Hot water vulcanization and subsequent annealing Prussian blue analog (PBA) precursor treatment were based on NiFe. NIFES catalyst with similar morphology to sea urchin showed the second bottom of 10 mA cm^{-2} to 200 mV. This is an conductive, complex atomic state, efficient material transfer, and Ni-rich active website exposure. As Wen and his colleagues pointed out, the abnormally porous Co sulphide-nanocube (A-$CoS_{4.6}O_{0.6}$ porous nanocubes [PNCs]) is based on the ion exchange reaction replacing Na_2S on $[Fe(CN)_6]^{3-}$ (Figure 5.6a,b). PNC A-$CoS_{4.6}O_{0.6}$ shows good OER performance, which is much higher than CoS_2. It can be compared with the latest RuO_2 catalyst (Figure 5.6c). In the OER process, the adsorption of *O intermediate is used by mixing two kinds of atomic O and pendants with Co-S, which has a very enhanced catalytic OER activity. According to the investigation, non-noble transition metal is also an effective OER electrocatalyst. For example, Ni Fe PBA nanocomposites were prepared by Nai et al. as the site-selective ammoniation process of Ni Fe PBA nanocomposites. As shown in Figure 5.6, the glazing time of ammonia plays a decisive role during the structure evolution from the shape of the nanocube to the nanostructure. After selenization of the Ni–Fe PBA nanocages in Se vapor, the Ni–Fe mixed diselenide nanocages can be formed in this self-templating

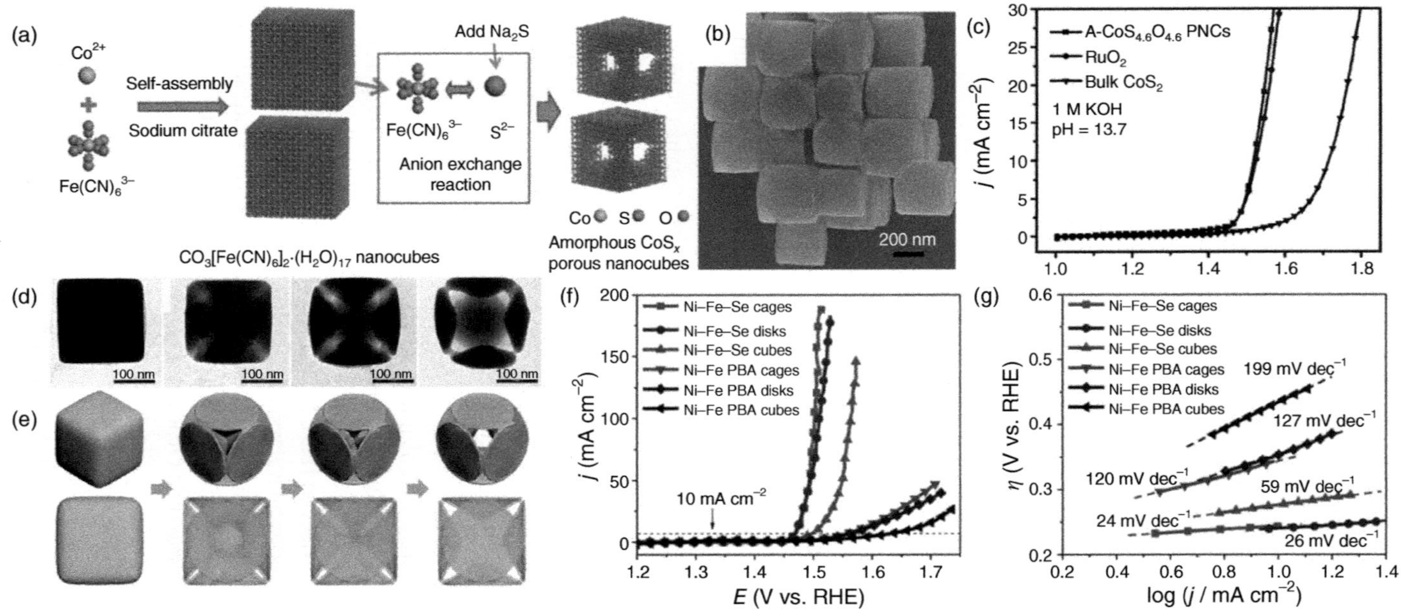

Figure 5.6 (a) Manufacturing process summary of A-$CoS_{4.6}O_{0.6}$ PNC. (b) A-$CoS_{4.6}O_{0.6}$ PNC is a representative image of field-emission scanning electron microscopy (FESEM). (c) OER curves of other catalysts were measured at 1.0 M KOH [19]. Source: Reproduced with permission of Wiley. In order to manufacture. (d) Ni Fe PBA nano meter, the TEM image of the product was observed in 0, 1, 3, and 5 minutes according to the etching time of ammonia water. (e) The corresponding sketch map of the structural evolution process in nano Cairo in Ni Fe PBA or no cube. (f) OER polar curve. (g) 1.0 M KOH electrolyte, six different catalysts are corresponding to Tafel slope [20]. Source: (a–c) Cai et al. [19]. © 2020, John Wiley & Sons, (b) Cai et al. [19]. Reproduced with permission of John Wiley and Sons, (d, e) Nai et al. [20]. Reproduced with permission of WILEY-VCH Verlag GmbH & Co. KGaA, Weinheim, (g) Nai et al. [20]. © 2017, John Wiley & Sons.

approach, which shows superior electrocatalytic performance toward the OER. It gets 10 mA cm^{-2} current density, and corresponding Ni Fe PBA obtained a small plate tilt (24 mV dec^{-1}) equivalent to Ni Fe see (Figure 5.6).

As mentioned above, iron, cobalt, and nickel materials have received the most research from OER because of their unique characteristics of O intermediate adsorption. Various nanostructures of Fe, Co, and Ni catalysts have recently been developed. Through a simple cation exchange strategy, Tong Zhang et al. reported the synthesis of NiO nanorods (NRs) with a large number of O vacancies [21]. During the synthesis process, carbon fiber paper (CFP) was used for in situ growing NiO nanorod arrays via a simple cation exchange method in which the sacrificial templates employed ZnO nanorod arrays. The conversion of ZnO NRs into NiO NRs takes place after the processes that Ni^{2+} cations are completely replaced with Zn^{+} cations. Through DFT calculations and experimental research, the author proved that the O-vacancy engineering facilitates the optimization of electronic and chemical characteristics of NiO NRs. The electron conductivity was significantly enhanced by the O vacancies on the surface of NiO NRs to simultaneously enhance the reaction kinetics of hydrogen evolution. The outstanding hydrogen evolution reaction (HER) catalytic activity/durability in alkaline solution was found on the resultant NiO NRs. In addition, these specially NiO NRs in situ grown on CFP substrates can be used directly as OER electrocatalysts, with higher performance compared to noble metal RuO_2 electrocatalysts.

In the work of Zhu et al., they first prepared CO_3O_4 nanowire arrays and then processed these nanowires under N_2 plasma at room temperature to prepare cobalt nitride nanowires [22]. Due to the better conductivity and larger retention of electrochemically active surface area (EASA), the obtained CoN nanowire array on foamed nickel exhibits excellent OER activity, with an overpotential as low as 290 mV and a current density of 10 mA cm^{-2}. The OER performance of CoN nanowire arrays with high density and excellent durability at different current densities confirms that metal nitrides are a kind of promising non-precious metal catalysts to supplement the various nanostructured metal oxides technology.

The preparation of metal catalysts guided by the MOF template is a general strategy to form nanostructures. Jian Zhou et al. reported the strategy that has been used to prepare layered CO_3O_4@X (X = CoP C, CoS, and CO_3O_4) catalysts for effective OER [23] in which the CO_3O_4@X is prepared from cobalt carbonate@ZIF-67, which is defined as CCH@ZIF-67. The synergistic effects of the obtained catalyst and unique hierarchical structures endow it with masses of active sites exposed on the surface, convenient for conductivity and ion diffusion, which are beneficial to improve its catalytic activity. Therefore, these CO_3O_4@X products exhibit high-efficiency OER catalytic activity together with excellent catalytic durability, far better than Ir/C and cobalt-based electrocatalysts reported previously. In particular, at a current of 10 mA cm^{-2}, the CO_3O_4@CoP catalyst only shows a low overpotential at 238 mV and has the highest electrocatalytic performance. In addition, CO_3O_4 X could effectively electro-oxidize some other small organic molecules (for example, ethanol, methanol, and glycerol). This approach is expected to be expanded into the manufacturing of more composite catalyst electrodes, which have a layered structure to achieve more effective water splitting. On the other hand, Zhou et al. prepared H_3LCoCN_{800} by direct pyrolysis of the precursor of layered phosphonate-based MOF from the phosphonic acid ligand 4-HOOC–C_6H_4–CH_2–$NHCH_2PO_3H_2$ $(H_3L)_{36}$ under N_2 atmosphere [23]. The catalyst on Ni foam is demonstrated to be an efficient alkaline OER electrocatalyst with the overpotential of only 215 mV at the current of 10 mA cm^{-2}. Combining the high-resolution XPS with X-ray absorption spectroscopy, the origin of the outstanding performance was revealed. These findings indicate that cobalt diphosphate with a longer distance between Co–O and Co–Co and distorted metal coordination geometry mainly contributed to the performance.

Compared with single transition metal catalysts, double or multiple transition metal catalysts show greater flexibility in the adjustment of their atomic/electronic structure and are moving toward a more effective OER. For example, FeNi and FeCo-based catalysts have been widely reported. A novel two-step method was used by Xu Zou et al. for the synthesis of the Ni–Fe–OH@Ni_3S_2/NF catalyst, which involves the steps of preparing Ni_3S_2 nanosheet arrays (defined as Ni_3S_2/NF), and also ultra-fast in situ growing Ni–Fe bimetal Ni_3S_2/hydroxide on nickel foam (NF) with amorphous structure (defined as Ni–Fe–OH) [24]. The Ni–Fe–OH@Ni_3S_2/NF composite nanomaterials exhibit high-efficiency OER catalytic activity at high currents (e.g. 1000 mA cm^{-2}). Ni–Fe–OH@Ni_3S_2/NF catalyst also has impressive OER electrocatalytic stability in 30 wt% KOH and 1 M KOH electrolytes. More results experimental measurements show that the outstanding structural stability, excellent electron conductivity, and high electrocatalytic activity are effectively integrated in the single catalyst system, making Ni–Fe–OH@Ni_3S_2/NF electrode to have significant OER electrocatalytic ability under high current density.

Shulin Zhao et al. reported that by controlling the annealing atmosphere, the NiFe-LDH/graphene oxide precursor was subjected to temperature-programming annealing and nitriding treatment to fabricate DR-Ni_3FeN nanocrystals and N-doped graphene (NG) nanohybrid (DR-Ni_3FeN/NG) [25]. In the nanohybrid, the DR-Ni_3FeN nanocrystals are anchored on the NG, and apart from about 10% stacking faults, they also mainly exhibit double crystal defects. This kind of nanohybrid can effectively catalyze OER in an alkaline medium with a small overpotential (250 mV) to reach the current of 10 mA cm^{-2} and a higher switching frequency (0.46 s^{-1}), which is better than the similar products (nearly defect-free) Ni_3FeN/NG, commercial IrO_2 and the most advanced OER catalyst. In addition to excellent activity, they also have better durability than similar products. As revealed by microstructure, spectroscopy, and electrochemical analysis, the enhanced OER performance of DR-Ni_3FeN/NG nanohybrid is derived from a large number of double crystal defects in the active phase of Ni_3FeN and the strong interaction between DR-Ni_3FeN and NG effect.

In the Zhang et al. study, amorphous ultra-thin oxygen-bound Ni–Fe–S (NFS) nanosheets loaded onto the Ti board are synthesized by lightly calcining Ni–Fe–S ultra-thin nanosheets (denoted as O-NFS) on the Ti board NFS, which is followed with electrochemical tuning (ECT) serving to be an effective catalyst for OER [26]. This strategy of combining ECT and O successfully achieved the coordinated adjustment of structural (active site) and electronic characteristics. With the study on the effects of ECT and O incorporation toward the OER performance, oxygen-incorporation Ni–Fe–S electrochemical tuning (O-NFSECT) catalysts provide new possibilities for meeting the industrial needs of high-efficiency electrocatalysts. The result indicates a small Tafel slope of 39 mV dec^{-1}, high currents of 3000 and 500 mA cm^{-2} at the overpotentials of only 435 and 300 mV, respectively, and excellent stability of 1100 mA cm^{-2} for 12 hours, and a high TOF value of 0.76 s^{-1} in 1 M KOH electrolyte. The unique open 3D structure was revealed to be responsible for these outstanding performances. The 3D structure has consisted with ultra-thin amorphous-like nanosheets. Therefore, a more active site could be exposed, and conductivity could be enhanced, and benefits from oxygen incorporation and ECT appropriate electronic structure.

The bimetal (Fe_xNi_{1-x}) 2P nanoarray was prepared by Zhang et al.; it is a very effective electrocatalyst for OER in alkaline and neutral media [26]. The iron/nickel ratio-dependent oxidation of nickel under alkaline and neutral conditions and its influence on OER kinetics are also discussed. As an integrated non-noble metal high-performance catalyst for OER, the optimized $(Fe_{0.5}Ni_{0.5})_2P$/NF electrode has excellent performance for OER in the neutral and alkaline environment. It can provide currents as high as 10 and 500 mA cm^{-2} with minimal overpotentials of 156 and 255 mV in a 1 M KOH electrolyte, respectively. Besides, the $(Fe_{0.5}Ni_{0.5})_2P$/NF electrode could reach the current of 10 mA cm^{-2} with the overpotential of 396 mV in 0.1 M phosphate buffer (KPi, pH = 7). It also showed excellent stability in 0.1 M KPi and 1 M KOH electrolyte. Compared to other non-noble OER catalysts studied until now, the neutral and alkaline performance proves that $(Fe_{0.5}Ni_{0.5})_2P$/NF has excellent catalytic activity.

As for FeCo-based materials for OER, Zhuang et al. reported a simple strategy for preparing iron–cobalt oxide nanosheets with abundant O defects using $NaBH_4$ as a reducing agent. Such catalyst could thereby offer abundant surface sites for the electrocatalysis of OER (Figure 5.7) [27]. In a 0.1 M KOH solution, the optimized Fe_1Co_1–ONS can achieve 10 mA cm^{-2} with the overpotential of only 308 mV. Besides, the Tafel slope is only 36.8 mV dec^{-1}, which is the best in reported data (Figure 5.7). The characterization proved that the excellent performance of Fe_1Co_1-ONS could be attributed to its abundant oxygen vacancies and thin atomic plates (Figure 5.7), which can greatly enhance the reaction activity, as well as the number of active surface sites. On the other hand, Ye Zhou et al. reported that a strong correlation has been established the relationship between Co–O covalent and the activity for oxygen evolution on the substituted spinel $ZnFe_xCO_{2-x}O_4$. It indicated that the increases in the valences of O 2p and Co 3d are the reasons to the enhanced performance brought about by the substitution of iron by 10–30 at% because it promotes the injection/extraction of electrons from oxygen [28]. The best-performing $ZnFe_{0.4}Co_{1.6}O_4$ oxide is better than the standard IrO_2 catalyst, has good cycle stability, and can be used as a promising low-cost electrocatalyst. Although the decoupling of proton–electron transfer was indicated by the dependence of pH with OER activity on $ZnFe_{0.4}Co_{1.6}O_4$ during oxygen evolution processes, it is not considered to be a better way for the lattice O participation. A

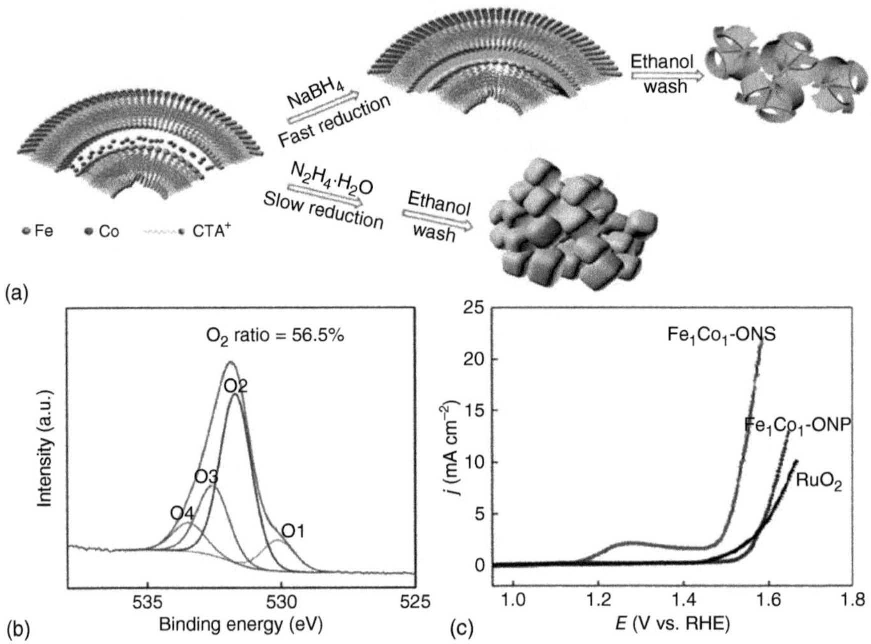

Figure 5.7 (a) Schematic diagram of the preparation of Fe_1Co_1-ONS and (b) Fe_1Co_1-ONP, O 1s spectra, and (c) CVs of Fe_1Co_1-ONS, Fe_1Co_1-ONP, and RuO_2. Source: Zhuang et al. [27]. © 2020, John Wiley & Sons.

larger energy gap between the Fermi level and the O p band center was generated in the spinel oxide lacking cations, which may hinder the activation of lattice oxygen. The work shows that the metal–oxygen covalent relationship is related to the basic understanding on oxygen evolution for oxide electrocatalysts at various defective chemistries. Town et al. prepare lots of $La_{0.6}Sr_{0.4}Co_{1-x}Fe_xO_{3-\delta}$ ($0 < x < 1$) nanofibers with porous structures and a diameter of 54–71 nm. Appropriate polymers were used to synthesize these materials by electrospinning at various parameters [29]. In the OER test at 10 mA cm^{-2}, the electrocatalyst of $La_{0.6}Sr_{0.4}Co_{0.6}Fe_{0.4}O_{3-\delta}$ (LSCF6464) nanofibers with the BET area of 24.2 m^2 g^{-1} only shows the potential of 647 mV vs. Ag/AgCl. This value is obviously lower compared to the commercial powdered LSCF (786 mV), as well as the latest IrO_2 catalyst (660 mV). LSCF6464 nanofiber catalyst also has excellent durability, at a current density of 10 mA cm^{-2}. In the case of mA cm^{-2} disks, there is almost no observable potential change for more than three hours. On the contrary, in the same measurement, the performance of the IrO_2 electrocatalyst continuously decreases.

In addition to iron, cobalt, and nickel-based materials, other transition metal catalysts for OER have also been reported. Han Xu et al. reported Cu_2O/Cu dendrites-consisted 3D hybrid foam with a large surface area for alkaline oxygen evolution. The excellent catalytic activity for OER was obtained on the hybrid Cu_2O/Cu foam, together with their stable stability and good reaction kinetics. The excellent OER catalytic performance is arisen from the special porous 3D foam, which can offer a short diffusion path and rapid transportation for the solution. It also facilitates the removal of the generated oxygen gas bubbles. Besides, the synergy and high-speed electronic transmission network were offered by the core–shell Cu_2O/Cu dendrites for the electrocatalytic reactions. The above results not only show that low-cost metal oxides have the potential as eco-friendly OER catalysts with high catalytic activity/durability in alkaline solution, but also the design will be used to develop more effective OER electrocatalysts. Zhao et al. reported another copper-based catalyst [29]. In this work, the authors found that the Co-doped Cu_7S_4 nanodisk with modified active site electronic structure showed enhanced electron transfer between Co and Cu sites and improved electrocatalytic kinetics. As expected, compared with bare Cu_7S_4, the co-designed Cu_7S_4 nanodisk exhibits an overpotential as low as 270 mV at a current of 10 mA cm^{-2}, as well as a low Tafel slope and increases the switching frequency. In addition to Co doping, Zhiguo Ye et al. also prepared various metal ions (V, Fe, Ni, and Co) doped with ultra-thin MnO_2 nanosheets, which were electrodeposited onto CFP. It is easy to use the anode co-deposition approach [30]. On the surface of carbon fiber, there were abundant nanoclusters at high density, which were composed of the doped ultra-thin MnO_2 nanosheets. These nanosheets have an average thickness of about 5 nm. Metal elements such as V, Fe, Ni, and Co were proven to dope in the MnO_2, thereby causing the increase in the MnO_2 conductivity. In the OER test, to reach the current of 10 mA cm^{-2} in 1 M KOH, the overpotentials of 390 and 245 mV were obtained on the doped ultra-thin MnO_2 nanosheet/CFP electrode and IrO_2/CFP electrode, respectively. A significant lower overpotential was maintained on the doped ultra-thin MnO_2 nanosheet/CFP electrode at a current density of 20 mA cm^{-2} during the long-term test, compared to the case in MnO_2 catalyst electrode.

Jingjie Wu et al. reported the high acidic OER catalytic activity on the two-dimensional transition metal dihalides (TMDs) is reported, taking semi-conductor from 2H-MoS_2 and conductor from 2H/1T-TaS_2 [31]. The 2D TMDs nanosheets were prepared by exfoliation of Li intercalation chemistry meditation in the liquid phase. The OER electrocatalytic activity of TMD nanosheets on carbon paper was tested in acid solution (such as 0.5 M H_2SO_4) and evaluated first from linear sweep voltammetry (LSV). The data indicate that 1T-MoS_2 has the best effective electrocatalytic performance. The benchmark current of 10 mA cm^{-2} on 1T-MoS_2 catalyst only requires an overpotential of 0.42 V. In contrast, for the same cases in 2H-MoS_2, 1T-TaS_2, and 2H-TaS_2, 0.48, 0.45, and 0.54 V of overpotentials were required, respectively. OER performance of the catalysts was also tested by the Tafel plot with the equation of $\eta = b \times \log(j/j_0)$, in which j_0 represents exchange current density and b stands for Tafel slope. High Tafel slope values were found in the Tafel graph, in which 1T TaS_2, 1T MoS_2, 2H TaS_2, and 2H MoS_2 had the Tafel slope of c. 282, 322, 255, and 361 mV dec^{-1}, respectively. However, these values of the Tafel slope are much lower compared to some other OER catalysts with non-noble metal components in acidic media, such as MnO_x samples with the Tafel slope of 653 mV dec^{-1}, or B-doped diamond samples with the Tafel slope of 680 mV dec^{-1}. The authors found that according to DFT calculations, the active sites of OER are located on the edges, not on the surface of the substrate. In addition, it is revealed that the 1T polymorph has the same tendency to have higher activity than the 2H counterpart. At the same time, MoS_2 quantum dots were prepared by Bishnupad Mohanty et al. through the one-step hydrothermal synthesis technology [32]. Compared with the prior art catalyst IrO_2/C, the synthesized MoS_2 quantum dots (MSQD) has higher OER activity and lower Tafel slope. The potential cycle of MSQD can activate the surface and improve OER electrocatalytic activity. The MSQD vertices are active, as demonstrated by DFT calculations, especially for the vacancies on edges. These results suggest that smaller flakes which have abundant defects on edges are more effective for OER. The adsorption of intermediates during oxygen evolution would be influenced by the existence of carbon quantum dot (CQD), resulting in significant inhabitation of the OER activity on MSQD.

Previously, an excellent Ta_2O_5 electrocatalyst for OER was exploited by the control over the O environment to induce and form a highly active surface for OER [33]. The stable surface structure of Ta_2O_5 and its initial overpotential to OER are calculated by system DFT simulation. The most active site for OER was demonstrated to be the O site on stable Ta_2O_5 (200) surface, and the starting overpotential is calculated to be only 0.25 V. Using simulated O-enriched conditions and oxygen-controlled pulsed laser deposition, the authors successfully have grown Ta_2O_5 nanolayers with high (200) exposure on the carbon cloth (CC). The obtained (200) plane shows excellent electrocatalytic activity for OER, with an initial potential of 0.385 V at the current of 10 mA cm^{-2} as well as the low initial overpotential at 0.29 V according to the simulation calculations. The novel strategy illustrated by such adjustment of O with the control over the surface toward the activation of electrocatalytic performance facilitates rapid development for transition metal oxide electrocatalysts for efficient energy conversion.

5.1.3 Other Catalysts

In addition to transition metal (including precious metals) materials, other most commonly used OER catalysts also include MOF complexes and metal-free materials, such as carbon and polymer materials. MOFs are the first OER catalytic systems to be widely studied. The periodic coordination bonds between the metal atom and organic ligands with a well-defined structure could consist of MOF. MOF can offer some unique advantages as compared to heterogeneous and homogeneous materials on the basis of its structure natures. Similar to molecular catalytic materials, MOFs with porous structures are easy to access on the active sites and possess a clear physical/chemical structure. Moreover, MOF is a highly crystalline solid material, which can be easily recycled and durable under chemical and physical attack. Recently, Shenlong Zhao et al. developed ultra-thin NiCo bimetallic organic framework nanosheets [34]. The mixture of phthalic acid (BDC) and Ni^{2+} had been used to synthesize NiCo bimetal-MOF (NiCo-UMOFNs) which show excellent catalytic OER performance. The foamed Cu was used to support the as-prepared NiCo-UMOFNs which exhibit an overpotential of 189 mV at the current of 10 mA cm^{-2} and also a low onset potential of 1.39 V under alkaline conditions. For at least 200 hours, a highly stable current density can be obtained at the stable overpotential of 250 mV, and the related Faraday efficiency (FE) is as high as 99.3%. The author found that the ultra-thin MOF sheet has coordination-unsaturated surface atoms serving as open sites for reactant adsorption, which is proved via various measurements (involving DFT calculations and X-ray spectroscopy). These results indicate that the main active centers are coordination-unsaturated metal atoms, and the cobalt and nickel coupling is essential for regulating catalytic performance.

In addition to bimetallic organic frameworks, trimetallic MOFs have also been reported. Feilong Li et al. study lots of trimetallic MOFs, which were based on Fe/Ni components (corresponding to the material of Lavoisier Institute, Fe/Ni/Co(Mn)-MIL-53) and were prepared by solvothermal synthesis. These catalysts were used directly as high-efficiency catalysts [35]. The prepared Fe/Ni/Co(Mn)-MIL-53 catalyst exhibits a volcanic OER performance that varies with composition. The optimal current density of $Fe/Ni_{2.4}$ is 20 $mA\,cm^{-2}$/$Co_{0.4}$-MIL-53 by reaching the low Tafel value at 52.2 $mV\,dec^{-1}$ at an overpotential of 236 mV. In addition, growing directly on NF can further improve the OER performance of these MOFs. Therefore, the tri-metal MOF had enhanced OER performance; for understanding the reasons, the authors further investigated the possible factors which may influence the performance. CV was then used for the calculation of C_{dl}. This value represents somewhat the ECSA. The results indicated that all the electrocatalysts have almost the same ECSA. Therefore, it indicates that these electrocatalysts possess the same or a similar number of surface-active sites on solid–liquid–gas reaction interfaces.

One of the biggest disadvantages of MOF materials is poor conductivity. Li Zhao et al. reported that the 2D cobalt 1,4-benzenedicarboxylate (CoBDC) and $Ti_3C_2T_x$ (MXene phase) nanosheets were hybridized in situ by an assisted process of interdiffusion reaction [36]. The obtained hybrid catalyst is used for OER. The potential was transformed relative to the RHE. It is observed that when the applied potential was set at 1.64 V, the anode current reaches 10 $mA\,cm^{-2}$, and a value of 48.2 $mV\,dec^{-1}$ was calculated for the Tafel slope in 0.1 M KOH. These results are superior to the results obtained with benchmark IrO_2 catalysts and are in the same level with or even greater compared to the obtained with the latest metal-based electrocatalysts previously reported. Although a large active surface area and a highly porous structure were provided by the CoBDC layer, the hydrophilic and conductive $Ti_3C_2T_x$ nanosheets enable the fast transfer of charges and ions the well-defined CoBDC–$Ti_3C_2T_x$ interfaces, then further promote the interaction of the aqueous solution with the contact surface for reaction-active CoBDC. For the fabrication of air cathodes in rechargeable zinc–air batteries, the hybrid nanosheets are further used. The fabricated Zn–air batteries were further used successfully in light-emitting diodes.

Carbon-based materials are another important type of catalyst used in OER. Changlin Zhang and others prepared an ordered porous carbon–graphene framework with N-doping. This catalyst can be used as a dual-function catalyst for OER and ORR [37]. The obtained N-doped mesoporous carbon frameworks (N-MCFs) electrocatalyst displayed oxygen reduction reaction (ORR) performance compared to the benchmark Pt/C. Furthermore, the OER performance is excellent, with the overpotential of 324 mV at 10 mA cm^{-2}, better compared to the most of the C-based catalysts, transition metal oxides, and their hybrids as reported. The performance is even comparable to IrO_2 and RuO_2 electrocatalysts. The mass transfer resistance was very small on the obtained N-doped mesoporous graphene frameworks (N-MGFs) catalyst, with excellent dual functionality in ORR/OER and also the good durability. Due to its ordered pores, high surface area, excellent intrinsic electrical conductivity, and internal diffusion characteristics, the electrocatalytic performance has been significantly improved. Similarly, in the presence of phosphonic acid aniline monomer, conductive carbon paper was used for directly grown layered mesoporous P and N co-doped carbon nanofibers directly grown on its surface via the processes of electrochemistry-induced polymerization [38]. The obtained material shows strong stability (little decrease in activity after 12 hours of continuous operation) and high activity in generating electrocatalytic oxygen and has a low overpotential for electrocatalytic oxygen (310 mV, at a current of 10 mA cm^{-2}). Its performance is in the same level with precious metal oxides (e.g. IrO_2). More abundant active sites and higher active surface areas were induced by double-doping with P and N in comparison with original carbon counterparts or single-doped ones. DFT calculations show the dopants of nitrogen and phosphorus. It can jointly activate the C atoms next to it and induce a synergistic enhancement of OER activity.

Kong gang Qu et al. have incorporated S into the carbon framework by a new method. S and N co-doped porous carbon nanosheets were obtained on the basis of polydopamine-graphene oxide (GD) hybrid [39]. S was conjugated with the GD hybrids after the pyrolysis at 800 °C, and the percentage of 16.7% of S doping reaches an efficiency as high as 6.1%. Such efficiency is significantly better compared to most of the S doping materials, as reported. Based on the bifunctional effects for ORR/OER, the obtained N and S co-doped porous C nanosheets show excellent performance, good dynamics, and excellent durability, which are better than other nonmetal elements-doped C as reported, and most of the transition metals catalysts and even noble metal electrocatalysts are much better. It is believed that a rich porous structure, high concentration of multiple dopants, and an excellent electron transfer rate can obviously accelerate the ORR/OER processes. In view of the physicochemical structural tunability and versatility of polydopamine (named as PDA), the employment of PDA-based carbon materials with heteroatom doping was offered with this general platform toward high-efficiency OER. In addition to a single carbon form, multiple carbon forms can be combined to enhance OER activity. In Abdelhamid M. El-Sawy et al. study, doping S in carbon nanotube (CNT)–graphene nanoleaf was reported by a continuous two-step process [40]. This bidding strategy introduces a stable sulfur–carbon active center. Scanning transmission electron microscopy electron energy loss spectroscopy (namely STEM-EELS) plots and spectra, combined with X-ray absorption near-edge spectroscopy (namely XANES), have shown that the fluorescence emission at the edge of sulfur K confirms the heterocyclic sulfur in the carbon ring of CNTs. The incorporation is increased with an extremely high mass-specific activity of 1000 mA cm^{-2}, and under a 400-mV overpotential, the TOF greater than (20% Ir/C) the TOF, which improves the diploid sulfur carbon The OER activity of the nanotube–graphene nanoleaf is at 570 mV. In addition, it reveals the unique connection between the graphene nanoleaf and the CNT wall and also confirms the uniform distribution of sulfur on the CNT. XANES and XPS help explain the excellent activity and stability by showing that a large amount of sulfur is incorporated inside the ring of CNT.

Zhang You et al. reported a conceptually novel general strategy. The catalytic performance of nonmetal CNTs for OER can be improved by this strategy through facile coating by polymer without having to any edge defects, functional groups, and heteroatom dopants. It is a graphite structure [41]. The strategy is simple, effective, green, and easy to expand and it is coated with specific types of electrochemically inert polymers with polar oxygen-containing groups (for example, poly(ethylene glycol), poly(vinyl acetate), poly(ethylene-maleic acid), poly(vinyl alcohol)). After pure CNTs (–O–, $-COOCH_3$, –COOH, –OH) through the interaction of non-covalence, it is easy to obtain many excellent nonmetal composite membrane electrocatalysts, resulting in an unexpected and extremely high electrocatalytic OER performance. Compared to the commercially available precious metal RuO_2 electrocatalyst, the electrocatalytic OER activity of pure CNTs is very poor. Combined experimental and computational studies have shown that the high OER performance observed can be attributed to the synergistic effect of the coated polymer layer as a co-catalyst as well as the intrinsic topological defects of CNTs (as active centers), for optimization of intermediate adsorption to improve OER.

In addition to carbon/polymer-based metal-free materials, phosphorus-based materials have also been reported. In Xiaohui Ren et al. study, the electrocatalytic OER performance of two-dimensional-layered black phosphorus nanosheets (BP) were prepared through a method of liquid exfoliation processes [42]. The BP nanosheets have great improvement in the electrocatalytic activity. From the studies of electrocatalytic measurement, it is found that the Tafel slope and the OER onset potential of BP nanosheets are $88\,mV\,dec^{-1}$ and 1.45 V, respectively. These data show that the obtained BP nanosheets with exfoliated processes exhibit good nanostructures, have high-efficiency electrocatalytic performance, and have excellent electrochemical OER performance and long-term stability. In addition, the authors obtained BP nanosheets at different thickness distributions by selective centrifugation treatment. It was found an interesting phenomenon that there is a relationship between thickness and OER performance on BP nanosheets. Among them, with the decrease in thickness of BP nanosheets by centrifugation, the tested OER performance was improved. Thus, it is an effective way to increase the specific surface area and produce more surface-active sites by decreasing the thickness of BP nanosheets toward excellent OER electrocatalytic activity. The authors further predicted the real active sites to offer some basic understanding related to the obtained OER activity on the nanosheets. At the same time, BP nanosheets with few layers are promising for the application of efficient OER electrochemical devices. On the other hand, Ranjith Prasannachandran et al. proposed a novel strategy that uses a one-step electrochemical exfoliation process to control the synthesis and in situ surface functionalization of phosphorescent quantum dots (PQDs) [43]. Phosphorescent quantum dots functionalized with nitrogen-containing groups (FPQD) show effectiveness and OER stability in terms of low Tafel slope of 48 mV dec^{-1}, the low potential at only 1.66 V at the current of 10 mA cm^{-2}, and excellent stability of the electrocatalytic activity. In addition, we observed that the electron transfer kinetics of FPQD to Fe^{2+}/Fe^{3+} redox probe was enhanced compared with the original PQD.

5.2 Photo-Assisted H_2O Oxidation

Light-assisted electrochemical water oxidation is an essential half-reaction in artificial electrochemical fuel synthesis. The lack of sustainable, low-cost, and efficient catalysts with photoelectric properties is the key factor limiting their applications of photo-assisted electrochemical water oxidation for energy conversion. Various metal compounds and heterostructure catalysts were exploited over the past decades. In this part, we will discuss progress in material types.

5.2.1 Metal Compound-Based Catalysts

The use of photoelectrochemical cells to generate H_2 serving as a clean-burning energy has recently attracted great attention. Until the discovery of effective, scalable, and stable photosensitive catalysts water oxidation by the photo-electrochemical approach, this method remains elusive. So far, many materials have been studied, such as TiO_2, $BiVO_4$, WO_3, Bi_2WO_6, TaON, $BaTaO_2N$, and other photoanodes. Among the reported catalysts, the hematite (named α-Fe_2O_3) had received widespread study interest because of its non-toxicity, availability, narrow band gap, and easy mass production [44]. In the work of Amira Y. Ahmed et al., a hematite nanoparticle photoanode with controlled size was successfully prepared from Sr^{2+} and Ba^{2+} ions by the simple method of chemical bath deposition processes [44]. Characterization of synthetic electrodes by intensity-modulated photocurrent spectroscopy (IMPS), scanning electron microscope (SEM), XRD, (P)EIS, inductively coupled plasma-optical emission spectroscopy (ICP-OES), and XPS were conducted. It had been observed that by adding Ba^{2+} and Sr^{2+} ions to the chemical bath, hematite nanoparticles showed a decrease in the average diameter by 13% and 35%, respectively. In agreement with the data, the Sr- and Ba-modified hematite photoelectrodes showed an increased ECSA by 3.2 and 2.4 times, respectively, compared to the bare hematite. At the standard photoelectrochemical test conditions with the potential of 1.23 V vs. RHE, the photocurrent generated by the Sr- and Ba-modified hematite photoanode is 3 times and 2.6 times higher, respectively, compared to the bare hematite. Furthermore, Sr- and Ba-modified hematite photoanodes have a shift in the onset by c. 220 and 150 mV, respectively, for water oxidation. On the basis of the data, there is a desired correlation between electroactive surface area and photocurrent found under large electrode potential (i.e. 1.3 V vs. RHE), indicating that the increased electroactive surface area can render the increased photocurrent. Under low bias potential, the surface state density of Sr- and Ba-modified hematite photoanode is higher, the charge transfer rate constant is higher compared to the bare hematite, the surface state density is higher, and the charge transfer rate constant is higher; thus, the authors can clarify the cathode shift of the OER onset potential and draw conclusions from (photo)electrochemical impedance spectroscopy ((P)EIS) and IMPS measurements.

The main challenge of catalyzing the oxidation of water to O_2 is the excessive overpotential (η), which is needed, and another challenge is related to the high (basic) or low pH (acidic) solution of the (photo)electrocatalyst at high anode potential. The persistent kinetic slow four electron together with four-proton continuous transfer processes under the anode potential. As mentioned earlier, high-priced metals have been shown to have high activity on OER. Therefore, high-valent metal compounds are expected to be used in (photo)electrocatalytic OER. Recently, Yin Wang had prepared FeOOH nanotube arrays directed vertically on conductive CC via a template electro-etching strategy through one-step electro synthesis [45]. In $FeCl_2$ solution, the controlled electrodialysis method works for 400 seconds at 1.5 V vs. Ag/AgCl. During the first 200 seconds, the vertically directed Co-based nanowire array is converted to CoO@FeOOH. After that, electroetching of Co ions takes place to form FeOOH nanotubes in the following 200 seconds. During the heat treatment, the phase change from β-FeOOH to amorphous can be prevented with the robust nanotube morphology. LSV was used to test the OER performance on the FeOOH nanotubes (named CC@FeOOH-NTs) in a 1.0 M KOH aqueous solution. Firstly, the author studied the electrocatalytic activity influenced by the FeOOH form. Compared with CC@FeOOH-NRs and CC@FeOOH-NPs, the CC@FeOOH-NTs has lower resistance and thus a lower overpotential to OER. Second, the electrocatalytic activity of CC@FeOOH-NTs catalyst with the heat treatment was compared. Compared with CC@FeOOH-NTs catalyst, the overpotential of CC@FeOOH-NTs catalyst –240 °C is reduced from 377 to 352 mV (at a current of 10 $mA\,cm^{-2}$) by 25 mV. Nevertheless, CC@FeOOH-NTs-300 °C has poor activity, while CC@Fe_2O_3-NTs catalyst has not even any activity. The Tafel diagram was used to check the OER kinetics. The CC@FeOOH-NTs-240 °C has a small Tafel slope at 47 $mV\,dec^{-1}$, obviously lower compared to CC@Fe_2O_3-NTs (65 $mV\,dec^{-1}$), CC@FeOOH-NTs (62 $mV\,dec^{-1}$), and the other controls samples. In addition, the photoactive OER electrocatalyst employed the CC@FeOOH-NTs-240 °C was further studied. In this test, a lower initial overpotential of only 280 mV and a smaller Tafel slope were found on CC@FeOOH-NTs-240 °C, under visible light conditions (light source uses 300 W xenon lamp with wavelength 420 nm $\leq \lambda$) (42 $mV\,dec^{-1}$) (≤700 nm filter). With the irradiation by visible light, when the current is 10 $mA\,cm^{-2}$, the overpotential is reduced by 24 mV. In comparison, the CC@Co nanowire array (NWA) did not show enhanced photoactivity in terms of OER activity. The CC@FeOOH-NTs-240 °C has a semicircle diameter in the EIS analysis, obviously lower than the light irradiation. In general, radiation will excite the semiconductor FeOOHs to generate electrons and holes. Electron transfer will be promoted by these electrons and holes within the photoelectrocatalyst, and the OER kinetics will be enhanced by the accelerated electrons. Additionally, the authors also collect the time-photovariable I–t curve of the CC@FeOOH-NTs-240 °C catalyst under chopped lighting and show the reproducible and rapid current response to the on–off lighting cycle. The dependence on the wavelength of incident photon-current conversion efficiency (IPCE) was examined (365–800 nm). Under visible light irradiation and at the overpotential of 343 mV, the current density of CC@FeOOH-NTs-240 °C reaches 16 $mA\,cm^{-2}$, which is 2.7 times greater compared to the current density under the corresponding dark conditions. Under visible light irradiation and at the potential of 1.58 V, the obtained TOF of CC@FeOOH-NTs-240 °C was calculated to be 0.022 s^{-1} relative to RHE. This value is 3.7 times compared to the dark environment (0.0059 s^{-1}). The OER FE is higher than 97%. In addition, the FeOOH nanotube morphology as well as the current were kept under irradiation for 15 hours to react.

Fe can also be used as a doping element combined with N to promote photoelectrochemical OER performance of the catalyst. Jiaying Lu et al. reported that trace amounts of Fe (N, Fe–CoS_2) were included in the N-doped CoS_2 nanorod arrays, which was synthesized on a stainless steel photoanode [46]. During the precursor synthesis process, the substrate contains some which will be incorporated in the material. As the main dopant, the intrinsic OER performance was enhanced by the N atom via the control over the electronic structure and charge distribution of surface-active sites. Besides, the design of the effective bandgap of the semiconductor can be achieved by the introduction of N in a lattice structure, which can also by forming an intermediate energy level between the band gaps. Thus, the excitation energy can be reduced for effectively extending the photo-responsive area to the visible region. The oxidation of the Co site causes the production of oxidized Fe impurities, which are significant for the promotion of OER performance. The promotion mechanism was explained by reduced absorption energy for O-based intermediates, enhanced electrical conductivity, and possible more charges to be involved in OER (Figure 5.8). As expected, after the modification with Fe and N element, the CoS_2 anode showed obviously higher electrocatalytic OER activity and also enhanced photocoupled OER activity. Under dark conditions, a large

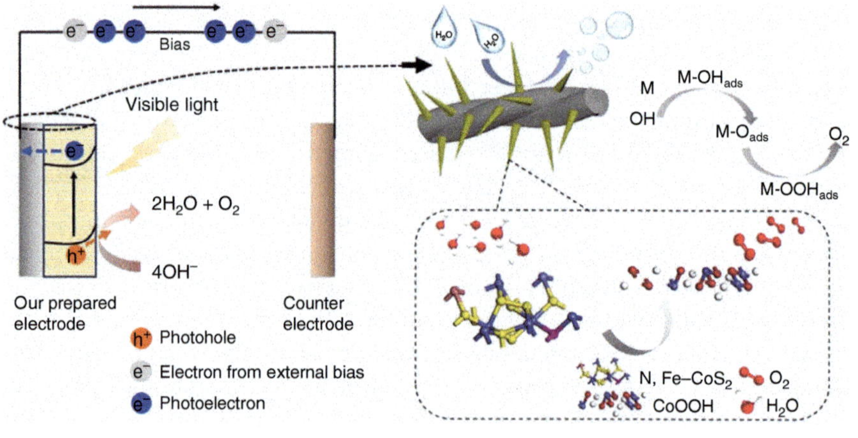

Figure 5.8 Schematic illustration of the photo-coupled electrochemical OER mechanism of N, Fe–CoS_2/SS. Source: Lu et al. [46]. © 2019, American Chemical Society.

current density at 770 mA cm^{-2} at the overpotential of 350 mV was achieved on the N, Fe–CoS_2/stainless steel (SS) anode. This value was about four times higher compared to that of the Fe–CoS_2/SS sample. Such an outstanding OER activity is arisen from the increase in the number of surface-active sites caused by the co-doping of N and Fe, the increase in electrical conductivity, and the increase in intrinsic activity. It is worth noting that as it is exposed under visible light and near-infrared (>420 nm), and a significant light response was observed on the photoanode. And such a response is more than 10 times compared to Fe–CoS_2.

$SrTiO_3$ is a kind of double-oxide material for various applications, which should be expected to be used in photocatalysis, ferroelectricity, and photoluminescence (PL). Kateřina Minhová Macounova et al. studied the role of surface size in photoelectrochemical water oxidation. On $SrTiO_3$ nanocubes of different sizes [47], the band gap illumination of the strontium titanate electrode causes an anode current. The photocurrent appears at a bias that approximately relative to the flat band potential of 220 mV. The bias voltage required to record the anode photocurrent increases with the increase in pH, reflecting the changes in the protonation of surface oxygen atoms. The photoelectrochemical activity of $SrTiO_3$ nanocubes is related to the size and increases with the increase in the particle size. A semi-quantitative analysis of the observed photocurrent and mass spectrometry of the reaction product showed that the contact of water with the luminescent $SrTiO_3$ nanocubes resulted in the formation of oxygen, hydrogen peroxide, and ozone. Oxygen and ozone are the main products of water oxidation on the [100]-oriented $SrTiO_3$ surface, and their fractions increase with increasing particle size. Hydrogen peroxide is simultaneously generated at low-dimensional parts (crystal edges and vertices) through oxygen reduction, and its content increases with decreasing particle size.

In the work of R. J. Kriek et al., revealing that Eu_2TeO_6 as a photocatalyst exhibits continuous light charging, which leads to an enhanced "continuous" dark current (after the illumination is terminated) [48] which is greater than the current obtained only by applying a potential. The processes via Pechini sol–gel method with calcined at 400 and 900 °C were used to synthesize telluride oxide samples, which were used to analyze the selective physic-chemical natures of the tellurium oxide. Appropriate methods were used employing this electrocatalysts and photoelectrochemical catalysts, and photocharged catalysts. A comparative evaluation for OER was conducted in alkaline media. Amorphous characteristics were found in the sample calcined at a high temperature of 400 °C, while the material which suffers from calcination at 900 °C was characterized as Eu_2TeO_6. An electrochemical cell with modification of a lamp was used for the photoelectrochemical evaluation. A glass carbon electrode holder was used for the deposition of the catalyst instead of on an indium-doped or fluorine-doped SnO_x glass plate. The band gap for Eu_2TeO_6 (calcined at 900 °C) is 4.31 eV, and better OER activity was found under the continuous UV irradiation. After UV irradiation for a period of time, a sustained activity up to 40 minutes was observed. Under the potential of 1.8 V vs. RHE and after 150 minutes of light irradiation, the increased photocurrent from 0.44 to 0.91 mA cm^{-2} was observed, with an increase of 107%. Unexpectedly, after irradiation for 150 minutes, a dark current continued for 40 minutes under the condition of 0.86 mA cm^{-2} (the measurement was terminated at this point), which continued to increase by 95% in comparison with the initial sample without irradiation. The phenomena may help to improve the understanding of light charging and ultimately help to develop solar energy conversion/storage with high efficiency.

Besides the properties of the catalytic materials, another key influencing factor is the electrolyte that also affects OER processes and also the performance of catalysts. In the electrolyte factor, the pH effect was also investigated. At a high pH, it was found that MnO_2 catalytic material shows better performance. According to reports, the water potential on the electrodeposited MnO_x is too low (about 150 mV), which can only be achieved in a certain organic salt (such as *n*-butylammonium sulfate) electrolyte [49]. In this case, water can be mainly oxidized to hydrogen peroxide (H_2O_2) at a lower overpotential, and H bonding with ammonium ions can stabilize the H_2O_2. For this reason, this reaction pathway was taken as the formation of *n*-butylammonium-assisted H_2O_2 electrocatalytic reaction. Such a pathway offers a promising way of water splitting under a low overpotential. In spite that a lot of energy has been invested in MnO_x research, most of the research has focused on dark electrocatalytic activity. In fact, MnO_x may be a potential catalyst for photoelectrochemical OER, as suggested by the more positive valence band (VB) energy from the MnO_x band structure compared to the H_2O/O_2 reaction potential. Fengling Zhou et al. showed the results that the photoelectrochemical OER activity of the MnO_x film was obviously enhanced by employing an amine ionic liquid-contained buffered aqueous electrolyte [49]. Buffer conditions and amine salts that maintain constant proton activity are both essential to achieving high photocurrent. In the Bi-buffered electrolyte with *n*-butylammonium nitrate (namely BAN) at a pH value of 9, compared to saturated calomel electrode (SCE) ($\eta = 540$ mV), the photocurrent obtained at 1.0 V reached at 4.5 mA cm^{-2} O_2 and O_2. It is found that the IPCE is greater than 3% at 400 nm and greater than 4% at 370 nm. The two H_2 produced at the same time in this system, and the H_2 that may be decomposed later accelerate the decomposition of H_2, and O_2 forms oxygen. The oxygen under light is proposed to explain the improvement of photocurrent.

5.2.2 Metal–Metal Heterostructure Catalysts

Among the various metal oxides explored in the past several years, one of the most potential materials as a photoanode is $BiVO_4$. This metal oxide belongs to the n-type semiconductor, and it is composed of low-cost elements with non-toxicity. In addition, three kinds of crystal systems were provided: a tetragonal phase (s-t), a zirconia structure with a tetragonal phase (z-t), and a scheelite structure with a monoclinic phase (s-m). Compared with the oxidation potential of water, due to the appropriate level of the VB and conduction band (CB), as well as the low band gap at approximately 2.4 eV ($\lambda < 516$ nm), the s-m system has a visible crystal structure. It shows the best O_2 release activity under irradiation, and thus this is a promising system currently for photoelectrochemical application [50]. In spite the presence of these excellent advantages, a lower rate for charge separation occurs on $BiVO_4$, and the kinetics for transfer of photo-generated carriers to the electrode back and also throughout the interface of electrolyte–semiconductor is slow. This case results in a large volume and more surface reorganization. In terms of photo-generated carrier transport, it is known that it can limit the electron transport kinetics by the distance between the holes transport throughout the interface of electrolyte–semiconductor and the VO_4 tetrahedrons of $BiVO_4$ crystal where the V-3d orbital of CB is located here. Therefore, the formation of heterostructures is an important method to improve the efficiency of $BiVO_4$. Simelys Hernández et al. have photoelectrically deposited a cobalt phosphate (CoPi) catalyst on the best $BiVO_4$ electrode [50]. With optimized technology, under 1 solar radiation (100 mW cm^{-2}) and neutral pH, the photocurrent at 1.23 V is as high as 3 mA cm^{-2} compared to RHE. The authors also quantify the charge transfer kinetics of $BiVO_4$ photoanode without and with CoPi.

Going forward, Zhang Xueliang et al. reported that a heterojunction photoanode consisted of a conformal $BiVO_4$ layer/WO_3-nanoplate array (NPA), which was further decorated by cobalt phosphate (CoPi). This photoanode can serve as an oxygen evolving electro-catalyst (OEC) and can significantly improve the performance of photoelectochemical cell (PEC) [51]. The processes of coating a thin conformal $BiVO_4$ layer on the pre-synthesized WO_3 NPA surface were used for the synthesis of $BiVO_4/WO_3$ nanocomposite via a stepwise method of spin coating, and the light-assisted electrodeposition method could realize the CoPi OEC. A maximum photocurrent of 1.8 mA cm^{-2} was achieved at the potential of 1.23 V on the best CoPi@$BiVO_4/WO_3$ heterojunction photoanode under the air mass (AM) 1.5G solar radiation conditions in phosphate buffer electrolyte relative to RHE. This value is 5–12 times better compared to the bare $BiVO_4$ and WO_3 photoanodes. PL spectroscopy measurements, EIS, and UV–vis absorption spectroscopy show that the better PEC performance was arisen from the fact that the $BiVO_4/WO_3$ heterojunction has the nature of type II and enhanced carrier separation/transmission. By modifying the CoPi promoter, the oxidation kinetics and light stability of water are improved.

In the work of Drialys Cardenas-Morcoso et al., the surface of the semiconductor $BiVO_4$ photoanode has been modified with a highly porous cobalt oxide (CoO_x) promoter derived from MOF. In its synthesis, first, the solvothermal route reported previously was used to prepare ZIF-67 [52]. In order to make ZIF-67 grow uniformly on $BiVO_4$ and adhere firmly, the $BiVO_4$ surface was further modified by a monolayer of trimellitic acid. After that, simply mixing 2-methylimidazole and cobalt nitrate in methanol solution was used to grow ZIF-67 under the room temperature, thereby uniformly covering the surface of the semiconductor with MOF particles (called $BiVO_4$-ZIF-67). Thereafter, the synthesized $BiVO_4$-ZIF-67 sample was calcined in the air at 400 °C for three hours; during this time, the conversion from MOF nanoparticles to CoO_x promoter occurs (called $BiVO_4$–CoO_x). Compared with the bare $BiVO_4$ electrode, the detailed photoelectrochemical characteristics of the photoanode modified by the cocatalyst showed a significant improvement in the catalytic performance (four times higher photocurrent, and 300 mV cathode shift of the catalytic starting potential). In addition, measurements using fast hole scavengers and EIS analysis show that the promoter of MOF conversion can cause the obvious acceleration in water oxidation kinetics rather than acting as a layer for passivation on the surface.

In addition to $BiVO_4$, due to its unique structure, electrical and optical properties, metal nitrides (such as gallium nitride (GaN)) have become a new generation of solar water-splitting materials. The latest development of molecular beam epitaxy has led to the controlled synthesis of single-crystal GaN nanowire arrays with high-quality interfaces on planar Si, greatly reducing manufacturing costs [53]. These grown GaN nanowire arrays have a defect-free structure and large carrier mobility, so that carriers can be efficiently extracted from the Si substrate. In addition, the structure of the nanowire array has the benefits for the exposure of high-density surface-active sites and enhancing sunlight absorption.

Similarly, in the work of Guanyu Liu et al., it is opposed to using GaN-based semiconductors for solar energy decomposition [54]. The valence band edges and conduction band of GaN cross the reduction and oxidation potentials of hydrogen/oxygen evolution and change of mole fraction of In in $In_xGa_{1-x}N$ could lead to the adjustment of the band gap in the whole solar spectrum, which is expected to achieve a multi-band structure for efficient H_2 generation by solar approach [54]. However, until now, the research related to the use of promoters to improve the GaN-based photoelectrodes toward more efficient OER is still lacking. In the obtained tapered nanowires, the length is about 1.5 μm and the size in the average diameter is about 200 nm. The GaN nanowires were used for the deposition of CO_3O_4 sub-monolayers, with the deposition time of 5, 10, and 15 seconds and annealing to form a solid nano-island form. Potential measurement and photocurrent density were used to evaluate the photoelectrochemical properties of the CO_3O_4–GaN photoanodes. It is worth noting that all the photoanodes loaded with CO_3O_4 showed a cathode shift of about 100 mV, and the photocurrent increased significantly when the potential was lower than 0.6 V vs. RHE. GaN-loaded five seconds CO_3O_4 had shown the highest activity. Compared with the bare GaN anode, the photocurrent density is increased by 240% under the condition of 0.4 V. This is an excellent promoter performance of photoelectrochemical systems. The saturation current on the bare GaN is observed to be larger compared to the currents on 10 and 15 seconds CO_3O_4–GaN. However, this current is still slightly smaller compared to that of five seconds CO_3O_4–GaN. The interaction between surface charge transfer efficiency (η_{trans}) and effective light absorption (η_{abs}) could be the reason for this result. In general, the following equation can be used for the evaluation of the actual OER photocurrent (J_{water}): $J_{water} = J_{max} \times \eta_{abs} \times \eta_{sep} \times \eta_{trans}$. In this equation, η_{sep} represents the efficiency for charge separation, J_{max} represents the maximum photocurrent density from theoretical saturation (J_{max}), and η_{trans} can be obtained from the percentage of surface contact holes during OER. In some semiconductors (like hematite and $BiVO_4$), due to surface reorganization, η_{trans} will be restricted even in the high potential region, which reduces the amount for surface contact holes that are involved in water oxidation. In some reports previously, integration of OER catalysts with a light absorber leads to an increase in η_{trans}; thus, the saturation photocurrent density could also be increased. Due to the high transparency of five seconds CO_3O_4 nano-islands, in regard to the sample of five seconds CO_3O_4–GaN, the value of $\eta_{abs} \times \eta_{sep}$ is nearly the same with the value of a bare GaN electrode. For the time being, the bare GaN electrode shows a low density in photocurrent, which was arisen from the higher rate for surface charge recombination compared to CO_3O_4–GaN, as well as its slow degradation without a promoter. After adding the CO_3O_4 promoter, in the Nyquist's plot, the arc radius of impedance is greatly reduced, indicating that the electrolyte–electrode interface has an enhanced charge transfer kinetics. Nevertheless, the samples of 15 and 10 seconds CO_3O_4 photoelectrode had smaller saturation photocurrent density compared to bare GaN. This is because the light absorption is reduced compared with increasing surface area, and the CO_3O_4 sub-monolayers had a lower transparency. The sample of five seconds CO_3O_4 nano-island sub-monolayer film has the highest performance, arisen from excellent OER activity, and the high optical transparency. In addition, the light stability characterization showed that, compared with pure GaN, the CO_3O_4 nano-islands greatly reduced the photoanode photo-corrosion. It is temporarily arisen from the hole scavenger by CO_3O_4, which can quickly collect photo-generated holes, thereby inhibiting the photo-oxidation corrosion and charge recombination process on the GaN surface.

Fe, Co, and Ni-based materials with obvious OER activity are also used in photoelectrochemical water oxidation due to their heterogeneous structure. It is proved that the in situ growth of the ultra-thin FeF_x nanolayer on the Fe_2O_3 photoanode can be used as an efficient OER co-catalyst to accelerate hole transfer, thereby improving the activity of PEC [55]. With its ultra-thin structure and high electronegativity, the ultra-thin FeF_x not only promotes rapid hole transfer between the anode/cocatalyst interface but also maximizes the contact area between the electrode and the reactant. As expected, this photoanode exhibits a remarkably improved photocurrent density of $2.4\,mA\,cm^{-2}$ at 1.23 V (reversible hydrogen electrode, AM 1.5G). This value is 12 times larger compared to the original Fe_2O_3 sample. In addition, a significant negative shift (0.1 V vs. RHE) of the starting potential was achieved at the same time. Since CoO_x nanoparticles are known to be excellent water oxidation catalysts, Xiubing Huang et al. introduced a simple strategy to attach sub-nano cobalt oxide catalysts to the surface of WO_3 [56]. Co-modified WO_3 (called WO_3-CoM-2) is prepared in two steps: the mixture of WO_3 and Co-ethylene diamine tetraacetic acid (EDTA) is annealed at 350 °C in flowing argon, and then a second step is performed at 500 °C in flowing air. The second calcination greatly improves the WO_3 performance in OER and PEC, increases the O_2 release activity by 13 times, and significantly improves the photocurrent density. In the Soltani et al. study, a three-dimensional nuclear-branched CoP@NiFe-OH nanohybrid supported on nickel foam (CoP@NiFe-OH) phosphorylated on the surface was prepared as a high-performance dual-functional electrocatalyst for alkaline water electrolysis [57]. Theoretical calculations and electrochemical tests show that the hybridization of CoP and NiFe-OH can produce a strong synergistic catalysis, which leads to rapid hydrogen and oxygen-release reaction kinetics. At the same time, the hierarchical core branch array and self-supporting architecture also result in CoP@NiFe-OH with multiple exposed active sites, faster ions, electron transport behavior, and longer service life. Due to these typical advantages, the dual-function

CoP@NiFe-OH electrode can accelerate the total water decomposition and only provide currents of 100, 50, and 10 mA cm^{-2} at 1.68, 1.63, and 1.53 V, respectively, exceeding the performance of precious metals (+) IrO_2/C//Pt/C (−) and most dual-function electrocatalysts (Figure 5.9). In the Francesco Malara et al. study, the multiple effects of the amorphous nickel hydroxide (NiOH) (NiOOH) thin layer were deposited by light on the hematite (α-Fe_2O_3) photoanode [59]. Compared with RHE, the reduction of the overpotential resulted in an advancement of the photocurrent starting potential of 150 mV, and the photocurrent increased by about 50% at 1.23 V. In order to explain these phenomena, we conducted in-depth electrochemical studies through EIS spectroscopy and CV. The results are found that the electrolyte was injected with the charge because of the increased with the amount of NiOOH deposition, as well as the resistance reduction for charge transfer on the electrolyte–electrode interface. The benefit of depositing NiOOH is attributed to the fact that it facilitates removing holes away from the surface trap by hematite. Since a part of the photo-generated holes is consumed in the Ni redox cycle, above the potential higher than 1.25 V, this action is alleviated.

Yilmaz et al. reported a conceptualized collaborative self-templating method guided by the pattern of carbonate hydroxide, enabling the synthesis of 3D interpenetrating hierarchical nano-architectures, and then thermal selenization treatment to prepare porous interconnection transition metals with defects

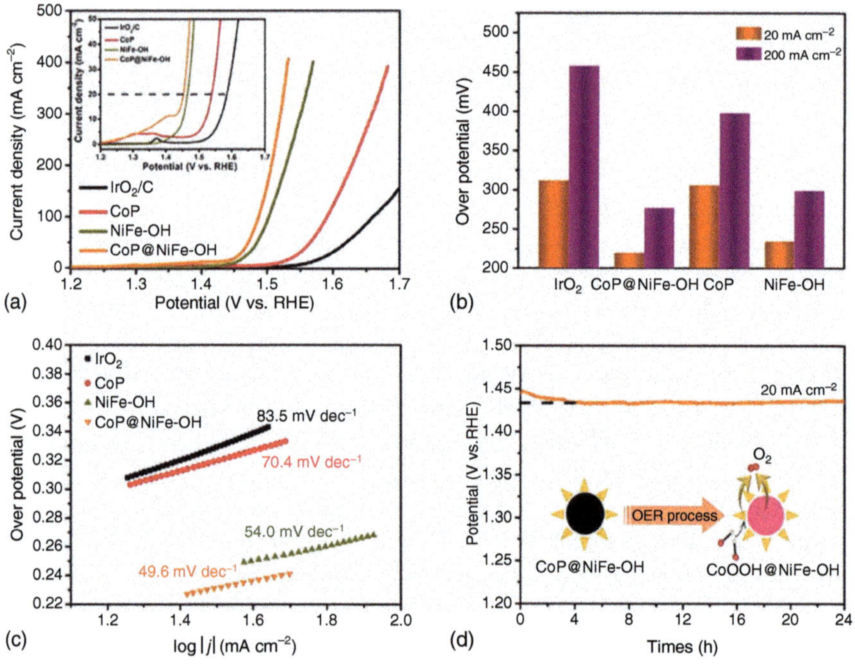

Figure 5.9 Polarization curves; comparison plots; corresponding Tafel slopes; chronopotentiometry curves of CoP@NiFe-OH at 20 mA cm^{-2} for 24 hours, Inset: Schematic illustration of the evolution process of the CoP@NiFe-OH during OER. Source: Jiang et al. [58]. © 2019 Elsevier.

selenide heterostructure [60]. The sacrificial MOF compounds (NiFe-PBA) and NiOH templates were explicitly developed to induce NiOH at high temperatures and pressure. Thermal decomposition to complete in situ defects and pore formation will subtype selenization into Ni–Fe selenide hybrids (named as NiFeSe@NiSeO@CC). Thus, an interconnected porous conductive network was produced with the microscopic level, and it can atomically expose abundant edges of the reactive crystal boundary. The systemically electrochemical tests show that the NiFeSe@NiSeO@CC sample has excellent electrochemical and photochemical properties. It has ultra-high potential and excellent stability for OER and HER and alkaline overall water splitting and has highly efficient and stable electrocatalytic activity. When p-type silicon was incorporated into the PEC for water splitting, HER activity can be further enhanced due to the catalytic properties as well as the established heterojunction. In the Huachuan Sun et al. study, by coupling ruthenium oxide (RuO_2) and NiFe-LDH on T-shaped NiTe, heterogeneous catalysts of NiTe@RuO_2 and NiTe@NiFe-LDH were prepared [61]. The manufactured NiTe@RuO_2 and NiTe@NiFe-LDH catalysts have excellent OER and HER activities, respectively. Theoretical calculations and experimental measurements show that the electron transfer on the interface during the HER process is beneficial to the adsorption of H*. The authors use NiTe@NiFe-LDH for anode and NiTe@RuO_2 for the cathode to construct a complete electrochemical cell for overall water splitting. The cell can provide a current of 200 $mA\,cm^{-2}$ at the cell voltage of 1.63 V. The whole battery works well under a voltage of 1.5 V and can continuously produce hydrogen/oxygen.

In addition to Fe/Co/Ni materials, copper-based materials are also used for photoelectrochemical water oxidation. In the Dong Chen et al. study, Au and Pt nano-layers with spatial spacing were deposited on the bottom and surface of Cu_2O nanoparticles, respectively [62]. The resulting new Au/Cu_2O/Pt photoelectrode includes a plasmon photosensitizer (Au) for enhancing light collection and generating hot electrons from the surface plasmon resonance (SPR) effect and an electron collector layer (Pt) used to promote photoelectron transfer. Many photoelectrochemical measurements, especially surface photovoltage (SPV) measurements, have well supported this synergistic effect. As expected, the photocurrent density–voltage curve shows that the Au/Cu_2O/Pt composite material reaches a photocurrent of $-3.55\ mA\,cm^{-2}$. This current value is almost 4.63 times that of pure Cu_2O at 0 V. On the other hand, Lei Ran reported the acceleration of transfer kinetics and charge separation on the aligned Cu_7S_4/TiO_2/CoCr LDH nanorod array [63]. The modification of CoCr-LDH and Cu_7S_4 on the basis of TiO_2, due to the highly matched band arrangement, makes the great enhancement in the efficiency of charge separation and photoanode's ultraviolet light absorption. It is an effective approach to prevent Cu_7S_4 from photo-corrosion by forming a heterojunction structure as demonstrated by attaching Cu_7S_4 with a protective layer. In addition to the increased active surface area arisen from the layered morphology, more importantly, it also offers pores for gas escape as well as sufficient active sites for the water oxidation process. Due to the fact that the three components have special energy band arrangement, it participates in the water-splitting process and has multiple reaction interfaces, because the Cu_7S_4 and CoCr-LDH have been distributed with the photo-induced holes facilitating OER. Therefore, the photocurrent density (2.04 mA cm^{-2} at the potential of 1.23 V) and photoconversion efficiency (0.58% at the potential of 0.6 V) were greatly enhanced by the synergistic effect of the photoanode consisted of ternary Cu_7S_4/TiO_2/CoCr-LDH heterogeneous.

5.2.3 Metal–Nonmetal Heterostructure Catalysts

Many non-metallic materials are also used to combine metallic materials to form heterogeneous structures. Among alternative nonmetal materials, carbon materials have been extensively studied. Generally, carbon materials are used as electronic mediators. Their robust structure, high electrical conductivity (graphene is 108 S m^{-1}), and large surface area greatly enhance the transport of photo-generated carriers inside the semiconductors, thereby inhibiting the recombination of photo-induced electron–hole pairs. The sp^2-bonded carbon network with conjugated structure has a large number of delocalized electrons, which improves the conversion efficiency for photocurrent of the system.

Teyyebah Soltani et al. reported the use of BVO to prepare highly stable nanocomposites consisted of BVO and reduced graphene oxide (rGO) at two weight ratios for rGO (10% and 5%). The synthesis uses a method by visible light-assisted photocatalytic reduction GO [57]. Then, an fluorine-doped tinoxide (FTO) substrate was used to drop-casting these synthesized BVO/rGO nanocomposite materials to study the effects from rGO in the electron transport insides the synthesized BVO/rGO film. During the water-splitting test by PEC experiment, compared with BVO/rGO 5% (377.9 μA cm^{-2}) and BVO (111.7 μA cm^{-2}), the highest photocurrent on BVO/rGO-10% photoelectrode was obtained 554.4 μA cm^{-2} at the potential of 1.2 V vs. Ag/AgCl. Compared with BVO, the BVO/rGO electrode shows not only the largest carrier density but enhanced conductivity and extended lifetime. The data indicated that BVO/rGO photoanode had excellent PEC performance. With the solar sunlight, the BVO/rGO film showed a negative shift in the open circuit potential (OCP) on the basis of the promoted separation of holes/electrons insides the film. BVO/rGO electrodes and BVO's photocurrent yield would also influence by the solution pH. In the NaOH solution (1 M, pH = 13.5) with entire measuring potential range (relative to Ag/AgCl, 0 to +1.2 V), the photocurrent density is higher than that of Na_2SO_4 solution (0.1 M, pH = 6.5) (Figure 5.10). This study proved the effects from coverage of rGO in BVO/rGO film, which would largely enhance the transfer of electrons through rGO. Besides, Palyam Subramanyam et al. prepared $BiVO_4$ on rGO nanocomposite material doped with molybdenum (Mo) and evaluated it as a photoanode for PEC water splitting [64]. By introducing Mo-dopant and rGO in $BiVO_4$, this composite material enhances the water oxidation performance of PEC because they help reduce charge recombination and enhance photocurrent. As a result, the Mo–$BiVO_4$@rGO composite photo anode exhibited a photocurrent density of 8.51 mA cm^{-2} at 1.23 V vs. RHE, which is two and four times greater than that of Mo–$BiVO_4$ (5.3 mA cm^{-2} at 1.23 V vs. RHE) and pristine $BiVO_4$ (2.01 mA cm^{-2} at 1.23 V vs. RHE) photoactive electrodes. In addition, the

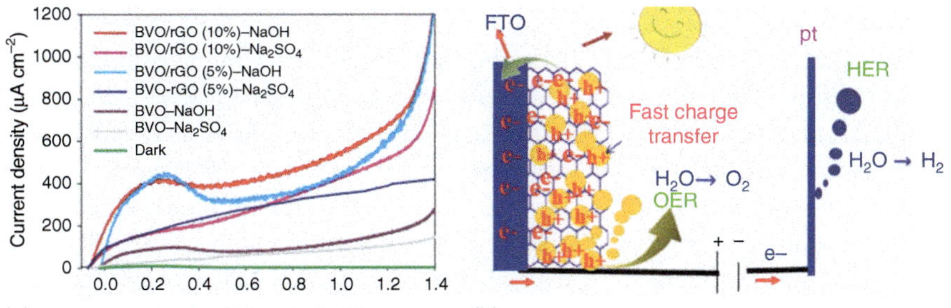

(a) (b)

Figure 5.10 (a) LSV curve scanned from −0.2 to 1.4 V vs. Ag/AgCl for BVO and BVO/rGO of 5% and 10% in Na_2SO_4 (0.1 M, pH = 6.5) and NaOH (1 M, pH = 13.6) electrolytes. (b) Possible mechanism for PEC water-splitting process catalyzed by rGO in the BVO/rGO photoelectrode under simulated solar light irradiation. Source: Soltani et al. [57]. © 2018 Elsevier.

ternary nanocomposite has good light conversion efficiency, low charge transfer resistance, and good external quantum efficiency (EQE). These studies show that the improved $BiVO_4$@rGO can improve the water-splitting activity of PEC and show that this method can be used to design more effective PEC water-splitting photo anodes.

Ashwani Kumar et al. developed the NiFe alloy NPs supported on N- and S-doped porous C matrix prepared from biomass (duckweed, DW) as effective OER/HER electrocatalysts [65]. In the OER test, an overpotential of only 267 mV was required for the annealed catalyst to reach 10 mA cm^{-2} in 1 M KOH. For total water splitting, the catalyst only need a battery voltage of 1.61 V to deliver 10 mA cm^{-2} and continuously release O_2 and H_2 bubbles for more than 200 hours. After integration with perovskite solar cells, the homologous DW electrolyzer exhibits overall water splitting by a solar-driven approach, and the value of 9.7% was reached for the solar-to-hydrogen conversion efficiency. Santosh K. Singh et al. reported a high-performance OER electrocatalyst, which consists of large-surface area N-doped porous graphene for supporting the NiCo alloy particles (named as NiCo/pNGr (75 : 25)) [66]. XPS, XRD, transmission electron microscopy, Raman, and DFT calculations have clearly studied the importance of N doping to achieve effective interaction of size-controlled NiCo particles and their uniform dispersion. NiCo/pNGr (75 : 25) has been further studied by DFT calculations combined with the electrochemical measurements, which displays that for reaching 10 mA cm^{-2}, the overpotential is 260 mV, with the small Tafel value of ≈87 mV dec^{-1}, and the long catalytic life. The calculation studies were performed to examine the NiCo particle with pNGr defect sites and doped N, which can maintain longer catalytic durability. In addition, NiCo/pNGr (75 : 25) serving as a water oxidation electrocatalyst can make the electrolyzer to operate at a rather low potential of 1.5 V in an alkaline medium.

Silicon oxide and titanium oxide are the other two types that effectively combine precious metal materials to create heterogeneous structures. Ivan A. Moreno-Hernandez et al. report pointed out that after forming a thin chemically oxidized SiO_x layer, applying a tin oxide (SnO_x) layer to the n-Si wafer can passivate the Si surface, and at the same time, produce a voltage of ≈620 mV of light under standard simulated sunlight (1G) [67]. The SnO_x layer forms ohmic contacts with Ni, Ir, or Pt films, which act as a precatalyst for the OER in 1.0 M H_2SO_4(aq) or 1.0 M KOH(aq). Using Ni in 1.0 M KOH(aq) or using Pt/IrO_x layer as OER catalyst in 1.0 M H_2SO_4(aq), the regenerative solar energy conversion efficiency to O_2(g) was obtained by 4.1% and 3.7%, respectively. For electrodes with patterned catalyst layers, stable photocurrents of >100 hours can be obtained in both 1.0 M H_2SO_4(aq) and 1.0 M KOH(aq). Seungtaeg Oh et al. reported a controllable heterojunction of CoO_x/SiO_x/*n*-Si prepared by atomic layer deposition (ALD) for solar-driven OER/HER [68]. By changing the deposition temperature, a phase conversion of the ALD CoO_x film occurs from CoO to CO_3O_4. In addition, the study of the system shows that its electrochemical and energy properties change greatly along with the change of phase. A high photovoltage was generated on the CoO, which was formed at 150 °C by establishing an ideal hole-selective heterojunction with *n*-Si, while higher catalytic performance for OER occurs on CO_3O_4 grown at 300 °C. In order to resolve the competitive relationship, a double-layer (DL) ALD CoO_x film was exploited by the authors, which has the advantages from CO_3O_4 and CoO. A high photocurrent of 3.5 mA cm^{-2} was generated from the DL ALD CoO_x/SiO_x/Si heterojunction photoanode without a buried junction, and it still kept a saturation of 32.5 mA cm^{-2} for 12 hours under simulated 1 solar radiation in 1 M KOH as the current density without significant degradation. In the work of Zion Tachan et al., the possibility of photoelectrochemical water oxidation using the TiO_2-catechol type II charge transfer complex has been demonstrated [69]. First, the formation of the TiO_2–catechol complex extends the spectral response of the system to the visible region. Secondly, the resulting complex acts as a hole reservoir, which can electrocatalyze the OER while reducing the required overpotential by 500 mV s^{-1}. In addition, the fact that type II charge transfer complexes can be formed similar to other types of metal oxides opens the way for the development of various new photocatalytic systems.

References

1 Feng, J., Lv, F., Zhang, W. et al. (2017). *Adv. Mater.* 29: 1703798.
2 Shi, Q., Zhu, C., Du, D., and Lin, Y. (2019). *Chem. Soc. Rev.* 48: 3181–3192.
3 Stoerzinger, K.A., Rao, R.R., Wang, X.R. et al. (2017). *ChemMedChem* 2: 668–675.
4 Rao, R.R., Kolb, M.J., Halck, N.B. et al. (2017). *Energy Environ. Sci.* 10: 2626–2637.
5 Laha, S., Lee, Y., Podjaski, F. et al. (2019). *Adv. Energy Mater.* 9, 1803795.
6 Lin, Y., Tian, Z., Zhang, L. et al. (2019). *Nat. Commun.* 10: 162.
7 Wang, L., Zhou, Q., Pu, Z.H. et al. (2018). *Nano Energy* 53: 270–276.
8 Ping, Y., Nielsen, R.J., and Goddard, W.A. 3rd, (2017). *J. Am. Chem. Soc.* 139: 149–155.
9 Akbashev, A.R., Zhang, L., Mefford, J.T. et al. (2018). *Energy Environ. Sci.* 11: 1762–1769.

10 Alia, S.M., Shulda, S., Ngo, C. et al. (2018). *ACS Catal.* 8: 2111–2120.
11 da Silva, G.C., Perini, N., and Ticianelli, E.A. (2017). *Appl. Catal. B* 218: 287–297.
12 Yang, H., Wu, Y., Lin, Q. et al. (2018). *Angew. Chem. Int. Ed.* 130: 15702–15706.
13 Park, J., Sa, Y.J., Baik, H. et al. (2017). *ACS Nano* 11: 5500–5509.
14 Oakton, E., Lebedev, D., Povia, M. et al. (2017). *ACS Catal.* 7: 2346–2352.
15 Kwon, T., Hwang, H., Sa, Y.J. et al. (2017). *Adv. Funct. Mater.*: 27.
16 Strickler, A.L., Escudero-Escribano, M.A., and Jaramillo, T.F. (2017). *Nano Lett.* 17: 6040–6046.
17 Hou, Y.H., Liu, Y.P., Gao, R.Q. et al. (2017). *ACS Catal.* 7: 7038–7042.
18 Wu, Z.-P., Lu, X.F., Zang, S.-Q., and Lou, X.W. (2020). *Adv. Funct. Mater.* 30, 1910274.
19 Cai, P., Huang, J., Chen, J., and Wen, Z. (2017). *Angew. Chem. Int. Ed.* 56: 4858–4861.
20 Nai, J., Lu, Y., Yu, L. et al. (2017). *Adv. Mater.* 29, 1703870.
21 Zhang, T., Wu, M.Y., Yan, D.Y. et al. (2018). *Nano Energy* 43: 103–109.
22 Zhu, D.D., Liu, J.L., and Qiao, S.Z. (2016). *Adv. Mater.* 28: 3423–3452.
23 Zhou, J., Dou, Y.B., Zhou, A. et al. (2017). *Adv. Energy Mater.* 7, 1602643.
24 Zou, X., Liu, Y., Li, G.D. et al. (2017). *Adv. Mater.* 29, 1700404.
25 Zhao, S.L., Li, M., Han, M. et al. (2018). *Adv. Funct. Mater.* 28, 1706018.
26 Zhang, J., Hu, Y., Liu, D. et al. (2017). *Adv. Sci. (Weinh.)* 4, 1600343.
27 Zhuang, L., Ge, L., Yang, Y. et al. (2017). *Adv. Mater.* 29, 1606793.
28 Zhou, Y., Sun, S., Song, J. et al. (2018). *Adv. Mater.* 30: e1802912.
29 Zhao, S., Guo, S., Zhu, C. et al. (2017). *RSC Adv.* 7: 1376–1381.
30 Ye, Z.G., Li, T., Ma, G. et al. (2017). *Adv. Funct. Mater.* 27, 1704083.
31 Wu, J.J., Liu, M.J., Chatterjee, K. et al. (2016). *Adv. Mater. Interfaces* 3, 1500669.
32 Mohanty, B., Ghorbani-Asl, M., Kretschmer, S. et al. (2018). *ACS Catal.* 8: 1683–1689.
33 Xiao, W., Huang, X.L., Song, W.D. et al. (2016). *Nano Energy* 25: 60–67.
34 Zhao, S.L., Wang, Y., Dong, J.C. et al. (2016). *Nat. Energy* 1: 1–10.
35 Li, F.L., Shao, Q., Huang, X., and Lang, J.P. (2018). *Angew. Chem. Int. Ed.* 57: 1888–1892.
36 Zhao, L., Dong, B., Li, S. et al. (2017). *ACS Nano* 11: 5800–5807.
37 Zhang, C.L., Wang, B.W., Shen, X.C. et al. (2016). *Nano Energy* 30: 503–510.
38 Zhu, Y.P., Jing, Y., Vasileff, A. et al. (2017). *Adv. Energy Mater.*: 7.
39 Qu, K.G., Zheng, Y., Dai, S., and Qiao, S.Z. (2016). *Nano Energy* 19: 373–381.
40 El-Sawy, A.M., Mosa, I.M., Su, D. et al. (2016). *Adv. Energy Mater.* 6.
41 Zhang, Y., Fan, X.L., Jian, J.H. et al. (2017). *Energy Environ. Sci.* 10: 2312–2317.
42 Ren, X.H., Zhou, J., Qi, X. et al. (2017). *Adv. Energy Mater.*: 7.
43 Prasannachandran, R., Vineesh, T.V., Anil, A. et al. (2018). *ACS Nano* 12: 11511–11519.
44 Ahmed, A.Y., Ahmed, M.G., and Kandiel, T.A. (2018). *Appl. Catal. B* 236: 117–124.
45 Wang, Y., Ni, Y.M., Wang, X. et al. (2018). *ACS Appl. Energy Mater.* 1: 5718–5725.
46 Lu, J., Cai, L., Zhang, N. et al. (2019). *ACS Appl. Mater. Interfaces* 11: 44214–44222.

47 Macounova, K.M., Nebel, R., Klusackova, M. et al. (2019). *ACS Appl. Mater. Interfaces* 11: 16506–16516.

48 Kriek, R.J., Iqbal, M.Z., Doyle, B.P., and Carleschi, E. (2019). *ACS Appl. Energy Mater.* 2: 4205–4214.

49 Zhou, F., McDonnell-Worth, C., Li, H. et al. (2015). *J. Mater. Chem. A* 3: 16642–16652.

50 Hernández, S., Gerardi, G., Bejtka, K. et al. (2016). *Appl. Catal. B* 190: 66–74.

51 Cheng, Y., Guo, Y., Zhang, N. et al. (2019). *Nano Energy*: 63.

52 Cardenas-Morcoso, D., Ifraemov, R., Garcia-Tecedor, M. et al. (2019). *J. Mater. Chem. A* 7: 11143–11149.

53 Zhou, B., Kong, X., Vanka, S. et al. (2018). *Nat. Commun.* 9: 3856.

54 Liu, G.Y., Karuturi, S.K., Simonov, A.N. et al. (2016). *Adv. Energy Mater.* 6, 1600697.

55 Feng, C.C., Wang, L., Fu, S.R. et al. (2018). *J. Mater. Chem. A* 6: 19342–19346.

56 Huang, X.B., Zhao, G.X., and Wang, G. (2017). *J. Mater. Chem. A* 5: 24631–24635.

57 Soltani, T., Tayyebi, A., and Lee, B.K. (2018). *Sol. Energy Mater. Sol. Cells* 185: 325–332.

58 Jiang, H., Gu, J., Zheng, X. et al. (2019). *Energy Environ. Sci.* 12: 322–333.

59 Malara, F., Minguzzi, A., Marelli, M. et al. (2015). *ACS Catal.* 5 (9): 5292–5300.

60 Yilmaz, C., Tan, C.F., Lim, Y.F., and Ho, C.W. (2019). *Adv. Energy Mater.* 9, 1802983.

61 Sun, H.C., Yang, J.M., Li, J.G. et al. (2020). *Appl. Catal. B* 272, 118988.

62 Chen, D., Liu, Z.F., Guo, Z.G. et al. (2018). *J. Mater. Chem. A* 6: 20393–20401.

63 Ran, L. and Yin, L.W. (2019). *Adv. Mater. Interfaces* 6.

64 Subramanyam, P., Vinodkumar, T., Nepak, D. et al. (2019). *Catal. Today* 325: 73–80.

65 Kumar, A., Chaudhary, D.K., Parvin, S., and Bhattacharyya, S. (2018). *J. Mater. Chem. A* 6: 18948–18959.

66 Singh, S.K., Kumar, D., Dhavale, V.M. et al. (2016). *Adv. Mater. Interfaces* 3.

67 Moreno-Hernandez, I.A., Brunschwig, B.S., and Lewis, N.S. (2018). *Adv. Energy Mater.* 8.

68 Oh, S., Jung, S., Lee, Y.H. et al. (2018). *ACS Catal.* 8: 9755–9764.

69 Tachan, Z., Hod, I., and Zaban, A. (2014). *Adv. Energy Mater.* 4, 1301249.

6

H_2O Reduction and Water Splitting Electrocatalytic Cell

6.1 Noble-Metal-Based HER Catalysts

Platinum (Pt) has a good hydrogen evolution reaction (HER) activity and is, to a large extent, preferred as HER active catalyst. Platinum materials have excellent electrocatalytic hydrogen production ability. In industrial applications, the main problems are the high cost of the corrosive solution and unstable catalysis. One way to solve the problem is to reduce the load of Pt while maintaining efficiency. Recently, people combine Pt and carbon materials to enhance the electrocatalytic performance. The control of C materials can improve the surface area of the catalyst and provide the basis for Pt loading.

In addition, when using Pt–Carbon material, the improvement of the conductivity of the substrate enables the electronic motion to be quickly used for the dynamics of HER. Excellent catalyst activity can be obtained by high interaction and synergy with metal species and carbon carriers so as to effectively prevent the aggregation and leaching of Pt NP in the electrochemical process. Sun et al. [1] cited the synergistic effect of the carbon matrix. Single platinum atoms and clusters supported on nitrogen-doped graphene nanosheets (NGNs) was fabricated for the HER, using the atomic layer deposition (ALD) technique, resulting in the utilization of nearly all the Pt atoms. The size and density of the Pt catalysts on the NGNs are precisely controlled by simply adjusting the number of ALD cycles. The Pt atoms and clusters on the NGNs show much greater activity for the HER in comparison with conventional Pt nanoparticles catalysts. After the Pt precursor was fixed to the NGN substrate, a single layer containing Pt was formed in an oxidizing environment.

After continuous immobilization and oxidation of the platinum catalyst, the surface reaction of ALD can control the particle size distribution that shows the activities of NGN, Pt/C commercial, and ADDPt/NGN about 50 and 100 ALD cycles. Compared with Pt/C, ALD50Pt/NGN and ALD100Pt/NGN showed higher HER activity, among which ALDPt/NGN had a Tafel slope as low as 29 mV dec^{-1}. The mass activity was obtained by considering the catalyst loading.

The results show that a single Pt atom and a single cluster can greatly improve the efficiency of Pt, rather than other NP, and decrease the integration cost of HER catalyst. Stability is an important factor in catalyst evaluation. ALD50Pt/NGN

Electrocatalysis in Balancing the Natural Carbon Cycle, First Edition. Yaobing Wang.
© 2021 WILEY-VCH GmbH. Published 2021 by WILEY-VCH GmbH.

shows a similar LSV curve before and after 1000 current measurement cycles. After the accelerated decomposition test, a scanning electron microscope image of ALDPt/NGN showed that the size of Pt particles increased slightly, but no apparent aggregation was found, and the high stability of ALD50Pt/NGN to HER was confirmed.

The metal alloy process is becoming another promising Pt-based HER catalyst upgrading process. Basic scientific research of base metal on bimetallic catalysts has shown that this change is essential to change the heterojunctions and unexpected properties of new nanostructures in bimetallic systems. The element provides Pt metal catalyst with excellent activity.

Ruthenium is also a good electrocatalytic hydrogen generator because of its metal hydrogen bond strength similar to that of platinum. Bick et al. studied the electrocatalytic performance of Ru in HER. A rare study reported the widespread use of Ru's pH in HER body. Bick et al. introduced Ru in a two-dimensional (2D) C_2N catalyst containing nitrogen (Ru@C_2N). It is made up of palm oil [2]. As a multi-purpose platform for adsorption, germination, and growth of Ru NP, C_2N scaffolds provide rich coordination sites for the distribution of small Ru nanoparticles in the C_2N layer. Prepared Ru@C_2N catalyst was tested in 0.5 m H_2SO_4 electrolyte. Ru@C_2N was similar to Pt/C (16 mV) and required a low overcapacity of 22 mV in order to provide 10 mA cm^2 of current density better than Co@C_2N (290 mV), Ni@C_2N (410 mV), Pd@C_2N (330 mV), and Pt@C_2N (60 mV) samples. A small table elevation indicates that Ru@C_2N follows the Volmer Table Reaction Mechanism. Ru@C_2N, which changed the tested solution from acid to alkali, was much higher over the low overload available from 17 mV to achieve the small table climb of the 10 mA cm^2 of current density. Calculating the density functional theory (DFT) of Ru@C_2N, the materials have different catalytic activities in different pH ranges. In acid solution, the catalyst has a similar value to Pt(111), which indicates that it has a good catalytic activity. This value is generally considered as a good binding energy for proton production and is generally used for adsorption, reduction, and gas desorption. At the same time, the reaction mechanism of the catalyst in the alkaline electrolyte is different from that in acid. Ru@C_2N can effectively adsorb H_2O, form H and OH, and transport protons in the next step. Thus, the efficiency loss in the volume step process due to the strong OH bond can be compensated. In the study of Huang et al., transition metal copper (Cu) and ruthenium form a channel-rich snowflake-like RuCu nanoplate (NS). In the nano wafer catalyst, the channel in the modified morphology can enlarge the surface area of the electrocatalyst, improve the electronic transmission capacity, and is conducive to the performance of the catalyst. DFT calculation demonstrated the surface electron activation in the channel-rich region. And also, amorphous copper played a decisive role in the synergistic effect for crystalline Ru species with good catalytic activity. The short-range disturbance reduces the average energy dissipation caused by the combined relaxation dissociation.

For the first time, precious metals have been used in hydrogen chemisorption experiments because of their low hydrogen binding energy. In the experiment of hydrogen adsorption/desorption, its (111) surface has good thermal stability and the ability of hydrogen adsorption and desorption. Similar to Pt(100), the location near HER top volcano station demonstrated that infrared is a significant alternative to platinum.

6.2 Non-Noble Metal Catalysts

Transition metal oxide (TMO) is familiar in electrocatalysis due to its stability, low cost, easy access as well as environmentally friendly characteristics. TMO has a good prospect in the field of electrocatalysis, but this kind of catalyst has the following problems: low conductivity, slow reaction kinetics, and lack of hydrogen adsorption sites because the above problems affect the performance of gas electrocatalysis. Scientists have found that TMOs based on molybdenum and tungsten, such as MoO_2 and WO_2, have high metal conductivity and good performance because they crystallize in a single clinical unit and have deformed rutile crystal structure. However, the aggregation of MoO_2 affects HER performance. In order to solve these problems and further improve the performance of TMO electrolytic catalyst, a variety of strategies are proposed, such as reducing the size of the structure, preparing porous structural materials, introducing surface defects or O defects by combining TMO with conductive carbon carrier, and doping other atoms. As discussed in the case of a metal catalyst, reducing the particle size of the catalyst will make the surface larger, thus making the surface reaction more active, thus increasing the electrocatalytic activity of TMO. For example, Cui et al. proposed that the activity of ultra-thin porous MoO_2 nanofibers (NFs) directly attached to the foamed nickel is much higher than that of the conventional dense MoO_2.

The mesoporous metal WO_2 carbon nanowires with better oxygen defects were produced by sintering the inorganic–organic WO_3 ethylene diamine hybrid. The resulting catalyst shows excellent HER performance, in which it has a low performance of 10.58 mV, and the Tafel slope as low as 46 mV dec^{-1}. This is due to the high concentration of oxygen defects, the mesoporous electronic structure, and the close interaction between NPs WO_2 and carbon matrix. O defects not only generate many active sites but also enhance electronic mobility because, with the reduction of the electronic state close to the Fermi level, the forbidden band provides the metal characteristics of WO_2 synthesis. A series of sub-chemically determined WO_3-x nanostructures with a high concentration of O gap were also manufactured, which increased HER activity, while HER activity was higher than its original counterpart. It is worth noting that the C-doped WO_x is fixed on the C carrier (WO_x@C/C) and is manufactured via carbonizing the tungsten phosphate coated with polypyridine. Its catalytic performance is excellent. The Pt-type catalytic performance is 36 MW and shows a 19 mV dec^{-1} ultra-small plate slope (Figure 6.1). The thick carbon shell results in the excellent activity of WO_x@C/C, which provides good conductivity, thus accelerating load transfer and changing its ΔGH^{**} value to a level suitable for H^+ adsorption.

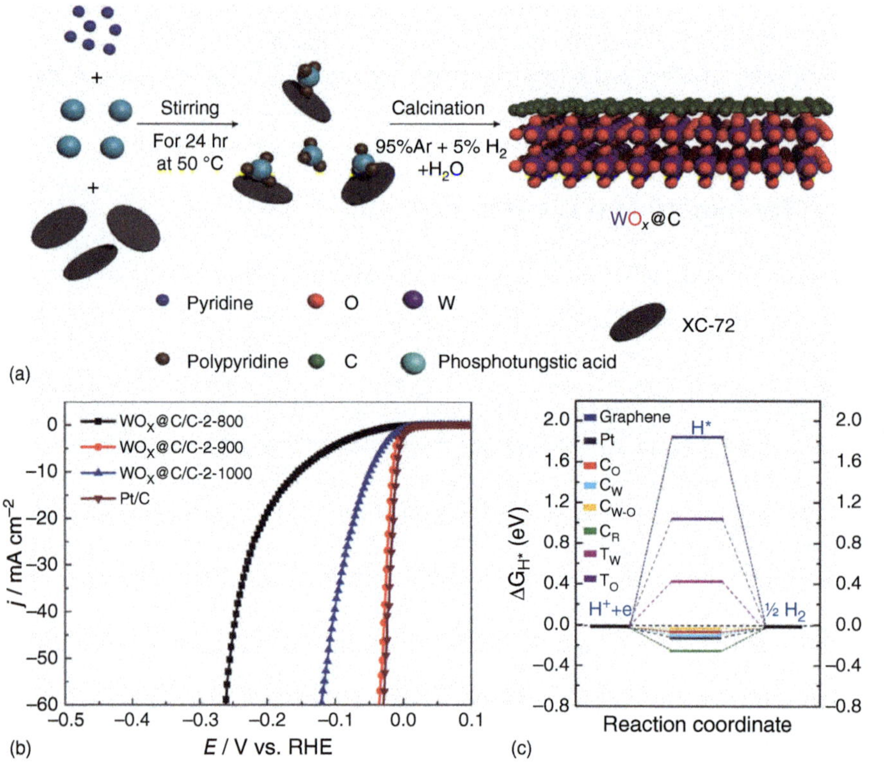

Figure 6.1 (a) Schematic diagram for WO_x@C/C. (b) Polarization curve. (c) Calculates H* of Gibbs free energy at the adsorption position in the WO_3@C model, the only WO_3 and graphene, and serves as a reference. Source: Jing et al. [3]. © 2018, John Wiley & Sons.

In recent years, transition metal carbides (TMCs) have many advantages like its low cost, good conductivity, good chemical stability, and thermal stability. The electric metal electronic structure allows the active TMC to absorb and activate hydrogen, which makes the intrinsic electrocatalyst Pt similar. In 2012, Hu and other scientists found that MO_2C had a good electrocatalytic activity for HER. The findings provide an opportunity for the materials with TMC-based structures. These M materials include Co, Fe, Ni, Mo, W, Ta, V, etc. Among them, molybdenum carbide and tungsten carbide, which are used in electrochemistry, are more prominent.

Among the alternative non-precious metal materials, Co-, Ni-, and Fe-based materials have received extensive research interest. For example, Fan Haosen et al. reported nickel cyanide coordination polymer nanosheets with 2D ultra-thin structure via a simple co-precipitation method [4]. After carburizing treatment in Ar/H_2 atmosphere, 2D uniform hybrid nanosheets were successfully obtained, in which N-doped carbon nanosheets were embedded with uniform Ni_3C nanodots. Hybrid nanosheets exhibit a thickness of about 10 nm and a lateral size of about 200–300 nm and have uniformly dispersed Ni_3C nanodots of 3–5 nm. When Fe is incorporated, excellent catalytic activities for both oxygen evolution reaction (OER) and HER were obtained on Ni_3C-based nanosheets. Specifically, Ni_3C nanosheets doped with 2 at.% Fe (atomic percentage) showed a small Tafel slope as low as 41.3 mV dec^{-1} and an overpotential as low as 292 mV for HER in alkaline medium. It also has excellent OER catalytic performance, with a small Tafel slope as low as 62 mV dec^{-1} and a low overpotential of 275 mV in alkaline medium.

For the exposure of more surface-active sites, electrocatalysts were usually reduced to single atoms fixed on specific substrates, clusters, or nanoparticles on supports to maximize the atom utilization. In addition, it may greatly change the inherent activity of transition metals. Zhang Longzhou et al. reported a simple and inexpensive strategy. This method is used to prepare highly stable, atom-dispersed Ni catalysts on defect graphene (DG) (A-Ni@DG) via incipient wetness impregnation process followed with acid-leaching process, in which Ni loading is as high as 1.24 wt% [5]. According to the direct observation results from the linear combination fitting (LCF) analysis of the X-ray absorption near-edge structure (XANES) and the high-angle circular dark-field scanning transmission electron microscope (HAADF-STEM) image, the authors inferred the Ni–C configuration. Single-atom Ni was trapped in the defects of graphene according to the HAADF-STEM image, forming a Ni@defect. As suggested by the DFT calculations related to three kinds of possible Ni–C configurations, it was found that the activities for OER/HER are different on the different Ni–C coordination. This result is completely distinct from metal nanoparticles for electrocatalysis. The catalytic mechanism assumes that Ni@defect provides activities for HER or OER. Unsurprisingly, high activity was obtained on A-Ni@DG, which is comparable to the benchmark Pt/C catalyst. Compared with RHE without attenuation in the catalytic process, its turnover frequency (TOF) at 100 mV is calculated to be 5.7 s^{-1}. This value is among the best ones reported so far. Of great interest, compared to RHE, to achieve the current density of 10 mA cm^{-2}, the A-Ni@DG electrocatalyst only requires the potential of 1.5 V for OER, which is far higher compared to the benchmark of Ir/C electrocatalyst. Moreover, the stable HER and OER performance prove the firm fixation of Ni on the surface of the DG substrate. Such fixation provides the basis for the good tolerance of long-term electrocatalytic reactions. The specific configuration of atomic Ni in the defect leads to the excellent performance of A-Ni@DG, and it changes the electronic structure of trapped Ni, thereby minimizing the energy barrier of OER/HER. In addition to defect anchoring, heteroatoms can also be used to anchor single-atom metal catalysts. Jie Yin et al. prepared monoatomic Ni–C–N nanosheet catalysts for HER [6]. They used a two-step process to prepare the Ni–C–N NSs. Firstly, the bulk Ni–C–N was obtained by annealing Ni-polyoxometalate metal–organic framework (Ni-POMMOF) under the NH_3 atmosphere at 380 °C for three hours and washing with HCl. Then, the combination of ultrasonic and wet chemical peeling processes in formamide was used for peeling off a large amount of Ni–C–N into Ni–C–N NSs. The obtained Ni–C–N NSs possess metallic properties and could be used as an efficient electrocatalyst for HER in 1.0 M PBS, 1.0 M KOH, or 0.5 M H_2SO_4. The Ni–C–N NSs material has electrocatalytic properties similar with ordinary Pt catalysts and can be used for long-term electrocatalysis for more than 70 hours at all pH values without significant current decay. The work shows the possibility to enhance HER performance by hybridizing Ni_3N with NiC to the levels of Pt.

6.3 Water Splitting Electrocatalytic Cell

Recently, new separators or water systems have been introduced [7]. Direct conversion of solar energy to hydrogen through the photolysis of water is a promising way to generate clean energy [8, 9].

In terms of limitations, a lot of ways were developed to combine the whole water distillation with solar cells to establish a new effective way for low-cost fractionation [10, 11]. Among these, short-circuit photoelectric density is 0.76 mA cm^{-2}, open-circuit photoelectric density open circuit voltage is 1.06 V. Due to these benefits, the solar energy conversion efficiency reaches 17.3%. They chose Ni-Fe-layered double hydroxide (LDH)/Ni foam as an effective double catalyst, which was combined with perovskite series cells in the aberration solar cell composition. In Figure 6.2, perovskites solar cells are arranged side by side and combined by wires to simulated solar radiation that provides energy to separate water [13]. These two facts make the photoelectric density of water in the separator about 10 mm, and the power supply square meter, equivalent to 12.3% of the solar hydrogen efficiency.

In another way, Dai's group gave an ideal image proposal, electrolysis/photolysis of water, and consumption of H_2. The water passing through the H_2 fuel cell must be in a closed cycle, and the electrolyte capacity of anode and cathode must match, so the efficiency of the whole water distillation system may reach a very high level. Gong et al. [13] reported a combination of electrocatalyst Ni@Cr_2O_3-NiO (CrN) HER and high-activity Ni-Fe-LDH-OER in 1 M KOH electrolyte (Figure 6.3). The initial voltage of the whole water separation system is 1.46 V and the current density is 200 mA cm^{-2} at 1.57 V. Alkaline metal catalyst can provide about 15% high yield when using GaAs solar cell to separate water.

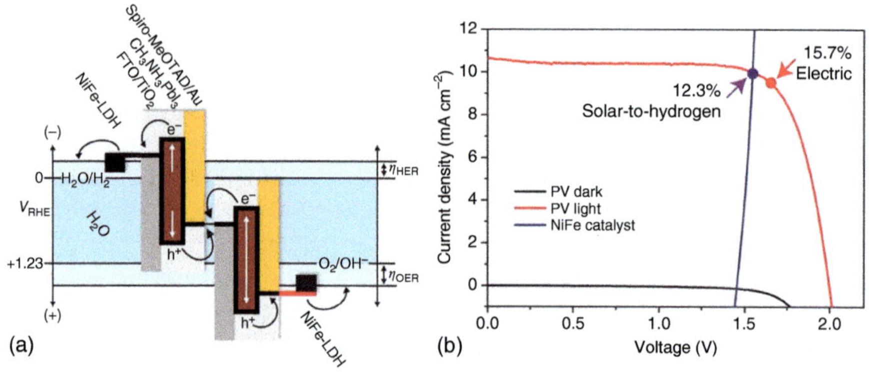

Figure 6.2 The combination of a perovskite tandem cell with a Ni-Fe-LDH/Ni foam electrode for sunlight-induced water splitting. (a) Schematic of the device and the relevant energy levels. (b) $J-V$ curves of the perovskite tandem cell under dark and simulated AM 1.5 G of 100 mW cm^{-2} illumination, and of the NiFe/Ni foam electrode based water-splitting cell in a two-electrode configuration. Source: Long et al. [12]. © 2016, Elsevier.

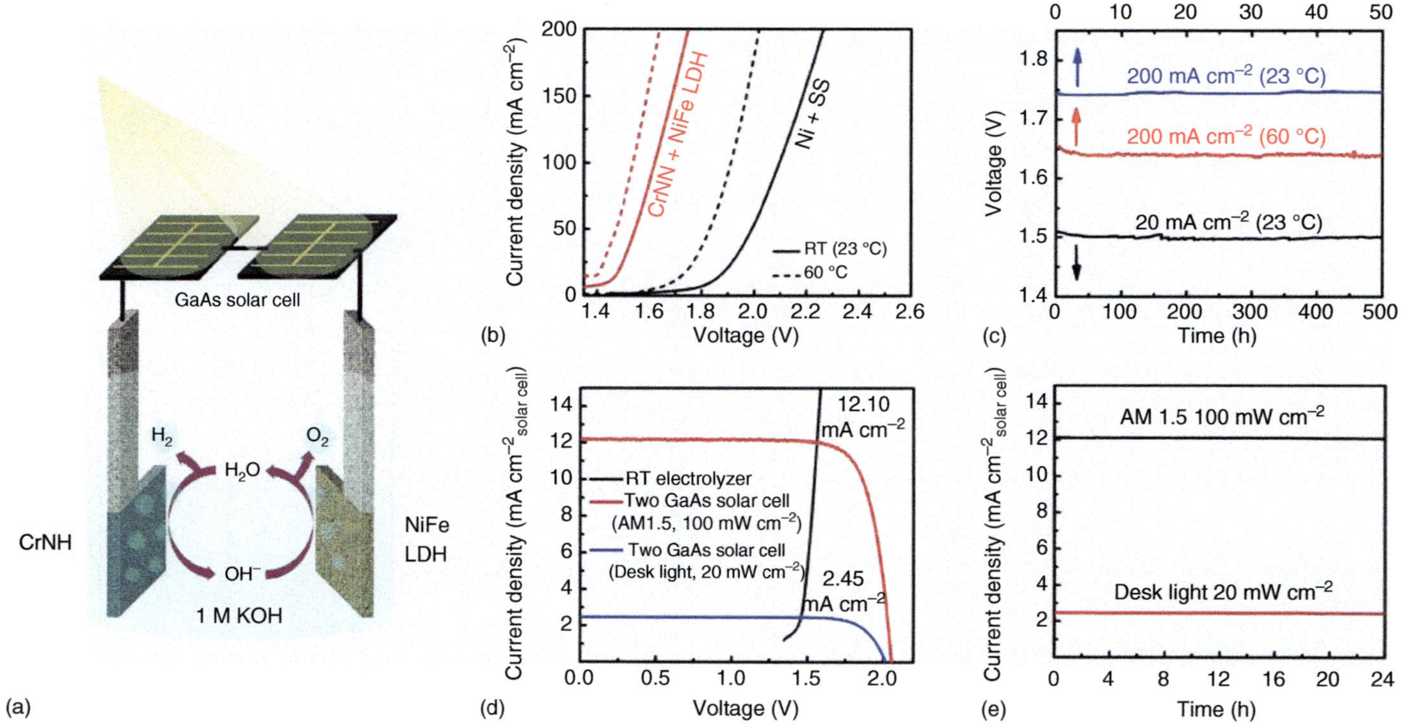

Figure 6.3 (a) Schematic representation of solar water decomposition with two gallium arsenide solar cells in series and alkaline electrolysis cells with CrNN and Ni-Fe-LDH catalysts; (b) use of CrNN catalysts as cathodes (24 mg cm^{-2},30 wt% nickel powder) and Ni-Fe-LDH as the anode. (c) No iR compensation for constant current density chronopotential curves of CrNN and Ni-Fe-LDH catalyst electrolyses. (d) The battery $I-V$ curve of the GaAs solar cell under simulated AM 1.5 G. (e) Water distribution of the current density over time driven by gas solar cells. Source: Gong et al. [13]. © 2015, John Wiley & Sons.

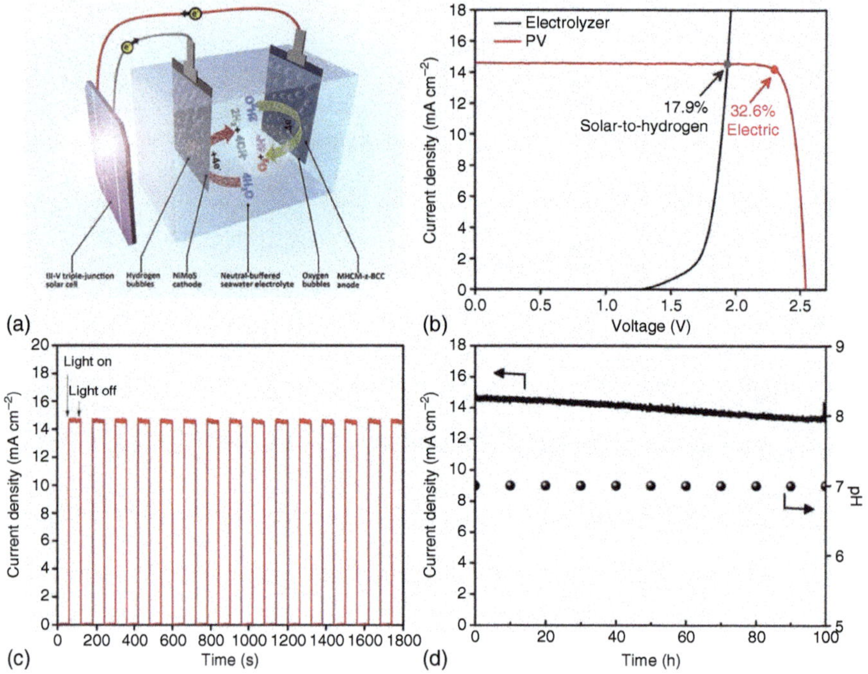

Figure 6.4 (a) Neutral pH photolysis of seawater, a schematic diagram of a photovoltaic seawater dividing device. (b) Simulate the $J-V$ curve (100 MW cm^{-2}) of the triple solar cell and the polarization curve of the MHCM-z-BCC/NiMos seawater separation device in the two-electrode structure (c) The current density time curve of the integrated seawater separator under AM 1.5 G radiation lighting. (d) The current density time curve of the seawater separator integrated under AM 1.5 G of continuous illumination and the registered buffer seawater electrolyte pH value [14]. Source: Hsu et al. [14]. © 2018, John Wiley & Sons.

Recently, the Qiao's group demonstrated a photoelectrolysis system for seawater (Figure 6.4). This system was based on CO_2 [$Fe(CN)_6$] nH_2O and NiMoS; they are regarded as anode and cathode. Furthermore, they are operated by a single commercial photovoltaic cell. This integrated system achieves the spontaneous and efficient production of high-purity H_2 and O_2 in neutral with a solar to hydrogen efficiency of 17.9% and about 14.6 mA cm^{-2} [14].

In conclusion, further development of bifunctional catalysts should pay attention. HER and OER response mechanisms need for further understanding [15, 16]. The following study needs to provide useful guidance for the design of catalysts that can be done by modifying crystalline and electronic structures. It is also important to examine the mechanism reduction during the cycle process because it provides solid information on improving the design and manufacture of catalysts. The design of the morphology and the composition of the catalysts could increase the number of exposed catalytic-active sites and encourage the diffusion of the reagents in the electrolyte. Technical problems should be taken into account when developing the new catalyst. First, consideration should be given to how the developing electric catalyst can be applied on a large scale. Several bifunctional catalysts with nanostructures endowed with hetero-atoms have shown excellent performance for global water separation [17, 18]. Also, it is important to develop in situ characterization techniques with theoretical calculations and analyses in order to understand the functioning mechanisms. Therefore, different in situ techniques such as Raman in situ operando, Fourier transform infrared (FTIR), X-ray absorption spectroscopy (XAS), and light microscopy characterization techniques [19–26] can be used to investigate important and valuable information about catalysts such as structural changes [7, 27–30].

References

1 Cheng, N., Stambula, S., Wang, D. et al. (2016). *Nat. Commun.* 7: 13638.

2 Mahmood, J., Li, F., Jung, S.M. et al. (2017). *Nat. Nanotechnol.* 12: 441–446.

3 Jing, S., Lu, J., Yu, G. et al. (2018). *Adv. Mater.* 30: e1705979.

4 Ross, M.B., Dinh, C.T., Li, Y. et al. (2017). *J. Am. Chem. Soc.* 139: 9359–9363.

5 Zhang, L.Z., Jia, Y., Gao, G.P. et al. (2018). *Chem. US* 4: 285–297.

6 Yin, J., Fan, Q.H., Li, Y.X. et al. (2016). *J. Am. Chem. Soc.* 138: 14546–14549.

7 Zhou, D.J., Li, P.S., Xu, W.W. et al. (2020). *ChemNanoMat.* 6: 336–355.

8 Encrenaz, T. (2008). *Annu. Rev. Astron. Astrophys.* 46: 57–87.

9 Kanan, M.W. and Nocera, D.G. (2008). *Science* 321: 1072–1075.

10 Hagiwara, H., Inoue, T., Kaneko, K., and Ishihara, T. (2009). *Chem* 15: 12862–12870.

11 Liu, Y.B., Zhou, B.X., Bai, J. et al. (2009). *Appl. Catal. B Environ.* 89: 142–148.

12 Long, X., Wang, Z., Xiao, S. et al. (2016). *Mater. Today* 19: 213–226.

13 Gong, M., Zhou, W., Kenney, M.J. et al. (2015). *Angew. Chem. Int. Ed. Engl.* 54: 11989–11993.

14 Hsu, S.H., Miao, J., Zhang, L. et al. (2018). *Adv. Mater.* 30: e1707261.

15 Reece, S.Y., Hamel, J.A., Sung, K. et al. (2011). *Science* 334: 645–648.

16 Brillet, J., Yum, J.H., Cornuz, M. et al. (2012). *Nat. Photonics* 6: 823–827.

17 Jacobsson, T.J., Fjallstrom, V., Sahlberg, M. et al. (2013). *Energy Environ. Sci.* 6: 3676–3683.

18 Pinaud, B.A., Benck, J.D., Seitz, L.C. et al. (2013). *Energy Environ. Sci.* 6: 1983–2002.

19 Cardenas-Morcoso, D., Ifraemov, R., Garcia-Tecedor, M. et al. (2019). *J. Mater. Chem. A* 7: 11143–11149.

20 Kriek, R.J., Iqbal, M.Z., Doyle, B.P., and Carleschi, E. (2019). *ACS Appl. Energy Mater.* 2: 4205–4214.

21 Laha, S., Lee, Y., Podjaski, F. et al. (2019). *Adv. Energy Mater.* 9, 1803795.

22 Kibria, M.G., Edwards, J.P., Gabardo, C.M. et al. (2019). *Adv. Energy Mater.* 9: 1803795.

23 Boyd, M.J., Latimer, A.A., Dickens, C.F. et al. (2019). *ACS Catal.* 9: 7578–7587.

24 Lu, J., Cai, L., Zhang, N. et al. (2019). *ACS Appl. Mater. Interfaces* 11: 44214–44222.

25 Macounova, K.M., Nebel, R., Klusackova, M. et al. (2019). *ACS Appl. Mater. Interfaces* 11: 16506–16516.
26 Ran, L. and Yin, L.W. (2019). *Adv. Mater. Interfaces* 6, 1800970.
27 Cheng, Y., Guo, Y., Zhang, N. et al. (2019). *Nano Energy* 63, 103897.
28 Yin, Z., Peng, H., Wei, X. et al. (2019). *Energy Environ. Sci.* 12: 2455–2462.
29 Sun, H.C., Yang, J.M., Li, J.G. et al. (2020). *Appl Catal B* 272, 118988.
30 Li, F.L., Shao, Q., Huang, X., and Lang, J.P. (2018). *Angew. Chem. Int. Ed. Engl.* 57: 1888–1892.

Part V

H_2 Oxidation/O_2 Reduction and Device

As two of the basic reactions in the electrochemical energy cycle, H_2 oxidation reaction (HOR) and O_2 reduction reaction (ORR) have received extensive study. Coupling of these two reactions could comprise H_2 fuel cell; meanwhile, many metal-air batteries involve the utilization of ORR. H_2 can be continuously generated by water powered by renewable energy such as solar and wind power while O_2 is one of the main ingredients in the atmosphere. Therefore, these new energy technologies utilizing H_2 and O_2 redox have great promise to solve the current issues related to a shortage of energy resources, environment pollution, and climate change. In this part, the discussion on HOR and ORR basics and their applicable devices will be introduced.

Electrocatalysis in Balancing the Natural Carbon Cycle, First Edition. Yaobing Wang.
© 2021 WILEY-VCH GmbH. Published 2021 by WILEY-VCH GmbH.

7

Introduction

H_2 oxidation reaction (HOR) is the two-electron transfer process by converting H_2 into H^+ or OH^-, while O_2 reduction reaction (ORR) refers to the 4-e^- transfer processes to transfer O_2 into H_2O or OH^- [1, 2]. The sluggish of these two half-reactions is always a bottleneck that limits the overall performance of the new energy devices, hindering their progress of commercialization [3, 4]. A great research effort was thus made for the development of highly efficient and stable electrocatalysts and electrodes for driving such electrochemical processes.

Previous metal Pt is regarded as the only catalyst with the highest activity for both HOR and ORR [5, 6] and has received considerable attention for the past several decades. To further improve the Pt activity and lower its cost, a variety of strategies have emerged, such as tuning the composition of Pt-based nanoparticles, forming hybrid nanostructures, constructing core–shell structures, altering the surface crystalline structure, etc., [6, 7] to develop cost-effective and highly efficient HOR/ORR electrocatalysts. Meanwhile, various Pd and Ru–Ir alloy electrocatalysts were also developed for HOR/ORR [7, 8]. In spite of these progresses, it is still a big challenging to lower the cost of previous metal electrocatalysts. Therefore, researchers have turned their attention to the exploitation of previous metal-free electrocatalysts [9, 10]. Such innovative approaches are accompanied with an additional requirement: the previous metal-free electrocatalysts are usually urged to use in alkaline electrolytes. It is noted that the HOR rate in acid medium would be higher than that in alkaline medium, while ORR proceeds fast under alkaline conditions [11, 12]. By rational design, the device performance in alkaline condition could be comparable to that in acid medium, e.g. hydroxide exchange membrane fuel cells (HEMFCs); its maximum power density is similar to that of proton exchange membrane fuel cells (PEMFCs) [13]. Similar to previous metal control, the design toward non-previous metals also involves tuning the atomic, nano, and micro structures, components, size, and so on.

In this part, we will first give the general electrochemical parameters for determining the performance of HOR and ORR, summarize the mechanisms for HOR and ORR, and discuss the research advances of electrocatalysts from previous metals to non-previous metals, and to metal-free materials. We will also introduce the configurations of H_2 fuel cell and typical Zn–air battery and discuss their research progress.

Electrocatalysis in Balancing the Natural Carbon Cycle, First Edition. Yaobing Wang.
© 2021 WILEY-VCH GmbH. Published 2021 by WILEY-VCH GmbH.

7.1 Electrocatalytic Reaction Parameters

7.1.1 Electrochemically Active Surface Area (ECSA)

7.1.1.1 Test Methods

There are usually two method sets for the determination of electrochemically active surface area (ECSA) in metal electrocatalysts [14]: one hand is to test the surface Faraday reactions' Coulomb charge, for instance, hydrogen deposition underpotential (HUPD) [15, 16], metal underpotential deposition (UPD) [14], CO-stripping method [15], and surface metal oxidation/reduction [17]. The other hand is extracted double-layer capacitors (C_{dl}) [17, 18].

HUPD is according to H atoms' adsorption/desorption on metal surfaces. For example, when a single layer of hydrogen atoms is adsorbed on the cathode on the Pt surface, desorption at the anode can be represented by the cyclic voltammetry (CV) peak relative to reference hydrogen electrode (RHE), ranging from ~0.05 to ~0.4 V. The desorption and adsorption regions of hydrogen atoms are compounded after deducting bilayer current. Coulomb charge is averaged in a shaded area and then divided by the particular charge (for a single layer of HUPD on the Pt surface, the specific charge is $210\,\mu C\,cm^{-2}$ Pt), and the ECSA of Pt can be obtained. HUPD has been popular applied to determine the ECSA of certain precious metals (e.g. Pt, Rh, and Ir), especially Pt/Pd-based alloys used for a variety of electrocatalysis research.

CO-stripping method through noting a single-layer charge transfer of CO stripped from the metal surface to quantify the surface area of the metal, e.g. CO stripping on Pt. A moderate reduction potential was used in the working electrode, and CO gas is blown into the electrolyte to trigger the CO monolayer adsorption at Pt. The electrolyte should be cleansing with Ar gas to separate dissolved CO before stripping CV measurement. CO stripping is represented by an oxidation peak at ~0.7 V relative to RHE, which is scanned at the same scan rate as the baseline integration (CV in Ar-saturated electrolyte).

The specific charge is $420\,\mu C\,cm^{-2}$ (used to form a CO monolayer on Pt) so as to convert the CO stripping charge to the metal ECSA. Similar to HUPD, CO stripping is popular applied to precious metal-based catalysts.

The surface area can be quantified by the surface metal redox reaction for metals that do not have a CV peak for HUPD or CO stripping, which depends on the interaction between oxide and metal atoms on the surface. It involves forming a single-layer metal oxide anode, integrating its reduced CV peak, and converting it to the surface area with a specific charge. For example, Ni, Ag, and Cu can use this method to determine the surface area.

Metal oxide electrocatalyst surface area measurement technology.

7.1.2 Determination Based on the Surface Redox Reaction

In theory, via calculating the Coulomb charge under the CV peak of a metal oxide with known specific charge oxidation/reduction to obtain ECSA. However, the use of me2tal oxide CV peaks in ECSA quantitative analysis is still questionable, which is unlike metal catalysts.

7.1.3 Determination by Electric Double-Layer Capacitance Method

Bilayer capacitance (C_{dl}) is the common way to measure metal oxide ECSA currently. Taking MnO_3 as an example, the specific method is as follows: in the 0.1 M KOH saturated with Ar, record CV curve (10, 20, 50, 100, and 200 mV s^{-1} voltage window: 1.1–1.2 V) at different scan rates; there should not be no Faraday current in this voltage window. Taking the average value of anode current or cathode current as the electric double-layer capacitance current I_c, plot the curve of I_c vs. scan rate vs. I_c that complies with an ideal capacitor of the ideal linear behavior. When $I_c = vC_{dl}$, the slope of the straight line is C_{dl}. It should be noted that in the electrocatalytic test, in order to ensure conductivity, like carbon black, a conductive agent is often added during electrode coating, so it is difficult to test the intrinsic activities of pure metal oxide powder. In order for C_{dl} to be entirely contributed by oxides, conductive agents such as carbon black should not be used when preparing electrodes by the drop-coating method because high surface area carbon black can cause illegal Faraday currents and interfere with ECSA calculations.

Another method of measuring C_{dl} is the electrochemical impedance spectroscopy (EIS), which establishes the equivalent resistance by fitting to determine the electric double-layer capacitance value.

The electrochemical performance of electrocatalysts is usually investigated via CV in alkaline solutions saturated with N_2 or Ar. The ECSA has been determined for the Pt group metal (PGM) electrocatalysts in most literatures based on the hydrogen desorption area after bilayer modification from CV. Other methods, including CO stripping, Cu_{upd}, Au–O, Pd–O reduction voltammetry, and OH desorption (for non-precious metal electrocatalysts), can be applied to measure the ECSA of some basic HOR electrocatalysts as well. The specific test conditions for ECSA for various basic HOR electrocatalysts are listed in Table 5.1. Obviously, the measurement conditions are very different from each other [19].

To quantify the ECSA of metal catalysts, there are two techniques: the surface redox reaction to derive Coulomb charge, including HUPD, CO stripping, surface reduction of metal and UPD, and illegal C_{DL} [20].

7.1.4 Kinetic and Exchange Current Density (j_k and j_0)

7.1.4.1 Definition

In addition to the ECSA value, two kinetic parameters can also determine the basic HOR activity of the electrocatalyst. One is j_0, which is corresponding to the electron transfer rate in the reversible case, and the other is j_k at a given potential. These two parameters can be calculated by the polarization curves with EIS testing corrected the internal resistance. Ignoring iR correction will cause a serious underrated HOR activity in alkaline media and will mislead the mechanism. The iR correction formula is listed.

Table 7.1 Summary of test conditions for ECSAs of HOR electrocatalysts in various media.

Electrocatalyst	Electrolyte	Temperature	Sweep rate ($mV\,s^{-1}$)	Method and charge density	ECSA ($m^2\,g^{-1}$)	Ref.
Pt/C	0.1 M $HClO_4$	293 K	100	HUPD region ($210\,\mu C\,cm^{-2}$)	120	[21]
Pt/Cu NWs	0.1 M KOH		Pt: 20 Cu: 50	Pt: HUPD region ($210\,\mu C\,cm^{-2}$) and CO stripping ($420\,\mu C\,cm^{-2}$) Cu: Cu oxidation ($288.4\,\mu C\,cm^{-2}$)	Pt: 35.9 Cu: 0.4	[22]
Ir/C	0.1 M KOH	293 K	50 and 20, respectively	HUPD region ($218\,\mu C\,cm^{-2}$) and CO stripping ($420\,\mu C\,cm^{-2}$)	64 and 71, respectively	[23]
Au/C	0.1 M KOH		100	AuO reduction region ($390\,\mu C\,cm^{-2}$)	5.8	[24]
Pd/C	0.1 M KOH; 0.1 M $HClO_4$		50	Pd-O reduction ($424\,\mu C\,cm^{-2}$)	KOH: 74.9 $HClO_4$: 74.0	[25]

The polarization curve of the HOR activity of electrocatalysts in alkaline media is usually investigated by testing on a rotating disk electrode (RDE), which is located in a saturated with H_2 alkaline solution within a given potential range and rotation speed. In Table 7.1, experimental conditions of the HOR electrocatalyst polarization curve in alkaline media are collected. Obviously, polarization curves are recorded under different conditions even if it is the equal electrocatalyst; it is impossible to acquire different polarization curves, which directly affects the basic HOR active.

7.1.4.2 Calculation

Nernstian diffusion overpotential equation modified Koutecky–Levich equation.

7.1.5 Overpotential HUPD

Information about electrode reactions is usually obtained by testing the current functioned as potential (by obtaining an iE curve). Certain terms are usually corresponding to the characteristics of the curve. When a cell possesses a defined equilibrium potential, then the potential is a key reference point in the system. When the Faraday current passes, the phenomenon that the electrode/battery potentials deviated from the equilibrium value is called polarization. The polarization degree is tested by the overpotential η (Figure 7.1).

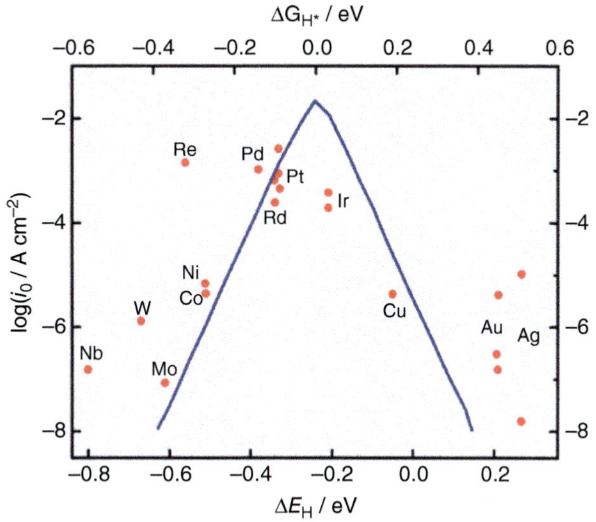

Figure 7.1 Experimentally measured HER/HOR exchange current densities (marked as dots) over different metal surfaces in acid solutions, plotted as a function of calculated HBE. The simple kinetic model (solid line) is plotted either as a function of the free energy for hydrogen adsorption according to the equation (17) or as a function of HBE. Source: Davydova et al. [2]. © 2018, American Chemical Society.

In theory, most hydrogen bonding energy (HBE) values previously reported were determined, except for some studies, in which the HUPD and HOR intermediates were assumed to be the same, and the thermodynamic parameters were calculated based on the HUPD peak potential. Although the binding energy value will be dispersed according to the nature of the H adsorption site (face-centered cubic [FCC], top), material type (pc, bulk, and facet), monolayer (ML), and in H_{ad} (θ). In the case of covering the surface, the HBE value is related to the atomic number of major components, as shown in Figure 5.1. The main components are made up of 18 various elements (Cr, Ti, Fe, V, Cu, Ni, and Co) in period 4, period 5 (Rh, Pd, Mo, Ag, and Ru), and period 6 (Ir, W, Re, Os, Au, and Pt) may be considered as HOR electrocatalyst due to its high H_2 adhesion possibility (or adhesion factor, i.e. contact with the surface of the ratio of the number of particles) and hydrogen [2].

Generally, a standard three-electrode system can be used to test the HOR/hydrogen evolution reaction (HER) activity of catalysts in an electrochemical cell. H_2SO_4 or $HClO_4$ is used as an electrolyte to simulate acidic conditions, and KOH or NaOH is used to simulate alkaline conditions. The exchange current density (i_0), the kinetic current density (i_k) at a fixed overpotential, or the overpotential required to achieve a certain current density is generally used as a parameter to judge the activity of different catalysts. CV or the linear sweep voltammetry (LSV) is employed to record the electrochemical behavior of the catalysts. In general, the RDE technique is required for HOR measurement due to the low solubility of H_2 in the electrolyte. The mass transport can be improved by the RDE technique, and the kinetics of the catalysts can be obtained from the mixed kinetic-diffusion control region. Generally, because of the H_2 low solubility in the electrolyte, the HOR measurement requires a RDE technology. RDE technology could improve mass transfer, and the area of hybrid kinetic diffusion control can acquire the kinetics of the catalyst [26].

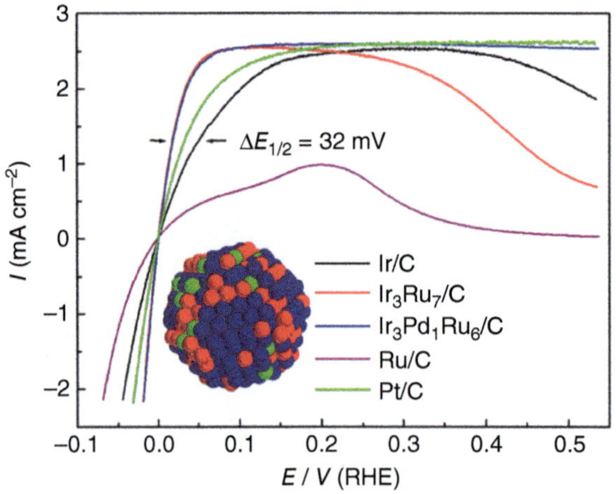

Figure 7.2 RDE voltammograms of Ru/C, Ir/C, and $Ir_{10-x}Ru_x$/C in 0.1 M KOH saturated with H_2. Rotation rate: 1600 rpm, Scan rate: 5 mV s^{-1}. The catalyst loading was 3.5 μg metal cm^{-2}. The compositions of the catalysts are indicated in the figure. Source: Wang and Abruna [27]. © 2017, American Chemical Society.

7.1.6 Tafel Slope

This parameter is mainly used to discuss the mechanism and determination of rate-determining step (RDS). Its value is associated with the nature of the electrode material, the state of the electrode surface, the solution composition, and the temperature. The difficulty values of the electron transfer steps can be compared according to different electrode reactions.

7.1.7 Halfwave Potentials

The half-wave potential refers to the electrode potential value on the current–electrode potential curve when the electrolytic current achieves half of the diffusion current.

Under certain experimental conditions, the half-wave potential ($E_{1/2}$) is only related to the inherent information of the ion, not the concentration, and is a characteristic constant of the ion, which can be used as the basis of qualitative analysis. The factors affecting half-wave potential include type and concentration of auxiliary electrolyte, temperature, the acidity of the solution, and complex formation.

The halfwave potentials for Ir_9Ru_1/C, Ir_7Ru_3/C, and Ir_3Ru_7/C were 32 and 15 mV negatively shifted when compared to Ir/C or Pt/C, respectively. Similarly, all studied Ir_9Pd_1/C and $Ir_{9x}Pd_1Ru_x$ alloy catalysts exhibited a higher activity than Ir/C, Ru/C, and Pd/C. The half-wave potentials for $Ir_6Pd_1Ru_3$/C and $Ir_3Pd_1Ru_6$/C were also 32 and 15 mV negatively shifted when compared to Ir/C or Pt/C, respectively. Compared to $Ir_{10x}Ru_x$/C catalysts, $Ir_{9-x}Pd_1Ru_x$/C catalysts were active over a wider potential region, due to the addition of Pd, which inhibits oxide formation [27] (Figure 7.2).

References

1 Kulkarni, A., Siahrostami, S., Patel, A., and Norskov, J.K. (2018). *Chem. Rev.* 118: 2302–2312.

2 Davydova, E.S., Mukerjee, S., Jaouen, F., and Dekel, D.R. (2018). *ACS Catal.* 8: 6665–6690.

3 Hong, W.T., Risch, M., Stoerzinger, K.A. et al. (2015). *Energy Environ. Sci.* 8: 1404–1427.

4 Bruce, P.G., Freunberger, S.A., Hardwick, L.J., and Tarascon, J.-M. (2012). *Nat. Mater.* 11: 19–29.

5 Zhao, Z., Chen, C., Liu, Z. et al. (2019). *Adv. Mater.* 31, e1808115.

6 Durst, J., Simon, C., Siebel, A. et al. (2014). *ECS Trans.* 64: 1069–1080.

7 Mao, J., He, C.T., Pei, J. et al. (2020). *Nano Lett.* 20: 3442–3448.

8 Bhowmik, T., Kundu, M.K., and Barman, S. (2016). *ACS Catal.* 6: 1929–1941.

9 Sheng, W.C., Bivens, A.P., Myint, M. et al. (2014). *Energy Environ. Sci.* 7: 1719–1724.

10 Shao, M., Chang, Q., Dodelet, J.P., and Chenitz, R. (2016). *Chem. Rev.* 116: 3594–3657.

11 Lu, S.F., Pan, J., Huang, A.B. et al. (2008). *Proc. Natl. Acad. Sci. U. S. A.* 105: 20611–20614.

12 Tang, D.P., Pan, J., Lu, S.F. et al. (2010). *Sci. China Chem.* 53: 357–364.

13 Pan, Z.F., An, L., Zhao, T.S., and Tang, Z.K. (2018). *Prog. Energy Combust. Sci.* 66: 141–175.

14 Łukaszewski, M., Soszko, M., and Czerwiński, A. (2016). *Int. J. Electrochem. Sci.* 11: 4442–4469.

15 Binninger, T., Fabbri, E., Kötz, R., and Schmidt, T.J. (2014). *J. Electrochem. Soc.* 161: H121–H128.

16 Lohmann-Richters, F.P., Abel, B., and Varga, A. (2018). *J. Mater. Chem. A* 6: 2700–2707.

17 Cossar, E., Houache, M.S.E., Zhang, Z.H., and Baranova, E.A. (2020). *J. Electroanal. Chem.* 870, 114246.

18 Watzele, S., Hauenstein, P., Liang, Y.C. et al. (2019). *ACS Catal.* 9: 9222–9230.

19 Cong, Y.Y., Yi, B.L., and Song, Y.J. (2018). *Nano Energy* 44: 288–303.

20 Wei, C., Rao, R.R., Peng, J. et al. (2019). *Adv. Mater.* 31, e1806296.

21 Durst, J., Siebel, A., Simon, C. et al. (2014). *Energy Environ. Sci.* 7: 2255–2260.

22 Alia, S.M., Pivovar, B.S., and Yan, Y. (2013). *J. Am. Chem. Soc.* 135: 13473–13478.

23 Zheng, J., Sheng, W., Zhuang, Z. et al. (2016). *Sci. Adv.* 2, e1501602.

24 Mahoney, E.G., Sheng, W.C., Yan, Y.S., and Chen, J.G.G. (2014). *Chemelectrochem* 1: 2058–2063.

25 Zheng, J., Zhou, S.Y., Gu, S. et al. (2016). *J. Electrochem. Soc.* 163: F499–F506.

26 Lu, S.Q. and Zhuang, Z.B. (2016). *Sci. China Mater.* 59: 217–238.

27 Wang, H. and Abruna, H.D. (2017). *J. Am. Chem. Soc.* 139: 6807–6810.

8
Hydrogen Oxidation Reaction (HOR)

8.1 Mechanism for HOR

This section describes the mechanism of hydrogen oxidation reaction (HOR) and mentions the complete details. At least for inorganic materials, this is recognized. At least in inorganic media, HOR is carried out via combining the following basic processes: (i) dihydrogen dissociative adsorption, (ii) transfer an electron from dihydrogen to catalyst, and (iii) release the adsorbent H atoms. We will discuss these three steps in Sections 8.1.1–8.1.2.

Here, the hydrogen adsorption site is represented by * on the catalyst, and adsorbed hydrogen is defined as H_{ad}. Adsorbing hydrogen without electron transfer is exhibited by the Tafel reaction. The adsorption of H_2 accompanied by transferring a proton and an electron is depicted as the Heyrovsky reaction, which produces an adsorbed hydrogen unit. The process of discharge of the H_{ad} unit and the release of protons is called as the Volmer reaction. Thus, the HOR can proceed through the Tafel–Volmer or Heyrovsky–Volmer mechanism [1].

8.1.1 Hydrogen Bonding Energy (HBE)

Some researchers claim that hydrogen bonding energy (HBE) is a descriptor enough to fully determine HOR dynamics. In alkaline media, corresponding to the two models of HOR pathways, two descriptors were proposed to correlate the HOR activity of electrocatalysts in alkaline media directly. One is the HBE, where only OH participates in basic HOR, and the other is hydroxyl binding energy or oxygen affinity energy, where OH_{ad} participates in basic HOR. Rheinlander indicated that HBE plays an important work in HOR in alkaline media. Relative to the hydrogen pressure base on HBE, it could be inferred that the exchange current density exports the order of reaction with 0.5 on the polycrystalline Pt [2].

HBE is known as the only description of hydrogen evolution reaction (HER)/HOR in acids. The relationship between volcanic type and reactivity is related to the hydrogen adsorption energy of metals. Gibbs free energy change and hydrogen adsorption energy with ideal HBE equal to zero can predict the highest HER/HOR activity [3]. Therefore, adjusting the HBE of the target material is a design principle for the research of new materials. Taking the same way for acid, in the Sheng et al. [4]

Electrocatalysis in Balancing the Natural Carbon Cycle, First Edition. Yaobing Wang.
© 2021 WILEY-VCH GmbH. Published 2021 by WILEY-VCH GmbH.

study, the volcanic relationship between HER exchanged current density on alkali metals is established by theoretical calculation of HBE, which shows that HBE remains the critical code of HER/HOR in alkali. The NiMo catalyst with the co-doping of HER/HOR in alkali has been developed under the circumstances; the results showed that since its weak HBE caused HER/HOR activity, it is superior to Ni [5].

8.1.2 Underpotential Deposition (UPD) of Hydrogen

The position of the hydrogen deposition underpotential (HUPD) peak in the Pt cyclic voltammetry (CV) has something to do with the strength of the hydrogen bond on the Pt surface [3, 6]. HOR had similar considerations. Some people think that the adsorption of OH directly in the alkaline medium will promote the whole reaction through the reorganization of H_{ad} and OH_{ad} [7]. Based on the study of HOR on PtRu NPs, Strmcnik et al. believe that since Pt is a hydrogen adsorption site, Ru, as a more sensitive oxygen element, has an adsorption effect on OH. The more oxophilic sites on Ir (defects) and PtRu material (Ru atoms) electrodes facilitate the adsorption of OH_{ad} species. Those then react with the hydrogen intermediates (H_{ad}) that are adsorbed on more noble surface sites. Therefore, in this regard, it is necessary to simultaneously expose the Pt site and the oxophilic site on the catalyst surface. However, via investigating the removal of surface of Ni from acid-leached PtNi and Pt, Lu and Zhuang [8] found PtNi and pickling PtNi have alike HOR activity, which indicates that the OH surface initially adsorbed on the Ni site has no effect on the activity of HOR.

8.2 Catalysts for HOR

The active properties of HOR catalysts using Pt-based materials, Pt-based alloy materials, and non-Pt group metal (PGM)-based materials were studied. In this section, we focus on the influence of catalyst synthesis, size, morphology, pH, and structure on electrocatalytic performance. In the volcano diagram for the activity of HOR/HER and PGM (Pt, Ir, Pd, etc.), it has the best activity corresponding to the optimized energy of the hydrogen binding area. Nevertheless, they cannot be applied due to the expensive cost of these PGMs. Benefiting from the latest developments in nanotechnology, certain catalysts with higher activity or lower prices (i.e. non-noble metals) were developed to reduce the amount of PGM used in fuel cells and/or electrolyzer. The new HOR/HER catalysts will be discussed in Sections 8.2.1–8.2.6.

8.2.1 Pt-based Materials

Pt-based materials: Pt/C in alkaline media, Otyama et al. have studied the size effects of HOR activity [9]. Heat-treating Pt/C from 100 to 750 °C (Tanaka Kikinzoku Kogyo, 2 nm) synthesizes different particle sizes of Pt/C catalysts.

The results indicated Pt/C with a lot of activity achieves a maximum parameter at of about 3 nm in the 0.1 M KOH aqueous solution, while with the increasing particle size, the specific activity increases and finally achieves an almost stabilized parameter.

With decreasing size, the HOR-specific activity decreases because the activity of edge atoms is lower than the increasing proportion of edge atoms on the end face. On the contrary, Durst et al. [10] studied that in alkaline media, the specific activity of HOR does not change with the change of particle size from 2 nm to the surface of the expansive body (polycrystalline Pt) but decreases with increase in particle size since a lot of activity increases. The difference in the impact of Ohyama and Durst study on the size of alkaline HOR may be due to different test systems. Durst studied only the basic HOR activity of Pt/C with 2 nm size. The experimental data of Pt/C with a size of 4 nm and polycrystalline Pt come from the reference [11]. It may be inappropriate to compare the HOR activity of polycrystalline Pt, 2 nm Pt/C, and 4 nm Pt/C in alkaline media in this way.

Pt/C-based materials. Comprehending how pH influences HOR activity and the HER is critical to the development of a stable, reasonably priced HOR/HER catalyst for electrolyzer and hydroxide exchange membrane fuel cells. “In a wider pH range (0–13), it indicates the metals are not related to pH dependence for HBE. We also found a general relationship between exchange current density and HBE on these four metals, suggesting that they may be the same basic steps and also rate-determining steps, and HBE is the main descriptor for HOR/HER activity. As the pH value increases, the initial dissolution potential of CO on the four metals decreased, indicating strong adsorption of OH and the enhanced effect of OH adsorption on HOR/HER [12].

Pt-CTF-based materials are the materials with the structure of covalent triazine skeleton with atomic Pt atoms (0.29 wt%). In their synthesis, via polymerizing 2,6-dicyanopyridine in the existence of conductive carbon particles synthesizes Pt-CTF, and then Pt is immersed in a K2[$PtCl_4$] solution. From the mechanism study, the current related to HOR begins to flow at 0 V (that is, there is no overcharge). Faraday current will not be generated by covalent triazine framework (CTF) w/o Pt; it is explained that Pt mounted on CTF is used as the electrocatalyst for HOR. That fits the conjecture, detecting no Pt-weighted aggregates in k3 EXAFS (extended X-ray absorption fine structure) oscillations Fourier transform. In summary, these results clearly indicate that the HOR acts on a single Pt atom in Pt- covalent triazine framework (PT-CTF), requiring less activated energy [13].

Pt-NPs-based materials are the most frequently used materials for HOR. We have developed a method using a nanoscale scanning electrochemical microscope (SECM) to study the geometrical properties and catalytic activity of a single Pt NP in the HOR. For the synthesis, electrodeposition of Pt NP with a three-electrode battery was used, where highly oriented pyrolytic graphite (HOPG), mercury sulfate electrode (MSE), and Pt disk electrode are a working electrode, a reference electrode, and a counter electrode, respectively [14].

Pt-NWs-based materials were studied by Scofield et al. [15] with better HOR performance in alkaline media compared to the commercial Pt/C and suggest that

1D nanomaterials were beneficial to enhance alkali metal HOR activity due to the presence of crystal plane segments and fewer potentially deteriorated defective parts and high aspect ratio. The morphological effect of Ru@Pt core–shell electrocatalyst on the activity of basic HOR was also studied [16]. The HOR activity in alkaline medium displays in the order of Pt/C $\leq$ $Ru_{0.6}Pt_{0.4}$ alloy < Ru_1@$Pt_{0.5}$ < Ru_1@Pt_1 (Tanaka Kikinzoku Kogyo). Ru@Pt is under the Pt load of 200 μg cm^{-2}, and the enhancement activities of Ru_1@$Pt_{0.5}$ and Ru_1@Pt_1 were 1.92 and 1.76, respectively, in comparison with $Ru_{0.6}Pt_{0.4}$ alloy.

Pt_7M_3 alloy NW (M = Au Cu, Co, Fe, Ru) [15] was synthesized to correlate the composition with the activity of HOR in alkaline media. Among the samples tested, in alkaline media, Pt_7Ru_3 nanowires (NWs) showed the highest activity of HOR. In detail, the synthetic Pt NW is much lower than the specific activity of 0.493 mA cm^{-2} (Figure 8.1) at the potential of 0.05 V vs. reference hydrogen electrode (RHE). In addition, both Pt_7CO_3 and Pt_7Fe_3 NW have much higher activity than Pt NW (Figure 8.1a,c). In contrast, Pt_7Cu_3 and Pt NW showed higher activity than Pt_7Au_3 NW (Figure 8.1b,c). The electronic interaction between the second metal and Pt causes the high HOR activity of Pt_7M_3 NWs (M = Ru, Fe, Co). The copper oxide formation on the Pt surface and the increased HBE, respectively, cause the lower HOR activity of Pt_7Cu_3 and Pt_7Au_3 NWs in alkaline media. Chemical vapor deposition causes the formation of Ru_xPt_y/C and Ru_xPd_y/C alloys [17]. In alkaline medium, the electrocatalysts' HOR activity has a decreasing order from Ru_xPt_y/C, Pt/C, Ru_xPd_y/C, to Pd/C (Table 8.1).

Crystalline ultrathin Pt–M alloy NWs (M = Cu, Co, Ru, Au, Fe) were also prepared. It was proved that compared with the commercial Pt standard and the synthesized Pt NWs, the synthesized alloy NW catalyst could increase the activity of the hydrogenation reaction (HOR). In the study of a structure–effect relationship, the electronic effect involves the contribution of "ligand effect" and reveals the "lattice strain effect". The overlap of M and Pt in the d-band is changed by ligand effect" because it relates to a decrease or increase of the electron density for Pt d-band, and the electronic structure of local Pt is changed with the effect. In comparison, the "lattice strain effect" is arisen from the tensile strain or compressive strain of surface, that is, the "disturbance" of the Pt lattice has an effect on the electronic structure because it has been demonstrated to influence its d orbital position. We have proven the existence of the synergistic effects by analyzing the Pt 4f regional displacement of each binary catalysts and the X-ray photoelectron spectroscopy (XPS) shift found in the CV hydrogen/oxide region. The change in the d-spacing value of all alloy structures is confirmed compared to the use of individual Pt [15].

As to the three highly active single metals: Pt, Pd, and Ir carbon-supported electrocatalysts, the pH effect on the rate of hydrogenation and release reaction (HOR/HER) was solved for the first time. It was shown that HOR/HER activity at high pH has the order of Pt > Ir > Pd, and the current exchange density is much larger than previously measured. We also show and calculate for the first time that from low (pH = 0) to high pH (pH = 13), the exchange current on the carbon-supported Pt, Ir, and Pd electrodes is reduced by about 2 orders of magnitude [18] (Figure 8.2).

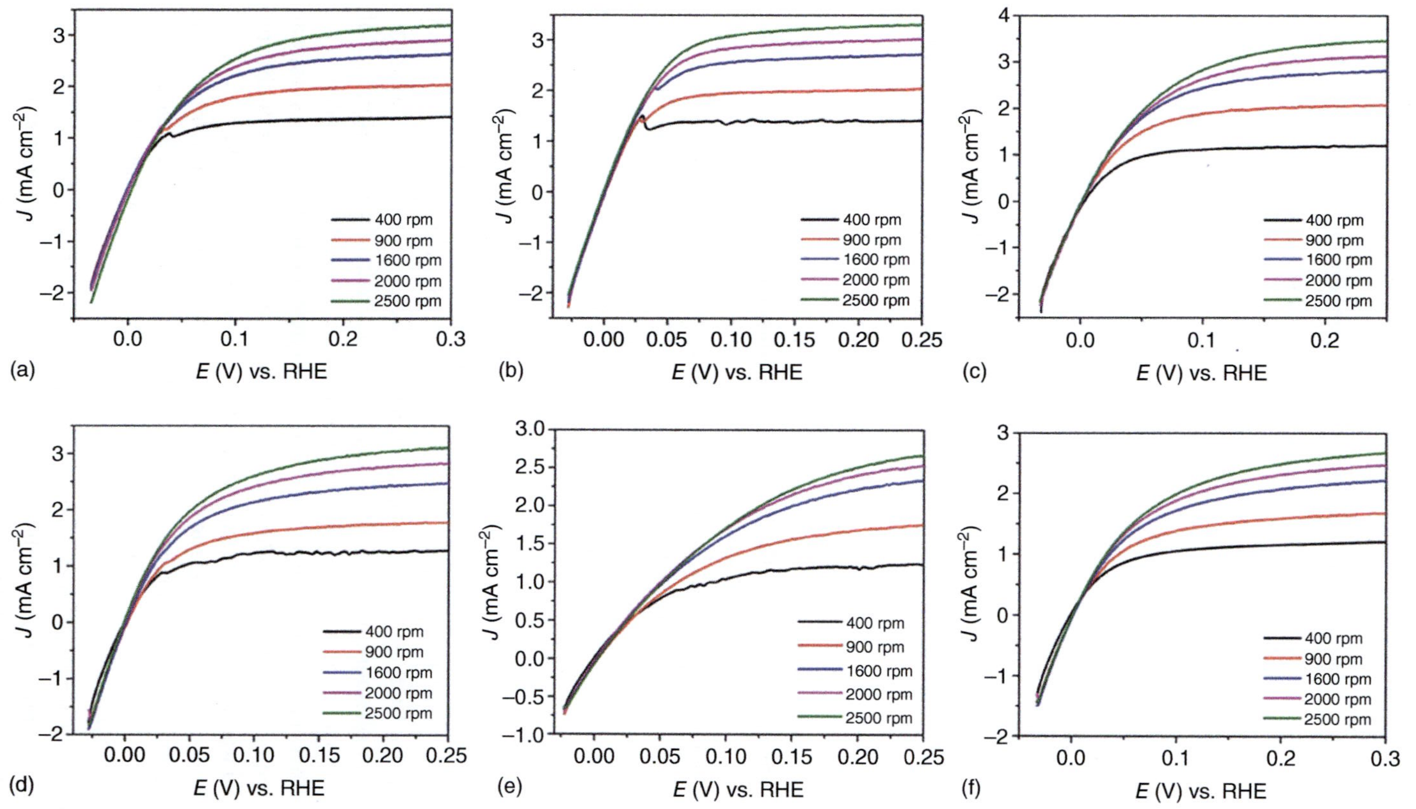

Figure 8.1 Hydrogen oxidation reaction curves obtained in a hydrogen-saturated 0.1 M KOH electrolyte at varying rotation speeds (i.e. 400, 900, 1600, 2000, and 2500 rpm): (a) Pt NWs; (b) Pt_7Ru_3 NWs; (c) Pt_7Fe_3 NWs; (d) Pt_7CO_3 NWs; (e) Pt_7Cu_3 NWs; (f) Pt_7Au_3 NWs. All curves are normalized to the geometric surface area of the electrode. Source: Scofield et al. [15]. © 2016 American Chemical Society.

Table 8.1 Computed Pt electrochemical surface areas calculated from the HUPD regions and CO stripping regions, respectively.[a)]

	Calculated Pt electrochemical surface area (cm^{-2})			
Catalyst	**HUPD region**	**CO stripping region**	**Activity meads at 0.05 V vs. RHE (mA $cm^{-2}{}_{geo}$)**	**Exchange current density (mA $cm^{-2}{}_{real}$)**
Pt	0.27	0.271	1.38	0.229
PtRu	0.176	0.273	2.2	0.493
PtFe	0.159	0.285	1.68	0.459
PtCo	0.168	0.267	1.64	0.394
PtCu	0.129	0.134	0.97	0.191
PtAu	0.55	0.227	1.23	0.162

a) Corresponding activities at 0.05 V vs. RHE normalized to the geometric surface area of the electrode. These have been compared with normalized HOR exchange current densities for all catalysts, determined from a combination of data from Tafel plots as well as from calculated HUPD charges.

Pt/Cu NW-based materials were also used for HOR. Platinum (Pt)-coated copper (Cu) nanowires, defined as Pt/Cu NWs, were synthesized by partial electrical displacement of Cu NW, with a diameter of 100 nm and a length of 25–40 μm. Pt/Cu NWs as HOR catalyst in alkali and Pt nanotubes templated by Cu (PtNT (Cu)), 5% Cu monolayer (5% ML (monolayer) Cu/BPPt) on bulk Pt electrode with the polycrystalline structure were studied), bulk polycrystalline Pt electrode (BPPt), and Pt on carbon (Pt/C). By comparing these catalysts, it indicates that the addition of Cu may benefit the HOR activity in Pt/Cu NWs via applying compressive strain onto Pt. The surface of copper further assists in the adsorption of hydroxyl groups, thus increasing the HOR activity for Pt/Cu NWs, which exceeds 3.5 times and 1.9 times the mass and area exchange current density of Pt on carbon. In the study of a structure–effect relationship, it has been proven that the addition of copper may benefit the HOR activity of platinum–copper nanowires by applying compressive strain on platinum; the surface of copper further helps for the adsorption of hydroxyl groups, thereby increasing the HOR activity of Pt/Cu NWs. Nørskov et al. displayed that copper has similar hydroxide and oxygen binding energy as iridium and ruthenium [19]. At 0.35 V relative to RHE, copper can be oxidized initially, and its potential was obviously lower than that of Pt (c. 0.8 V vs. RHE). In spite that bulk oxidation was not useful during the onset of HOR, it was shown experimentally that oxophilicity increased compared to Pt. Overall, the area and mass exchange current density for Pt/Cu NWs were 3.5 times and 1.9 times higher than that of Pt/C, respectively [20] (Figure 8.3).

PtRu-based materials are one typical, highly active catalyst for HOR. It has been suggested that more oxygen-philic sites and PtRu materials (Ru atoms) promote the adsorption of OH_{ad} substances. Then, those react with hydrogen intermediates (H_{ad}), which are adsorbed on more expensive surface parts [7]. For the synthesis of

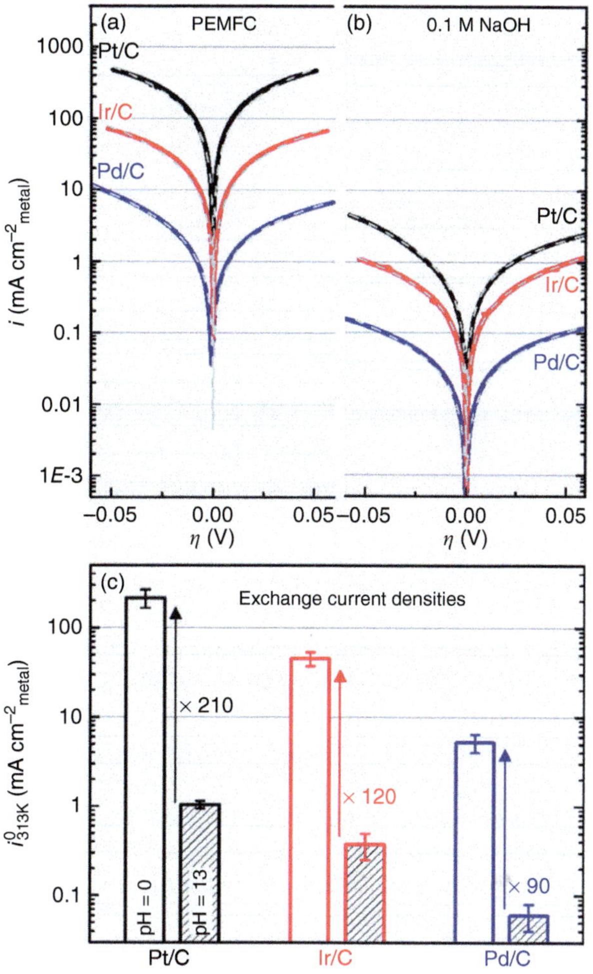

Figure 8.2 Upper part: HOR/HER Tafel plots of the specific current densities on Pt/C (black lines), Ir/C (red lines), and Pd/C (blue lines) in: (a) PEMFC and (b) 0.1 M NaOH electrolyte. The Butler–Volmer fits are indicated as gray dashed lines. Lower part: averaging according to Butler–Volmer and micropolarization equations, HOR/HER exchange current density (i_0 = 313 K) in PEMFC (empty column) and 0.1 M NaOH (stripe column). Data recorded in ambient pressure and 313 K H_2-saturated atmosphere. Source: Durst et al. [18]. Licensed under CC BY 3.0.

Pt-shell/Ru core-based materials, a Pt shell was deposited onto the Ru core, with removing any surface oxide on the Ru nanoparticles. After cooling the solution to room temperature, we added 6 ml of 50 mM H_2PtCl_6 ethanol solution (Pt:Ru atomic ratio 1.0) with vigorous stirring to ensure its uniform dispersion. The mixture was kept at an oil bath temperature of 80 °C for two hours, and then NaOH was added dropwise in an amount equal to 0.5 times the molar amount of Pt (0.5 × 300 μmol). After another half-hour, the colorless mixture was cooled to room temperature, then filtered, washed, and dried in vacuo. This Ru_1@Pt_1 sample has a metal weight percentage of 31% Pt, Ru is 16%, and the remaining carbon load is 53%. Single-crystalline Ru@Pt core–shell nanoparticles with two-monolayer-thick Pt shells, which doubles the activity on Pt–Ru alloy nanocatalysts. Using the atomic ratio of Pt:Ru of 0.5, our carbon-supported nanocatalysts were made with 1.0 and 1.0 Pt shells, respectively, named Ru_1@$Pt_{0.5}$ (1ML) and Ru_1@Pt_1 (2ML) catalysts. From the performance test, the HOR/HER activity in 1 M KOH saturated with hydrogen was measured using

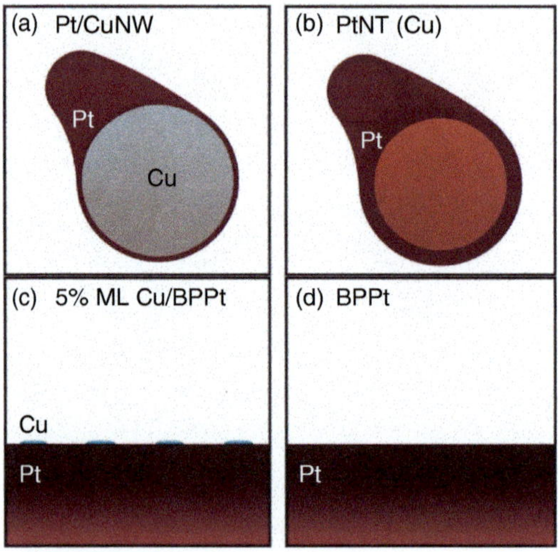

Figure 8.3 Schematic of (a) Pt/Cu NWs, (b) Pt NTs (nanotubes) (Cu), (c) 5% ML Cu/BPPt, and (d) BPPt. Source: Alia et al. [20]. © 2013, American Chemical Society.

the suspended gas diffusion electrode (GDE) method. The smaller the charge transfer resistance (CTR), the higher the activity; therefore, the performance is in the order of $Ru_1@Pt_1$ (2ML) > $Ru_1@Pt_{0.5}$ (1ML) > $Ru_{0.6}Pt_{0.4}$ alloy > Pt. Compared with Pt catalysts, the three Ru-containing catalysts showed enhanced performance. The Pt mass activity of $Ru_1@Pt_1$ (2ML) is 4.0, $Ru_1@Pt_{0.5}$ (1ML) is 3.7, and $Ru_{0.6}Pt_{0.4}$ alloy is 2.1. The mechanism study doubles the activity of $Ru_{0.6}Pt_{0.4}$ alloy nanocatalyst and confirms the benefits of placing all Pt atoms in the top two atomic layers. From the temperature-dependent activity, we identified a clear activation barrier. For the Δ of the Pt catalyst, the E_a value is 355 meV or 34.3 kJ mol^{-1}, and the ΔE_a value of the $Ru_1@Pt_1$ (2ML) catalyst is 315 meV or 30.4 kJ mol^{-1}, that is, the activation free energy close to the Volmer reaction, $\Delta G^{-*} = 0$ V, which is the rate-determining step in the main route [21]. This value is consistent with a boost of peak power density from 0.6 to 1.0 W cm^{-2} (about 1.67 times higher) in the AEM hydrogen fuel cell by using Ru-Pt alloy nanoparticles with the same 0.4 mg cm^{-2} total metal loading as with 0.4 mg cm^{-2} Pt nanoparticles. However, this significant improvement is almost impossible. It is interpreted as an oxophilic effect because in the K_t solution, the reactive hydroxyl species may generate a negative potential greater than that generated on the surface of PtRu in some parts of the Pt surface. Instead the doping of Ru, the input has an electronic effect to weaken Pt–H_{ad} interaction, as uncovered from the voltammetric behavior together with density functional calculations, and therefore is beneficial to the desorption of H_{ad} in oxidative processes, which is the rate-determining step for alkaline HOR [21] (Figure 8.4).

Here, we introduce the kinetic study on HOR in a representative catalytic system with an alkaline medium. Electrochemical tests indicate that HOR rates for PtRu/C and Ru/C systems are decoupled from the corresponding HBE. Thus, it challenges the currently popular HBE mechanism. The combination of in situ spectroscopic with electrochemical data validates another bifunctional mechanism, which provides convincing evidence that there are hydroxyl groups on the Ru surface of the

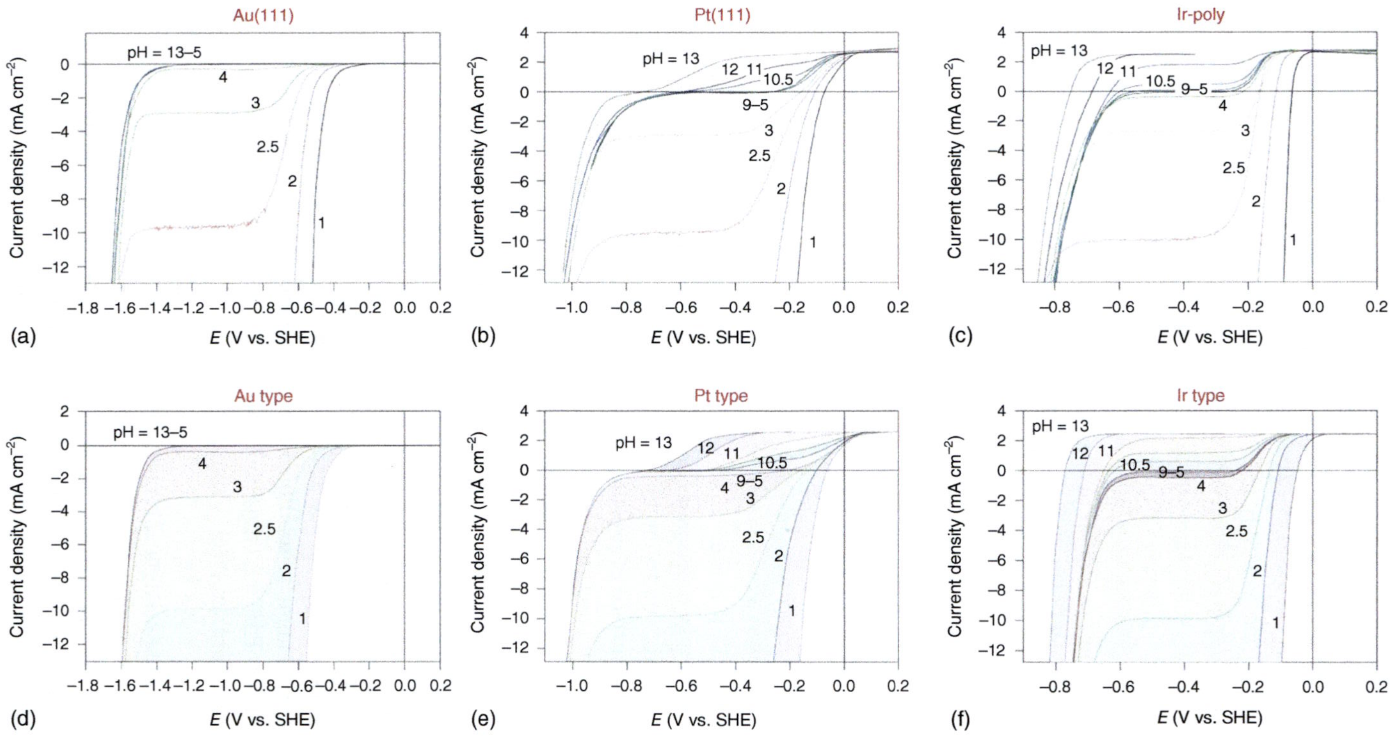

Figure 8.4 Current–voltage behavior was compared vs. standard hydrogen electrode (SHE), which, for the H^+/H_2 couple, is by definition $E^0 = 0.0$ V. (a–c), Measured pH-dependent current–potential polarization curves at rotation rates of 1600 rpm and a sweep rate of 50 mV s^{-1} for Au(111) (a), Pt(111) (b), and Ir-poly (c). (d–f) Simulated current–potential polarization curves for the HER/HOR Au-type electrode ($K_{1anodic} = 0$, $K_{2anodic} = 0$, $K_{1cathodic} = 10^{-9}$, and $K_{2cathodic} = 10^{-6}$) (d), Pt-type electrode (both anodic and cathodic, $K_1 = 10^{-4}$, $K_2 = 10^{-3}$) (e), and Ir-type electrode (both anodic and cathodic, $K_1 = K_2 = 1$) (f). Simulation of the polarization curves was used to determine the role of pH-dependent (H_3O^+, OH^-) and pH-independent (H_2, H_2O) reactant concentrations in the overall current–potential relationships. The presence of the proton branch and water branch for the HER was explained. Two distinct mass-transport-controlled plateaux at intermediate pH values (pH 9) for the HOR are noted for Pt and Ir. From comparing the experimental and simulation results for the HOR at pH > 9, we determine that the OH^- species plays a critical role in the overall HOR rates. Source: Strmcnik et al. [7]. © 2013, Springer Nature.

HOR potential range, and it plays an important role in enhancing the Volmer step with the rate-determining step. This conclusion provides an important reference for designing and selecting HOR catalysts [22].

Pt/NbTiO_2-based materials were used for HOR. Significant progress in breaking the limits of the mass activity of PGM is expected to realize an ultra-low PGM loading on membrane electrode assembly (MEA) below 0.125 g kW^{-1} while still gaining the performance as required. However, due to the very low load/thickness of these next-generation catalyst layers, durability is expected to be an important challenge. Specifically, in the processes of the Air/Air Start Shutdown (start-up [SU]/shutdown [SD]) event, the cathode may reach an extremely high potential of >1.5 V, resulting in apparent oxidation of the catalyst support. In work, a new solution to overcome the SU/SD loss is presented, which depends on the new mechanism to impart selectivity of the hydrogenation reaction (HOR) to the Pt/NbTiO_2 catalyst. The rotating disk electrode (RDE) test was used to investigate the effect and help clarify its mechanism while performing an in situ MEA test to prove that the same effect could be realized at the MEA level. This novel discovery can provide key enabling technologies for the next-generation MEA designs with high durability [23].

8.2.2 Pd-Based Materials

HOR is a counterpart of HER; it is also consisted of three main steps, which can occur via the Volmer–Heyrovsky or Volmer–Tafel reaction pathway or a combination pathway of both [24, 25]. Due to the high reversibility of both two reactions, the activity of HOR on the surface of precious metals almost follows the trend of HER. Due to the rapid kinetics and the interference of H absorption, it is not easy to reveal the HOR kinetics on Pd [26]. However, it is generally believed that in both acidic and alkaline media, the HOR activity on Pd is much lower compared to that of Pt. For example, on the basis of the results from RDE and electrochemical impedance spectroscopy (EIS) in H_2SO_4 solutions, Pronkin et al. [27] conclude that the Pd/C nanoparticles (2.6 and 4.0 nm) have slower HOR kinetics that is more than 2 orders of magnitude lower compared to the Pt/C HOR kinetics, and the HOR kinetics are independent of particle size. Durst et al. [28] studied that in the H_2 pump experiment conducted in proton exchange membrane fuel cells (PEMFCs), this was thought to minimize the mass transfer resistance. The exchange current density on Pd/C (2.8 nm) obtained at 40 °C is ~3 mA cm^{-2}. This value is about 40 times smaller than Pt/C (~120 mA cm^{-2}), as measured in the same conditions.

It turns out that Pd-based materials are promising catalysts for both HER and HOR because of their lower cost compared to Pt/C and have excellent CO tolerance. The stability of the Pd-based catalyst should be further improved. In a hydrogen environment, due to the absorption of hydrogen on Pd, the Pd lattice will expand, and the structure may be broken. The absorption of H in Pd also affects the evaluation of electrochemical-specific surface area (ECSA) based on H adsorption/desorption charges.

8.2.3 Ir-Based Materials

The active site of HOR/HER can be then determined via correlating activity with its key characteristics (such as the number of sites similar to HBE). Strmcnik et al. indicated that the polarization curve of polycrystalline Ir for HOR in 0.1 M NaOH basically overlaps with the concentration overpotential curve, suggesting that the HOR/HER kinetics on polycrystalline Ir is fast [7]. Therefore, the mechanical understanding of HOR/HER's activity on Ir has aroused great interest because it is a cost-effective candidate to Pt (the average price of Pt and Ir from November 2013 to November 2014 is US$ 1413.53 per ounce and 542.16 US dollars/ounce), but more importantly, it can clarify the guidelines for the design of effective HOR/HER catalysts based on non-precious metals. In comparison with the extensive research for Pt, the HOR/HER kinetics research for Ir is limited: most of the work is carried out in acidic electrolytes, and its activity may be underestimated [29–31]. Study on mechanism and pH effect in an alkaline environment should be focused [7, 18].

Due to its high stability and low price in acidic media, Ir has attracted great attention as a promising alternative for Pt catalysts. However, its electrocatalytic activity is obviously lower than Pt. Control over the morphology of the Ir-based nanocatalysts is of great significance in practical applications because both the electrocatalytic activity and stability are closely related to the size and shape of the nanocatalysts. Another method to enhance the HOR catalytic activity of Ir-based nanocatalysts is to appropriately change the electronic structure of Ir alloy combining certain transition metals (e.g. Co and Fe). Our previous reports indicate that Ir-based alloy materials produced via the solvent evaporation hydrogenation reduction (SE-HR) process are highly active for HOR catalysis. However, the deterministic determination of the effect of alloying is still out of reach [32].

8.2.4 Rh-Based Materials

Rhodium is a well-known precious metal that has high catalytic activity for many important industrial processes, such as automotive engineering (synthesis of catalytic converters), petroleum refining processes, fine chemical production, and energy generated by fuel cells [33]. The nanostructure can be easily adjusted by adjusting the size and shape of the metal nanostructure. Although Rh metal has good activity on HER/HOR, its activity is 20–30 times lower than that of Pt metal [28]. Several groups have also studied rhodium-based catalysts such as RH_2P nanocubes [21], Rh/Si NPs [16], Rh–MoS_2 [34], Rh tetrahedra, and concave tetrahedral/nanosheets toward various applications in electrocatalytic energy conversion and storage area. Although the part of HER/HOR applications in the acidic medium is already available [35], however, studies on HOR/HER activity in basic media are still very rare [12, 36].

8.2.5 Ru-Based Materials

Ruthenium-based catalysts are considered as candidates for HER/oxygen evolution reaction (OER) bifunctional catalysts, supported by two recent publications [37].

Table 8.2 Summary of alkaline HOR activity of noble metal-based electrocatalysts.

Electrocatalyst	Loading (μg_{metal} cm^{-2})	$j_{0.05V}$ a) (vs. RHE)	$j_{0.5}$ ($mAcm^{-2}{}_{metal}$)	$j_{0.m}$ ($Ag^{-1}{}_{metal}$)	References
Ru/C (3.1 nm)	10		0.063	82	[9]
Commercial Pt/C	10		0.05	60	[9]
Commercial Pt/C b)	2.5		1.0 ± 0.1	~1200	[18]
Commercial Ir/C b)	5		0.37 ± 0.12	~218	[18]
Commercial Pd/C b)	5		0.06 ± 0.02	~63	[18]
Pd-CN_x b)	43		0.037	~18.5	[34]
Commercial Ir/C b)	10		0.21 ± 0.02	128.6 ± 18.8	[32]
Ir/C-300C b), c)	10		0.22 ± 0.07	110.3 ± 13.1	[32]

a) charge density, charge density of reduction or oxidation for one monolayer of adsorbed species.
b) commercial Pt/C electrocatalyst.
c) r.t., room temperature.

The dual-function catalyst may simplify the water electrolysis system because a single catalyst covers both electrodes [38]. In the conventional asymmetric electrolyzer, the HER catalyst is different from the OER catalyst. The OER electrode compartment is exposed to an oxidizing environment, while the HER compartment experiences a reduction potential. Long-term oxidation and reduction loads on the electrodes may cause the components of the electrolytic cell to deteriorate or corrode. On the other hand, a symmetric electrolysis cell based on a single bifunctional catalyst can be operated by periodically switching the electrodes. Therefore, the water electrolysis operation is expected to be more durable [39, 40] (Table 8.2).

8.2.6 Non-noble Metal Materials

One of the potential advantages of hydroxide exchange membrane fuel cells (HEMFCs) compared to PEMFC is that owing to the highly stable types of non-noble metals in alkaline media, non-noble metal electrocatalysts can act as alkaline HOR electrocatalysts. The main non-noble metal materials involve Ni-, Co-, and Cu-based catalysts. Since Raney nickel alloy and NiB were used as HOR catalysts in alkaline fuel cells (AFCs), great attention has been drawn to Ni as a HOR electrocatalyst in HEMFC. Unfortunately, so far, Ni has not been used as a good basic HOR electrocatalyst in HEMFC. Thus, many efforts have been made to promote the alkaline HOR performance of nickel-based catalysts to replace noble metals.

Nickel-based materials are the most commonly used non-noble metal catalysts for the production of H_2 (HER) by combining bimetallic catalysts, including Cr, ternary Ni–Mo–Fe alloys, and HOR in AFC. Ti, La, Cu, several Ni-based, and Mo have been exploited to enhance HOR activity. However, compared to Ni, HOR has very limited reinforcement on Ni alloys, and the mechanism behind it is unclear. Previous studies have shown that decoration of Ni with Cr can adjust the Ni d-band

Table 8.3 Alkaline HOR activity of PGM-free electrocatalysts.

Electrocatalyst	Loading (mg_{metal} cm^{-2})	$j_{k0.05v}$ (vs. RHE)a) (mA $cm^{-2}{}_{disk}$)	$j_{0.5}$ (mA $cm^{-2}{}_{metal}$)	$j_{0.m}$ (A g^{-1})	Peak current density (mA cm^{-2})	References
Ni/Cr (anode)	5.0 (anode)	NA	NA	NA	50 (in H_2–O_2)	[41]
Ag (cathode)	1.0 (cathode)	NA	NA	NA	40 (in H_2–O_2)	
Ni/W (anode)	17.5 (anode)	NA	NA	NA	NA	[42]
CoPPYb)	2.0 (cathode)	NA	NA	NA	NA	
Ni/N-CNTc)	0.25	2.33	0.028	3.54	NA	[43]
Ni/N-CNTc)	0.25	0.48	0.0092	0.98	NA	[43]
Nic)	0.25	0.07	0.0013	0.15	NA	[43]
Nic)		0.002		NA	NA	[5]
CoNiMoc)		0.044	0.015	NA	NA	[5]
Pt disk		1.44	0.61	NA	NA	[5]
$Ni_{0.95}Cu_{0.05}$/C	0.25	NA	0.014	2.5	NA	[44]

NA: Not applicable/not found.
a) Activity obtained at 0.05 V vs. RHE.
b) Carbon-supported Co-polypyrrole.
c) Kinetic parameters obtained after iR correction.

electron structure, thereby weakening the Ni—O bond, thereby making the Ni metal surface available for the proceeding of HOR. The authors prepared NiMo and CoNiMo catalysts by electrodeposition and further tested the HOR activity using a temperature-controlled rotary disk electrode approach in 0.1 M KOH. Compared with the electroplated Ni catalyst in the KOH solution, the electroplated NiMo and CoNiMo multimetal catalysts show significantly enhanced HOR activity. In order to understand qualitatively trends of HOR activity, the authors conducted H_2-temperature program desorption (H_2-TPD) experiments as well as density functional theory (DFT) calculations on alloy systems with simpler structures. These systems facilitated for DFT calculations and TPD experiments. The results indicate that weak HBE is an important factor to improve the HOR activity on multimetal of both CoNiMo and NiMo electrocatalysts [5] (Table 8.3; Figure 8.5).

Ni/N-CNT was synthesized by two-step processes [43]. First, by reducing the nickel salt in the aqueous solution, Ni is selectively grown on the lightly oxidized multi-walled carbon nanotubes (CNTs). Next, hydrazine and ammonia are added, and then a hydrothermal reaction is performed at 150 °C. This process leads to better crystallization of Ni component, partial reduction of the oxidized CNTs, and, moreover, doping N into CNTs. In the performance test, Ni/N-CNT has a much higher current density than that of Ni or Ni/CNT, and the displayed starting potential is only 0 V. The nitrogen-doped carbon nanotube (N-CNT) has little HOR catalytic activity, indicating that the better catalytic activity on Ni/N-CNT was arisen from the synergistic effects in the combined structure. Ni/N-CNT

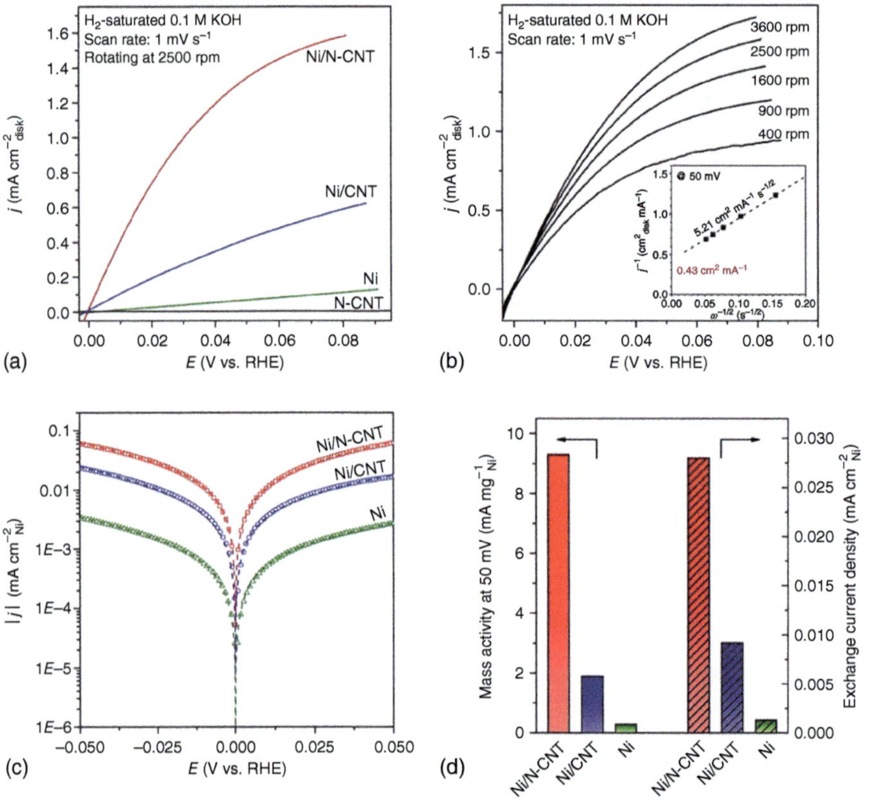

Figure 8.5 HOR performances. (a) Polarization curves of Ni/N-CNT, Ni/CNT, Ni (all of the three catalysts with the loading of 0.25 mgNi cm^2), and N-CNT (0.1 mgC cm^2) catalysts in H_2-saturated 0.1 M KOH at a scan rate of 1 mV s^{-1} and rotating speed of 2500 rpm. (b) Polarization curves of Ni/N-CNT in H_2-saturated 0.1 M KOH at a scan rate of 1 mV s^{-1} and various rotating speeds. Inset is the Koutecky–Levich plot at an overpotential of 50 mV. (c) HOR/HER Tafel plots of the specific current density on Ni/N-CNT, Ni/CNT, and Ni in H_2-saturated 0.1 M KOH. The dashed lines indicate the Butler–Volmer fitting. (d) Mass activity at 50 mV (unpatterned) and exchange current density (patterned) of the Ni/N-CNT, Ni/CNT, and Ni, respectively. Source: Zhuang et al. [43]. Licensed under CC BY 4.0.

electrocatalyst was also measured in electrolyte saturated with Ar, and there was no anode current, demonstrating the electrocatalytic reaction of HOR. The slope is calculated to be 5.21 cm^2 mA^{-1} s$^{-1/2}$, which is close reasonably with the theoretical value (4.87 cm^2 mA^{-1} s $^{1/2}$ for HOR's 2-e transfer). The intercept of the extrapolated line is corresponding to the reciprocal of the pure dynamic current density. At a 50-mV overpotential, the j_k of Ni/N-CNT is 2.33 mA cm^{-2}. The exchange current density for Ni/N-CNT is as high as 0.028 mA cm^{-2}, which is three times and 21 times higher compared with Ni/CNT and Ni. The exchange current density for Ni/N-CNT is one of the best PGM-free HOR electrocatalysts and can be compared with some PGM electrocatalysts (such as Pd).

Bimetallic nickel–molybdenum catalyst was prepared by thermal reduction of molybdenum and nickel precursors onto the surface of Ketjen Black (KB) [45].

In the performance test, the activity of NiMo/KB was tested using a thin-film method, in which catalyst suspension droplets in isopropanol (1.0 mg ml^{-1}) were casted onto the polished substrate of a glassy carbon rod (diameter is 5 mm) and dried on nitrogen flow continued for 60 minutes. After being immersed in an electrochemical cell, NiMo/KB is adjusted by cycling in the potential range of −0.20 to 0.40 V at a scan rate (ν) of 20 mV s^{-1}. To reduce the types of Ni oxides which were formed during catalyst storage in air and passivation, 0.1 M NaOH electrolyte is used. The ECSA of Ni was evaluated by the CV test, which was obtained in a potential range of −0.06 to 0.40 V at $\nu = 20\,mV\,s^{-1}$. After subtracting the charge associated with the Ketjen Black carbon carrier, the entire anode charge value corresponding to the formation of a single layer of α-Ni$(OH)_2$ and the conversion factor of 0.514 mC cm^{-2} were considered. The HOR exchange current comes from the anode scan of the linear scan voltammogram, which is micropolarized in a potential interval from −0.06 to 0.40 V in the H_2-saturated 0.1 M NaOH solution with a scan rate of 5 mV s^{-1}. The linearization of the area is obtained.

To further understand the origin of the promotion of HOR/HER activity on NiMo catalysts relative to polycrystalline nickel, the Tafel–Heyrovsky–Volmer mechanism was used to microdynamically model the hydrogen electrode reaction and the two-step oxidation mechanism of nickel. Figure 8.6 compares the current–potential curve of the experiment with the current–potential curve simulated with the parameters listed in Table S1. In the potential range of −0.06 to 0.15 V relative to RHE, a good agreement can be achieved between the experimental CV curve in the N_2 and H_2 atmospheres and the simulated CV curve (Figure 8.6), while at higher potentials, the simulation current is much worse than the simulated current at higher potentials than the experimental current (for more details, please refer to the supplementary information for readers). Compared with polycrystalline nickel electrodes, the Gibbs energy ($\Delta_{ad}G_0$) of hydrogen intermediate adsorption has increased significantly since about 2000. The value −0.1/−0.15 eV is approximately for pure Ni. NiMo/KB electrocatalyst must be assumed to be 0.04 eV in order to reproduce the key characteristics of the current–potential curve. Therefore, the increase in

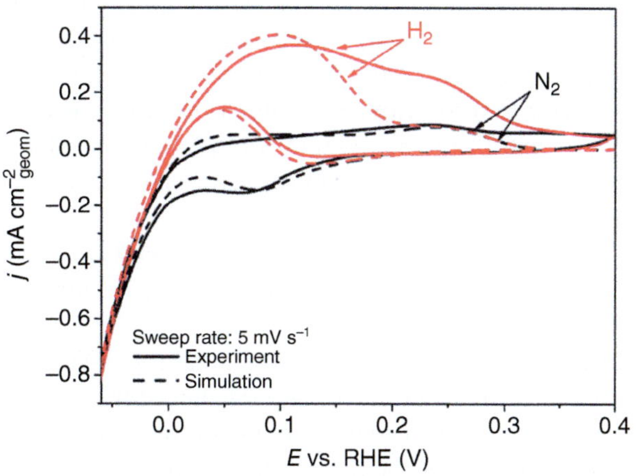

Figure 8.6 Comparison of the experimental (solid line) and simulated (dashed line) cyclic voltammograms for NiMo/KB (with a loading of 100 µg cm^{-2}) in N_2-(black) or H_2-saturated (red) 0.1 M NaOH at a sweep rate of 5 mV s^{-1}. All the curves are presented after the subtraction of the KB carbon contribution. Source: Kabir et al. [45]. © 2017, Royal Society of Chemistry.

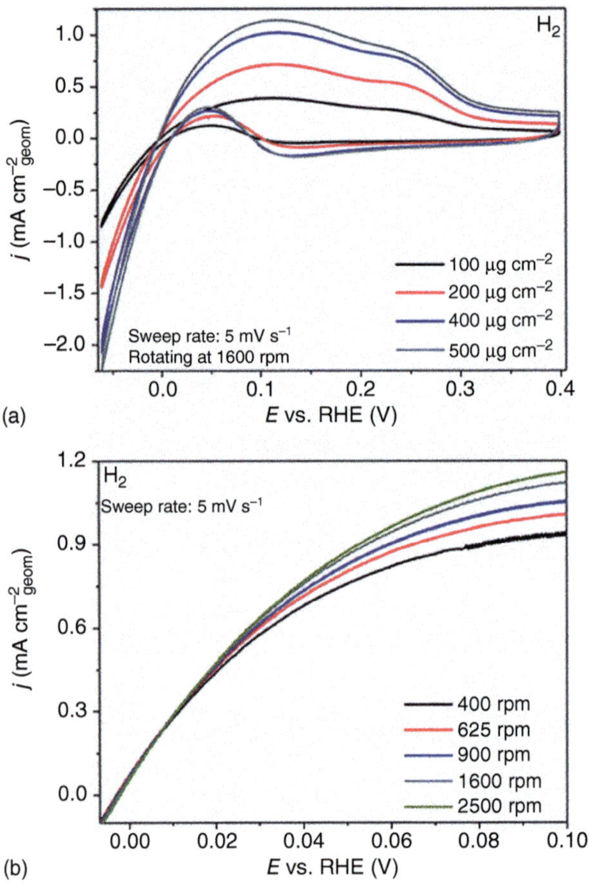

Figure 8.7 (a) Cyclic voltammograms obtained for NiMo/KB with various loadings in H_2-saturated 0.1 M NaOH at a sweep rate of 5 mV s^{-1} and rotating speed of 1600 rpm; (b) linear sweep voltammograms obtained for NiMo/KB (with a loading of 500 μg cm^{-2}) in H_2-saturated 0.1 M NaOH at a sweep rate of 5 mV s^{-1} and various rotating speeds. Source: Kabir et al. [45]. © 2017, Royal Society of Chemistry.

HOR/HER activity obtained on NiMo/KB was due to the increase in ($\Delta_{ad}G_0$), which is equivalent to a reduction of the binding energy of Ni–H_{ad} in NiMo/KB by about 20%. Compared with the pure Ni electrode, the value is 0.14/0.2 eV, which is too strong for hydrogen absorption. The modification of the binding energy of Ni–H_{ad} with the presence of Mo is consistent with the work of Paloukis et al. The author explained this through the electronic effect of Mo, which caused the Ni_0 peak of Ni 2p3/2 to move from 852.4 to 852.6 eV. It is worth noting that the Ni0 peak in the Ni 2p3/2 XPS spectrum of NiMo/KB is located at 852.5 eV. Similarly, Mo oxide (observed by XPS on the Ni surface) can also improve the HOR activity on Ni by promoting the Volmer step (Figure 8.7).

Ni_3M (M = Co, Fe, Cu, Mo) was synthesized by reducing the salt precursor in a sodium borohydride-based solution; a lot of catalysts consisted of bimetallic Ni_3M nanoparticles supported on carbon (VXCMAX22, Cabot, USA) were prepared, where M includes Cu, Co, Fe, and Mo. Carbon-supported single-metal Ni nanoparticles were also synthesized as a reference [46]. Further details can also be found in the authors' previous publications. In the performance test, first, the electrochemical stability for Ni_3M/C catalyst to dissolution was investigated by (scanning flow

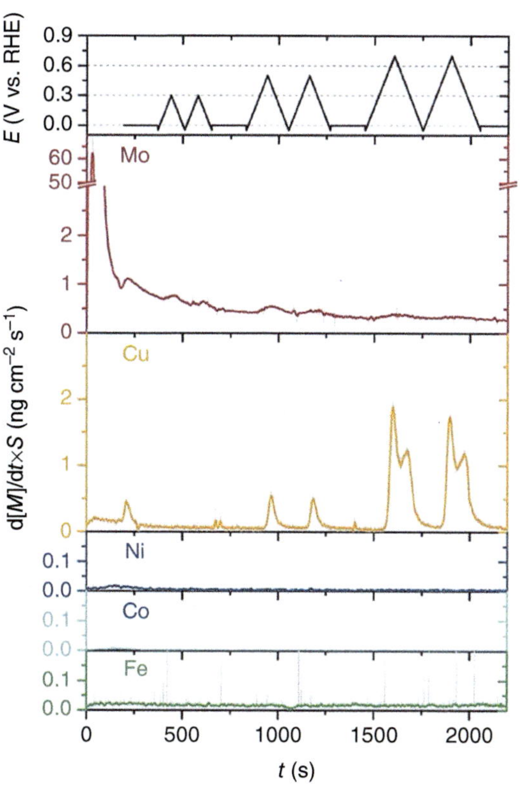

Figure 8.8 Online dissolution data during the potentiodynamic cycling protocol (shown in the upper pane) in 0.05 M KOH within the potential window of −0.05 to 0.7 V_{RHE} for Fe (green), Co (light blue), Ni (dark blue), Cu (orange), and Mo (red). The light gray traces represent the original dissolution data, and the colored components were smoothed for clarity. Source: Davydova et al. [46]. © 2019, American Chemical Society.

cell inductively coupled plasma) SFC-ICP-MS. The results are shown in Figure 8.8. As shown in Figure 8.8, after a certain degree of dissolution owing to contact with the electrocatalyst, the dissolution signals of Co, Ni, and Fe are still below the detection limit of inductively coupled plasma mass spectrometry (ICP-MS) (about 1, 10, and 300 pg cm^{-2} s^{-1}). The potential upper limit (UPL) is independent of the CV cycle. Fe and Co exhibited small dissolution peaks when contacted with electrolyte at open circuit voltage (OCP) (22 and 3.4 ng cm^{-2}, respectively). In potential cycles in the range of −0.05 to 0.7 V vs. RHE, no dissolution of Ni, Fe, and Co was observed. In contrast, the dissolution behavior of Cu is absolutely different. First, the amount of dissolved copper (37.5 ng cm^{-2}) was obviously higher than the tested values of Co, Ni, and Fe (Table 8.4).

CeO_2/Ni was prepared by modifying Ni with CeO_2 on XC-72 via a sequential process including solvothermal synthesis of CeO_2 and Ni precursor reduction by $NaBH_4$ with the existence of CeO_2 and XC-72 [47]. In the performance test, it can be clearly seen that CeO_2(r)–Ni/C-1 exhibits the highest catalytic performance. The HOR polarization curve of CeO_2(r)–Ni/C-1 in 0.1 M KOH saturated with H_2, which is a function of the rotation speed from 2500 to 400 rpm, and the Koutecky structure constructed at the overpotential of 0.05 V. The slope of the Levich diagram is calculated, which is 13.8 cm^2 mA^{-1} rpm1/2, which is in good agreement with the theoretical value of 14.8 cm^2 mA^{-1} rpm1/2. According to the Koutecky–Levich

Table 8.4 Electrochemical stability data for the Ni_3M/C electrocatalysts.

Catalyst	As-prepared		Online ICP-MS dissolution of M[a] ($ng\ cm^{-2}$)		i_0 ($mA\ cm^{-2}$ Ni) after AST[b]		i_0 ($mA\ cm^{-2}$ Ni) after corrosion test[c]
	i_0 ($mA\ cm^{-2}$ Ni)	ECSA (m^2 Ni g^{-1} Ni)	$\eta = 0.3\ V_{RHE}$	$\eta = 0.7\ V_{RHE}$	$\eta = 0.3\ V_{RHE}$	$\eta = 0.7\ V_{RHE}$	
Ni/C	0.064 ± 0.003	3.1 ± 0.084	<d.l.[d]	<d.l.	0.062 ± 0.003	negl.[e]	negl.
Ni_3Co_1/C	0.058 ± 0.005	1.93 ± 0.33	<d.l.	<d.l.	0.058 ± 0.010	negl.	negl.
Ni_3Cu_1/C	0.065 ± 0.009	4.40 ± 0.68	2.91	265.1	0.060 ± 0.008	negl.	negl.
Ni_3Fe_1/C	0.089 ± 0.012	1.89 ± 0.39	<d.l.	<d.l.	0.085 ± 0.010	negl.	negl.

a) Dissolution stability.
b) Long-term cycling stability.
c) Chemical stability.
d) < **d.l.**: below detection limit.
e) **negl.**: the value is negligibly low.
Source: Davydova et al. [46]. © 2019, American Chemical Society.

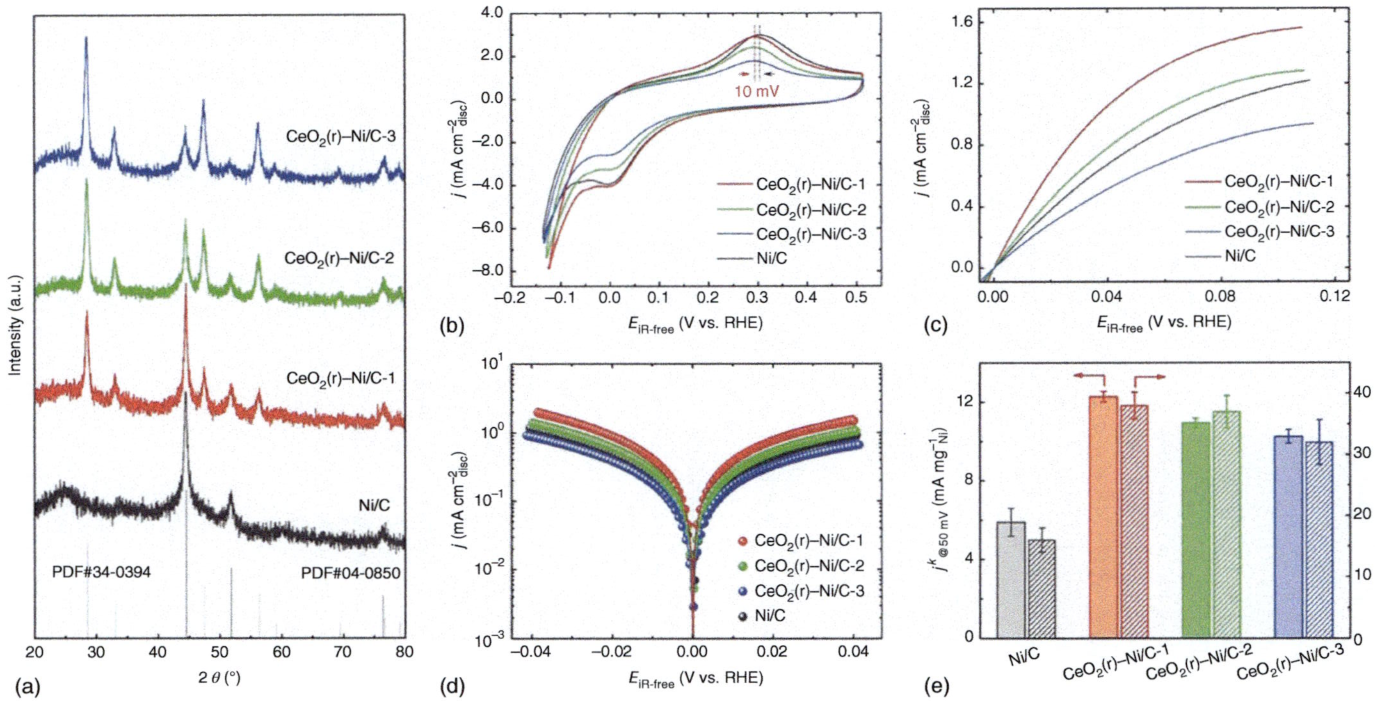

Figure 8.9 X-ray powder diffraction (XRD) patterns (a) and electrochemical tests (b–e) of Ni/C, CeO_2(r)–Ni/C-1, CeO_2(r)–Ni/C-2, and CeO_2(r)–Ni/C-3. CVs (b) in Ar-saturated 0.1 M KOH at a scan rate of 50 mV s^{-1}. HOR polarization curves (c) in H_2-saturated 0.1 M KOH at a rotating speed of 2500 rpm with a scan rate of 5 mV s^{-1}. Tafel plots (d) derived from (b). The lines indicate the Butler–Volmer fittings. (d) Comparison of the mass activities and the exchange current densities of the four catalysts. Source: Yang et al. [47]. © 2019, John Wiley & Sons.

equation, the dynamic current density (j_k) value measured based on the intercept is 1.73 mA cm^{-2}. In addition, the accurate loading of Ni in these catalysts through inductively coupled plasma-atomic emission spectrometry (ICP-AES) was 52.3, 46.1, 36.9, and 25.1 wt, for Ni/C, CeO_2(r)–Ni/C-1, CeO_2(r)–Ni/C-2, and CeO_2(r)Ni/C-3, respectively. Hence, after normalized by the mass of Ni on the working electrode, the mass activity of CeO2(r)–Ni/C-1 is around 12.28 mA mg Ni^{-1} at 50 mV, which is higher than that of Ni/C (5.89 mA mg Ni^{-1}), CeO2(r)–Ni/C-2 (10.95 mA mgNi^{-1}), CeO2(r)–Ni/C-3 (10.26 mA mgNi^{-1}), and most of the other Ni-based catalysts, as shown in Figure 8.9.

In the performance test, the current density of Ni/NiO/C-700 increases along with increasing rotation speed, which is because the mass transfer is carried out faster at a higher rate. This is achieved by fitting j^{-1} and $w^{-1/2}$ at the overpotential of 0.05 V, and a slope of 5.78 cm^2 mA^{-1} s$^{-1/2}$. Chemie can be obtained with a current density of j_k 1.59 mA cm^{-2} (0.32 mg cm^{-2}, close to the theoretical values and reported theoretical values of other Ni-based HOR electrocatalysts). The calculated kinetics of Ni/NiO/C-700 is a disk when h = 50 mV. Considering its Ni mass load, according to the thermogravimetric analysis (TGA) results shown in the support information, the Ni content is 63.2 wt%. At h = 50 mV, the mass activity of Ni/NiO/C-700 is 5.0 mA/mg Ni correspondingly, which makes Ni/NiO/C-700 to be one of the most active HOR electrocatalysts composed of non-noble metal elements. In fact, the HOR activity for Ni/NiO/C-700 is an order of magnitude at least higher than the HOR activity of pure Ni-based catalysts.

In summary, the optimization and identification of electrocatalysts require a very accurate and careful assessment of the activity. This chapter aims at providing detailed experimental procedures to guide new readers and researchers to reliably measure the specific and mass activities of hydrogen electrocatalysis (HER and HOR). Accurate evaluation of specific activity needs a reliable estimation of the surface area of the electrocatalyst. We first discussed the method of testing the metal catalyst ECSA: the coulomb charge of the surface redox reaction, including HUPD, CO stripping, and the metal underpotential deposition (UPD). The ambiguity of the ECSA methods was then evaluated using a Pt-based catalyst as an example model to remind the reader that each method only offers an approximation for the surface area, and in fact, no method could be perfect. Next, we discuss the surface area evaluation of non-metallic catalysts. The commonly used ECSA method is double layer capacitance (CDL), but it is a big challenge to achieve accurate estimation.

References

1 Lu, S.Q. and Zhuang, Z.B. (2016). *Sci. China Mater.* 59: 217–238.

2 Cong, Y.Y., Yi, B.L., and Song, Y.J. (2018). *Nano Energy* 44: 288–303.

3 Tian, X., Zhao, P., and Sheng, W. (2019). *Adv. Mater.* 31: e1808066.

4 Sheng, W.C., Myint, M., Chen, J.G.G., and Yan, Y.S. (2013). *Energy Environ. Sci.* 6: 1509–1512.

5 Sheng, W.C., Bivens, A.P., Myint, M. et al. (2014). *Energy Environ. Sci.* 7: 1719–1724.

6 Karlberg, G.S., Jaramillo, T.F., Skulason, E. et al. (2007). *Phys. Rev. Lett.* 99: 126101.

7 Strmcnik, D., Uchimura, M., Wang, C. et al. (2013). *Nat. Chem.* 5: 300–306.

8 Lu, S. and Zhuang, Z. (2017). *J. Am. Chem. Soc.* 139: 5156–5163.

9 Ohyama, J., Sato, T., Yamamoto, Y. et al. (2013). *J. Am. Chem. Soc.* 135: 8016–8021.

10 Durst, J., Simon, C., Siebel, A. et al. (2014). *ECS Trans.* 64: 1069–1080.

11 Sheng, W.C., Gasteiger, H.A., and Shao-Horn, Y. (2010). *J. Electrochem. Soc.* 157: B1529–B1536.

12 Zheng, J., Sheng, W., Zhuang, Z. et al. (2016). *Sci. Adv.* 2: e1501602.

13 Kamai, R., Kamiya, K., Hashimoto, K., and Nakanishi, S. (2016). *Angew. Chem. Int. Ed.* 55: 13184–13188.

14 Kim, J., Renault, C., Nioradze, N. et al. (2016). *J. Am. Chem. Soc.* 138: 8560–8568.

15 Scofield, M.E., Zhou, Y.C., Yue, S.Y. et al. (2016). *ACS Catal.* 6: 3895–3908.

16 Scofield, M., Zhou, Y., Yue, S. et al. (2015). *ACS Catal.* 5: 6764–6772.

17 St. John, S., Atkinson, R.W. III,, Unocic, R.R. et al. (2015). *J. Phys. Chem. C* 119: 13481–13487.

18 Durst, J., Siebel, A., Simon, C. et al. (2014). *Energy Environ. Sci.* 7: 2255–2260.

19 Nørskov, J.K., Rossmeisl, J., Logadottir, A. et al. (2004). *J. Phys. Chem. B* 108: 17886–17892.

20 Alia, S.M., Pivovar, B.S., and Yan, Y. (2013). *J. Am. Chem. Soc.* 135: 13473–13478.

21 Elbert, K., Hu, J., Ma, Z. et al. (2015). *ACS Catal.* 5: 6764–6772.

22 Li, J., Ghoshal, S., Bates, M.K. et al. (2017). *Angew. Chem. Int. Ed.* 56: 15594–15598.

23 Banham, D., Ye, S.Y., O'Toole, A. et al. (2016). *Nano Energy* 27: 157–166.

24 Wang, J.X., Springer, T.E., Liu, P. et al. (2007). *J. Phys. Chem. C* 111: 12425–12433.

25 Conway, B. and Tilak, B. (2002). *Electrochim. Acta* 47: 3571–3594.

26 Rau, M.S., Quaino, P.M., de Chialvo, M.R.G., and Chialvo, A.C. (2008). *Electrochem. Commun.* 10: 208–212.

27 Pronkin, S.N., Bonnefont, A., Ruvinskiy, P.S., and Savinova, E.R. (2010). *Electrochim. Acta* 55: 3312–3323.

28 Durst, J., Simon, C., Hasche, F., and Gasteiger, H.A. (2015). *J. Electrochem. Soc.* 162: F190–F203.

29 Montero, M.A., Fernandez, J.L., de Chialvo, M.R.G., and Chialvo, A.C. (2013). *J. Phys. Chem. C* 117: 25269–25275.

30 Trasatti, S. (1972). *J. Electroanal. Chem.* 39: 163–184.

31 Zoski, C.G. (2003). *J. Phys. Chem. B* 107: 6401–6405.

32 Zheng, J., Zhuang, Z.B., Xu, B.J., and Yan, Y.S. (2015). *ACS Catal.* 5: 4449–4455.

33 Kundu, M.K., Mishra, R., Bhowmik, T., and Barman, S. (2018). *J. Mater. Chem. A* 6: 23531–23541.

34 Bhowmik, T., Kundu, M.K., and Barman, S. (2016). *ACS Catal.* 6: 1929–1941.

35 Masud, J., Nguyen, T.V., Singh, N. et al. (2015). *J. Electrochem. Soc.* 162: F455–F462.
36 Montero, M.A., de Chialvo, M.R.G., and Chialvo, A.C. (2015). *J. Power Sources* 283: 181–186.
37 Zheng, Y., Jiao, Y., Zhu, Y. et al. (2016). *J. Am. Chem. Soc.* 138: 16174–16181.
38 Wang, J., Cui, W., Liu, Q. et al. (2016). *Adv. Mater.* 28: 215–230.
39 Park, H.S., Yang, J., Cho, M.K. et al. (2019). *Nano Energy* 55: 49–58.
40 Martindale, B.C.M. and Reisner, E. (2016). *Adv. Energy Mater.* 6: 1502095.
41 Lu, S.F., Pan, J., Huang, A.B. et al. (2008). *Proc. Natl. Acad. Sci. U.S.A.* 105: 20611–20614.
42 Hu, Q.P., Li, G.W., Pan, J. et al. (2013). *Int. J. Hydrogen Energy* 38: 16264–16268.
43 Zhuang, Z., Giles, S.A., Zheng, J. et al. (2016). *Nat. Commun.* 7: 10141.
44 Cherstiouk, O.V., Simonov, P.A., Oshchepkov, A.G. et al. (2016). *J. Electroanal. Chem.* 783: 146–151.
45 Kabir, S., Lemire, K., Artyushkova, K. et al. (2017). *J. Mater. Chem. A* 5: 24433–24443.
46 Davydova, E.S., Speck, F.D., Paul, M.T.Y. et al. (2019). *ACS Catal.* 9: 6837–6845.
47 Yang, F., Bao, X., Li, P. et al. (2019). *Angew. Chem. Int. Ed.* 58: 14179–14183.

9

Oxygen Reduction Reaction (ORR)

9.1 Mechanism for ORR

The oxygen reduction reaction (ORR) mechanism is very complicated, involving many basic and irreversible steps. If you do not consider some details in the reaction processes, the mechanism of ORR on the electrode can be divided into 2e and 4e reduction; the difference is whether there is a peroxide intermediate in the solution. Regardless of the 2e reaction process, the difference is that either O_2 or H_2O_2 of produces is formed (H medium). The H_2O_2 intermediate oxidation products may decompose in the reverse reaction, thereby forming the O energy conversion efficiency of the battery system and damaging the electrode.

9.1.1 Battery System and Damaged Electrodes

H_2O (acidic medium) or OH 4e will be formed in the alkaline state, thereby decreasing the efficiency for the energy conversion of the battery. On the other hand, intermediate oxygenated products would damage the battery systems as well as the electrodes. The reduction is a direct (alkaline medium) O_2 one-step reduction process. However, on the surface of most catalysts and electrodes, ORR is easy to follow the process of combining $2e^-$ and $4e^-$, while direct $4e^-$ process is relatively difficult. This is because of the high dissociation energy of the O—O bond in the oxygen molecule, which is about 494 kJ mol^{-1}. The 2e-reaction pathway would directly produce H_2O_2 or HO_2, which only requires 146 kJ mol^{-1}. This case would reduce the activation energy for ORR and increase the possibility for the reaction to occur. Therefore, a good ORR electrocatalyst requires high selectivity toward the 4e-reduction process. At the same time, the ORR should be carried out at as high a potential region as possible to achieve the 4e transition of O_2, increase the output voltage of the catalyst, and improve the efficiency of the catalyst. It can be said that the cathode ORR electrocatalyst controls indirectly the energy conversion rate and also affects the wide application of catalyst. Therefore, a better understanding of the ORR mechanism is required for the design of an inexpensive and effective ORR catalyst [1].

Electrocatalysis in Balancing the Natural Carbon Cycle, First Edition. Yaobing Wang.
© 2021 WILEY-VCH GmbH. Published 2021 by WILEY-VCH GmbH.

9.1.2 Intermediate Species

In general, the four-electron pathway is preferred due to higher reaction kinetics and efficiency [2]. To date, in practical applications, platinum-based materials are considered to be ideal ORR electrocatalysts with four electron paths.

Reports of ORR electrocatalysis are presented according to the mechanism, with separate sections for proceeding catalysis by initial outer- and inner-sphere electron transfer processes to O_2 [3]. $4e^-$ process and $2e^-$ process have a difference in the electrochemical energy barriers. The key point to choose the 4e or 2e process is to tune the dissociation of O—O bonds in adsorbed *OOH.

9.2 Catalysts in ORR

Recently, many groups have published many excellent comments on the design and synthesis of Pt-based or non-Pt catalysts for ORR. Most of the reports focus on different methods and approaches for preparing specific types of ORR catalysts. Given the rapid development of Pt-based and Pt-free electrocatalysts, there is an urgent need to update this field in a timely manner, focusing on novel design concepts and major breakthroughs in ORR electrocatalysts on Pt-based and Pt-free.

9.2.1 Noble Metal Materials

ORR plays a vital role in the current new electrochemical energy conversion technologies, and it is the key factor for total output as well as commercial applications. In particular, the ORR process involves the transfer of O_2 molecules and the product accumulation, which may inhibit the Pt active site. Additionally, due to the decisive issues related to oxygen adsorption on the catalyst surface, O—O bond activation/cracking, and product removal, ORR itself has slow kinetics. Despite the high requirements for advanced catalysts in the applications, the important requirements are still challenging tasks.

Continuous research on the accurate design and intentional synthesis of excellent ORR electrocatalysts has been conducted, and a variety of effective ORR electrocatalysts were exploited. Among various candidates, platinum (Pt) has been clearly demonstrated to be an essential component of $4e^-$ transfer during ORR. However, the sluggish kinetics in the ORR forces a relatively large amount of Pt material loading, resulting in an obvious increase in the overall cost of fuel cells.

For Pt-based materials, three aspects of the research were focused: (i) reduce the size of Pt particles, increase the Pt dispersion, and increase the specific surface area; (ii) develop an approach for preparing Pt particles with specific orientation surfaces; (iii) various physical and chemical approaches are used to combine other non-metal/metal elements to the Pt electrocatalyst to alloy it or to disperse Pt component into other metal alloys, transition metals, and oxides to form a mixture or alloy in a specific ratio, e.g. core–shell structure.

For pure Pt electrocatalyst, the structure effect on bulk single crystals was studied. The ORR behavior of low-index Pt surfaces (e.g. Pt(100), Pt(111), and

Pt(110)) has been widely studied. These results are summarized in some review reports and will not be focused here. The conclusion is that in weakly adsorbed electrolytes (e.g. $HClO_4$ electrolyte), the ORR activity follows the order of Pt(111) ≈ Pt(110) ≫ Pt(100). Feliu et al. and Hoshi et al. systematically studied high refractive index Pt the structural effect of the surface, and it is concluded that the ORR activity is dependent highly on the direction of the surface steps and the platform. It is demonstrated that in the high refractive index plane *n*(*hkl*)-(*mno*), the activity increases with the density of the platform (or the width of the platform) decrease and increase, apart from the *n*(110)-(111) surface (*n* represents the number; (*hkl*) and (*mno*) represents the ladder atomic row, respectively, represents the structure of the ladder and the stairs). In later research, Hoshi et al. confirmed the general trend on the (111) platform, except that the number of order atomic rows on the surface is $n = 2$. ORR activity did not rely on the (100) platform step density. It is proposed that the active site on the high refractive index surface with (111) platform is located on the edge of the (111) platform and its adjacent platform row. However, the mechanism for high ORR activity is not fully understood. These results are possibly an important guide for the development of highly active Pt electrocatalysts, which should have stepped surfaces rather than a smooth (111) surface because Pt(111) is not as active as its adjacent stepped surface. The activity of Pt nanoparticles (NPs) with (221) = 4(111)-(111) and (331) = 3(111)-(111) facets with high refractive index is expected to be higher than (111) more than three times higher. Attempts have been made to prepare this high refractive facet Pt nanoparticles. The poor stability of the specially designed structures, especially during potential circulation, is a big issue.

Particle size/shape effect was also studied. Inspired by the trend of ORR activity, which was gained on single-crystal Pt, namely, high refractive index plane > (111) > (100), lots of efforts were made to prepare shape-controlled nanostructures with optimization of the structural effect. It is found to be combined with the (100) plane. Compared with the aspect of ZnO, octahedral Pt particles surrounded by (111) planes have higher activity in ORR, which is consistent with the overall work of single crystal. Pt particles with high refractive index plane include tetrahedron (*hk*0), oblique hexahedron (*hkk*), and trioctahedron (*hhk*), and at least one Miller index is >1; it has been proved that its activity is higher than (111) or (100). The activity enhancement is arisen from the steps, ledges, and buttons, the high-density low-coordination atoms on the junction. The main problem with shape-controlled Pt particles is the stability under ORR conditions because they tend to develop into thermodynamically balanced shapes.

Single-atom Pt is a novel, highly efficient catalyst for the ORR. For example, two Pt with different weight percentages were deposited on TiC and TiN nanoparticles with acid treatment using the initial wet impregnation method [4]. For monoatomic catalysts, 0.2 wt% of Pt was deposited, and for nanoparticle catalysts, 5 wt% Pt has been deposited. In the performance test study, ORR has been conducted on both Pt1/TiC and Pt1/TiN. The ORR on Pt nanoparticles usually generates H_2O through the four-electron reaction pathway. However, the strong O—O the double bond cannot be dissociated on the active site where atoms are dispersed. Therefore, the

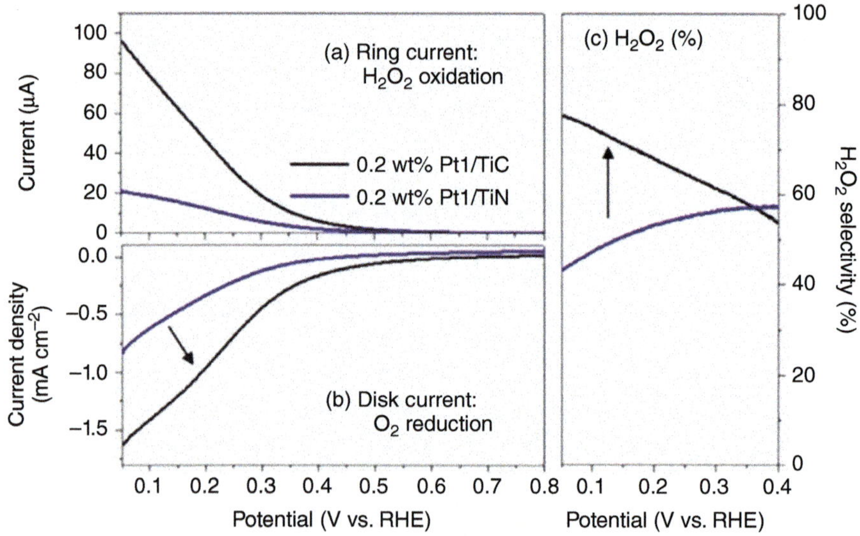

Figure 9.1 Oxygen reduction reaction polarization curves on single-atom Pt: (a) ring currents measured concurrently during ORR with a potential held at 1.2 V; (b) ORR polarization curves of single-atom Pt in O_2-saturated 0.1 M $HClO_4$ solution with a scan rate of 0.01 V s^{-1}; (c) H_2O_2 selectivity calculated from disk and ring currents. Source: Yang et al. [4]. © 2017, American Chemical Society.

monoatomic platinum produces H_2O_2 as the main product. Pt1/TiN and Pt1/TiC produce a large amount of H_2O_2, which is due to the atomic active sites (Figure 9.1). The disk current in Figure 9.1b indicates simultaneous generation. In all potential ranges, the ORR current density on Pt1/TiC is almost twice that of Pt1/TiN. The ringing current in Figure 9.1a shows the H_2O_2 generated on the disk. It is oxidized. Pt1/TiC shows a higher current for oxidation. The selectivity of H_2O_2 can be estimated from the disk current and the ringing current, as shown in Figure 9.1c. The Pt1/TiC electrocatalyst shows a better activity and a greater effect on H_2O_2, as well as high selectivity. At 0.2 V (vs. reversible hydrogen electrode [RHE]), ORR activity of 0.2 wt% Pt1/TiC is −0.96 mA cm^{-2}, H_2O_2 selectivity is 68.0%, and 0.2 wt% Pt1/TiN-R4 activity. It is −0.34 mA cm^{-2} for oxygen reduction activity and 53.2% for H_2O_2 selectivity, respectively.

Pt nanoparticles and nanocrystals are commonly used for ORR [5]. For the synthesis of Pt nanoparticles, it has been recently reported that monodisperse Pt nanocubes can be prepared by reducing $Pt(acac)_2$ with the existence of oleylamine–oleic acid and trace amounts of $Fe(CO)_5$. Compared with other Pt nanoparticles, the ORR overpotential is reduced by nearly 0.05 V on the 7 nm Pt nanocube, which can be seen from the ORR potential occurring at approximately 200 °C. For 7 nm Pt, NP is 0.92 V; 3 and 5 nm Pt, NP is approximately 0.87 V. Near the half-wave potential (about 0.75 V), the current generated by 7 nm Pt NP is about four times higher than that of 3 or 5 nm Pt NP. This ORR enhancement of the nanocube dominated by Pt(100) agrees with that observed on the ORR activity of Pt(*hkl*) on the extended electrode surface. The current density of the 7 nm platinum nanocube measured from rotating disk electrode (RDE) is four times higher than that of the 3 nm polyhedron (or 5 nm cubic) Pt NP, indicating the significant effect of Pt NP shapes on the ORR under proton exchange membrane fuel cell (PEMFC) operation conditions. In the mechanism study, the ORR activity with shape dependence can be attributed to the different adsorption of sulfate ions onto Pt(111) and (100) surfaces. Sulfate ions may bind to Pt surface atoms in the electrolyte and hinder ORR performance. Due to the triple symmetry of the 3 equiv O atoms in the sulfate and the Pt atoms on the (111) layer, the trend of the bonding strength is (111)@(100). For ORR, the situation is in reverse, and in H_2SO_4, the (100) plane has more active sites compared to the (111) plane. Nanocubes have a main (100) face and therefore have high ORR performance. The similar low activity found from 3 to 5 nm NPs agrees with recent reports on the size-independent Pt NPs on the activity of fuel cell. The shape-dependent electrocatalysis demonstrated further that trace iron in the Pt NPs will not contribute to the increasing activity in Pt NP catalysts.

For the synthesis of Pt nanocrystals [6], they can be prepared by reducing Pt precursors, decomposing organometallic complexes, or a combination of these two pathways (for example, the hydrogenation decomposition of $Pt(acac)_2$ with the existence of a blocking agent.) Figure 9.2 shows the polarization curve of 20 nm Pt polyoctahedron and Pt/C catalyst at room temperature. The half-wave potentials of 20 nm Pt/C catalyst and Pt polyoctahedron were 0.840 and 0.843 V, respectively. The illustration in Figure 9.2 compares the specific activity and mass activity at the potential 0.85 V. Although a similar mass activity was observed, the specific

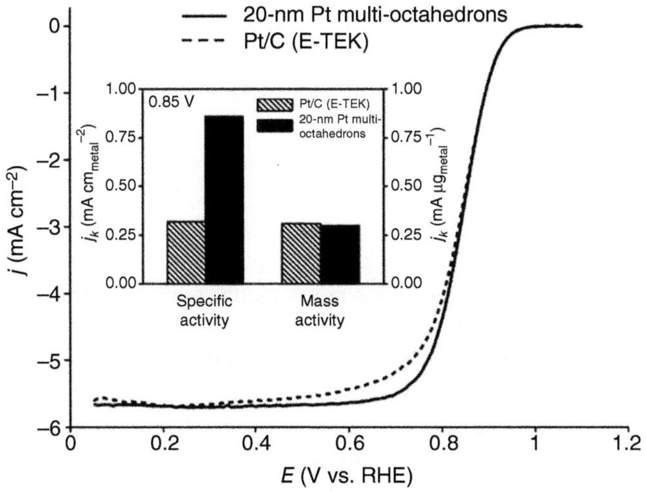

Figure 9.2 ORR polarization curves for 20-nm Pt multioctahedrons and Pt/C catalyst recorded at room temperature with a sweeping rate of 10 mV s^{-1} in O_2-saturated $HClO_4$ solution (0.1 M). The inset compares the specific activity (left) and mass activity (right) for these two catalysts at 0.85 V. The kinetic currents were calculated using mass-transport correction from ORR polarization curves. The loading of 20-nm Pt multioctahedrons and Pt/C catalyst on the rotating disk electrode was the same, 15 g (Pt) cm^{-2}. Source: Chen et al. [6]. © 2008 Elsevier.

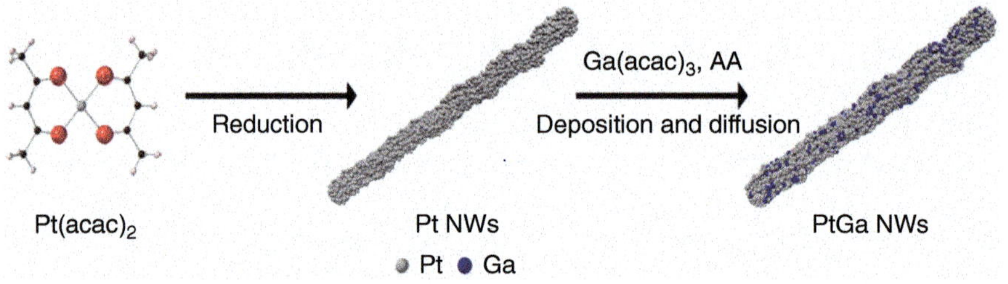

Figure 9.3 Structural and compositional characterizations of PtGa alloy NWs. Schematic illustration showing the evolution process of PtGa NWs. Source: Gao et al. [7]. © 2019, American Chemical Society.

activity of Pt polytetrahedron with the size of 20 nm was 2.7 times its specific activity. The Pt/C catalyst is probably because of the preferential exposure of 111 faceted steps on its surface. In this study, it was found that the electro-oxidation efficiency per unit surface area of Pt THHs (tetrahexahedrons) is about two times higher. Compared with Pt spheres and the commercially available 3.2 nm Pt/C catalyst, the electrocatalytic efficiency of ethanol is increased by about four times. The mechanism study has revealed that similar to the shape effect on catalysis, atoms at the edges, steps, and knots are more catalytically active; thus, it is expected that high refractive index facets will show higher electrocatalytic activity than the normally generated low refractive index facets. For example, Pt THHs having high indices (730), (520), and/or (210) facets acquired from electrochemical treatment were shown to be used as electrocatalysts for the electro-oxidation of ethanol and formic acid, respectively. In this study, compared with Pt nanospheres and the commercially available 3.2 nm, it was found that the electro-oxidation efficiency of Pt THHs for ethanol is about two times per unit surface area and the oxidation efficiency for formic acid is about four times.

9.2.1.1 Platinum/Carbon Catalyst

The results clearly demonstrated that Pt THHs has a high catalytic performance per unit surface area for small organic molecules oxidation (Figure 9.3).

Pt alloy catalysts can not only enhance the ORR performance but also reduce the cost. PtGa ultrathin nanowires were prepared via a two-step method [7]. The electrochemical surface areas (ECSAs) for Pt NWs (nanowires) /C, $Pt_{4.31}Ga$ NWs/C, and commercial Pt/C catalysts are 83.4, 53.4, and 65.7 $m^2\ g^{-1}$ Pt. Figure 9.4a exhibits the positive ORR polarization curve of the catalyst tested in the O_2-saturated 0.1 M $HClO_4$ solution which was obtained at room temperature using the RDE method. The kinetic current of the catalyst on the basis of the Koutecký–Levich equation was further calculated. The specific and mass activities were acquired via normalizing the dynamic flow with the Pt loading or ECSA of the corresponding catalyst (Figure 9.4b,c). In detain, the $Pt_{4.31}Ga$ NWs/C nanocatalyst exhibited the highest specific area and mass activity at 0.9 V vs. RHE of 3.28 mA cm^{-2} and 1.89 A mg^{-1} Pt, respectively, which is 12.1 and 10.5 times higher than the Pt/C catalyst (specific activity is 0.27 mA cm^{-2}; mass activity is 0.18 A mg^{-1} Pt). It has been suggested that

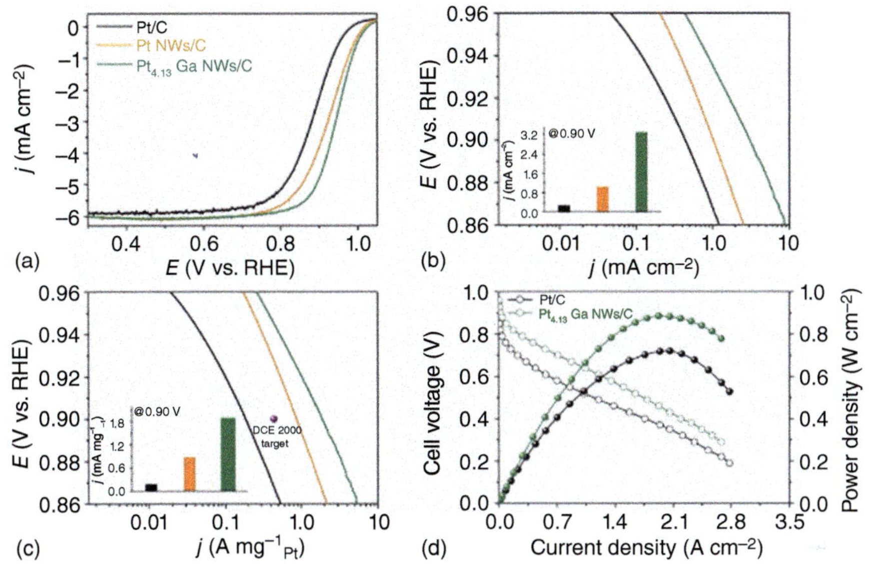

Figure 9.4 Electrocatalytic performance of $Pt_{4.31}Ga$ NWs/C, Pt NWs/C, and commercial Pt/C catalysts. (a) ORR polarization curves were recorded at room temperature in O_2-saturated 0.1 M $HClO_4$ solutions at a sweep rate of 10 mV s^{-1} and a rotation rate of 1600 rpm. Tafel plots of (b) specific activity and (c) mass activity. (d) H_2O_2 fuel cell *i*–*V* polarization and power density plots of prepared membrane electrode assembly (MEA) with the cathode Pt loading of 0.12 mg_{Pt} cm^{-2} for $Pt_{4.31}Ga$ NWs/C and 0.14 mg_{Pt} cm^{-2} for commercial Pt/C. The color scheme in panel (a) applies to panels (a–c). Source: Gao et al. [7]. © 2019, American Chemical Society.

p–d hybridization interaction could enhance the ORR activity. Specifically, since the electron structure determines the electrocatalytic performance, understanding the changes in the electronic structure caused by alloyed Ga is the key to explaining the promoted ORR activity. The ratio of $ECSA_{CO}$/ECSA H_{upd} is 1.5; this is because the hydrogen adsorption strength on the platinum surface is reduced (thus leading to an underestimation of ECSA H_{upd}). Therefore, we reasonably conclude that the $Pt_{4.31}Ga$ NWs in the electrocatalyst present Pt skin surfaces. Another point worth to be mentioned is the change in the initial potential of CO electro-oxidation. It was observed that the $Pt_{4.31}Ga$ alloy NWs/C catalyst showed the largest negative start potential (0.75 V vs. RHE), while the Pt/C catalyst commercially available showed the largest positive start potential (0.83 V vs. RHE). The negative migration of the potential for CO stripping of the $Pt_{4.31}Ga$ alloy NWs/C catalyst indicates that the energy for adsorption is weakened, and the center of the d band is reduced, which actually proves the experimental evidence that the activity and CO tolerance are improved.

PtNiRh trimetallic nanowires were also prepared as a promising catalyst for ORR [8]. The ECSA of the trimetallic PtNiRh NWs/C catalyst was 106.4 m^2 g^{-1} Pt. The positive ORR polarization curve of the catalyst at room temperature is further recorded in 0.1 M $HClO_4$ solution saturated with O_2, as shown in Figure 9.5.

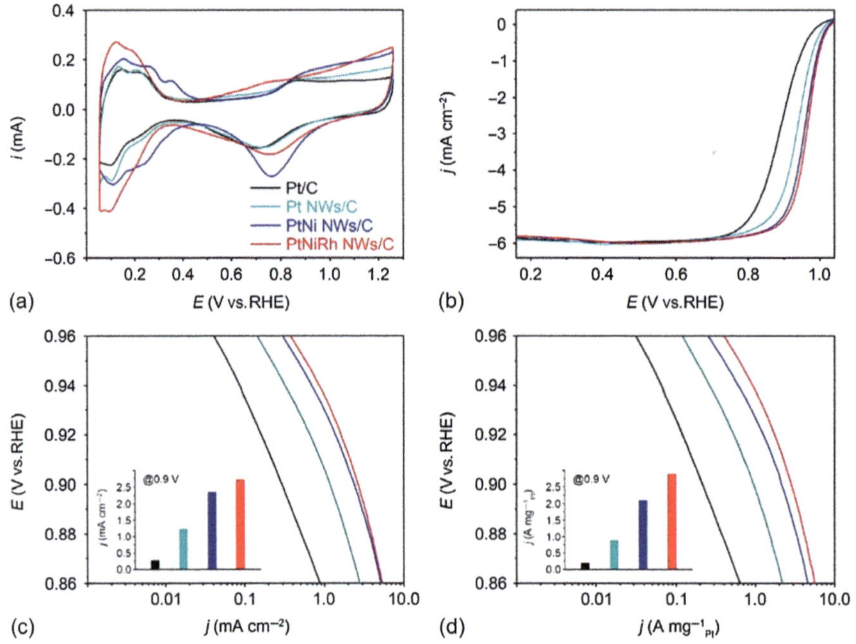

Figure 9.5 Electrocatalytic performance of different catalysts. (a) CV curves were recorded in N_2-saturated 0.1 M $HClO_4$ solutions at room temperature with a sweep rate of 50 mV s^{-1}. (b) Positive-going polarization curves recorded in O_2-saturated 0.1 M $HClO_4$ solutions with a sweep rate of 10 mV s^{-1} and a rotation rate of 1600 rpm. (c) Specific activities at 0.9 vs. RHE. (d) Mass activities at 0.9 V RHE. The color scheme in panel (a) applies to all other panels. Source: Li et al. [8]. © 2018, American Chemical Society.

According to the polarization curve, the kinetic current of the catalyst can be evaluated from the Koutecký–Levich equation, with normalization to the ECSA and Pt masses to derive specific area and mass activity, respectively (Figure 9.5c,d). At 0.9 V vs. RHE, the trimetallic PtNiRh NWs/C catalyst has the best specific activity (2.71 mA cm^{-2}).

Pt–Ni rhombic dodecahedra have been prepared with a start from the formation of a Pt-rich branch, and induced Ni deposition gradually takes place around this branch [9]. In short, the three-dimensional element distribution in Pt–Ni solid rhombic dodecahedra (SD) has been successfully controlled, resulting in structural differences after Ni corrodes into excavated or hollow nanoframes (NFs). The excavated NF appears to extend inward from the edge of Pt enrichment, and hollow NF has a gap inside the edge of Pt enrichment. These differences result in a significant improvement in the ORR activity of the excavated NF. The cyclic voltammogram (CV) in Figure 9.6a shows the H_{upd} process used to test the ECSA, for H-NF which is 60.2 m^2 g^{-1} Pt, and the ECSA for E-NF is 48.1 m^2 g^{-1} Pt, both lower than the commercial Pt/C ECSA (76.1 m^2 g^{-1} Pt), normalized at 0.95 V vs. RHE on ECSA H_{upd} and Pt mass loads, giving specific activity and mass activity, and compared with the earlier study of Pt_3Ni NF using Pt skin (Figure 9.6). Among the catalysts reported, E-NF has the highest specific activity and mass activity; they are about 10 times and

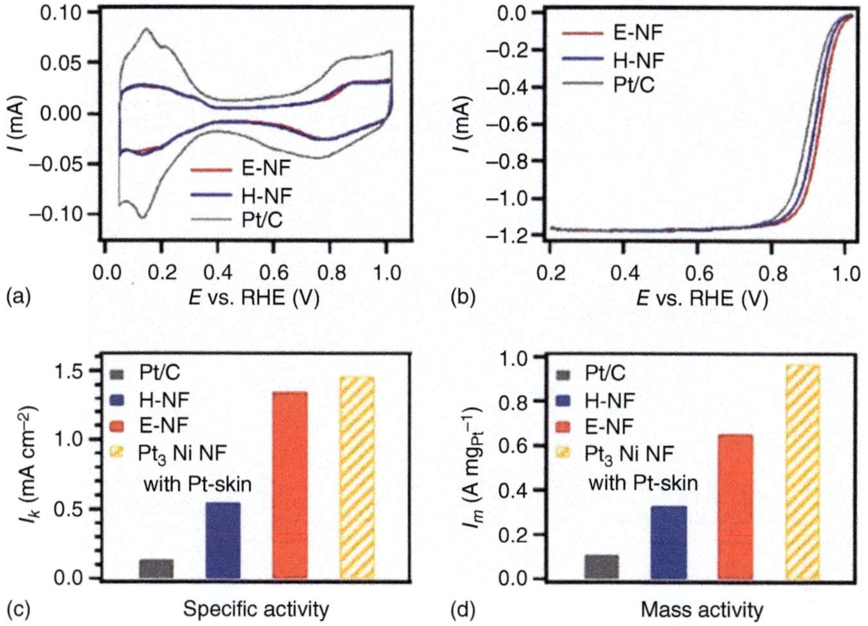

Figure 9.6 (a) Cyclic voltammograms recorded at room temperature in Ar-saturated 0.1 M $HClO_4$ solution with a sweep rate of $50\,mV\,s^{-1}$. (b) ORR polarization curves recorded at room temperature in O_2-saturated 0.1 M $HClO_4$ solution with a sweep rate of $20\,mV\,s^{-1}$ and a rotation rate of 1600 rpm. (c) Specific activity and (d) mass activity at 0.95 V vs. RHE for Pt/C, H-NF, and E-NF as compared to Pt_3Ni NFs with Pt skin. Source: Becknell et al. [9]. © 2017, American Chemical Society.

6 times higher than the commercially available Pt/C. The specific activity of E-NF is $1.35\,mA\,cm^{-2}$, which is 145% higher than that of H-NF ($0.55\,mA\,cm^{-2}$). There is no change in the morphology after the electrochemical test.

PtPd alloy nanosheets showed an ECSA of c. $40.12\,m^2\,g^{-1}$ (c. $58.98\,m^2\,g^{-1}$ metal, close to Pt/C metal) [10]. In addition to ECSA, the electrochemical double-layer capacitance (Cdl) was further calculated. Figure 9.7 exhibits that the Cdl on PtPd NSs (nanosheets) ($8.26\,mF\,cm^{-2}$) was slightly bigger compared to that of Pt/C ($7.23\,mF\,cm^{-2}$), which further indicates that PtPd NSs have much more active sites exposed on the surface and possibly higher current density, tested in a 0.1 M KOH solution. Obviously, the half-wave potential ($E_{1/2}$) of PtPd NSs (0.879 V) was more correct than the positive wave of Pt/C (0.836 V). The former has more favorable dynamics. More specifically, the as-obtained PtPd NSs display 2.4 times enhancement in mass activity (MA) ($382.10\,mA\,mg^{-1}$) and 3.5 times enlargement in specific activity (SA) ($0.95\,mA\,cm^{-2}$) at 0.80 V, rivaling those of Pt/C catalyst ($156.60\,mA\,mg^{-1}$ and $0.27\,mA\,cm^{-2}$, Figure 9.8). As an important evaluation of electrocatalysts, the stability of PtPd NSs was studied by continuously scanning Pt/C as a solvent 1000 times in 0.1 M KOH solution. Compared.

A series of Pt_3M (M = Ni, Co, Fe, Ti, V) [12] have been exploited, and the basic relationship between the electrocatalytic tendency on the surface of Pt_3M

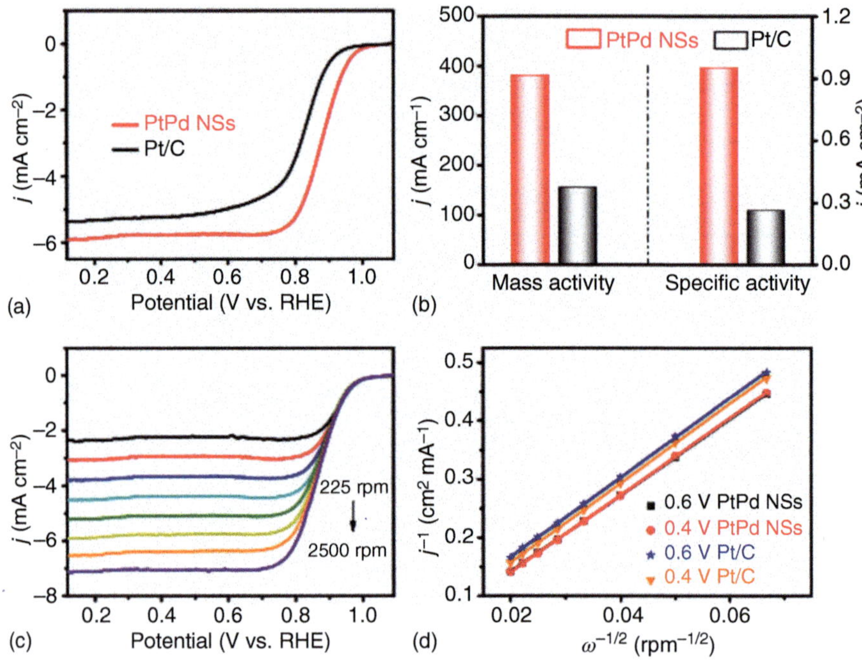

Figure 9.7 (a) ORR polarization curves of the catalysts in O_2-saturated 0.1 M KOH at a rotation rate of 1600 rpm and a scan rate of 5 mV s^{-1}. (b) The MA and SA values at 0.80 V. (c) ORR polarization curves of PtPd NSs with different rotation rates. (d) The K–L plots for ORR at 0.4 and 0.6 V. Source: Chen et al. [10]. © 2019 Elsevier.

(M = Ni, Co, Fe, Ti, V) between the as-determined surface electronic structure (d-band center) and oxygen reduction activity was revealed. This relationship shows a "volcano-type" shape, where the maximum electrocatalytic activity depends on the balance between adsorption energy for reactive intermediates and the surface coverage of bystanders (blockers). The establishment of electrocatalytic trends for extending the surface can be used to understand the activity mode of Pt_3M nanocatalysts and offers a basis to the catalytic enhancement of cathode electrocatalysts. With a combination of both simulation and experimentation to find surfaces with the activity as desired, advanced concepts in nanoscale catalysts engineering have been exploited. The average energy of the d-state is different, and the displacement is reduced on Pt modified electronically by the subsurface, which indicates that the OH_{ad} fractional coverage on these surfaces decreased compared to metallic Pt, the same as the recent Pt–Co's theoretical research is consistent with other Pt–M systems. In spite that these results provide useful information, it is impossible to determine the truly systematic experimental trend. By summarizing the above-mentioned Pt_3Fe surface and Gad other Pt_3M surfaces, the authors also considered how to simultaneously control the center position of the d-band (1-Θad term). To visually show the overall trend, under a constant overpotential, the dynamic current is drawn according to the center position of the d-band on the surface of the Pt skin (and Pt skeleton) determined by experiment. At this time, it is

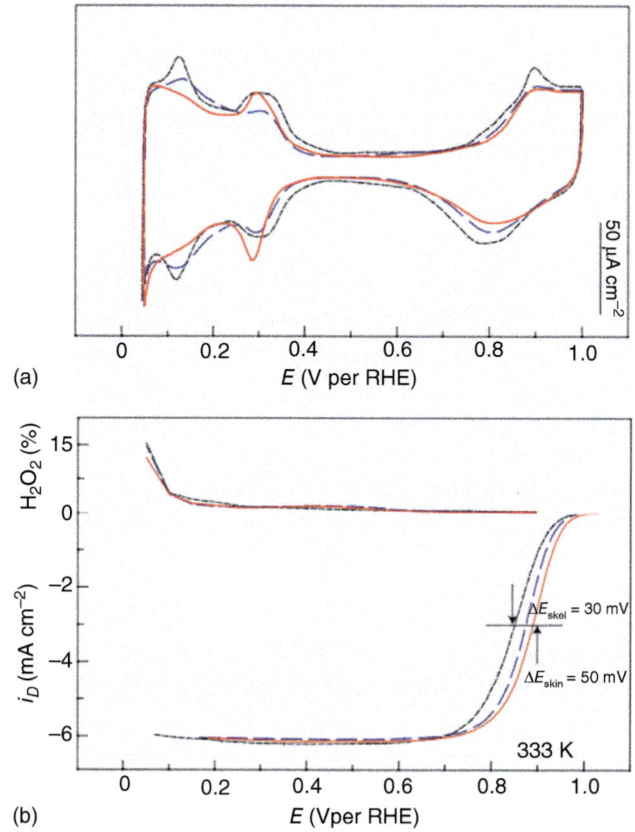

Figure 9.8 Electrochemical properties of Pt_3Fe surfaces. (a) Cyclic voltammograms of the Pt-skeleton (blue line) and Pt-skin (red line) electrodes as well as for a pure polycrystalline Pt (black line) electrode in 0.1 M $HClO_4$. (b) Polarization curves for the ORR on the disk electrode and the corresponding fraction of peroxide detected on the ring electrode. Source: Zhao et al. [11]. © 2019, Royal Society of Chemistry.

essential to get the summary relationship. As the average energy of d-state on the surface atoms is systematically changed throughout the periodic table of elements, the uniqueness is that the correlation between the surface electronic properties and reactivity determined by the experiment is established. The central position of the Pt skin surface is volcanic and has the largest catalytic activity for Pt_3Co. The Pt framework also has the same trend *t* surface, although the center of the d-band of Pt framework surface measured in the electrochemical environment could deviate from the value obtained in ultrahigh-vacuum (UHV) due to the dissolution of the 3d element, which also makes the surface morphology distinct from Pt skin surface.

Pt can also be used to combine with the rare earth element. Pt_3Sc and Pt_3Y were prepared in bulk and polycrystalline forms [13]. At 0.9 V, the specific activity of Pt_3Sc is 50% higher compared to that of Pt, while the specific activity of Pt_3Y is increased by six times. At 0.87 V, the activity of Pt_3Sc increased by 80%, while the activity of Pt_3Y increased by order of magnitude. Since the polarization curve of Pt_3Y has a steep gradient (or Tafel slope), at a potential lower than 0.87 V, the increase in Pt activity should be more significant. The Pt_3Y catalyst electrode exhibits the highest ORR activity tested in the following conditions applicable to any bulk polycrystalline metal surface. The activity of the Pt_3Y sample is closer to the activity of polycrystalline Pt_3Ni than that of single-crystal $Pt_3Ni(111)$, while the activity of polycrystalline Pt_3Ni is not so strong.

Surface modification is a way to improve the ORR activity of Pt catalysts. For example, Mo is obtained by further growing Mo on a pre-formed octahedral Pt_3Ni/C catalyst [14]. The octahedral Pt_3Ni/C and Mo-Pt_3Ni/C electrocatalysts showed similar ECSA, at 66.6 and 67.5 $m^2\ g^{-1}$ Pt, respectively, which is equivalent to the commercial Pt/C electrocatalyst (75.6 $m^2\ g^{-1}$ Pt). The authors calculated the kinetic current from the ORR polarization curves with considering mass transfer correction. To compare different catalysts' activities, the kinetic current was normalized relative to the ECSA and metal Pt loading. As shown in Figure 9.9, it is shown that the specific activity of octahedral Mo-Pt_3Ni/C at a voltage of 0.9 V is 10.3 $mA\ cm^{-2}$ compared to the RHE. In contrast, the specific activity for the undoped Pt_3Ni/C electrocatalyst is only ~2.7 $mA\ cm^{-2}$. According to the mass load of Pt, the mass activity of the Mo-Pt_3Ni/C catalyst is 6.98 $A\ mg^{-1}$ Pt. The specific activity for Mo-Pt_3Ni/C electrocatalyst is 81 times higher than that of the Pt/C catalyst. The mass activity for Mo-Pt_3Ni/C electrocatalyst has increased by a factor of 73. Mo-Pt_3Ni/C exhibits the best ORR performance, with a specific activity of 10.3 $mA\ cm^{-2}$ and a mass activity of 6.98 $A\ mg^{-1}$ Pt, which is comparable to the commercial Pt/C catalyst (0.127 $mA\ cm^{-2}$ and 0.096 $A\ mg^{-1}$ Pt). In the mechanism study, theoretical calculations show that Mo would exist on the subsurface position near the edge of the particles under vacuum conditions and prefers the surface

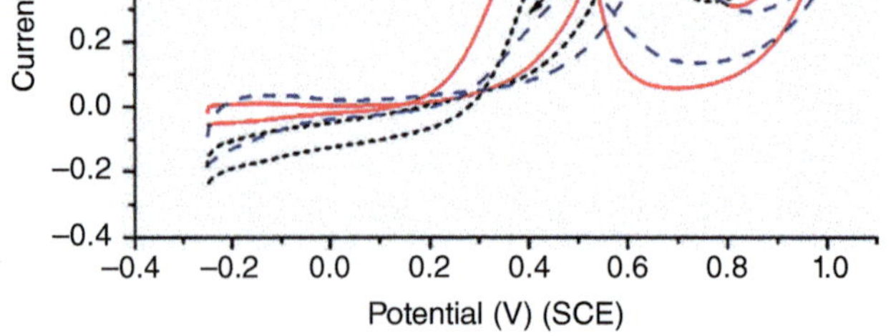

Figure 9.9 CV curves for the electro-oxidation of methanol by the as-prepared Pt–Au alloy trisoctahedra, Pt–Au alloy octahedra, and Pt black. The methanol oxidation was recorded in 0.1 M $HClO_4$ + 0.1 M CH_3OH solution at a scan rate of 50 $mV\ s^{-1}$ between −0.25 and 1.0 V. Source: Wang et al. [15]. © 2016, Royal Society of Chemistry.

apex/edge position under oxidation conditions. Therefore, Mo can both improve the performance of Pt_3Ni catalysts and improve its stability.

Core–shell structure Pt catalysts could achieve both the low-content Pt and high performance by electron modulation/strain control [14]. Pd@$Pt_{1.8}$Ni core–shell nanocrystals were synthesized using octahedral Pd nanocrystals as seeds and by controlling the overgrowth of Pd nanocubes according to previous work. In the comparison of Pt mass activity between Pd@Pt–Ni/C and Pt/C. It can be seen that at 0.9 V, the mass activity for Pd@Pt–Ni/C is 0.79 A mg^{-1} Pt, which is 4.9 times that of Pt/C (0.16 A mg^{-1} Pt).

9.2.1.2 Pd and Pt

The mass activity of Pd@Pt–Ni/C is 0.48 A mg^{-1} Pd + Pt. To reveal the ORR durability of Pd@Pt–Ni/C electrocatalyst, the authors applied a voltage of 1.05 V at a scan rate of 50 mV s^{-1} in a 0.1 M $HClO_4$ solution saturated with N_2 from 0.6 to 5. After 4000 cycles, Pd@Pt–Ni/C showed a little loss of only 6.7% in ECSA, while Pt/C showed a big loss of 22%. CV measurements showed that Pd@Pt–Ni/C decreased by only 10% in ECSA after 6000 cycles, while Pt/C dropped significantly by 27%, which indicates that Pd@Pt–Ni/C has a higher ratio than Pt/Ct–Ni/C. The improvement in catalytic performance was arisen from two factors: (i) the Pt and Ni alloy increase the Pt active site, which may weaken the oxygen-containing audience's combination of Pt, 26, and (ii) the two Interface interaction. Thin shells and Pd cores may change the electrons and geometry of shell atoms (Figure 9.10).

On the other hand, Pd@Pt core–shell concave decahedra were prepared by controlling the Pt deposition onto pre-formed Pd decahedral seeds [17]. Pt atoms are deposited initially onto the apex of decahedral seed crystal and then diffuse to other areas along the edge/ridge toward the surface, and then diffuse across the plane. Unlike the coating of Pd icosahedral seeds, even when deposited at 200 °C, Pt atoms would probably remain on the vertices and ridges/edges of decahedral seeds, which naturally form a core–shell coated with a concave surface structure. Compared to the commercial Pt/C catalysts, Pd@Pt concave decahedrons exhibit significant improvement in both catalytic activity and durability to ORR. For a concave decahedron with a weight percentage of 29.6%, its ratio (1.66 mA cm^{-2} catalyst 0.36 mA cm^{-2} Pt) as well as mass (1.60 A mg^{-1} Pt) ORR activity was increased by 4.4 and 6.6 times compared to Pt/C, respectively (0.32 A mg^{-1} Pt). After 10 000 cycle tests, the mass activity of the concave decahedron is still 0.69 A mg^{-1} Pt, which is over twice higher than that of the original Pt/C catalyst.

9.2.2 Transition Metal Catalysts

The commercialization of fuel cells depends on cost reduction, in which reducing or removing the use of platinum group metals (PGMs) in electrocatalysts is a particularly important path. Compared with anodes, cathodic oxygen reduction (ORR) has a higher kinetic barrier and requires more PGM catalysts. If PGM can be replaced with earth-rich elements through a simple synthetic route, the cost of fuel cells will be greatly reduced. Jasinski used cobalt phthalocyanine as

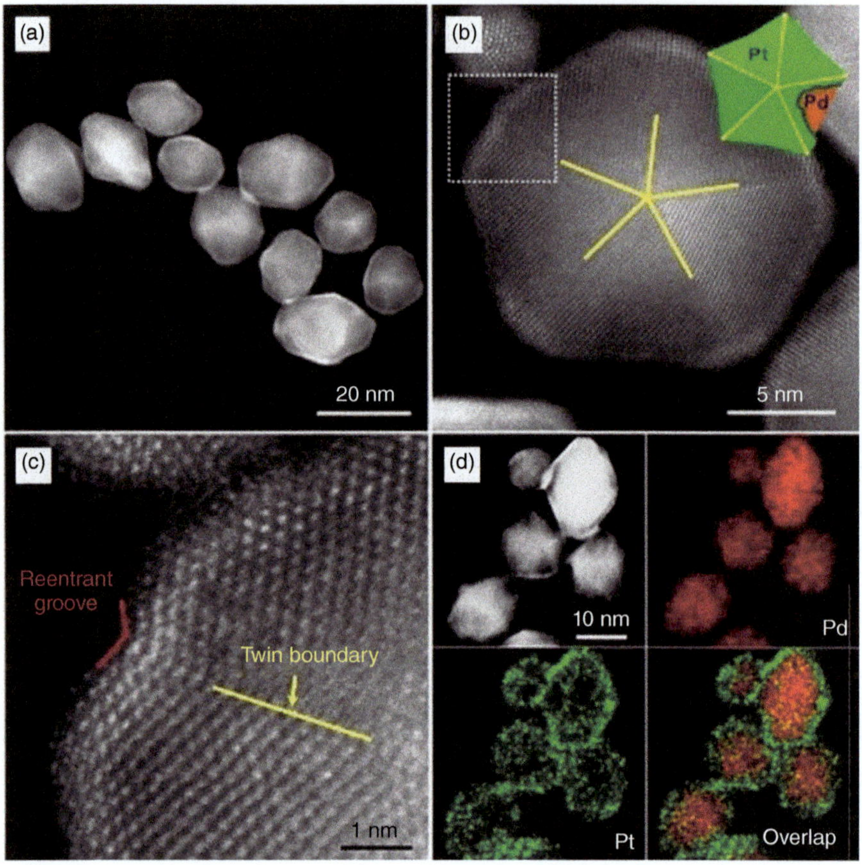

Figure 9.10 Structural and compositional analyses of the Pd@Pt core–shell concave decahedra with 29.6 wt% Pt. (a) High-angle annular dark field (HAADF)-scan transmission electron microscope (STEM) image. (b) Atomic-resolution HAADF-STEM image and a model of the decahedron viewed along its fivefold axis (inset). (c) Atomic-resolution HAADF-STEM image is taken from the corner marked by a box in (b), showing the presence of a re-entrant groove at a vertex and twin boundary along a ridge. (d) HAADF-STEM image of the concave decahedra and the corresponding energy dispersive X-ray spectroscopy (EDX) mapping of elemental Pd and Pt. Source: Wang et al. [16]. Reproduced with permission of American Chemical Society.

a precursor in 1964 and began to explore transition metal/nitrogen coordinated carbon (TM/N/C) as an ORR catalyst. Since then, research in this direction has attracted many followers. In particular, in recent years, Fe/N/C catalysts have received great attention owing to their promising catalytic activity.

Ideally, an effective ORR catalyst should have a high surface area and high-volume active site density that can be exposed through a high surface area without an inert carrier, such as the amorphous carbon used in traditional electrocatalysts. This method is exemplified using a metal–organic framework, such as a zeolite imidazolate framework (ZIF), as a precursor. Since the initial research on ZIF-based PGM-free ORR catalysts, the research on ZIF as a catalytic precursor has been rapidly developed and many encouraging results have been achieved, especially for complex iron or iron Fe/N/C catalyst synthesized from ZIF. As a precursor, ZIF has an adjustable structure, inherent porosity, high surface area, and diverse characteristics, so it is quite attractive [18].

Most of these electrocatalysts can only perform well in alkaline electrolytes; however, their activity/stability tends to decrease sharply in neutral/acidic media with the long-term operation. Therefore, to maximize electrocatalytic activity and also improve the durability under wide pH conditions, transition metals or compounds, and conductive heteroatoms (N, S, B, etc.), hybrid-doped nanocarbons have been developed. For example, Fe–N–C hybrid electrocatalysts have demonstrated to have high activity against ORR in alkaline/acid electrolytes. Similarly, the FeN_x configuration is uncovered as the essential active sites in Fe–N–C materials. However, FeN_x species are nearly undetectable in most metal oxide/carbon-based matrices, and these oxides also exhibit good activity for ORR. Keeping these issues in mind, it is needed to take action to uncover the active sites in new catalysts that have excellent ORR activity and durability in a wide pH media. In this way, general guidance can be offered to explore new feasible and advantageous catalysts [19]. Lots of metal oxides, especially metal oxides from Group IV and V, are chemically stable in acid solution and were proposed to be catalyst supports to replace carbons [20]. The problems associated with them are the lack of adsorption sites for oxygen species on the surface of metal oxides and also the low conductivity, resulting in extremely low ORR performance in their bulk form. To address these problems, surface modification, alloying, doping, and formation of highly dispersed nanoparticles are suggested.

Among various non-noble metal catalysts for ORR, Fe-based catalysts have received extensively studied [20–22]. For example, M–NCs (nitrogen carbon) catalysts (including Fe–NCs) are excellent electrocatalysts for ORR, which can be prepared by the pyrolytic reaction [23]. To demonstrate the potential application of M–NC SAC (single atom catalysts), the electrocatalytic ORR performance of Fe–NC SAC was evaluated and further compared to the commercial Pt/C catalyst and also other control samples (e.g. Fe@C–Glu, Fe@C–N, and C–N–Glu). As shown in Figure 9.11, the latest commercial Pt/C (20 wt%) shows excellent ORR activity with respect to the initial potential (as defined by a potential reached by the current density) at 0.1 mA cm^{-2} is 0.96 V and the half-wave potential reaches 0.85 V. Although the current Fe–NC SAC clearly shows a positive initial shift potential (0.98 V) and a half-wave potential (0.90 V), they are higher than Pt. The positive polarity of the C catalyst is 20 and 50 mV and is better than most non-previous metal ORR catalysts. The mass activity for Fe–NC-SAC at 0.90 V is 9.0 A g^{-1}. In addition, Fe–NC SAC exhibits a mass current as high as 100.7 A g^{-1} at 0.90 V, 65 A g^{-1} greater than that of Pt/C, which is 35.7 A g^{-1}. In contrast, the sample of Fe@C–N (without Glucose preparation), Fe@C–Glu (without melamine), and C–N–Glu (without iron) showed relatively poor ORR catalytic activity. The excellent ORR electrocatalysis activity of Fe–NC SAC is further confirmed by the lowest Tafel slope (48 relative to 69 mV dec^{-1} for Pt/C; Figure 9.11) and the least H_2O_2 yield (<3.5% at the entire potential range; Figure 9.11). These results suggest that the ORR process on Fe–NC SAC follows a 4-e reaction path and has high catalytic efficiency.

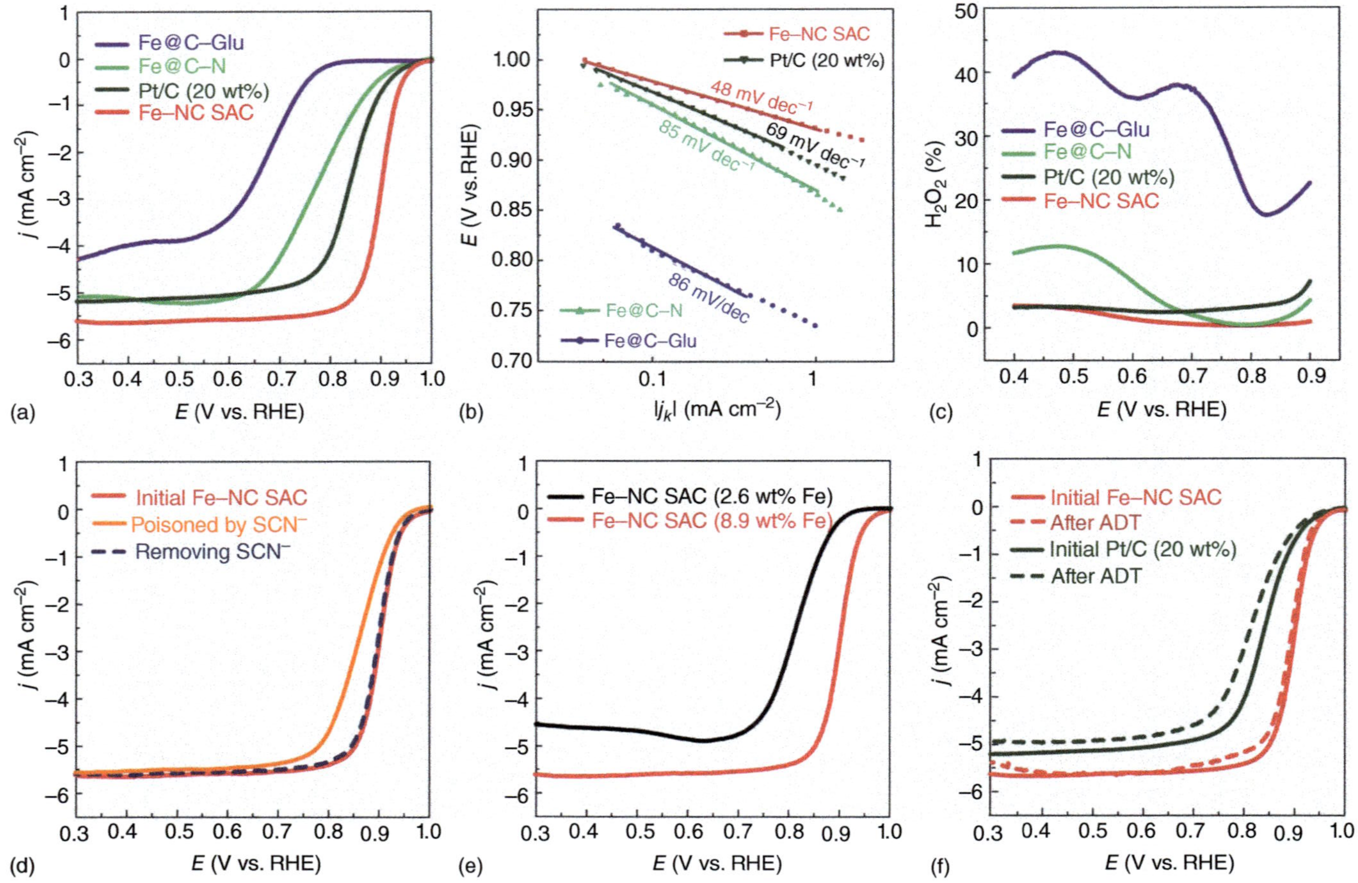

Figure 9.11 Evaluation of the electrocatalytic performance of Fe–NC SAC for ORR. (a) Steady-state ORR polarization curves of Fe–NC SAC, Pt/C, and control samples (Fe@C–N and Fe@C–Glu). (b) Corresponding Tafel plots. (c) Hydrogen peroxide yield. (d) Steady-state ORR polarization curves of Fe–NC SAC recorded in O_2 saturated 0.1 M KOH with or without poisoning by 0.01 M SCN^-, and one collected after removing SCN^-. (e) Steady-state ORR polarization curves of Fe–NC SAC with different Fe loading. (f) Steady-state ORR polarization curves of Fe–NC SAC and Pt/C before and after 5000 potential scanning cycles in O_2 saturated 0.1 M KOH. Source: Zhao et al. [23]. Licensed under CC BY 4.0.

Metal-based two-dimensional conjugated aromatic networks involving metal phthalocyanines (CAN-Pc(M), M = Fe, Co, and Fe/Co) were prepared [23]. First, transfer the 0.2 g of polyphthalocyanine and 15 ml of H_2O to a zirconia capsule containing a zirconia ball with a diameter of 0.5 mm. The vessel was fixed in a planetary ball mill and stirred for one hour. The resulting product was centrifuged to remove unstripped samples, collected carefully, and further dispersed in ethanol saturated with N_2. Determination of the concentration of the as-prepared CAN-Pc(M) solution was conducted via evaporation of a certain amount of the supernatant, which is calculated to be 0.5 mg ml^{-1}. In the performance test, constant current discharge curves at different currents clearly showed that the performance of zinc-air batteries with CAN-Pc (Fe/Co) is better than that of commercial Pt/C, especially at the high currents 100 mA cm^{-2}. When standardized by the weight of the zinc electrode consumed, the specific capacity of the zinc-air battery using CANPc (Fe/Co) is 717, 650, 610, and 559 mAh g^{-1} Zn and the current densities were 5, 20, 50, 100, and 150 mA cm^{-2}, respectively, with an energy density of 605 and 391 Wh kg^{-1} Zn. The authors also summarized the good ORR performance and Zn-air battery performance of CAN-Pc (Fe/Co) and highlighted its superiority over most previously reported non-noble metal catalysts. The good performance of CAN-Pc (Fe/Co) was due to the synergy, which is related to the abundant exposed active sites and effective mass transfer via the ultra-thin catalytic layers. In addition, mechanically charged batteries have operated for lots of cycles (Figure 9.12).

Fe-doped ZIF nanocrystal was prepared by partial replacement of Fe ions with Zn toward chemical bonding to the imidazolate ligand in the form of Fe–N_4 moieties [24]. It is different from the physical adsorption of iron ion. One-step activation can specifically generate atomically dispersed iron sites in porous carbons. The best ORR activity is tested on the catalyst with a 50 nm size. Further decreasing size to 20 nm will result in apparent particle agglomeration, thereby reducing activity. Employing a homogeneous Fe atom catalyst model, it clarified the processes of active site generation by correlating the tested ORR performance with changes in chemical bonds in the precursor by the thermal activation. The critical temperature at which the active site is formed is 800 °C, which is related to the novel Fe components, and its reduction–oxidation value (from Fe^{3+} to Fe^{2+}) may combine with FeN_4, which is embedded in the carbon plane. The further increasing temperature causes continuous enhancement of activity, which is related to the increase of graphite Fe–N and N species. The novel atomic iron catalyst obtained considerable ORR activity in an acidic medium (0.5 M H_2SO_4). Compared to RHE, its half-wave potential is 0.85 V and the gap with Pt/C (60 $\mu g_{Pt}/cm^2$) is 30 mV. The same catalyst can be used to obtain higher stability. The catalyst loses only 0.02 V after 10 000 potential cycles from 0.6 to 1.0 V in the acid saturated with O_2.

9.2.3 Metal-Free Catalysts

Recent studies have indicated that nitrogen-doped carbon materials (e.g. carbon nanotubes and graphene) can be used as a low-cost, efficient, and metal-free alternative to Pt for ORR. N-doped carbon materials co-doped with a second

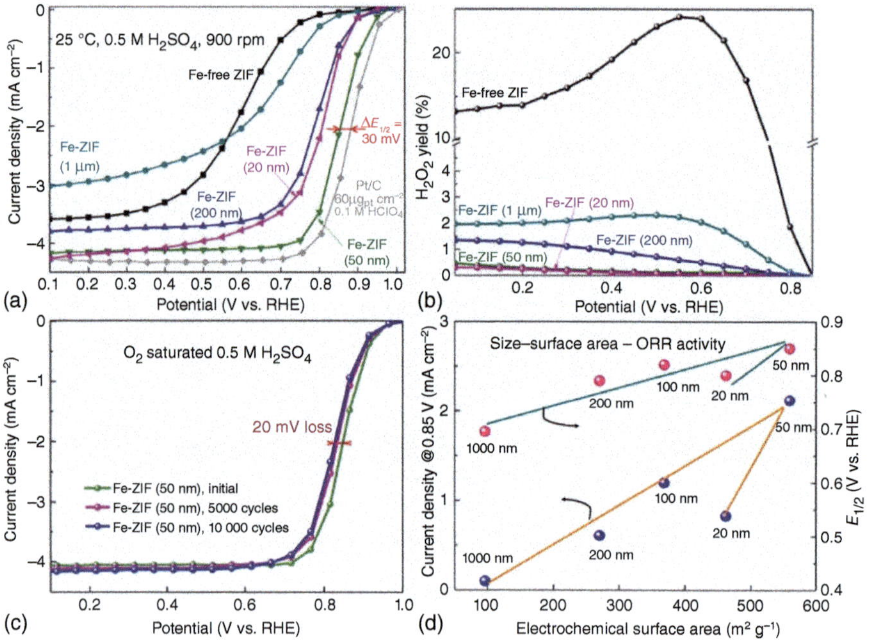

Figure 9.12 (a) ORR polarization plots for Fe-ZIF-derived catalysts in 0.5 M H_2SO_4 and Pt/C catalysts (60 μg_{Pt} cm^{-2}) in 0.1 M $HClO_4$ at 25 °C and 900 rpm. (b) Calculated H_2O_2 yield for Fe-ZIF catalysts as a function of particle size. (c) Stability AST by cycling the potential (0.6–1.0 V in O_2-saturated 0.5 M H_2SO_4). (d) Correlation between ORR activity and SA indicating a clear size dependence. Source: Zhang et al. [24]. © 2017, American Chemical Society.

element (such as S, B, or P) can adjust the electronic structure properties and surface polarity to further improve ORR performance. It has also been indicated to introduce heteroatoms, including PGMs, into carbon supports to improve the ORR performance of catalysts. When a heteroatom is bonded to a C skeleton, due to the difference in bond length and atomic size, it will bring about some defects at nearby sites, leading to uneven charge distribution. Additionally, since heteroatoms are normally covalently bonded within the C skeleton, they will not fade even after long-term working. Actually, N-doped C materials have demonstrated better durability even than commercial Pt/VC catalysts. These excellent properties encourage researchers to develop ways to combine the other heteroatoms (such as B, P, and S) into C materials. Particularly, Peng and coworkers reported that the first batch of P-doped graphite layers and multiwall carbon nanotube (MCNT) had a high electrocatalytic performance [25].

It is found that the vertical alignment of the iron phthalocyanine, which is an N-containing metal heterocyclic molecule, can generate N-doped carbon nanotubes (vertically aligned N-doped carbon nanotube [VA-NCNT]) with or without NH_3 vapors as an effective ORR catalyst [26]. In the performance test, the half-wave potential of the ORR on the nitrogen-free nonaligned carbon nanotube (NANCNT)/glass carbon electrode (GC) and VA-NCNT/GC electrode is

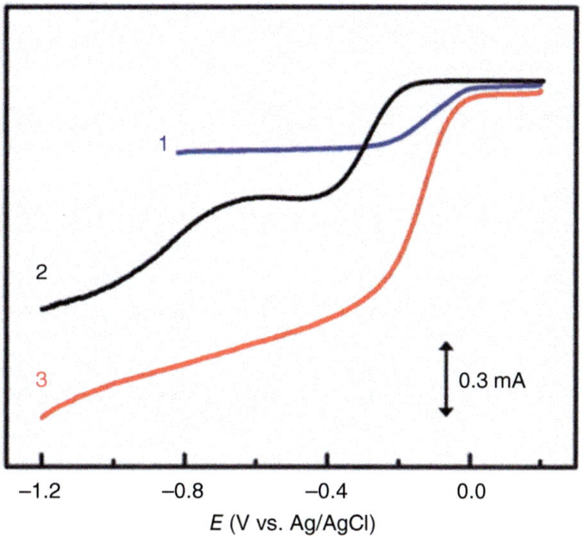

Figure 9.13 RRDE voltammograms for oxygen reduction in air-saturated 0.1 M KOH at the Pt-C/GC (curve 1), vertically aligned nitrogen-free nonaligned carbon nanotubes (VA-CCNT)/GC (curve 2), and VA-NCNT (curve 3) electrodes. Because of the technical difficulties associated with the sample mounting, amperemeter responses with the Pt ring electrode were not measured for the vertically aligned carbon nanotubes. Source: Yan et al. [27]. © 2015, Royal Society of Chemistry.

comparable to the half-wave potential on the Pt–C/GC electrode (−0.1 V), but relative to the Pt–C/GC electrode, a significantly enhanced steady-state diffusion current (~0.8 mA) was observed in the larger potential range of the VA-NCNT/GC electrode. Therefore, in alkaline solutions, the VA-NCNT/GC electrode is much better than ORR's Pt–C/GC electrode. To study the stability of the VA-NCNT/GC electrode to ORR, +100 000 cycles for VA-NCNT/GC electrode compared to Pt–C/GC electrode in 0.1 M KOH saturated with were performed and the voltage ranges in between 0.2 and −1.2 V. As can be seen from Figure 9.13, in addition to ORR, commercial Pt/C has clear deteriorated on, while VA-NCNT/GC electrode kept relatively stable.

Mesoporous N-doped carbons containing trace amounts of Fe (Fe–N/C) were prepared by a heat-removable nanoparticle template process [28]. Subsequent heat treatment at high temperature resulted in the evaporation and decomposition of FeO(OH) nanocrystals and the generation of mesoporous carbon, the surface area of which significantly increased to 934.8 $m^2\ g^{-1}$. The resulting Fe–N/C samples showed significant catalytic activity for ORR. Figure 9.14 displays the RDE voltammograms of four samples synthesized at different temperatures during pyrolysis. It can be seen that for the Fe–N/C-600 sample, when scanning the electrode potential negatively to reach +0.68 V (vs. RHE), a non-zero cathode current begins to appear. However, the initial potential is more positive at +0.87 V for Fe–N/C-700, while for Fe–N/C-800 and Fe–N/C-900, the initial potential is at +0.98 V and Time correction 0.96 V. At the same time, the limit current also showed a significant change between samples. For example, at +0.45 V, the limiting current of Fe–N/C-600 is 0.86 $mA\ cm^{-2}$, the limiting current of Fe–N/C-700 is 1.06 $mA\ cm^{-2}$, and it is 4.81 $mA\ cm^{-2}$ for Fe–N/C-800, is 2.92 $mA\ cm^{-2}$ for Fe–N/C-900, which indicates that the Fe–N/C-800 sample is the best ORR catalyst in this series. Consistent behavior was observed in cyclic voltammetry and rotating ring-disk

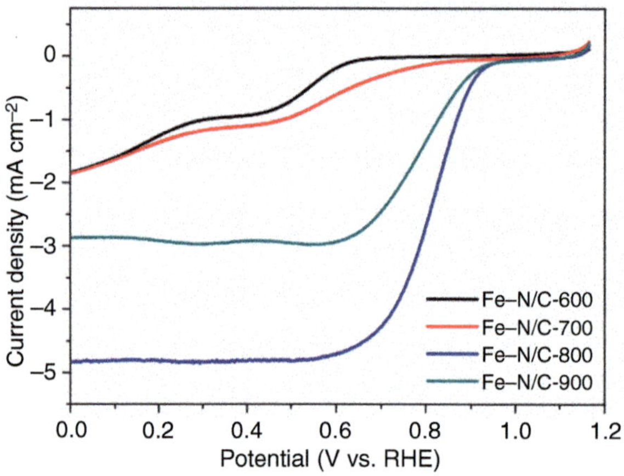

Figure 9.14 Rotating disk electrode (RDE) voltammograms of a glassy carbon electrode modified with 79.6 μg cm^{-2} of Fe–N/C-*T* (*T* = 600, 700, 800, and 900) in an O_2-saturated 0.1 M KOH solution at a rotation speed of 1600 rpm. Potential sweep rate 10 mV s^{-1}. Source: Niu et al. [28]. © 2015, American Chemical Society.

electrode (RRDE) measurements. Interestingly, this coincides with the highest graphite N concentration and surface area of the Fe–N/C-800 catalyst, indicating that graphite nitrogen may play a leading role in determining the ORR activity.

For the preparation of 3D mesoporous C foams doped with N and P [29], a one-step method, which involves the process of pyrolyzing polyaniline aerogel with the existence of phytic acid, was used. Aniline monomers are polymerized in the presence of phytic acid, and PANi hydrogel process [28] is produced by template-free hard gelation. A scalable and low-cost method to prepare 3D mesopores C doped with N and P (mesopores carbon doped with N and P [NPMC]) was developed. The linear scanning voltammogram (LSV) curve in Figure 9.15 confirms the electrocatalytic activity of NPMC-1000. The positive initial potential of the RHE is 0.94 V, and the half-wave potential of the RHE is 0.85 V. These obtained values are comparable with the values of Pt/C and are superior to most metal-free ORR catalysts reported previously, and even better than recently reported metal-containing C-based catalysts. In addition, the NPMC-1000 electrode had a much larger limiting current than that of the NPMC-900 and NPMC-1100 electrodes and is equivalent to the limiting current of Pt/C (Figure 9.15a).

Zigzag-edged graphene nanoribbons with a carbon nanotube backbone [30] (GNR@CNT) with zigzag on carbon nanotube (CNT) were synthesized and used as a cathode catalyst in H_2/O_2 PEMFC. The CNT's main chain in the middle of graphene nanoribbons (GNR) is combined with the carbon black spacer, effectively exposing the zigzag carbon of ORR. Although still lower than Pt/C (20 wt% Pt) in the acidic electrolyte, GNR@CNT exhibits considerable performance at an initial potential of 0.760 V and a half-wave potential of 0.63 V. It is worth noting that under acidic conditions, the initial potential and limiting current of GNR@CNT are higher than that of (nitrogen doped graphene nanoribbons) N-GNR@CNT, which is contrary to the trend in alkaline media. This phenomenon can be explained by the protonation of pyridine nitrogen of N-GNR@CNT in an acidic solution. This study demonstrates defective graphitic carbon to be promising for use in the PEMFC application.

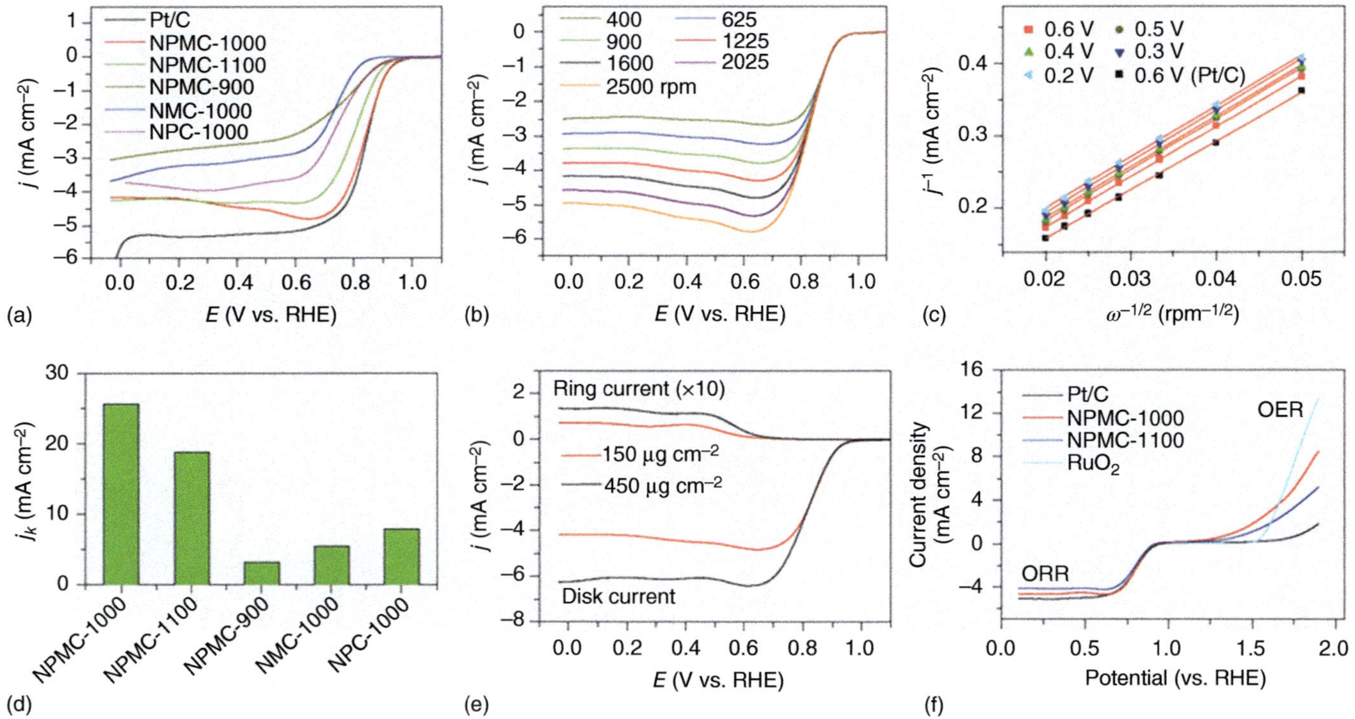

Figure 9.15 Electrocatalytic activity for ORR and oxygen evolution reaction (OER). (a) Linear scan voltammogram (LSV) curves for NPMC-900, NPMC-1000, NPMC-1100, (mesopores carbon doped with N) NMC-1000, (mesopores carbon doped with P) NPC-1000, and commercial Pt/C catalyst at an RDE (1600 rpm) in O_2-saturated 0.1 M KOH solution. Scan rate, 5 mV s^{-1}. (b) LSV curves of NPMC-1000 in oxygen-saturated 0.1 M KOH at various rotating speeds. (c) K–L plots for NPMC-1000 and Pt/C at various potentials. (d) The kinetic current of various samples for O_2 reduction at 0.65 V. (e) RRDE measurements (1600 rpm.) of ORR at an NPMC-1000 electrode with different catalyst loadings. (f) LSV curves of NPMC-1000, NPMC-1100, RuO_2, and commercial Pt/C catalyst on an RDE (1600 rpm) in 0.1 M KOH (scan rate, 5 mV s^{-1}), showing the electrocatalytic activities toward both ORR and OER. Source: Zhang et al. [29]. © 2015, Springer Nature.

9.3 Hydrogen Peroxide Synthesis

Hydrogen peroxide (H_2O_2) is a high value-added chemical that has been widely used, including chemical, military, medical, and environmental protection [31, 32]. It is evaluated that the quantity demanded of H_2O_2 will achieve 6000 kton in 2024 [33]. At present, the main method of industrial-scale production is anthraquinone oxidation. A high degree of automation is conducive to large-scale production. However, inevitable side reactions and complicated purification processes are urgent problems that need to be solved effectively [31]. As a simple and environmentally friendly method, the two-electron ORR at mild temperature and atmospheric pressure is a promising alternative method; it can also be produced on-demand and avoid the required H_2O_2 transportation and storage issues [34].

9.3.1 Catalysts Advances

The first example of H_2O_2 electrosynthesis published to date referred to the use of a Hg–Au electrode for ORR but only produced a few milligrams of H_2O_2 [35]. Significant progress has been made employing activated carbon electrodes with highly porous structures, greatly increasing the surface area, thereby increasing overall productivity [36]. The porous carbon cathodes were formed. As the field advanced, the substrate for electrocatalysts over many years [37, 38], electrode materials have become more complex, transitioning to doping with carbon, precious metal oxides, and alloys. The development of calculation methods for predicting the activity of H_2O_2 electrosynthesis materials promotes the further improvement of catalyst materials because the experimental paper continues to describe the materials predicted by density functional theory (DFT) calculations, which are closer to the peak of the Sabatier volcano map [39, 40].

For all electrocatalysts, an inverse relationship exists between current density and selectivity to H_2O_2. Some materials provide excellent selectivity to H_2O_2 production at lower overpotentials and lower overall rates, but selectivity deteriorates at higher overall rates. If the reaction is carried out at a low overpotential, even the simplest materials could show excellent H_2O_2 formation selectivity. However, even if the current density is about hundreds of milliamperes per square centimeter, the catalyst used in practical applications needs to maintain its selectivity. When comparing catalysts, it is important to take the consideration of the excessively high potential at which the given H_2O_2 selectivity is reported and the material selectivity should be determined within a certain range of overpotentials. Table 9.1 lists some of the main materials used for H_2O_2 electrosynthesis and the experimental conditions used to measure its selectivity. Now, we discuss some of these catalysts and emphasize their highly selective mechanism origin.

9.3.1.1 Pure Metals

Analyzing the H_2O_2 electrosynthesis activity of materials has become very common because the application for fuel cells requires ORR electrocatalysts to generate

Table 9.1 Electrode materials for the electrosynthesis of H_2O_2.

Material	Faradic efficiency (FE) (%)	*n*	Anodic/cathodic	Method	Solution	pH	*E* vs. RHE	*j* (mA cm^{-2})
Carbon nanotubes	90[a)]	1.8	Cathodic	RRDE	0.1 M KOH	13[b)]	—	—
Mesoporous carbon	95	2.1[a)]	Cathodic	GDE	0.5 M NaOH	13.7[b)]	0.43	−150
Doped carbon								
N–C	95	2.1	Cathodic	RRDE	0.1 M H_2SO_4	0.3	0.1	−1.2[c)]
Born-doped diamond	93	2.9[a)]	Cathodic	CA	0.5 M H_2SO_4	1[b)]	−1.7[c)]	—
Co(II) phthalocyanine	81.5	2.3	Cathodic	GDE	0.1 M K_2SO_4	2[b)]	0.23	−0.02[c)]
Fe(II) phthalocyanine	78.2	2.4	Cathodic	GDE	0.1 M K_2SO_4	2[b)]	−0.18	−0.01[c)]
Pt/S–C	96	2.1	Cathodic	RRDE	0.1 M $HClO_4$	1.3[b)]	0.1[c)]	−0.05[c)]
AuPd/C	95	2.1[a)]	Cathodic	RRDE	0.1 M $HClO_4$	1[b)]	0	−0.8[c)]
Sn_6Ni/C	88	2.2	Cathodic	RRDE	1 M NaOH	14[b)]	0.7[d)]	−0.34[d)]
CeO_2/C	95	2.1[a)]	Cathodic	RRDE	1 M NaOH	14[b)]	0.84	−0.15[c)]
WO_3/C	84	2.3[a)]	Cathodic	GDE	0.1 M K_2SO_4	0.3[b)]	−1.08	—
$BiVO_4$	95	2.1[a)]	Anodic	LSV	1 M $NaHCO_3$	8.3	2.8[c)]	15[c)]
$CaSnO_3$	76	2.5[a)]	Anodic	CA	2 M $NaHCO_3$	8.3	3.2	34[c)]
$BiVO_4/WO_3/Al_2O_3$	80	2.4[a)]	Photo-oxidative	PVC	2 M $NaHCO_3$	7.9	—	1
IrO_2/Ta_2O_5	79.3	2.4[a)]	Cathodic	CA	0.1 M Na_2SO_4	7	—	0.001

Note: The methods and apparatus used for evaluating these methods involve a rotating ring-disk electrode (RRDE), a gas diffusion electrode (GDE), linear sweep voltammetry (LSV), chronoamperometry (CA), and photovoltaic cell (PVC). When multiple compositions of doped/alloyed catalysts were focused, the best value recorded is shown. *E*, *j*, potential; *n*, number of electrons transferred; current density; RHE, reversible hydrogen electrode.

a) Value calculated assuming efficiency = $(2 - n/2) \times 100\%$ [41].
b) pH was not explicitly reported but approximated here on the basis of the electrolyte used.
c) Demonstrative potentials and current densities were read off steady-state voltammograms.
d) Demonstrative current densities and potentials are determined from a Koutecký–Levich plot.

minimal H_2O_2 even at a low overpotential. In fact, any H_2O_2 formed can be easily detected by either scanning electrochemical microscopy [42–46] or at the ring on the RRDE [47–51], or induced from a reduction in the number of transferred electrons (n) per O_2 substrate from an ideal ORR value (4) [52–56]. Therefore, ORR electrocatalysts that realize incomplete reduction would be less useful for fuel cells but may be greatly useful for H_2O_2 electrosynthesis, and the failed ORR catalysts could possibly be suitable for H_2O_2 generation. The selectivity for the generation of H_2O_2 on pure metal surfaces relies largely on the given potential. For ORR, scanning electrochemical microscopy showed that Ag, Cu, Hg, Au, Pt, and Pd generate part of H_2O_2 at relatively low overpotential. However, at all potentials, as studied, only Hg could keep high selectivity for production H_2O_2 [47].

Although it is encouraging that all catalysts can generate H_2O_2 while maintaining an ideal overpotential, the low overpotential requirement may lead to inherently low current density. Alternatively, the given catalyst can be adjusted by increasing the mass transfer rate to and from the electrode surface and/or from the electrode surface to facilitate the production of H_2O_2. In this case, H_2O_2 can quickly diffuse from the electrode before the further reaction can proceed. Increasing the mass transport rate was reached using microelectrodes [52, 53], RDEs [52, 55], nanoparticles [56, 57], and flow cells [58] by changing physical parameters including the particle shape/size, catalyst loading, rotation rate, electrode radius, or flow rate. The effect by employing small catalyst particles is not only physical; the smaller particles are beneficial in providing an end-to-end O_2 binding mode for H_2O_2. In fact, the size of Pt clusters on the inert carrier of indium tin oxide is inversely proportional to their selectivity for generating H_2O_2 from O_2 [59]. Although the increasing H_2O_2 production is considerable under large-scale transportation conditions, this material cannot provide 100% H_2O_2 selectivity ($n = 2$), so most of the work focuses on more practical solutions, e.g. alloys and composites.

9.3.1.2 Metal Alloys

A few of the best catalysts for electrosynthesis of H_2O_2 from O_2 are mostly derived from metal alloys, e.g. W–Au [60], Pd–Au [61, 62], and Pt–Hg [39, 63]. These electrocatalysts are usually designed by calculations that have already predicted the optimized strength toward O_2 adsorption [40, 64]. Theoretical studies have predicted correctly the activity of evenly dispersed and core–shell alloy [60] catalysts. The two alloyed metals need not be active separately for the electrosynthesis of H_2O_2. Indeed, alloying between active metal and another metal, which is relatively inert under the given potential, can yield discrete reaction sites and is coated by a relatively inert material [65]. For instance, in a Pd–Au alloy, the Pd site performs an initial electron transfer from $2e^-$ to O_2; while in the Pd–Au alloy, Pd transfers from $2e^-$ to O_2. Au can only bind O_2 weakly and cannot crack O—O bonds alone, but the reaction rate is too slow to be practically used without alloying Pd [66]. The reactivity of the alloy catalyst depends on the spacing/distance between the metal atoms, so fine modulating the ratio between inert metal and active metal can provide a volcanic type relationship between H_2O_2 selectivity and component.

In addition to Pd–Au, calculation studies showed that when Pd or Pt is alloyed with various metals, the provided catalyst has a high selectivity for H_2O_2, and $PtSn_2$ is close to the peak of theoretical volcanic activity. Unfortunately, the potential for dissolution of these many metals is more negative compared to the reduction potential for ORR, so operation conditions will cause them to dissolve from the alloy, rendering the formation of a Pd or Pt shell. Considering the stability, the best candidate is $PtHg_4$. The same concept of metal active sites on a relatively inert substrate was employed in developing single-atom catalysts [67], which feature an active metal single atom, like Pt isolated on (or in) inactive supports like TiN or S-doped C [68]. The Pt1 active site is more conducive to O_2 terminal binding than other binding methods, so the selectivity of H_2O_2 generation can reach 95%.

The fact of optimizing selectivity usually has an adverse effect on the reaction rate must never be ignored. For Pd_xAu_{1-x}, a lower value of x would benefit H_2O_2 selectivity, but the current density is lower. An ideal electrode catalyst will be a compromise-diluting active material, which will increase selectivity, yet will reduce the overall reaction rate. The current of electrocatalytic activity supposes that the system keeps its integrity at the operating conditions. Some catalysts are only stable in kinetics, and when PdAu nanoparticles suffered from oxidation or reduction potentials, partial separation of phase may occur, providing Au- or Pd-rich shells, respectively. These novel catalysts have different selectivity for the production of H_2O_2 [69], suggesting the requirement to not only measure activity but also the robustness of electrocatalysts at the operating conditions.

9.3.1.3 Carbon Materials

Metal alloys have encouraging electrocatalytic activity but are often consisted of precious metals that are not cost-effective and have limited availability. This stimulated the exploiting of carbon-based materials to serve as abundant and cheap electrocatalysts. The starting point for these materials can be one of various carbon allotropes, such as graphite [70–72], graphene [73, 74], and carbon nanotubes [26, 75, 76] as well as porous [41, 77, 78] and amorphous [79–81] carbon. Among these candidates, porous carbon materials possess large surface areas and thus facilitate mass transport efficiently and also favor the production of H_2O_2 by providing abundant defect sites compared to the uniform materials [82]. According to DFT calculations, some defect structures provide a material that reduces volcanic peaks by 2e-O_2, highlighting the key role as the active sites. Generally, the overall coordination saturated carbon sites have less reactivity compared to these defect sites. However, the optimal structure is not apparent; this is because of the catalyst with smaller pores to have a higher defect density. Since any H_2O_2 will not diffuse easily away from the electrode before further reaction, it may also lead to H_2O_2 yield reduce [83].

Besides changing the carbon electrode material structure, the performance of the H_2O_2 electrosynthesis catalyst can also be improved by adjusting the surface or body composition. For example, the current of an H_2O_2-generating carbon catalyst could be enhanced via oxidizing surface anodically [84, 85] or thermally [71] to offer more active surfaces decorated with OH and/or O groups; in the case of carbon, annealing can be used to significantly increase the selectivity to H_2O_2 formation to provide more hydrophobic surfaces, which facilitates the O_2 diffusion to electrode and assists to maintain a fast reaction rate. Phase boundary. At the gas diffusion electrode (GDE) [86], as mentioned earlier, it is expected that carbon oxide materials with multiple $-CO_2H$ and –C–OC–C– positions will have the best hydrogen peroxide activity [87].

The activity of carbon-based materials depends more on the effects of dopants than the microstructure and/or defects[74, 88]. Lots of dopants were studied, including boron [82, 89], nitrogen [90, 91], phosphorus [73, 74], sulfur [74, 92], fluorine [93], and numerous transition metals [94–97]. When doped with N atoms, both mesoporous carbon and graphite can display high selectivity for the production of H_2O_2 [98, 99]. For instance, N–C frameworks containing viologen and triazine groups have lots of redox-active sites. Even during the eight-hour electrolysis process, they can accept electrons and then transfer these electrons to the O_2 molecule to produce selectively H_2O_2 [100]. An exception has been observed for carbon nanotubes, where carbon doping leads to an increase in the apparent number of the transferred electrons ($n = 1.8–3.9$) [26].

Most dopants of transition metals are not suitable for the H_2O_2 electrosynthesis, and the selectivity for H_2O_2 on metal-doped carbon catalysts will reduce along with the increase of dopant concentration [95, 97]. Similarly, carbons with lower content of N and Fe–N dopant doping give greater selectivity [101]; an intuitive result is that higher concentrations of active sites favor the $4e^-$ ORR. One can acquire catalysts with selectivity by fixing [M(porphyrinato)] or [M(phthalocyaninato)] complexes onto carbon supports [102–104], but these catalysts would suffer from degradation with the existence of H_2O_2, causing them unstable during long-term use [97, 105].

9.3.1.4 Electrodes and Reaction Cells

Complementing the study on the development of catalysts with desired intrinsic properties, it is an important pursuit to design catalysts and other battery components into highly efficient devices. The desired electrode structures should have abundant adsorption sites, allowing H_2O_2 molecules to desorb quickly before subsequent reactions occur. When the interaction between metal and H_2O_2 is weak, the desorption of H_2O_2 is not only favored but also under the conditions of rapid mass transfer. The high surface area and rapid mass transfer will result in high current density, which is particularly ideal when promoting electrochemical cells from laboratory to industrial scale.

Some different designs on cell and electrode solve the same requirements with different approaches. A common choice for H_2O_2 electrosynthesis from O_2 is the GDE [86, 106], which is consisted of porous layers with hydrophobic components, e.g. impregnating carbon paper with the hydrophobic polymer. The electrode serves as a membrane in-between the liquid electrolyte and O_2. The catalytic material is deposited on GDE to form a solid–liquid interface, so when O_2 flows across the membrane, it could be reduced immediately after dissolution. Many studies related to this field have focused on the catalyst natures, and metal nanoparticles are the most common candidate. It should pay attention not to load too much electrocatalyst in the GDE, as this will hinder the gas flow physically and further reduce the current efficiency. GDE has many superiorities over ordinary electrodes because the gas phase facilitates mass transfer to be faster than the solution promotion, and researchers do not require to know the solubility limit of O_2 in the electrolyte.

It has been described how GDE fits for H_2O_2 electrosynthesis because of the ease of handling gaseous O_2. When people consider using anode H_2O_2 for electrosynthesis, the use of liquid flow cells may still have a similar effect (Figure 9.16). It could collect products in a container or circulate the intermediate so that it can be further achieved in the same volume of electrolyte to obtain a high-content product as desired. When an arrangement such as a flow system above a parallel plate electrode or a flow system using a mesh or channel electrode is used to change the solution flow rate, many different reaction environments are established [108]. These flow cells are very useful for the early development of new materials because they show a well-defined mass transfer method on the basis of flow velocity and geometry, so the establishment of the reaction is relatively simple and can be verified by using element analysis.

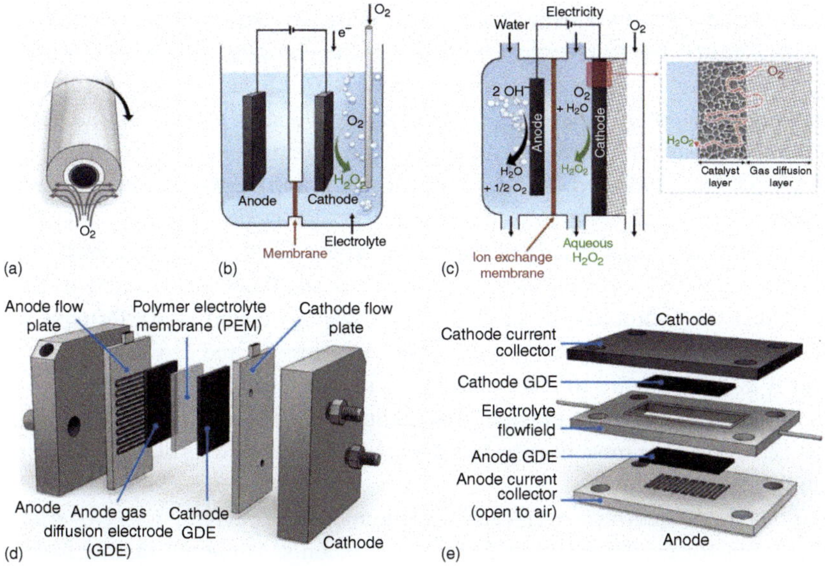

Figure 9.16 Schematic illustrations of different types of electrochemical cell for H_2O_2 production [107]. (a) RRDE setup to quantitatively measure the production of H_2O_2 from oxygen reduction. (b) Schematic of an H-cell configuration with a catalyst deposited on a solid substrate in a non-flowing electrolyte. (c) Schematic of a continuous flow cell with a catalyst deposited on a GDE. (d) Membrane-based flow reactor containing a MEA consisting of the anode and cathode electrodes on either side of a membrane. The MEA is assembled in between the anode and cathode current collectors, which have flow field patterns, and separates the respective OER and 2e$^-$ ORR for H_2O_2 production. (e) Microfluidic reactor consisting of a flow channel with liquid electrolytes. Source: Jung et al. [107]. © 2020, American Chemical Society.

Designs on other reactors were developed to increase the total current by maximizing active area rather than mass transfer. Highly porous electrodes (e.g. carbon felt or foam) can be used to provide a large surface area on which electrochemical reactions may occur, which can be achieved [109]. As scaled up, the concept adapts the form of a trickle bed reactor, where the electrode has placed the bed of catalyst nanoparticles in the chamber [110]. The electrolyte is fixed to the top and then dropped under gravity so that the waste water could be collected on the bottom. Combining a large surface area with a low flow rate can provide concentrated H_2O_2 electrolyte [111–114].

Another reactor of choice is the PEMFC [104, 115]. In the device, the cathode and anode are located in two compartments divided by the H^+ conductive polymer membrane. These batteries would be effective for the H_2O_2 electrosynthesis because of the separation of the two electrodes, which suggests that the H_2O_2 would be formed on the anode and cannot be degraded at the cathode, and vice versa. The electrosynthesis in fuel cells offers the possibility for collecting electrical energy as well as useful products at the same time, which has obvious energy advantages compared to the traditional synthesis that only consumes energy [116].

If the compartments of the electrochemical cell are separated, it is probable to generate some useful products at the anode and cathode at the same time and there will be no interaction between the products. Perhaps, the most famous example of the divergent paired electrochemical processes is the chloralkali processes, which produce Cl_2 and NaOH by oxidization and reduction of brine. A lot of examples of divergent processes are presented, such as the conversion of 4-(*t*-butyl)toluene into phthalide, methyl phthalate and 4-(*t*-butyl)benzaldehyde dimethylacetal [117], dienes and CO_2 into diols and diacids [118], L-cystine into L-cysteine and L-cysteic acid [119] and O_2 and H_2O into H_2O_2 and oxidants such as O_3 [120] or HO· [121]. An alternative way is the convergent paired electrochemical processes, where the same product can be produced in both compartments. This was applied to the electrosynthesis of hydroquinone [122, 123] and glyoxylic acid [124]. Of course, the cathode and anode H_2O_2 can also be produced from O_2 and H_2O at the same time, and it is encouraging to have obtained a patent for a complete electrolytic cell that operates at a Faraday efficiency exceeding 90%. These convergent pairing processes seem to be the most practical method for electrosynthesis of H_2O_2 toward the industrial scale.

Various reactor designs have been used in academic study and industry. H_2O_2 electrosynthesis reactor design has applied for many patents. These are mainly based on the generation of H_2O_2 from the O_2 cathode on gas diffusion cathode in the PEMFC, with the purpose of achieving a higher mass transfer rate and minimizing the problems related to the low solubility of O_2 in the aqueous electrolyte. By improving the design and orientation of the gas/liquid phases, these have been combined with stackable modular units, in which the gas/liquid phases flow through multiple sequentially stacked GDEs. It largely improves the whole output capacity. Besides, bipolar electrolysis cells similar to those used in the chloralkali process were used to produce useful products at both the anode and cathode, e.g. H_2O_2 and HOCl, or H_2O_2 at both cathode and anode.

References

1 Lang, P., Yuan, N.N., Jiang, Q.Q. et al. (2020). *Energy Technol. (Weinh.)* 8: 1900984.
2 Yang, W.L., Yang, X.P., Jia, J. et al. (2019). *Appl. Catal., B* 244: 1096–1102.
3 Pegis, M.L., Wise, C.F., Martin, D.J., and Mayer, J.M. (2018). *Chem. Rev.* 118: 2340–2391.
4 Yang, S., Tak, Y.J., Kim, J. et al. (2017). *ACS Catal.* 7: 1301–1307.
5 Wang, C., Daimon, H., Onodera, T. et al. (2008). *Angew. Chem. Int. Ed.* 47: 3588–3591.
6 Chen, J.Y., Lim, B., Lee, E.P., and Xia, Y.N. (2009). *Nano Today* 4: 81–95.
7 Gao, L., Li, X., Yao, Z. et al. (2019). *J. Am. Chem. Soc.* 141: 18083–18090.
8 Li, K., Li, X.X., Huang, H.W. et al. (2018). *J. Am. Chem. Soc.* 140: 16159–16167.
9 Becknell, N., Son, Y., Kim, D. et al. (2017). *J. Am. Chem. Soc.* 139: 11678–11681.
10 Chen, H.Y., Jin, M.X., Zhang, L. et al. (2019). *J. Colloid Interface Sci.* 543: 1–8.
11 Zhao, X., Yin, M., Ma, L. et al. (2011). *Energy Environ. Sci.* 4: 2736–2753.
12 Stamenkovic, V.R., Mun, B.S., Arenz, M. et al. (2007). *Nat. Mater.* 6: 241–247.
13 Greeley, J., Stephens, I.E., Bondarenko, A.S. et al. (2009). *Nat. Chem.* 1: 552–556.
14 Huang, X., Zhao, Z., Cao, L. et al. (2015). *Science* 348: 1230–1234.
15 Wang, J., Li, B., Yersak, T. et al. (2016). *J. Mater. Chem. A* 4: 11559–11581.
16 Wang, X., Vara, M., Luo, M. et al. (2015). *J. Am. Chem. Soc.* 137: 15036–15042.
17 Zhao, X., Chen, S., Fang, Z. et al. (2015). *J. Am. Chem. Soc.* 137: 2804–2807.
18 Liu, K., Wang, W., Guo, P.H. et al. (2019). *Adv. Funct. Mater.* 29, 1806300.
19 Li, Y., Huang, J., Hu, X. et al. (2018). *Adv. Funct. Mater.* 28: 1803330.
20 Shao, M., Chang, Q., Dodelet, J.P., and Chenitz, R. (2016). *Chem. Rev.* 116: 3594–3657.
21 Sharifi, T., Gracia-Espino, E., Chen, A.R. et al. (2020). *Adv. Energy Mater.* 10: 1902084.
22 Morozan, A., Jousselme, B., and Palacin, S. (2011). *Energy Environ. Sci.* 4: 1238–1254.
23 Zhao, L., Zhang, Y., Huang, L.B. et al. (2019). *Nat. Commun.* 10: 1278.
24 Zhang, H., Hwang, S., Wang, M. et al. (2017). *J. Am. Chem. Soc.* 139: 14143–14149.
25 Liu, Z.W., Peng, F., Wang, H.J. et al. (2011). *Angew. Chem. Int. Ed.* 50: 3257–3261.
26 Gong, K., Du, F., Xia, Z. et al. (2009). *Science* 323: 760–764.
27 Yan, Y., Miao, J., Yang, Z. et al. (2015). *Chem. Soc. Rev.* 44: 3295–3346.
28 Niu, W., Li, L., Liu, X. et al. (2015). *J. Am. Chem. Soc.* 137: 5555–5562.
29 Zhang, J., Zhao, Z., Xia, Z., and Dai, L. (2015). *Nat. Nanotechnol.* 10: 444–452.
30 Liu, S., Li, G.Z., Gao, Y.Y. et al. (2017). *Catal. Sci. Technol.* 7: 4007–4016.
31 Campos-Martin, J.M., Blanco-Brieva, G., and Fierro, J.L. (2006). *Angew. Chem. Int. Ed.* 45: 6962–6984.

32 Yang, S., Verdaguer-Casadevall, A., Arnarson, L. et al. (2018). *ACS Catal.* 8: 4064–4081.
33 Kim, H.W., Ross, M.B., Kornienko, N. et al. (2018). *Nat. Catal.* 1: 282–290.
34 Seh, Z.W., Kibsgaard, J., Dickens, C.F. et al. (2017). *Science* 355 (6321): eaad4998.
35 Traube, M. (1887). *Ber. Kgl. Akad.Wiss* 2: 1041–1050.
36 Berl, E. (1939). *Trans. Electrochem. Soc.* 76: 359–369.
37 Mizuno, S. (1948). *Bull. Tokyo Inst. Technol.* 13: 102.
38 Ignatenko, E.B. (1964). *Zh. Prikl. Khim.* 37: 2415.
39 Verdaguer-Casadevall, A., Deiana, D., Karamad, M. et al. (2014). *Nano Lett.* 14: 1603–1608.
40 Shi, X., Siahrostami, S., Li, G.L. et al. (2017). *Nat. Commun.* 8: 701.
41 Chen, S.C., Chen, Z.H., Siahrostami, S. et al. (2018). *ACS Sustainable Chem. Eng.* 6: 311–317.
42 Rodriguez, P. and Koper, M.T. (2014). *Phys. Chem. Chem. Phys.* 16: 13583–13594.
43 Maruyama, J., Inaba, M., and Ogumi, Z. (1998). *J. Electroanal. Chem.* 458: 175–182.
44 Markovic, N.M., Gasteiger, H.A., and Ross, P.N. (1995). *J. Phys. Chem.* 99: 3411–3415.
45 Zecevic, S., Drazic, D.M., and Gojkovic, S. (1989). *J. Electroanal. Chem.* 265: 179–193.
46 Shih, Y.H., Sagar, G.V., and Lin, S.D. (2008). *J. Phys. Chem. C* 112: 123–130.
47 Sanchez-Sanchez, C.M. and Bard, A.J. (2009). *Anal. Chem.* 81: 8094–8100.
48 Sanchez-Sanchez, C.M., Rodriguez-Lopez, J., and Bard, A.J. (2008). *Anal. Chem.* 80: 3254–3260.
49 Shen, Y., Trauble, M., and Wittstock, G. (2008). *Anal. Chem.* 80: 750–759.
50 Dobrzeniecka, A., Zeradjanin, A., Masa, J. et al. (2013). *Catal. Today* 202: 55–62.
51 Johnson, L. and Walsh, D.A. (2012). *J. Electroanal. Chem.* 682: 45–52.
52 Pletcher, D. and Sotiropoulos, S. (1993). *J. Electroanal. Chem.* 356: 109–119.
53 Birkin, P.R., Elliott, J.M., and Watson, Y.E. (2000). *Chem. Commun.*: 1693–1694.
54 Sheng, H., Ji, H., Ma, W. et al. (2013). *Angew. Chem. Int. Ed.* 52: 9686–9690.
55 Zhou, R.F., Zheng, Y., Jaroniec, M., and Qiao, S.Z. (2016). *ACS Catal.* 6: 4720–4728.
56 Chen, S.L. and Kucernak, A. (2004). *J. Phys. Chem. B* 108: 3262–3276.
57 Taylor, S., Fabbri, E., Levecque, P. et al. (2016). *Electrocatalysis* 7: 287–296.
58 Ilea, P., Dorneanu, S., and Popescu, I.C. (2000). *J. Appl. Electrochem.* 30: 187–192.
59 von Weber, A., Baxter, E.T., White, H.S., and Anderson, S.L. (2015). *J. Phys. Chem. C* 119: 11160–11170.
60 Antonin, V.S., Parreira, L.S., Aveiro, L.R. et al. (2017). *Electrochim. Acta* 231: 713–720.
61 Edwards, J.K. and Hutchings, G.J. (2008). *Angew. Chem. Int. Ed.* 47: 9192–9198.

62 Pizzutilo, E., Kasian, O., Choi, C.H. et al. (2017). *Chem. Phys. Lett.* 683: 436–442.
63 Siahrostami, S., Verdaguer-Casadevall, A., Karamad, M. et al. (2013). *Nat. Mater.* 12: 1137–1143.
64 Norskov, J.K., Rossmeisl, J., Logadottir, A. et al. (2004). *J. Phys. Chem. B* 108: 17886–17892.
65 Shao, M. (2011). *J. Power Sources* 196: 2433–2444.
66 Markovic, N.M., Adzic, R.R., and Vesovic, V.B. (1984). *J. Electroanal. Chem.* 165: 121–133.
67 Liu, J., Bunes, B.R., Zang, L., and Wang, C.Y. (2018). *Environ. Chem. Lett.* 16: 477–505.
68 Yang, S., Kim, J., Tak, Y.J. et al. (2016). *Angew. Chem. Int. Ed.* 55: 2058–2062.
69 Jirkovsky, J.S., Panas, I., Romani, S. et al. (2012). *J. Phys. Chem. Lett.* 3: 315–321.
70 Miao, J., Zhu, H., Tang, Y. et al. (2014). *Chem. Eng. J.* 250: 312–318.
71 Wang, Y., Liu, Y.H., Wang, K. et al. (2015). *Appl. Catal., B* 165: 360–368.
72 Zhou, L., Hu, Z.X., Zhang, C. et al. (2013). *Sep. Purif. Technol.* 111: 131–136.
73 Zhao, Z., Li, M., Zhang, L. et al. (2015). *Adv. Mater.* 27: 6834–6840.
74 Zhao, Z.H. and Xia, Z.H. (2016). *ACS Catal.* 6: 1553–1558.
75 Zhang, X.W., Fu, J.L., Zhang, Y., and Lei, L.C. (2008). *Sep. Purif. Technol.* 64: 116–123.
76 Kozlova, L.S., Novikov, V.T., Garaeva, G.R. et al. (2015). *Prot. Met. Phys. Chem. Surf.* 51: 985–989.
77 Sun, Y.Y., Sinev, I., Ju, W. et al. (2018). *ACS Catal.* 8: 2844–2856.
78 Valim, R.B., Reis, R.M., Castro, P.S. et al. (2013). *Carbon* 61: 236–244.
79 Lobyntseva, E., Kallio, T., Alexeyeva, N. et al. (2007). *Electrochim. Acta* 52: 7262–7269.
80 Perez, J.F., Llanos, J., Saez, C. et al. (2016). *Electrochem. Commun.* 71: 65–68.
81 Ilea, P., Dorneanu, S., and Nicoara, A. (1999). *Rev. Roum. Chim.* 44: 555–561.
82 Chen, S., Chen, Z., Siahrostami, S. et al. (2018). *J. Am. Chem. Soc.* 140: 7851–7859.
83 Park, J., Nabae, Y., Hayakawa, T., and Kakimoto, M.A. (2014). *ACS Catal.* 4: 3749–3754.
84 Potapova, G.F., Kasatkin, E.V., Panesh, A.M. et al. (2004). *Russ. J. Electrochem.* 40: 1193–1197.
85 Vlaic, C.A. and Dorneanu, S.A. (2015). *Stud. Univ. Babes-Bolyai, Chem.* 60: 141–150.
86 Perez, J.F., Saez, C., Llanos, J. et al. (2017). *Ind. Eng. Chem. Res.* 56: 12588–12595.
87 Lu, Z.Y., Chen, G.X., Siahrostami, S. et al. (2018). *Nat. Catal.* 1: 156–162.
88 Chai, G.L., Hou, Z.F., Ikeda, T., and Terakura, K. (2017). *J. Phys. Chem. C* 121: 14524–14533.
89 Coria, G., Perez, T., Sires, I., and Nava, J.L. (2015). *J. Electroanal. Chem.* 757: 225–229.
90 Xia, G.S., Lu, Y.H., and Xu, H.B. (2015). *Electrochim. Acta* 158: 390–396.

91 Iglesias, D., Giuliani, A., Melchionna, M. et al. (2018). *Chem* 4: 106–123.
92 Perazzolo, V., Durante, C., Pilot, R. et al. (2015). *Carbon* 95: 949–963.
93 Xie, J., Zhao, X., Wu, M. et al. (2018). *Angew Chem. Int. Ed.* 57: 9640–9644.
94 Nabae, Y., Malon, M., Lyth, S.M. et al. (2009). *Proton Exchange Membrane Fuel Cells 9*, vol. 25, 463–467. The Electrochemical Society.
95 Lefevre, M. and Dodelet, J.P. (2003). *Electrochim. Acta* 48: 2749–2760.
96 Nallathambi, V., Lee, J.W., Kumaraguru, S.P. et al. (2008). *J. Power Sources* 183: 34–42.
97 Bezerra, C.W.B., Zhang, L., Lee, K.C. et al. (2008). *Electrochim. Acta* 53: 4937–4951.
98 Fellinger, T.P., Hasche, F., Strasser, P., and Antonietti, M. (2012). *J. Am. Chem. Soc.* 134: 4072–4075.
99 Sidik, R.A., Anderson, A.B., Subramanian, N.P. et al. (2006). *J. Phys. Chem. B* 110: 1787–1793.
100 Peng, L.Z., Liu, P., Cheng, Q.Q. et al. (2018). *Chem. Commun. (London)* 54: 4433–4436.
101 Muthukrishnan, A., Nabae, Y., Okajima, T., and Ohsaka, T. (2015). *ACS Catal.* 5: 5194–5202.
102 Barros, W.R.P., Reis, R.M., Rocha, R.S., and Lanza, M.R.V. (2013). *Electrochim. Acta* 104: 12–18.
103 Silva, F.L., Reis, R.M., Barros, W.R.P. et al. (2014). *J. Electroanal. Chem.* 722: 32–37.
104 Yamanaka, I., Ichihashi, R., Iwasaki, T. et al. (2013). *Electrochim. Acta* 108: 321–329.
105 Schulenburg, H., Stankov, S., Schunemann, V. et al. (2003). *J. Phys. Chem. B* 107: 9034–9041.
106 Luo, H.J., Li, C.L., Wu, C.Q., and Dong, X.Q. (2015). *RSC Adv.* 5: 65227–65235.
107 Jung, E., Shin, H., Hooch Antink, W. et al. (2020). *ACS Energy Lett.* 5: 1881–1892.
108 Walsh, F.C. and de Leon, C.P. (2018). *Electrochim. Acta* 280: 121–148.
109 Gonzalez-Garcia, J., Banks, C.E., Sljukic, B., and Compton, R.G. (2007). *Ultrason. Sonochem.* 14: 405–412.
110 Oloman, C. and Watkinson, A.P. (1979). *J. Appl. Electrochem.* 9: 117–123.
111 Abdullah, G.H. and Xing, Y.C. (2017). *Ind. Eng. Chem. Res.* 56: 11058–11064.
112 Foller, P.C. and Bombard, R.T. (1995). *J. Appl. Electrochem.* 25: 613–627.
113 Lei, Y.M., Liu, H., Jiang, C.C. et al. (2015). *J. Adv. Oxid. Technol.* 18: 47–56.
114 Yamada, N., Yaguchi, T., Otsuka, H., and Sudoh, M. (1999). *J. Electrochem. Soc.* 146: 2587–2591.
115 Otsuka, K. and Yamanaka, I. (1990). *Electrochim. Acta* 35: 319–322.
116 Jirkovsky, J.S., Busch, M., Ahlberg, E. et al. (2011). *J. Am. Chem. Soc.* 133: 5882–5892.
117 Steckhan, E., Arns, T., Heineman, W.G. et al. (2001). *Chemosphere* 43: 63–73.
118 Matthessen, R., Fransaer, J., Binnemans, K., and De Vos, D.E. (2015). *ChemElectroChem* 2: 73–76.

119 Babu, K.F., Sivasubramanian, R., Noel, M., and Kulandainathan, M.A. (2011). *Electrochim. Acta* 56: 9797–9801.

120 Tatapudi, P. and Fenton, J.M. (1994). *J. Electrochem. Soc.* 141: 1174–1178.

121 Espinoza-Montero, P.J., Vasquez-Medrano, R., Ibanez, J.G., and Frontana-Uribe, B.A. (2013). *J. Electrochem. Soc.* 160: G3171–G3177.

122 Paddon, C.A., Atobe, M., Fuchigami, T. et al. (2006). *J. Appl. Electrochem.* 36: 617–634.

123 Ito, S., Katayama, R., Kunai, A., and Sasaki, K. (1989). *Tetrahedron Lett.* 30: 205–206.

124 Chen, R.Y., Huang, Z.X., Zheng, X., and Chen, Z. (2010). *Chem. Eng. Commun.* 197: 1476–1484.

10

Fuel Cell and Metal-Air Battery

10.1 H_2 Fuel Cell

Storage of electrical energy into chemical bonds, for example, water splitting [1–4] to form H_2 as the energy carrier, offers high energy densities compared to other energy conversion/storage technologies, including Li-ion batteries [5]. In the front of the environment and energy demand, a hydrogen fuel cell can be fabricated with hydrogen oxidation reaction (HOR) as the anode reaction and oxygen reduction reaction (ORR) as the cathode reaction Figure 10.1.

There are seven main parameters for H_2–O_2 fuel cell:

Theoretical specific capacities: In the conditions as specified in design (e.g. discharge rate, temperature, termination voltage, etc.), the minimum capacity of fuel cell/metal-air battery should be capable of discharging, and the unit is ampere hours, with the symbol *C* usually used. The theoretical capacity of the cell/battery can be determined accurately on the basis of the amount of fuel or active metal material in the battery/cell. Because of some possible side reactions occurring in the cell/battery and the special design needs, the true capacity of the cell/battery is normally lower relative to the theoretical capacity.

Rated voltage: The typical voltage of the cell/battery during operation at a given temperature is also named as the nominal voltage. It can be used for reference as selecting different types of cells/batteries. The actual voltage for the cell/battery varies along with the different use conditions during operating. The open-circuit voltage of the cell/battery is equal to the difference between the balanced electrode potentials between positive and negative electrodes. It is related only to the types of active materials, not to the amount of active materials. The cell/battery voltage is a DC voltage in essence; however, with some special conditions, some phase-forming film or the phase change of the catalysts caused by the redox reactions will lead to a small fluctuation of voltage. This phenomenon is called noise. The fluctuation amplitude is small, while the frequency range would be wide, so it can be easily distinguished from the self-excited noise.

Charge and discharge rate: There are two important scales that express this concept: magnification and time rate. First, the time rate is related to the charge–discharge

Electrocatalysis in Balancing the Natural Carbon Cycle, First Edition. Yaobing Wang.
© 2021 WILEY-VCH GmbH. Published 2021 by WILEY-VCH GmbH.

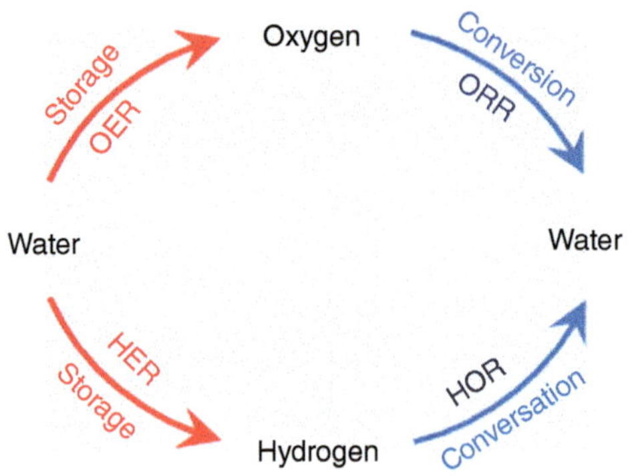

Figure 10.1 Hydrogen and oxygen cycle for energy storage and energy conversion. The two half-cell reactions for energy storage by water electrolysis are the oxygen evolution reaction (OER) and hydrogen evolution reaction (HER), and the two half-cell reactions for energy conversion are oxygen reduction reaction (ORR) and hydrogen oxidation reaction (HOR). Source: Wei et al. [5]. © 2019, John Wiley & Sons.

rate as expressed with the charge–discharge time and is equal numerically to the number of hours obtained by dividing the cell/battery's rated capacity (Ah) with a specified charge–discharge current. Magnification is another reprehensive charge/discharge rate, and its value is the inverse of the time rate. The discharge rate for the primary cell/battery is expressed by the time from the fixed resistance to the termination voltage. The discharge rate has a bigger impact on battery performance.

Impedance: The cell/battery normally possesses a large electrode–electrolyte interface area, so the cell/battery can be equivalent to a series circuit with small resistance, large capacitance, and inductance. However, the actual situation would be much more complicated, especially the impedance of the cell/battery will change with frequency and/or time. The tested impedance is valid only for a specific test state.

Battery life: Storage life involves the maximum storage time allowing for the battery to start for use. The total time including usage time and storage time is named as the battery's effective time. Battery life includes wet storage life and dry storage life. Cycle life is related to the maximum number of charge/discharge cycles that a cell/battery could achieve in the specified conditions. When specifying the cycle life, one should also specify the charge/discharge cycle test conditions, such as discharge depth charge/discharge rate and ambient temperature range.

Self-discharge rate: The rate at which the cell/battery loses its capacity during storage. Expressed as a percentage of the capacity lost from self-discharge per unit of storage time before storage.

Charge and discharge curve: After the battery is discharged, its operating voltage always changes with time. The curve drawn by the battery's operating voltage and discharge time or capacity is called the discharge characteristic curve. The curve is flat, indicating that the battery's operating voltage is stable. Measuring the discharge curve of a battery is one of the basic methods for studying battery performance. According to the discharge curve, you can determine whether the battery performance is stable and the maximum current allowed by the battery during stable operation.

The H_2–O_2 fuel cell main consisted of the separator, electrode (anode and cathode), electrolyte:

Separator: Recent work has shown that the performance of anion exchange membrane fuel cell (AEMFC) is almost the same as the latest proton exchange membrane fuel cell (PEMFC) [6]. However, much data has been collected when delivering pure oxygen or carbon-dioxide-free air to the cathode. In general, in order to avoid the adverse effect of CO_2 on the overall performance of AEMFC, CO_2 is usually removed from the oxidant, which is due to the carbonation caused by the reaction of CO_2 with OH-anions to generate HCO_3^- and CO_3^{2-}. When the OH-form of anion exchange membrane (AEM) is exposed to ambient air (containing about 400 ppm of CO_2), the OH-anions present in the membrane will undergo chemical conversion through an acid–base reaction, thereby generating bicarbonate and carbonate anions, according to Yaanagi and Fukuta, who found that for A201 AEM (Japan Tokuyama Corp., Japan), the OH concentration decreased by about 5%. CO_3^{2-} was formed 90% within the first 10 minutes of exposure to air. Thereafter, when the OH_2-concentration slowly increased, the results of Reaction 1 (forward) and Reaction 2 (backward) decrease HCO_3 and CO_3 decreases. The concentration is low, and the CO_2 concentration remains constant. As time passes, the concentration of OH continues to decrease until after 30 minutes. All OH was consumed. After two hours, the results showed that the anion content of AEM is a mixture of bicarbonate and carbonate, of which about 60% is HCO_3^- and 40% is CO_3^{2-}. In addition, AEMFC faces two more challenging issues [7]. One is to further improve the ionic conductivity, thereby obtaining better performance. Since the diffusion coefficient of OH^- is much lower than H^+, a higher ion exchange capacity (IEC) is needed so that a similar conductivity could be achieved. However, due to excessive polymer swelling, higher IEC results in sacrificing mechanical properties. The other is to continue to improve the stability of AEMFC in harsh environments and high temperatures. The polymer backbone and functional groups related to conductivity, mechanical, and chemical stability are easily attacked by OH′ and free radicals. Therefore, a lot of research has been conducted on improving the electrical conductivity of AEMFC and enhancing thermal and chemical stability.

Electrolyte: Recently, attention was paid to developing solid polymer electrolytes used as thin films with a thickness of several tens of microns, which mitigates the carbonate issue arisen from the use of alkaline electrolytes [7]. However, the carbonate issue cannot be eliminated completely because that alkaline anion exchange membrane fuel cell (AAEMFC) may be extremely sensitive to the carbon dioxide atmosphere. As a result, Pan et al. presented a new method for preparing an ultrathin composite membrane of alkaline polymer electrolyte (APE), which can avoid the leakage issue [7]. The APE ionomer quaternary ammonia polysulfone (QAPS) was impregnated into the porous polytetrafluoroethylene (PTFE) membrane and then performed a self-cross-linking process to generate a semi-interpenetrating network. The results displayed that the resulting membrane (QAPS@PTFE, 25 mm thick) showed low swelling degree (3.1% at 60 °C), low area resistance (0.09 V cm^2), and high mechanical strength (31 MPa). The peak power density at 550 mW cm^{-2} is used on both the anode and cathode of H_2/O_2 AAEMFC using Pt (0.4 mg cm^{-2}). The membrane uses humidified H_2 and O_2 at 60 °C at a rate of 120 sccm.

Electrode and device progress: Many researchers divide oxyhydrogen fuel cells into two categories on the basis of the nature of the membrane: AEMFC and proton exchange membrane (PEM). For more than a decade, AEMFC has received great attention, mainly because they have the potential to resort to electrocatalysts that do not contain Pt group metals (PGMs) and ultimately do not contain critical raw materials (CRMs) [8]. The latest achievements in the development of AEM with high hydroxide ion conductivity have enabled the manufacture of hydroxide exchange membrane (HEM)-charged AEMFC to achieve an impressive peak power density. Powered by air, the peak power density is 1.8 and 0.5 W cm^{-2}. However, recently PGM-based catalysts have been used on the two electrodes, normally Pt/C on the cathode and Pt or Pt/Ru on the anode. The alloy has achieved the most recent beginning-of-life (BoL) performance. This state can be explained by history, and the search for PGM-free catalysts for ORR in acidic medium has also greatly facilitated the development of metal nanocube (NC) catalysts for ORR in alkaline media. AEMFC and, therefore, PEMFCs that are resorting to catalysts that do not contain PGM and CRM, at this moment, are compromised due to the low HOR activity of PGM and catalysts in alkaline media. AEMFCs have attracted vast interest recently due to their potential advantages over current technologies based on PEM [6]. Among them, AEMFC's high pH environment can, in principle, use low-cost catalysts and bipolar plate materials. It also promotes the use of polyhydrocarbon-based chemicals as ionomer membranes, thereby greatly reducing fuel exchange in batteries. In addition, since the electrolyte is a polymer membrane instead of a concentrated KOH solution (as with alkaline fuel cell [AFC] technology), the problem of carbonate precipitation is completely avoided.

10.2 Metal-Air Battery

The metal-air battery could generate electricity through the oxidation/reduction reactions between the anode metals and the O_2 from the air in the porous cathode, analogous to the principle in fuel cells. The open structure in the cathode is the most important feature for the metal-air battery, which enables a continuous supply of O_2 from the air. This character directly gives its preference, especially for higher energy density. In addition, the open structure further gives the metal-air battery many advantages, such as lightweight and compact and cost-effective, because the cathode replaces the bulky and expensive components used in lithium-ion batteries [9].

Depending on the types of metals used in the anodes, metal-air batteries are also diverse. Among them, Li-air (oxygen) and Zn-air (oxygen) batteries are one of the most promising batteries [10, 11]. So far, due to the combination of its extremely high theoretical energy density (e.g. 5200 Wh kg^{-1} [involving oxygen]) and dangerous hazards, Li-air batteries are being frequently discussed but remain controversial batteries. The potential hazard originates from the explosive reactivity of metallic Li with air and/or water and may ignite the flammable organic electrolyte used for most batteries. Another obvious disadvantage is the high cost (approximately US\$60 lb^{-1}) and limited Li resources only in the natural mineral deposits in Chile and Australia [12]. Because of these economic and safety issues, all the aforesaid shortcomings of Li-air batteries restrain the practical commercialization toward a large scale.

Since innate safety stems from the use of abundant global assets of Zn mines (Zn is the fourth highest mineral in the earth's crust, about 300 times that of lithium) and noncombustible aqueous electrolytes, Zn-air batteries have strong application potential. Although the theoretical specific energy density of Li batteries is 1084 $Wh\,kg^{-1}$ (involving oxygen), which is smaller than Li-air batteries, they are still four times higher than current Li-ion batteries. In addition, low equilibrium potential, flat discharge voltage, low cost, long service life, and environmental friendliness, and other advantages further ensure the vigorous development of Zn-air batteries in a broad market energy demand. In the Zn-air battery, the most important and complicated part is the air cathode, which is integrated with the oxygen catalyst layer and the gas diffusion layer (GDL). It is also related closely to cost and performance, which are the most primary technical challenges that have to be addressed for commercialization. However, the structure of the air cathode and the wettability of the components are usually neglected, and even the use of highly active oxygen electrocatalysts directly leads to the deterioration of the performance of the entire battery [9].

Many companies, including Fluidic Energy, EOS Energy Storage, and Zinc Nyx Energy Solutions, also joined the study and contributed masses of outstanding works [13]. However, as one of the most promising candidates for energy conversion and storage technology, rechargeable Zn-air batteries are still in the early stages. Therefore, more and more researches are focused on Zn-air batteries with excellent electrochemical performance, and wearable and flexible electronic devices for rechargeable batteries for consumer batteries and power supplies are constantly evolving.

10.2.1 Metal-Air Battery Structure

A typical Zn-air battery consists four parts: a zinc electrode, an air electrode, an electrolyte, and a separator (Figure 10.2) [14]. In discharge processes, the oxygen in the atmosphere penetrates the surface of the electrocatalyst through the GDL and is reduced to OH-ions at the air electrode. Simultaneously, metallic Zn is released by an external circuit and oxidized to Zn^{2+}. The dissolved Zn^{2+} ions would coordinate with OH-ions to generate soluble $Zn(OH)_4{}^{2-}$. In charging processes, a reversible reaction occurs, Zn is plated on the anode, and O_2 is released from the surface of the electrocatalyst of the air electrode [15]. Each aforesaid part could affect the performance of the Zn-air battery individually.

According to the following chemical equation, in addition to hydrogen evolution reaction (HER) (which will be discussed later), the equilibrium potential for a Zn-air

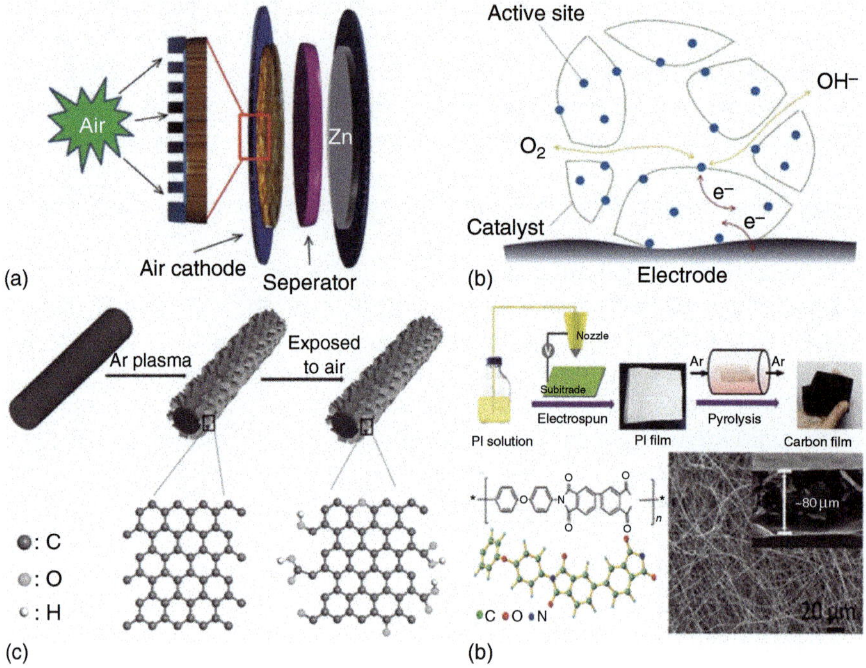

Figure 10.2 (a) Schematic illustration of a Zn–air battery. Adapted with permission. (b) Schematic illustration of the electrocatalysis process in ORR and OER. (c) In situ exfoliation for preparation of the edge-rich and oxygen-functionalized graphene/carbon fiber electrode. Reproduced with permission. (d) Nanoporous carbon nanofiber films (NCNFs), and their electron microscopy images. Source: Huang et al. [14]. 2019, John Wiley & Sons.

battery is 1.65 V, by using the formula of $E_{eq} = E_0$ (cathode) − E_0 (anode). However, the actual operating voltage for the Zn-air battery is lower than 1.65 V, which is because of the internal loss of the Zn-air battery caused by activation as well as the loss of ohm and concentration. It would be beneficial to understand the electrochemical behavior of Zn-air batteries via the polarization curves in the cathode and anode reactions.

$$\text{Anode: } Zn \rightarrow Zn^{2+} + 2e^-$$

$$Zn^{2+} + 4OH^- \rightarrow Zn(OH)_4{}^{2-}$$

$$(E^\circ = -1.25\ \text{V vs. normal hydrogen electrode [NHE]})$$

$$Zn(OH)_4{}^{2-} \rightarrow ZnO + H_2O + 2OH^-$$

$$Zn + 2H_2O \rightarrow Zn(OH)_2 + H_2\uparrow \quad \text{(also possible)}$$

$$\text{Cathode: } O_2 + 2H_2O + 4e^- \rightarrow 4OH^- \quad (E^\circ = 0.4\ \text{V vs. NHE})$$

$$\text{Overall reaction: } 2Zn + O_2 \rightarrow 2ZnO \quad (E^\circ = +1.65\ \text{V})$$

Figure 10.3 depicts a potential (v)–current (i) curve for effectively understanding the main origin for potential loss in Zn-air batteries using oxygen. Note that the

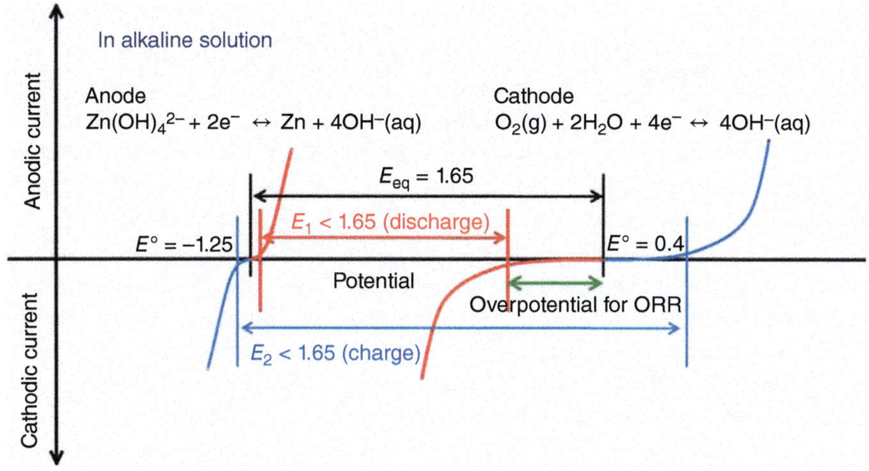

Figure 10.3 Schematic polarization curves of zinc-air cell. The equilibrium potential of the Zn-air battery is 1.65 V, but the practical voltage is lower than this value. Source: Zhu et al. [16]. © 2016, John Wiley & Sons.

overpotential on the zinc anode is smaller than that on the air cathode. Similarly, the cathode current occurs at the cathode, the air electrode. In contrast, during discharge (red line), an anode current appears on the anode (zinc electrode). Of course, when charging, each current direction is opposite (blue line). It can be seen from the polarization curve that a high overpotential is required for the generation of OH-ions through the ORR. This is the reason that the operating voltage of an actual Zn-air battery (E_1, Figure 10.3, red line) is much lower relative to the open circuit potential (OCV) (E_{eq}) of 1.65 V. Taking into consideration of the reverse reaction, that is oxygen evolution reaction (OER), charging requires a bigger potential (Figure 10.3). As can be seen from the brief discussion earlier, the use of ORR has positive and negative effects on Zn-air batteries. The punctuality gives the battery a high energy density due to the lack of oxygen active material. On the contrary, the negative point urges the Zn-air battery to have an obvious potential loss, which ultimately leads to a reduction in battery power density.

The different parts of the metal-air battery have different functions, and the bottlenecks and advances were discussed further:

Zinc anode: Zinc-air batteries use pure metallic Zn as the negative electrode active material. Zn will be oxidized during discharge. Although the corrosion rate of zinc in alkaline solution is slower than that of aluminum, following the formula of $Zn + 2H_2O \rightarrow Zn\ (OH)_2 + H_2\uparrow$, hydrogen gas would be generated from the corrosion process by HER. Thus, it is more rational to adapt the electrochemical behavior of metal Zn in alkaline solution, not just metal Zn in the gas phase. Since this HER should not take place in the discharge process, the method of inhibiting the hydrogen release reaction was regarded to be the most important research: it is hoped and possible to consume completely metal Zn in the discharge process, thereby ultimately increasing the capacity of the Zn-air battery.

Separator: The function of the separator in the Zn-air battery is to transfer the OH-ion from the air cathode to the Zn anode [10]. The basic requirements for a suitable separator involve high stability in alkaline electrolyte, suitable pore size and porosity for mass transport, high ionic conductivity, and no electron conductivity. Since Zn-air batteries use hydroxide ion migration rather than proton H^+ migration, the separator is prepared from polyvinyl alcohol, polyethylene, polyolefin, or polypropylene. However, the porosity of the separator makes Zn cations migration from anode to cathode possible, which leads to a reduction in battery capacity. Kiros has designed a new method to modulate the permeability of zincate ions by importing the inorganic matter $Mn(OH)_2$ into the microporous separator. The results showed that compared with commercially available separators, cationic polys (polymethylsulfonic acid-1,4-phenylenethio-1,4-phenylene trifluoromethanesulfonate) could effectively hinder Zn^{2+} ions from anode to cathode. In addition to the aforementioned basic requirements, a viable secondary Zn-air battery requires some other basic requirements. The separator should be inactive to oxidation and should be stable in charging and discharging processes. In addition, it requires high absorption content for the electrolyte and a fine pore structure to maintain the electrolyte in the pores and also prevent the growth of zinc dendrites.

Electrolyte: The alkaline electrolytes used for the Zn-air battery are mainly KOH, NaOH, and LiOH, except for neutral NH_4Cl; the rest are alkaline [10]. Among them, KOH is widely used in Zn-air batteries due to its excellent ionic conductivity. For example, ionic conductivity of K^+ is $73.50\,\Omega^{-1}\,cm^2/equiv$ compared to Na^+ ($50.11\,\Omega^{-1}\,cm^2/equiv$). In total, 30% KOH is usually used due to the maximum ionic conductivity of this concentration. In order to decrease the electrolyte resistance, the KOH concentration of the solution can be increased; however, too high a KOH concentration may cause the viscosity of the electrolyte to increase. In addition, following the reaction of $Zn(OH)_4{}^{2-} \rightarrow ZnO + H_2O + 2OH^-$, a high concentration of electrolyte will result in the formation of ZnO, which in turn increases the viscosity. The present issue of the KOH electrolyte is that it is very sensitive to atmospheric CO_2 when the zinc-air battery works in an alkaline solution and is exposed to air due to the reaction with OH^- ions to generate carbonates; during this process, the concentration of hydroxyl ions also naturally decreases. Therefore, the carbonation issue of alkaline electrolyte will cause decreased battery capacity.

Air cathode: Using O_2 in a Zn-air battery requires that the air cathode has both a suitable catalyst for ORR and an ideal porous structure [10]. When designing an air electrode, these two requirements should be considered at the same time. As mentioned earlier, a catalyst is required to reduce the high activation energy of ORR; in addition, the structure of the air cathode is also a vital factor determining the overall performance of the Zn-air battery. Actually, the air cathode only serves as a matrix for ORR. Taking into consideration the overall reaction processes of the Zn-air battery ($2Zn + O_2 \rightarrow 2ZnO$), the only materials for consummation are metal Zn and O_2. Since O_2 can be indefinitely supplied from the atmosphere, in principle, the air cathode can be reused until physical damage such as rupture of the air cathode after discharge. The highly porous structure of the air cathode forms a favorable diffusion path for O_2 and serves as a substrate with a high surface area for the catalyst supporting. Thus, carbon materials, including carbon nanotubes (CNTs) and activated carbon, can be utilized as the substrate of the air cathode. Generally, the air cathode is composed of a catalytic layer and a GDL and is fabricated by laminating them with a metal (e.g. Fe) grid as the current collector. Carbon materials and hydrophobic binders (e.g. PTFE) as a moisture-proof agent consisted the GDL.

Primary and mechanically rechargeable Zn-air batteries [14]: ORR catalysts in an air cathode are the key component of a primary as well as the mechanically rechargeable Zn-air battery [17]. Developing effective electrocatalysts to reduce O_2 to water is essential for improving the energy conversion efficiency of zinc-air batteries. Because of its catalytic activity comparable to or even better than commercial Pt/C electrocatalysts, various heteroatoms-doped carbons were prepared and used as efficient catalysts for Zn-air batteries.

Electrically rechargeable Zn-air batteries: Compared to the primary and mechanically rechargeable Zn-air batteries, rechargeable Zn-air batteries are easier to be designed in terms of battery configuration and, more importantly, recycling without the requirement for additional disassembly processes [14]. Rechargeable Zn-air batteries need an efficient bifunctional air electrode for the proceeding of both ORR and OER. These reactions are very slow due to the sluggish kinetic redox process and corrosive/oxidizing chemical and electrochemical environment. High-efficiency catalysts for air electrodes must not only be highly efficient and reversible ORR/OER but also to reduce the potential difference between charge and discharge processes.

Solid-state Zn-air batteries: Solid Zn-air batteries with special characteristics of good flexibility, stretch ability, and wear resistance can drive and promote the emerging next-generation portable electronic devices, e.g. electronic skins, flexible displays, and portable sensors [14, 18]. Generally, the solid electrolyte consisted of KOH in polyvinyl alcohol (KOHPVA) polymer is used, where polyvinyl alcohol (PVA) acts as an inert polymer matrix and host, and the KOH serves as an ion conductor to manufacture an all-solid-state zinc-air battery [13]. This polymer electrolyte can reduce the possibility of electrolyte leakage and manufacture a flexible battery coupling with a flexible electrode [18]. In order to enhance further the mechanical properties, a stretchable cross-linked poly(vinyl acetate)-poly(ethylene oxide)-KOH (PVA-PEO-KOH) water with a high ionic conductivity as high as $0.3\,S\,cm^{-1}$ was prepared. The gel polymer electrolyte can withstand a maximum strain of 300% [19]. The incorporation of a cross-linking agent (e.g. glutaraldehyde) can enhance the mechanical strength. The electrolyte of polyacrylic acid (PAA)-KOH exhibits good conductivity, ideal mechanical strength, and high water retention. In addition to the polymer electrolytes, the electrolyte membranes of hydroxide conductive cellulose were also reported, and the ionic conductivity was further improved by hydroxide conductive quaternary ammonium salts and functionalized graphen oxide (GO) [20].

The lithium-air battery is another type of advanced metal-based energy conversion system, which also utilizes O_2 from the air as the reactant in the air cathode. The lithium-air battery has high theoretical specific energy density as well as great potential in the applications of the emerging target, including remote electric vehicles. Over the past few years, lithium-air batteries have gained extensive research, which greatly promoted the basic understanding of the reaction in the batteries. Figure 10.4 exhibits the typical structure of a Li-air battery, which utilizes a catalyst with porous structure as the working electrode, lithium foil as the counter electrode, and a nonaqueous electrolyte-soaked separator in between [21]. The Li-air battery will generate Li_2O_2 on the cathode via ORR processes during the discharge. In the charging process, the reversible decomposition of Li_2O_2 via OER takes place. Bruce and coworkers [21] found that the electrolyte in the Li-air battery plays a key role during Li_2O_2 nucleation processes, and the number of donors (DN) of the solvent affects the ORR pathway greatly. In detail, the reaction would follow the solution path in a solvent of high DN, and the intermediate product of LiO_2 is dissolved in the electrolyte, resulting in the generation of large Li_2O_2 nanoparticles on the cathode, with excellent electrochemical activity. In comparison, solvents with low DN would render LiO_2 deposition onto the electrode surface, leading to the growth of Li_2O_2 films and poor performance of the battery.

The multilayer porous structure is also very vital in enhancing the catalytic activity of the air electrode materials. As suggested by Wang and coworkers, the wall density of 3D ordered mesoporous (3DOm) carbon with different specific surface areas and the pore size can significantly influence the generation and decomposition of Li_2O_2 as well as the accumulation of other by-products, leading to different battery performance (Figure 10.5) [22]. Similar conclusions were obtained from the

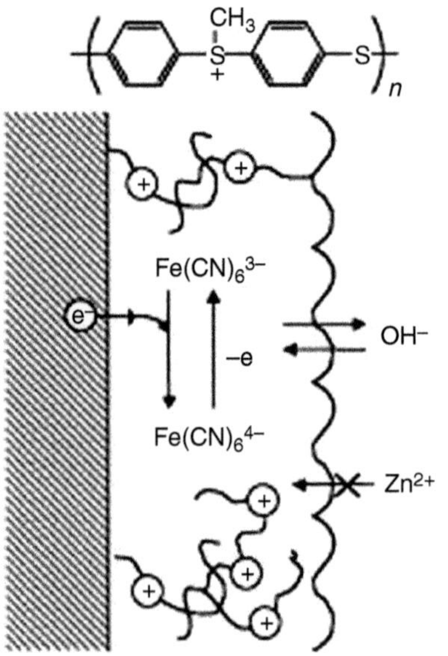

Figure 10.4 Scheme of the modified electrode using polysufonium. Source: Lee et al. [10]. © 2011, John Wiley & Sons.

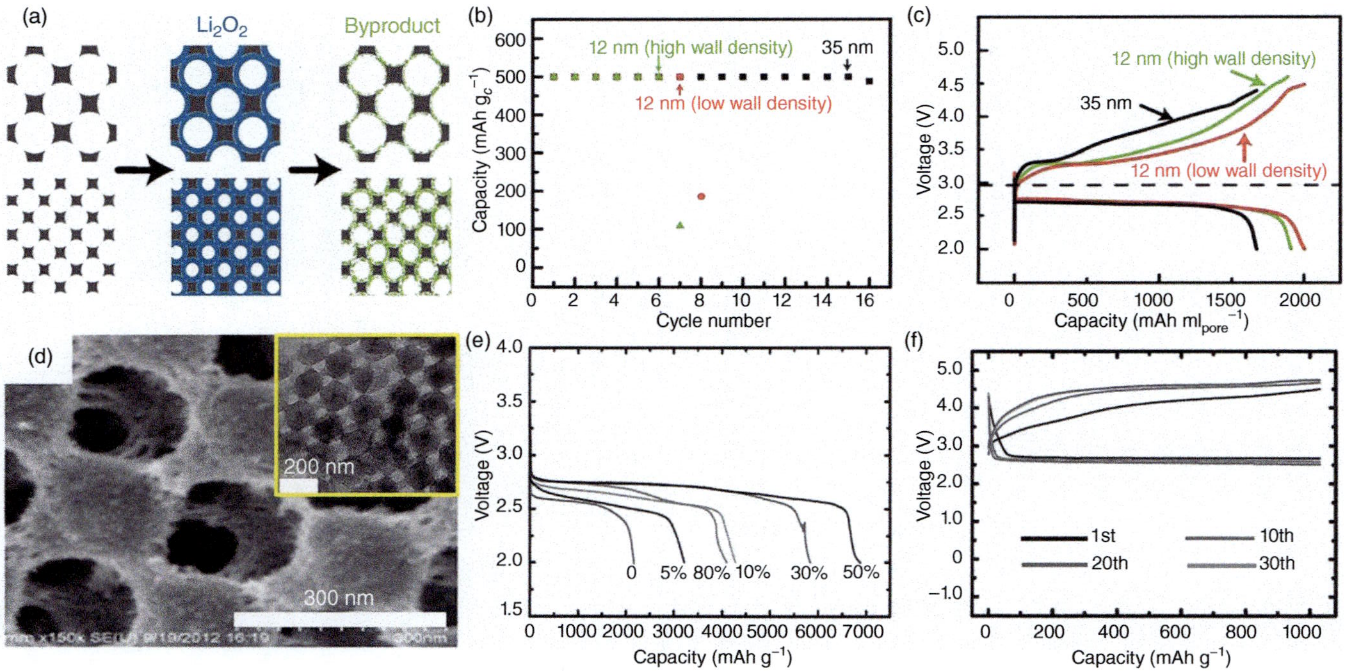

Figure 10.5 (a) Illustration of reaction products in 3DOm carbons with different pore structures. (b, c) Electrochemical performance of 3DOm carbons. Source: Xie et al. [22]. © 2015, John Wiley & Sons. (d) Mesoporous/macroporous carbon sphere array(MMCSA). (e) Battery performance using different amounts of the MMCSAs and (f) 30 wt% MMCSAs. Source: Guo et al. [23]. © 2013, John Wiley & Sons.

investigation of the air electrode containing 3D ordered mesoporous/macropore carbon ball sphere array (MMCSA) in Figure 10.5d [23]. The mesopores facilitate the wetting of the electrolyte and thus enhance the electron transfer and diffusion of Li^+. The macropores can offer ideal pathways for the diffusion of O_2 and offer enough space for Li_2O_2 storage. Based on such unique architecture, the cycle performance and discharge specific capacity of MMCSA are obviously enhanced (Figure 10.6e,f).

Dai and coworkers [24] developed porous graphene (holey graphene [hG]) wafer, which can be served as the air electrode in Li-O_2 battery, as depicted in Figure 10.6a. The hG cathode at a mass load of 5 mg cm^{-2} could offer an ultrahigh area capacity of c. 40 mAh cm^{-2} at the current density of 0.2 mA cm^{-2} (Figure 10.6b). Apparently, in the thick cathode structure on the air side, the discharge product of Li_2O_2 accumulates preferentially in an inclined manner, which may cause more Li_2O_2 storage and thus higher capacity. These results suggested that freestanding hG could be utilized as a potential high-quality load and highly active cathode in practical Li-O_2 batteries.

Through continuous thermal annealing of hG in air and NH_3 atmosphere, a low-cost yet highly efficient N-doped porous graphene (N-HGr) was synthesized and served as a cathode catalyst in Li-O_2 batteries [25]. Figure 10.6c exhibits the initial discharge/charge curves of the N-HGr electrode at the fixed capacity of 800 mAh g^{-1} at 40 mA g^{-1}. The voltage platforms for discharge and charge reached 2.93 and 3.47 V, respectively, thus realizing a round-trip efficiency of up to 85%. After more than 60 cycles, the terminal voltages in charge and discharge kept nearly unchanged in Figure 10.6d, while the r-Gr electrode displayed poor cycle stability over the course of 30 cycles in Figure 10.6e. As shown in Figure 10.6f, in the potential range of 2.2–4.4 V and at the current of 100 mA g^{-1}, the N-HGr electrode offered a high discharge/charge capacity of c. 17 000 mAh g^{-1}. It is worth noting that in similar test conditions, the electrocatalytic activity of the N-HGr electrode is obviously higher than that of the r-Gr electrode and even better than the electrocatalytic activity of composite electrodes consisting of transition metal oxides and/or precious metals reported thus far. This result is due to the unique, peculiar structure of r-Gr, as shown in Figure 10.6g. The pores on its surfaces can enhance the transportation of both Li^+ and O_2, conducive to the reversible formation/decomposition of Li_2O_2. In addition, the defects on the plane serve as active electrocatalytic sites, which enhance both ORR and OER through metal-free electrocatalysis, resulting in superior Li-air battery performance.

This part provides an overview of HOR/ORR parameters, mechanism, catalyst advances, and the corresponding device development of fuel cell, metal-air batteries. Although the general mechanism for HOR/ORR is becoming clear and vast advances on high-performance electrocatalysts have rendered this area to be approaching the practical applications, there are still some challenges/bottlenecks remaining unsolved, and the development direction for this research area is still in vague. Here we have given some summarization and perspective on these aspects.

Beyond previous metal catalysts for HOR, most of the nonprevious metal catalysts were designed based on nickel element by alloying with second or more components.

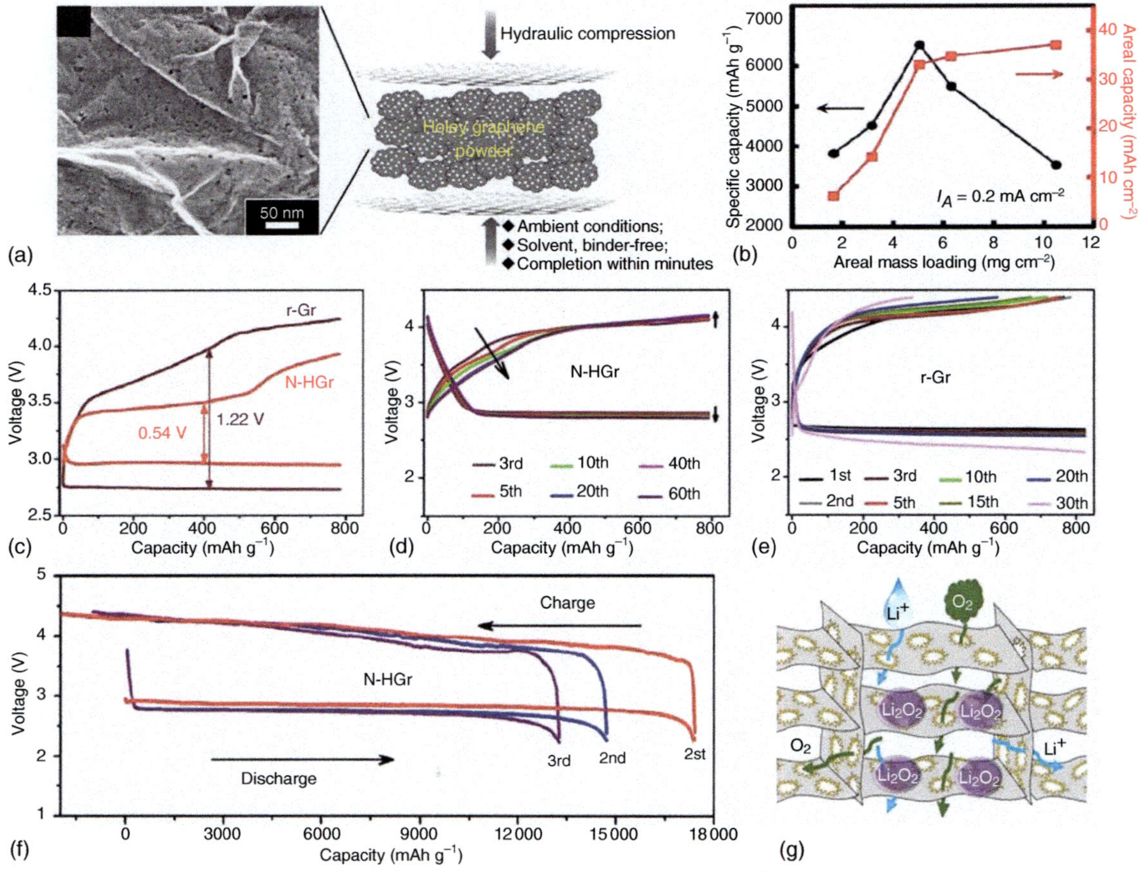

Figure 10.6 (a) SEM image of a freestanding hG electrode made by hydraulic compression. (b) Specific capacities and areal capacities of the hG electrodes based on areal mass loading. Source: Lin et al. [24]. Reproduced with permission of American Chemical Society. (c) Initial and (d, e) typical discharge/charge curves of the N-HGr electrode with a fixed capacity of 800 mAh g^{-1}. (f) Discharge/charge profiles of the N-HGr electrode within 2.2–4.4 V. (g) Schematic illustration of reactions on the N-HGr electrode. Source: Shui et al. [25]. © 2016, American Chemical Society.

The HOR mechanism on these kinds of catalysts is still not fully understood, and the design on the catalysts is still limiting. Generally, for mechanism study and catalyst design, the dissociative adsorption of H_2, electron transfer from H_2 to the catalyst, and discharge of adsorbed H atom could be the rate-determining steps and should be emphatically considered. In addition, the adsorbed species of H_2 are different in acid (only hydrogen) and alkaline (hydrogen and hydroxide) media, which would influence the reaction mechanism of HOR. The catalysts, which have a high affinity for H adsorption, the balance between H and OH adsorption, and high electron transfer ability, are recommended. The ultralow content of previous metal catalysts and the nonprevious metal catalysts, together with their scaling-up synthesis methods, are needed to develop.

There are already many nonprevious metal catalysts showing high ORR performance comparable to the previous metal Pt/C. Most of these electrocatalysts worked, preferably in alkaline media. There are three main kinds of highly active sites for ORR, namely M–N sites, C–N sites, and encapsulated metals. Among these active sites, the type Fe/N/C and Co/N/C showed great potential to replace Pt/C. The current challenges for the nonprevious metal catalysts rest in their stability and large-scale preparation. Some advanced characterization techniques such as in situ transmission electron microscopy (TEM), X-ray photoelectron energy spectroscopy (XPS), X-ray absorption spectroscopy (XAS), X-ray diffraction (XRD) are needed to analyze the degradation mechanism of these nonprevious catalysts, together with their active site's transformation. The efforts toward practical applications of nonprevious metal catalysts are suggested to focus on stability evaluation, which should consider many other factors, such as electrolyte types, catalyst loadings, atmospheric temperature/pressure, current densities/changes, and flow rate, etc.

As both HOR and ORR suggested applications of nonprevious metal catalysts in alkaline media, an alkaline electrolyte for the H_2–O_2 fuel cell is required. The AEMFC could greatly reduce the cost issue of PEMFC not only because of the use of cheap electrocatalysts but also the cost-effective and stable membrane materials and simple device configuration. Aqueous metal-air batteries are usually operated in alkaline media, which are excellent candidates for the next-generation battery technologies on the basis of the high specific energy density. With considering their applications in electric vehicles and smart grids, mechanical stability and the oxidation-tolerance ability during long-term operation, especially for the carbon-based catalysts, should be further improved. The electrocatalysts/electrode fabrication technology for metal-air batteries can refer to that of the H_2–O_2 fuel cell, taking into account of allowing for cost-effective and eco-friendly features.

Along with the great achievements in the catalysts, membrane, electrodes for H_2–O_2 fuel cell, and metal-air batteries, this research field is suggested to be prosperous with broad prospects. This chapter is expected to offer in-depth understanding and inspiration for designing various advanced electrocatalysts from synthesis, electrocatalytic performance to device applications.

References

1 Hwang, J., Rao, R.R., Giordano, L. et al. (2017). *Science* 358: 751–756.
2 Lewis, N.S. and Nocera, D.G. (2006). *Proc. Natl. Acad. Sci. U.S.A.* 103: 15729–15735.
3 Montoya, J.H., Seitz, L.C., Chakthranont, P. et al. (2016). *Nat. Mater.* 16: 70–81.
4 Seh, Z.W., Kibsgaard, J., Dickens, C.F. et al. (2017). *Science* 355 (6321): eaad4998.
5 Wei, C., Rao, R.R., Peng, J. et al. (2019). *Adv. Mater.* 31: e1806296.
6 Ziv, N., Mustain, W.E., and Dekel, D.R. (2018). *ChemSusChem* 11: 1136–1150.
7 Pan, Z.F., An, L., Zhao, T.S., and Tang, Z.K. (2018). *Prog. Energy Combust. Sci.* 66: 141–175.
8 Davydova, E.S., Mukerjee, S., Jaouen, F., and Dekel, D.R. (2018). *ACS Catal.* 8: 6665–6690.
9 Pan, J., Xu, Y.Y., Yang, H. et al. (2018). *Adv. Sci.* (Weinheim, Germany), 5: 1700691.
10 Lee, J.S., Kim, S.T., Cao, R. et al. (2011). *Adv. Energy Mater.* 1: 34–50.
11 Liu, Q.C., Li, L., Xu, J.J. et al. (2015). *Adv. Mater.* 27: 8095–8101.
12 Kim, S.W., Seo, D.H., Ma, X.H. et al. (2012). *Adv. Energy Mater.* 2: 710–721.
13 Fu, J., Cano, Z.P., Park, M.G. et al. (2017). *Adv. Mater.* 29: 1604685.
14 Huang, Y., Wang, Y., Tang, C. et al. (2019). *Adv. Mater.* 31: e1803800.
15 Zhang, J., Xia, Z., and Dai, L. (2015). *Sci. Adv.* 1: e1500564.
16 Zhu, A.L., Wilkinson, D.P., Zhang, X.E. et al. (2016). *J. Energy Storage* 8: 35–50.
17 Zhang, J., Zhao, Z., Xia, Z., and Dai, L. (2015). *Nat. Nanotechnol.* 10: 444–452.
18 Tan, P., Chen, B., Xu, H.R. et al. (2017). *Energy Environ. Sci.* 10: 2056–2080.
19 Xu, Y., Zhang, Y., Guo, Z. et al. (2015). *Angew. Chem. Int. Ed.* 54: 15390–15394.
20 Yin, J., Fan, Q.H., Li, Y.X. et al. (2016). *J. Am. Chem. Soc.* 138: 14546–14549.
21 Aurbach, D., McCloskey, B.D., Nazar, L.F., and Bruce, P.G. (2016). *Nat. Energy* 1: 16128.
22 Xie, J., Yao, X., Cheng, Q. et al. (2015). *Angew. Chem. Int. Ed.* 54: 4299–4303.
23 Guo, Z., Zhou, D., Dong, X. et al. (2013). *Adv. Mater.* 25: 5668–5672.
24 Lin, Y., Moitoso, B., Martinez-Martinez, C. et al. (2017). *Nano Lett.* 17: 3252–3260.
25 Shui, J., Lin, Y., Connell, J.W. et al. (2016). *ACS Energy Lett.* 1: 260–265.

Part VI

Small Organic Molecules Oxidation and Device

One of the essential parts in the artificial carbon cycle involves the oxidation of small organic molecules. In this part, we will discuss the typical and recent advances in electrocatalysts for small organic molecule oxidation, including methanol, methane, ethanol, formic acid, ethylene glycol, glucose oxidase, and glycerol. The reaction mechanism and promotion mechanism will be focused on. The configuration, operation, and advances of direct fuel cells (DFCs) will also be overviewed. The perspective for rendering the small organic molecule oxidation as a stronger and more feasible supplement in the electrocatalytic carbon cycle will be summarized.

Electrocatalysis in Balancing the Natural Carbon Cycle, First Edition. Yaobing Wang.
© 2021 WILEY-VCH GmbH. Published 2021 by WILEY-VCH GmbH.

11

Introduction

The alternative candidates of small organic molecules include ethanol, methanol, formic acid, glucose oxidase, ethylene, glycerol, glycol, and so on. The small organic molecules are much easier in storage and transportation than hydrogen [1]. Additionally, the devices converting these small organic molecules to electronic energy are called direct fuel cells (DFCs), which can be manufactured from the unit cells composed of a membrane electrode assembly (MEA) [2]. These DFCs have several advantages: first, they showed high-energy density as well as environmentally benign nature and are suitable for the applications of portable electronic devices and transportation [3]. For example, the complete ethanol oxidation produced $8.0\,kW\,h^{-1}\,kg^{-1}$ [4]. Second, compared to the regular multiple-step nature from chemical to thermal and then to electrical energy processes, the DFCs have the one-step nature directly from chemical to electrical energy via the electrochemical process, offering the unique basis of high efficiency [1]. Finally, with the development of perfluorinated cation exchange membranes, e.g. Nafion® proton exchange membranes and anion exchange membranes and the cathode catalysts for oxygen reduction reaction (ORR) [5], the DFCs have achieved an unprecedented maturity stage, making them a promising candidate as the future energy conversion devices.

Currently, there are some major obstacles restraining the rapid development and applications of DFCs, which includes the high cost of noble metal catalysts, especially for the anodes, the relatively low activity, stability, and unclear reaction mechanism [5–7]. For example, to convert ethanol into CO_2 through this complete oxidation reaction, 12-electron transfer processes are required [8]. The complex reaction would reduce the efficiency of the electrocatalytic processes. It was generally regarded that alkaline electrolyte enables faster reaction kinetics of small organic molecule oxidation [9]. Besides, alkaline electrolyte provides the basis for material stability and easy management [9]. Therefore, the DFCs employing alkaline polymer electrolyte membrane and electrolyte also have a great opportunity as promising low-temperature fuel cells in the future. However, from the basic point of view, several factors could be improved for solving these issues: the selection of small organic molecules with the type of catalyst, the membrane used, proper structures, and physical factors such as the concentration and temperature [10]. Among these alternative improved factors, the electrocatalysts and electrodes for

Electrocatalysis in Balancing the Natural Carbon Cycle, First Edition. Yaobing Wang.
© 2021 WILEY-VCH GmbH. Published 2021 by WILEY-VCH GmbH.

small organic molecule oxidation are the most crucial factors and thus attracted a huge amount of studies [5].

Thus far, noble metals Pt and Pd are still the indispensable components in electrocatalysts for small organic molecule oxidation reactions [11, 12], and the main goal of current studies is to improve the activity and stability while reducing their use amount so as to lower the cost of the catalyst and device. Compared to Pt, the Pd catalysts have received wide attention based on the high activity for formic acid, ethanol, and ethylene glycol, lower cost (half as much as Pt), and their intrinsic advantages by working in alkaline media with faster kinetics and minimizing corrosion of the materials [12]. With the development of nanotechnology and nanoscience over the past decades, the research mode for developing electrocatalysts has shifted gradually from the traditional trial-and-error methods to the accurate design and fabrication of nanocatalysts at atomic and molecular levels. For example, the Pt- and Pd-based catalysts were normally prepared by loading their nanoparticles on various supports with high surface area and conductivity. The nanocatalysts could be further controlled with precise composition, shape, size, nanostructure, and morphology. Various supports could provide tunable anchoring effects and electronic effects for improving the intrinsic activity and stability of the nanocatalysts [13]. Thus, the electrocatalytic properties of Pt- and Pd-based catalysts are usually greatly altered and different from the properties of bulk counterparts. The advances in synthesis protocols and characterization techniques also offer us new insights into the design principles for highly active and stable active sites for small organic molecule oxidation.

11.1 Primary Measurement Methods and Parameters

11.1.1 Primary Measurement Methods

Cyclic voltammetry (CV) has been widely used to obtain qualitative and quantitative information about various electrochemical reactions [14–18]. Power of CV arisen from the ability to fast offer abundant information about the thermodynamics about redox processes and the kinetics of electron transfer reactions and coupling chemical or adsorption reaction processes. CV is usually the first experiment in the electrical analysis. In particular, it provides a rapidly localized electroactive substance of redox potential and facilitates the evaluation of the influence of the medium during the redox processes [19].

CV involves linear scanning of potential at a fixed working electrode (usually in a solution without stirring) by using a triangular potential. On the basis of the information found, a single cycle or multiple cycles could be employed. In a potential scan, the potentiometer records the current value generated by applying the potential. The diagram of current functioned as potential is called a cyclic voltammetric diagram [20–27]. Cyclic voltammograms are complex, time-varying functions of a great deal of chemical and physical parameters. Figure 11.1 shows a typical CV response of reversible redox processes during a potential scanning period. Let us

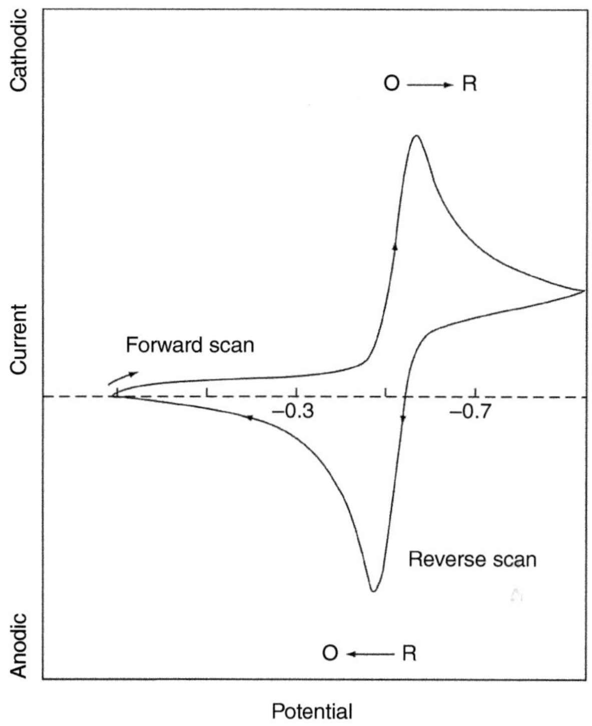

Figure 11.1 Typical cyclic voltammogram for a reversible O + $ne^- \leftrightarrow$ R redox process. Source: Ref. Wang [28]. © 2006, John Wiley & Sons.

assume that only the oxidation state O exists initially. For example, a potential scanning negatively was selected in the first half cycle, starting with values that did not decrease. When the applying potential gets close to the characteristic $E°$ of the redox processes, the cathode current increases until it reaches the highest point, namely the peak. After crossing the potential range where the reduction reaction occurs (exceeding the peak value by $90/n$ mV at least), the scanning direction is reversed. On reverse scanning, R molecules (produced in the first half of the cycle and deposited adjacent to the surface) would be oxidized back to O, forming an anode peak.

The characteristic peaks in the CV are formed by the diffusion layer adjacent to the electrode surface. The peaks could be better understood by examining the concentration–distance profile carefully during a potential scan. It is noted that continuous changes in the surface concentration are accompanied by an increase in the thickness of the diffusion layer (as one expected in a static solution). The obtained peak current reflects the continuous variation of concentration gradients over time. Thus, an increasing peak current reflects the realization of diffusion control, and a decreasing current (above peak) indicates a $t^{-1/2}$ dependence (which is independent from applied potentials). For these reasons, the reverse and forward currents have the same shape. The use of ultramicro electrodes, in which the mass transport is controlled in radial diffusion (rather than linear diffusion), produces an S-shaped cyclic voltammetry diagram.

The CV data are featured by several key parameters, which involve four observables, namely, two peak potential and two peak current. These observables offer the basis for electrochemical information acquirement developed by Nicholson and Shain [29] for understanding the CV response.

In a reversible system: the peak current for a redox couple is presented by the Randles–Sevcik equation (at 25 °C)

$$i_p = (2.69 \times 10^5)n^{2/3}ACD^{1/2}v^{1/2} \tag{11.1}$$

In which n represents the electron transfer number, A is electrode area (which is in the unit of cm^2), D is diffusion coefficient ($cm^3\ s^{-1}$), v is scan rate of potential ($V\ s^{-1}$), and C represents concentration ($mol\ cm^{-3}$). Correspondingly, the current should be proportional directly to increase with the square root of the scan rate as well as concentration. This dependence with the scan rate indicates an electrode reaction governed by mass transport. The forward-to-reverse peak current ratio, $i_{p,f}/i_{p,r}$, is unity for a simple reversible redox process. Particularly, this $i_{p,f}/i_{p,r}$, ratio can be influenced strongly by chemical reaction processes which are coupled to the redox processes. Current peaks are usually tested by extrapolating the baseline current curve.

The peak position (E_p) is correlated to the formal potential of redox processes. The formal potential for a reversible redox process is centered in the middle of $E_{p,a}$ and $E_{p,c}$:

$$E^{\circ} = \frac{E_{p,a} + E_{p,c}}{2} \tag{11.2}$$

The separation between the potentials of redox peaks in the reversible redox processes is given by

$$\Delta E_p = E_{p,a} - E_{p,c} = \frac{0.059}{n}\text{V} \tag{11.3}$$

Therefore, the peak separation is usually used for determining the electron transfer number and as a standard for the Nernstian behavior. Correspondingly, a quick 1e process should show a ΔE_p of c. 0.059 V. Both the anodic and cathodic peak potentials are uncorrelated with scan rate. On the other hand, it has a possibility to correlate the half-peak potential ($E_{p/2}$, the potential where current is half of the peak current) with the polarographic half-wave potential, $E_{1/2}$:

$$E_{p/2} = E_{1/2} \pm \frac{0.028}{N}\text{V} \tag{11.4}$$

In the reversible multielectron transfer reaction processes, the CV diagram consisted of some evident peaks if $E°$ for the individual reaction steps is successively higher and separated well.

The above situation would be very different as the reduction and oxidation reactions are slow, or they are coupled to chemical reactions. Actually, these non-ideal processes normally received the greatest chemical interest and for which the diagnostic power of CV is very useful. This information is normally acquired from comparing the experimental CV data with those obtained from theoretical ones.

In an irreversible and quasi-reversible system, the individual peaks could be reduced in size and widely separated (curve A). Totally, irreversible systems are usually characterized by the shift of peak potential along with scan rate as follows:

$$E_p = E^{\circ} - \frac{RT}{\alpha n_a F}\left[0.78 - \ln\frac{k^{\circ}}{D^{\frac{1}{2}}} + \ln\left(\frac{\alpha n_a F v}{RT}\right)^{1/2}\right] \tag{11.5}$$

where n_a is the charge transfer step, and α represents transfer coefficient. Thus, E_p takes place at the potentials higher than E°, accompanied with overpotentials related to α and k°. Regardless of the k° value, this peak displacement could be compensated by the proper change in the scan rate. The half-peak potential and peak potential will differ by $48/\alpha n$ mV (at 25 °C). Consequently, the CV diagram becomes more drawn- out as decreases of αn.

The peak current is obtained by

$$i_p = (2.99 \times 10^5)n(\alpha n_a)^{1/2}ACD^{1/2}v^{1/2} \tag{11.6}$$

should be proportional to bulk concentration but would be reduced in height, which is dependent on the α value. Assuming that a value of 0.5 is used, the current ratio of the reversible-to-irreversible peaks is 1.27 (e.g. for an irreversible process, the peak current is c. 80% of the peak current in a reversible process).

For quasi-reversible systems which have a k° between 10^{-1} and 10^{-5} cm s^{-1}, the current is governed over both the mass transport and charge transfer. The shape of the CV diagram should be functioned with $k^{\circ}/\sqrt{\pi a D}$ (where $a = nFv/RT$). With the increase in $k^{\circ}/\sqrt{\pi a D}$, the process will approach a reversible process. In the case of small values for $k^{\circ}/\sqrt{\pi a D}$ (e.g. at fast v), the system would show an irreversible behavior. Generally, the CV diagrams for quasi-reversible systems would be more drawn out and present larger separation in peak potentials relative to reversible systems [30–32].

Chronoamperometry method refers to step-up the applied potential on the working electrode from the potential value where the non-Faraday reaction occurs to the value in which effective concentration of the electroactive substance on the working electrode surface becomes zero. A fixed working electrode and a stationary solution should be employed. The resulting relationship of current and time is monitored. Since mass transport occurs only through diffusion at these conditions, the curves in current functioned with time represent the concentration gradient change adjacent to the electrode surface. As the reactants are consumed, the diffusion layer gradually expands, and over time the slope for the concentration curve decreases. Thus, the current (on the planar electrode) decayed over time, displayed by the *Cottrell* equation;

$$i(t) = \frac{nFACD^{1/2}}{\pi^{1/2}t^{1/2}} = kt^{-1/2} \tag{11.7}$$

where A, n, F, C, T, and D are surface area, electron number, Faraday constant, concentration, time, and diffusion coefficient, respectively. This $it^{1/2}$ constant is often referred to as Cottrell behavior. Deviations from this behavior take place over a long period of time (typically over 100 points) due to natural convective effects, coupled chemical reactions, and the use of non-planar or microelectrodes, which have high perimeter area ratios. For the latter case, a current independent of time, while proportional to the concentration, is acquired when $t > 0.1$ seconds because of a large contribution from radial diffusion. Similar considerations are used in spherical electrodes in which the current responses include time-dependent and time-independent terms along with the potential step. Also, recall that for a short value of T ($t < 50$ ms), the current signal includes an extra background contribution from current charging. This exponential decay in charge current stands for a major contribution for the obtained response without electroactive substances.

Chronoamperometry is usually adapted to measure the surface area of the working electrode and the diffusion coefficient of electroactive substances. Chronoamperometry can also be used to study the mechanism of the electrode process. Of particular interest to this task is the reversal of the two-step amperage experiment (the second process is used to detect the fate of the species produced in the first process).

To analyze the charge–time dependence, the potential-step experiment could be further used. This is achieved by integrating the current derived from the potential step and adding corrections with charge in consideration of the reaction with adsorbed species (Q_i) as well as double-layer charging (Q_{dl}):

$$Q = \frac{2nFACD^{1/2}t^{1/2}}{\pi^{1/2}} + Q_{dl} + Q_i \tag{11.8}$$

The procedure for charge measurement, regarded as chronocoulometry, is very useful for testing the quantity of adsorbed reactants; this is due to the capacity to separate charges which were generated by solution and adsorbed species. A curve of charge (Q) relative to $t^{1/2}$, called an Anson plot, produces an intercept at $t = 0$ corresponding to the total value of Q_{dl} and Q_i. The former one could be evaluated by deducting the intercept acquired in the same experiment conducted in a blank solution.

Tafel tests characterize the process information of the reactions tuned by electron transfer rate (e.g. reactions with mass transport fast enough). The relationship of current–potential in these reactions is distinct from those aforesaid-discussed mass-transport-controlled reactions.

For reconsideration of electron transfer reaction of $O + ne^- \leftrightarrow R$, electron transfer between the molecular orbital of O or R and conduction band of the electrode was involved in the true electron transfer step (for example, from the conduction band into an unoccupied orbital of O in a reduction process). The rate of the forward reaction V_f is first order in O:

$$V_f = K_f C_O(0, t) \tag{11.9}$$

while that of the reversed reaction V_b is first order in R:

$$V_b = K_R(0, t) \tag{11.10}$$

where k_b and k_f are the backward and forward rate constants, respectively. These constants are dependent on the applied potential following these exponential relationships:

$$k_f = k^{\circ} \exp[-\alpha nF(E - E^{\circ})/RT] \tag{11.11}$$

$$k_b = k^{\circ} \exp[(1 - \alpha)nF(E - E^{\circ})/RT] \tag{11.12}$$

where flutter represents the transfer coefficient, and $k°$ represents the standard rate constant. The $k°$ value ($\mathrm{cm\,s^{-1}}$) indicates the reaction information between electrode material used and specific reactant. The different values in dimensions (zero or one dimension) represent the symmetry of free energy curves, which are relative to the products and reactants. In regard to symmetric curves, the plant spacing approaches 0.5. Lateral energy is a representation of an energy fraction that enters the system for actually reducing the activation energy. Generally, Eqs. (11.11) and (11.12) show that changing in the potential used would exponentially affect k_f and k_b. Both negative and positive potentials thus accelerate redox processes. For the oxidation process, the energy from the electron in R's donor orbital should be higher than or equal to the energy from the electron on the electrode. For the reduction process, the energy of electrons on the electrode should be in turn higher than their energy in the R's acceptor orbit.

Since the net reaction rate can be depicted as:

$$V_{\mathrm{net}} = V_f - V_b = k_f C_{\mathrm{O}}(0, t) - k_b C_{\mathrm{R}}(0, t) \tag{11.13}$$

and as backward and forward currents are proportional to V_b and V_f, respectively

$$i_f = nFAV_f \tag{11.14}$$

$$i_b = nFAV_b \tag{11.15}$$

the total current can be given by the difference between the currents based on the backward and forward processes:

$$i_{\mathrm{net}} = i_t - i_b = nFA[k_t C_{\mathrm{O}}(0, t) - k_b C_{\mathrm{R}}(0, t)] \tag{11.16}$$

By replacing the expressions of k_f and k_b (Eqs. (11.10) and (11.11), respectively), one can obtain the *Butler–Volmer equation*:

$$i = nFAk^{\circ} \left\{ C_{\mathrm{O}}(0, t) \exp\left[-\frac{\alpha nF(E - E^{\circ})}{RT} - C_{\mathrm{R}}(0, t) \exp(1 - \alpha)nF(E - E^{\circ})/RT \right] \right\} \tag{11.17}$$

The relationship between reaction current and potential, which is controlled by the electron transfer rate, is described. It is noted that the net current relies highly on the concentration of each reduction/oxidation reaction species near the surface, as well as applied potential. For instance, Figure 11.2 displays the dependence on current-potential of $C_{\mathrm{O}}(0,t) = C_{\mathrm{R}}(0,t)$ and reuse = 0.50. A big negative potential can promote the charge motion in the cathode direction or slow it down toward the anode direction. Consequently, the anode current would become insignificant; meanwhile, the net current is combined with the cathode component. The acceleration/deceleration of the cathode current and anode current is not symmetrical necessarily. When the value is different from 0.5, the value will also be different.

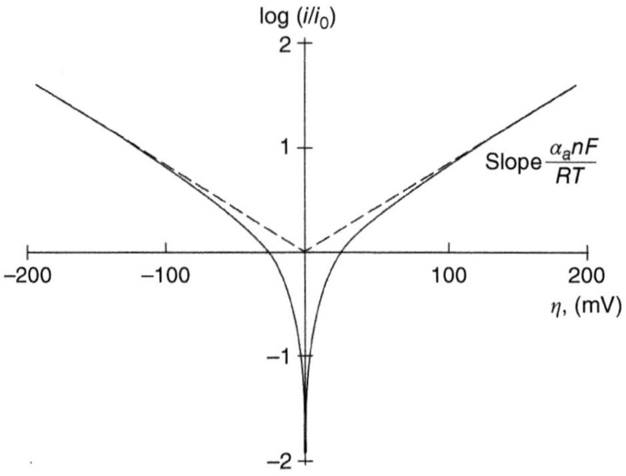

Figure 11.2 Tafel plots for cathodic and anodic branches of the current–potential curve. Source: Ref. Wang [28]. © 2006, John Wiley & Sons.

Similarly, the contribution of the cathode current is not found at the positive potential large enough.

No net current will flow when E is equal to $E°$. However, this case would be changed and kept dynamic, with carriers moving continuously in both directions and the anode and cathode current components being equal and opposite. At $E°$, the absolute value of these components was defined as the exchange current of i_o. This value is in direct proportion to the standard rate constant:

$$i_o = i_c = i_a = nFAk^{\circ}C \tag{11.18}$$

The value of 10^{-6} μA cm^{-2} is within the range of exchange current density for usual redox couples (25 °C). Thus, we can compose the Butler–Volmer equation as follows

$$i = i_o \left[\exp\left(-\frac{\alpha nF\eta}{RT}\right) - \exp(1-\alpha)\frac{nF\eta}{RT}\right] \tag{11.19}$$

in which $\eta = E - E_{\text{eq}}$ represents overvoltage. We can define this overvoltage in regard to a specific reaction, which has a known equilibrium potential.

As key kinetic parameters, i_o and α can be extracted from Eq. (11.19). In the exponential terms in Eq. (11.19), one can be ignored in comparison to the other in the case of overvoltages large enough (>118/n mV). For instance, in the case of big negative overpotentials like $i_c \gg i_{a,}$ and Eq. (11.19) will become

$$i = i_o \exp\left(-\frac{\alpha nF\eta}{RT}\right) \tag{11.20}$$

and hence, it can be obtained that

$$\ln i = \ln i_o - \alpha nF\eta/RT \tag{11.21}$$

Tafel [33] derive the relationship between logarithmic current and potential, as known as the Tafel equation. By plotting the $\log I$ oppose η tafel plot branch of cathode and anode curve between current and overvoltage. The graph is only linear if it is over potential, and as we approach zero, we observe a severe deviation from linearity. Extrapolating the linear parts of these curves to zero provides the understanding for $\log i_o$, and the slope gives you the value of the transfer coefficient. Equation (11.21) is rearranged to obtain another form of the Tafel equation:

$$\eta = a - b \log i \tag{11.22}$$

Tafel slope processes the value of $2.303RT/\alpha nF$. In the case of $n = 1$, $\alpha = 0.5$, the value is 118 mV (room temperature). Equation (11.22) exhibits that a small potential used, which only exceeds a little above equilibrium potential, could dramatically increase the current. However, the current cannot actually be increased to infinity due to the limits of the reactant transport rate to the electrode surface.

As the potential departure from $E°$ in a small value, linearization takes place in the exponential term of Eq. (11.20), with the current being in direct proportion to η:

$$i = i_0 nF\eta/RT \tag{11.23}$$

Hence, in a small potential region around $E°$, the net current is in direct proportion to the overvoltage.

It is also noted that the net current is zero in the case that $E = E_{eq}$, the equation of Eq. (11.17) can be obtained:

$$C_O(0,t)\exp[-\alpha nF(E - E^{\circ})/RT] = C_R(0,t)\exp[(1-\alpha)nF(E - E^{\circ})/RT] \tag{11.24}$$

An exponential form of Nernst equation would be generated from rearrangement of Eq. (11.24)

$$\frac{C_O(0,t)}{C_R(0,t)} = \exp[nF(E - E^{\circ})/RT] \tag{11.25}$$

The formal potential is thus correlated to equilibrium potential in a known reaction:

$$E_{eq} = E^{\circ} + (2.3RT/nF)\log Q \tag{11.26}$$

in which the equilibrium ratio function *is* Q.

11.1.2 Primary Parameter

Electrochemical surface area (ECSA) [34, 35]: as an important parameter, the ECSA can be used to judge mass and/or specific activity of electrocatalysts. For such calculation of ECSA, many methods can be used: using the hydrogen adsorption layer as so-called H_{upd} region in the potential scan, or using pre-adsorbed monolayer detachment oxidation (such as potentiodynamic), or employing a potential experimental step (by using chronoamperometric), as presented by, Kinoshita et al. previously. Alternatively, using Cu, which may be underdeposited, is also possible, but this has a major drawback because that the catalytic reactions would be greatly affected by metal ions and that it is really hard to clean the glass cells used. Therefore, advanced technologies must be used; one of them is differential electrochemical mass spectrometry (DEMS). However, such a technique is not cost effective and cannot be available for most research groups.

Traditional H_{upd} areas can be applied for electrocatalysts with the high surface area because of the convenience and speed of CV. However, these data are also used to determine surface area when studying co-oxidation reactions. The ECSA was calculated using H_{upd} and coextraction. Figure 11.3 shows the potential change curves of the three nanocatalysts under different loads. Resume catalyst shows the general features of Pt electrode in perchlorate solution, H_{upd} cost estimate false capacitance is usually a straight line in the double area after calibration, and Figure 11.3 below –CO- free curve record report CO-free and saturated solution of 0.05 V strip CO dissolution peak area between adjustment according to the first background and resume is considered in the potential region between 0.4 and 1.0 V anion adsorption (OH, and ClO_4^-). In the adsorption process of H_{upd} oxidation,

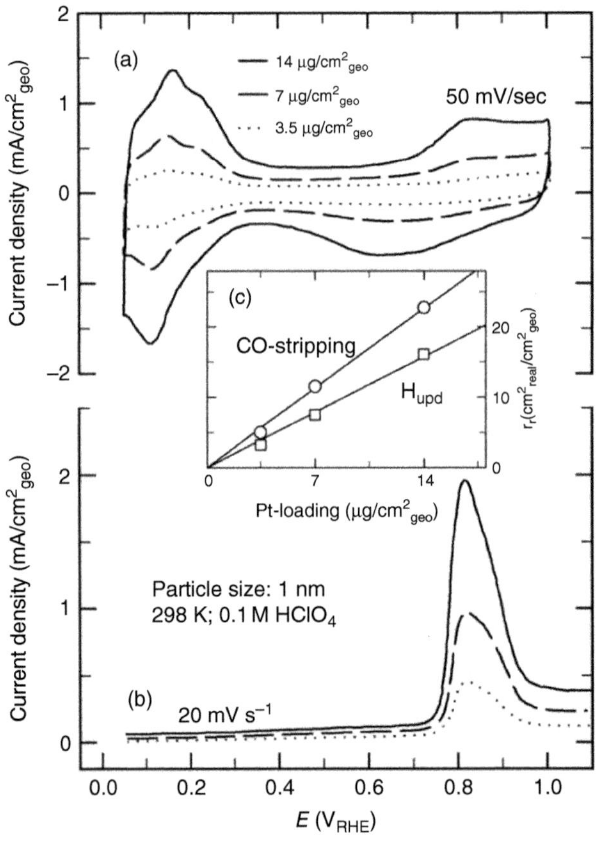

Figure 11.3 CV (a) and CO stripping (b) for the 1 nm catalyst for Pt loadings of 14, 7, and 3.5 µg cm^{-2}, respectively, in 0.1 M $HClO_4$ solution at room temperature. Inset (c): roughness factor η calculated from (a) and its dependence on the Pt loading. Source: Ref. Mayrhofer et al. [36]. © 2008, Elsevier.

only one electron is transferred, while CO oxidation to CO_2 requires 2e, so the charge should be reduced by half correspondingly.

The charges for Q_H and Q_{CO} can thus be presented as follows

$$Q_H = \frac{\int idE}{v} \quad \text{and} \quad Q_{CO} = \frac{\int idE}{2v} \tag{11.27}$$

with the scan rate v, the potential E, and the current i, the catalyst's roughness factor of r_f representing the ratio of the real surface area of Pt (A_{real}) and electrode geometric area of (A_{geo}) could be obtained as follows:

$$r_f = \frac{A_{real}}{A_{geo}} = \frac{(Q_H \text{ or } Q_{CO})/0.238\ \text{cm}^2}{195\ \mu C\ \text{cm}^{-2}} \tag{11.28}$$

In which 0.238 cm^2 is used for geometric electrode area, and 195 $\mu C\ cm^{-2}$ is used for the charge density of bulk polycrystalline Pt.

The linear relation of roughness factor with Pt loading in an 1 nm electrocatalyst is shown in Figure 11.3, which has an intercept of 1 initially in the region for H_{upd}- and CO dissolution. After linear approximation, this slope is approximately equal to the $m^2\ g^{-1}{}_{Pt}$ electrochemically active surface area (ECA) of the electrocatalyst, which has a big surface area. It should be noted that the surface area calculated by CO is 1.4 times that of H_{upd}, 154 – 112 $m^2\ g^{-1}{}_{Pt}$, respectively. At 2 nm, 5 nm, and nanostructured Pt film (NSTF), the rates for CO stripping and H_{upd} charge decreased along with the change of particle size by the value from 1.4 to c. 1.25, 1.1, and finally 1.0, respectively. One reason for the observed differences may be that of different sizes having different saturation coverage.

Besides, there is a systematic error in determining surface area by the amount of charge corrected by constant bilayer capacitance. This traditional background correction applies only to large Pt electrodes, while the carbon carrier has a significant effect on high surface area catalysts. To make it clearer, the authors propose a novel approach for evaluation and subtraction of the carbon carrier capacity before the traditional Pt bilayer correction. This process is shown in Figure 11.4. The CO volume oxidation curve of CV in the Ar purification solution is compared with that of the 1 nm catalyst. For the background correction, ordinary company peel curves can also be employed; however, it should anneal the surface, which can prevent the formation of co-oxidizing defects in the early ignition area lower than 0.8 V vs. RHE for the two cases; supported capability is not invariable along with potential but shows the slope areas lower than 0.4 V vs. RHE, and therefore cannot be ignored. Therefore, before traditional correction by double-layer for Pt As in, carbon carrier capacity correction (minus gray line) must be performed for CV in the H_{upd} region. During that process, the cost of the H_{upd} would approximately equal to that of the CO-stripping process.

The aforesaid two methods of measuring the platinum catalyst ECSA, that is, the CO-stripping and H_{upd} process, are both important and beneficial. However, it should also be noted that these two methods have somewhat systematic errors. For correcting these errors and accurately evaluate ECSA from H_{upd}, it should be deduced of the background, especially on carbon carriers.

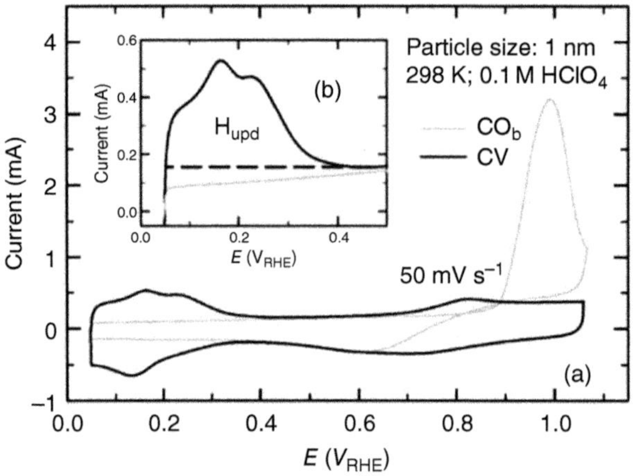

Figure 11.4 Using CO bulk oxidation (CO_b) correct the capacity caused by the carbon support. CV (black) in argon-purged solution and the CO_b curve (gray) (a). The inset (b) shows the general capacity of the H_{upd} area (black dots), and the appropriate corrections for the CO_b are shown in gray. In this case, the H_{upd}: co-stripping ratio will be 1.0 : 1.23 at the regular stripping and 1.0 : 1.0 at the appropriate background stripping. Source: Ref. Mayrhofer et al. [36]. © 2008, Elsevier.

The double layer can also represent the ECSA of a catalyst. A lot of other methods can be used to determine ECSA. However, only for the catalysts with similar physical and chemical properties can the method be applicable for comparison; it cannot be applied for comparison in different studies. Here, anode's and cathode's double-layer charging/discharging currents functioned as scanning rates are tested.

A linear line will be generated from the obtained plot between scan rate and the difference in double-layer charging/discharging currents ($\triangle j(j_a - j_c)$), and double-layer capacitance of nanomaterial can be depicted by the slope. Under most situations, this method is applicable to evaluate trends of electro-catalysis activity of nanomaterials. It is also shown a typical example of studying Co oxide films coated on fluorine-doped tin oxide (FTO) compared to the bare FTO substrate by using the C_{dl} method. The CVs of both Co_xO_y/FTO and FTO before activation and Co_xO_y/FTO after activation indicate an increase of C_{dl} after activation, which proves an increase of ECSA. The result here agrees well with the trend of electrocatalytic activity.

Some obvious drawbacks still exist even that the method, as mentioned above, has the ability to compare the activity trend of similar nanomaterials. First, ion adsorption and desorption processes in the non-Faraday are used to measure the relative ECSA, while Faraday processes refer to gas generation reaction processes. It is not proper to associate C_{dl}-derived ECSA to trends in the activity of pure Faraday electrocatalysts. Therefore, there is no need to worry if the relative ECSA found using the C_{dl} method is not consistent with the OER or HER activity trend of the catalyst. Instead, to make the correlation, other more reliable methods need to be used to determine ECSA. For most situations, when this difference in activity trends between ECSA and Faraday gas generation as measured by C_{dl}, it is best to use other electrical analysis techniques (such as EIS) [32, 37] to investigate the cause. Therefore, it can be concluded that nanomaterial's ECSA cannot be taken as a basic parameter for activity evaluation. However, it can be used in a study to evaluate the activity trends of electrocatalysts. In addition, to provide more accurate data for evaluation of active trend, employing other methods to determine ECSA is also a requisite.

References

1 Ong, B.C., Kamarudin, S.K., and Basri, S. (2017). *Int. J. Hydrogen Energy* 42: 10142–10157.
2 Zhang, Z.H., Liu, J., Gu, J.J. et al. (2014). *Energy Environ. Sci.* 7: 2535–2558.
3 Kamarudin, S.K., Achmad, F., and Daud, W.R.W. (2009). *Int. J. Hydrogen Energy* 34: 6902–6916.
4 Lamy, C., Rousseau, S., Belgsir, E.M. et al. (2004). *Electrochim. Acta* 49: 3901–3908.
5 Martínez-Huerta, M.V. and Lázaro, M.J. (2017). *Catal. Today* 285: 3–12.
6 Serov, A. and Kwak, C. (2009). *Appl. Catal., B* 90: 313–320.
7 Zhang, L.K., Wu, W., Jiang, Z., and Fang, T. (2018). *Chem. Pap.* 72: 2121–2135.
8 Zhou, W.J., Zhou, Z.H., Song, S.Q. et al. (2003). *Appl. Catal., B* 46: 273–285.
9 Antolini, E. and Gonzalez, E.R. (2010). *J. Power Sources* 195: 3431–3450.
10 Yu, E.H., Krewer, U., and Scott, K. (2010). *Energies (Basel, Switz.)* 3: 1499–1528.
11 Vigier, F., Coutanceau, C., Perrard, A. et al. (2004). *J. Appl. Electrochem.* 34: 439–446.
12 Yin, Z., Lin, L.L., and Ma, D. (2014). *Catal. Sci. Technol.* 4: 4116–4128.
13 Ren, M.J., Zou, L.L., Yuan, T. et al. (2014). *J. Power Sources* 267: 527–532.
14 Alcaide, F., Alvarez, G., Cabot, P.L. et al. (2011). *Int. J. Hydrogen Energy* 36: 4432–4439.
15 Arenz, M., Stamenkovic, V., Schmidt, T.J. et al. (2003). *Phys. Chem. Chem. Phys.* 5: 4242–4251.
16 Bianchini, C. and Shen, P.K. (2009). *Chem. Rev.* 109: 4183–4206.
17 Brankovic, S.R., Wang, J.X., and Adzic, R.R. (2001). *Electrochem. Solid-State Lett.* 4: A217–A220.
18 Bruno, M.M., Viva, F.A., Petruccelli, M.A., and Corti, H.R. (2015). *J. Power Sources* 278: 458–463.
19 Ishikawa, K., Ohyama, J., Okubo, K. et al. (2020). *ACS Appl. Mater. Interfaces* 12: 22771–22777.
20 Lu, Y.Z. and Chen, W. (2010). *J. Phys. Chem. C* 114: 21190–21200.
21 Maiyalagan, T. and Viswanathan, B. (2008). *J. Power Sources* 175: 789–793.
22 Matsuoka, K., Iriyama, Y., Abe, T. et al. (2005). *J. Electrochem. Soc.* 152: A729–A731.
23 Mitsushima, S., Koizumi, Y., Uzuka, S., and Ota, K.I. (2008). *Electrochim. Acta* 54: 455–460.

24 Neto, A.O., Franco, E.G., Arico, E. et al. (2003). *J. Eur. Ceram. Soc.* 23: 2987–2992.

25 Neto, A.O., Linardi, M., and Spinace, E.V. (2006). *Ionics* 12: 309–313.

26 Prabhuram, J., Manoharan, R., and Vasan, H.N. (1998). *J. Appl. Electrochem.* 28: 935–941.

27 Rincon, A., Perez, M.C., and Gutierrez, C. (2010). *Electrochim. Acta* 55: 3152–3156.

28 Wang, J. (2000). *Analytical Electrochemistry*. New York: Wiley VCH.

29 Nicholson, R.S. and Shain, I. (1964). *Anal. Chem.* 36: 706.

30 Rocha, R.S., Camargo, L.M., Lanza, M.R.V., and Bertazzoli, R. (2010). *Electrocatalysis* 1: 224–229.

31 Rocha, R.S., Reis, R.M., Lanza, M.R.V., and Bertazzoli, R. (2013). *Electrochim. Acta* 87: 606–610.

32 Spinner, N. and Mustain, W.E. (2013). *J. Electrochem. Soc.* 160: F1275–F1281.

33 Deng, J., Lin, S.C., Fuller, J. 3rd et al. (2020). *Nat. Commun.* 11: 3686.

34 Jafarian, M., Mahjani, M.G., Heli, H. et al. (2003). *Electrochem. Commun.* 5: 184–188.

35 Joglekar, M., Nguyen, V., Pylypenko, S. et al. (2016). *J. Am. Chem. Soc.* 138: 116–125.

36 Mayrhofer, K.J.J., Strmcnik, D., Blizanac, B.B. et al. (2008). *Electrochim. Acta* 53: 3181–3188.

37 Ma, M., Jin, B.J., Li, P. et al. (2017). *Adv. Sci. (Weinheim, Ger.)* 4: 1700379.

12

C1 Molecule Oxidation

12.1 Methane Oxidation

12.1.1 Reaction Mechanism

A typical electrolytic cell consists of a cathode, an anode, a membrane in between cathode and anode, and also an external circuit (Figure 12.1). The ion exchange membrane allows only proton or oxide ions to be transported. Generally, oxidation occurs on the anode surface, and reduction occurs on cathode surface.

To apply voltage or current, the external circuit should be used. Some possible reaction pathways and products for electrocatalytic methane conversion are shown in Table 12.1, in which the associated redox potential at 25 °C is relative to a normal hydrogen electrode.

Thermodynamically, on the basis of the fact that the redox potential needed for producing CO_2 is the lowest, thus the formation of CO_2 should be the easiest reaction (Table 12.1). There are also some other products with partial oxidation, such as CH_3OH, HCOOH, HCHO, and CO, and the formation of oxidation coupled products. Compared with CO_2, C_2H_6, and C_2H_4 have more redox potentials. This makes it difficult to realize the oxidative CH_4 coupling and electrocatalytic partial CH_4 oxidation [1].

12.1.1.1 Solid–Liquid–Gas Reaction System

Gas-phase methane, solid-phase electrodes, and liquid-phase electrolytes are the three key components for conducting solid–liquid–gas reactions. The acid–base nature of the media has a significant effect on the electrode and its surfactant. Therefore, two categories were involved in the reaction media, namely, acidic media and/or basic/neutral media. In this chapter, the research progress of the solid–liquid–gas electrocatalytic transformation of CH_4 based on two reaction media is reviewed.

12.1.2 Acidic Media

Electrophiles like Hg II and Pt II can be used for activation of CH_4 in highly acidic media [2–4]. The 2e oxidation of CH_4 can be tuned by electrophilic ions to generate methyl bisulfate (CH_3OSO_3H), and methyl bisulfate is not overoxidized due to

Electrocatalysis in Balancing the Natural Carbon Cycle, First Edition. Yaobing Wang.
© 2021 WILEY-VCH GmbH. Published 2021 by WILEY-VCH GmbH.

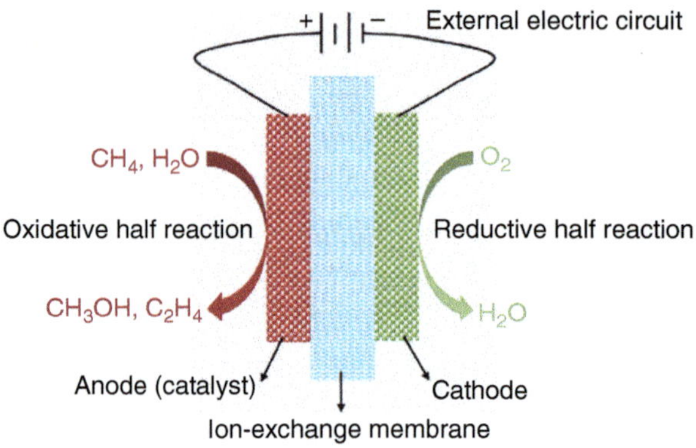

Figure 12.1 Schematic diagram of a typical electrocatalytic cell for methane conversion. Source: Xie et al. [1]. © 2018, Elsevier.

Table 12.1 Some reactions may be related to the electrocatalytic conversion of CH_4.

Reaction	*E* (V) vs. NHE (pH = 0, temperature = 298 k)
$CH_4(g) + H_2O(l) \rightarrow H_3OH(a) + 2H^+(a) + 2e^-$	0.58
$CH_4(g) + H_2O(l) \rightarrow HCHO(a) + 4H^+(a) + 4e^-$	0.46
$CH_4(g) + 2H_2O(l) \rightarrow HCOOH(a) + 6H^+(a) + 6e^-$	0.26
$CH_4(g) + H_2O(l) \rightarrow CO(g) + 6H^+(a) + 6e^-$	0.26
$CH_4(g) + 2H_2O(l) \rightarrow CO_2(g) + 8H^+(a) + 8e-$	0.17
$2CH_4(g) \rightarrow C_2H_6(g) + 2H^+(a) + 2e^-$	0.35
$2CH_4(g) \rightarrow C_2H_4(g) + 4H^+(a) + 4e^-$	0.44

the electron absorption for the HSO_4 group (Eq. (12.1)). Methyl hydrogen sulfate can be hydrolyzed to the product of methanol (Eq. (12.2)). H_2SO_4 is an oxidant and can be converted to SO_2 in such a reaction (Eq. (12.1)). Therefore, the reoxidation of SO_2 to H_2SO_4 is essential for forming such a complete cycle (Eq. (12.3)). Equation (12.4) represents the aforesaid net reaction.

$$CH_4 + 2H_2SO_4 \rightarrow CH_3OSO_3H + SO_2 + 2H_2O \tag{12.1}$$

$$CH_3OSO_3H + H_2O \rightarrow CH_3OH + H_2SO_4 \tag{12.2}$$

$$SO_2 + 1/2O_2 + H_2O \rightarrow H_2SO_4 \tag{12.3}$$

$$\text{Net: } CH_4 + 1/2O_2 \rightarrow CH_3OH \tag{12.4}$$

With the exception of the Pt catalyst, only high-valence forms of most electrocatalysts can be used to activate CH_4 for generating CH_3OH, which is reduced to a low-priced substance (Figure 12.2) [5]. To evaluate possible driving force for CH_4 oxidation functionalization, the potential for 2e redox of metal centers should be analyzed. Thus, it is expected that metal ions and high oxidation–reduction potential CH_4 will be used more rapidly, but at the same time may hinder the oxidants in

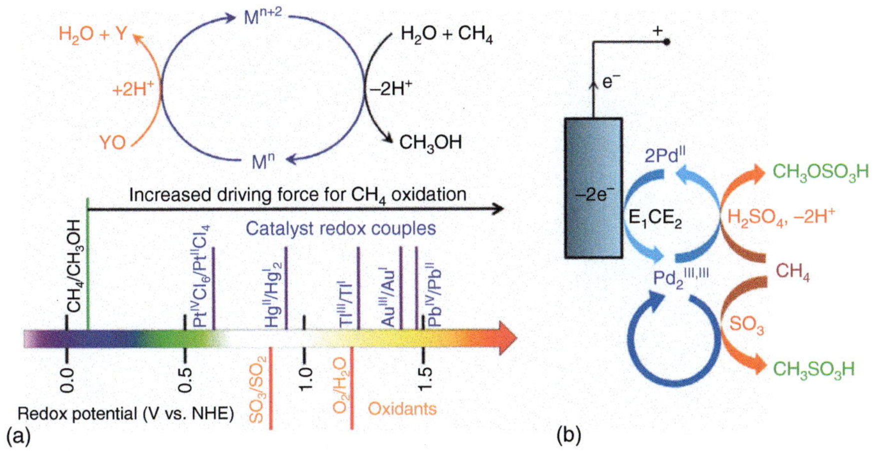

Figure 12.2 (a) Simplified catalytic cycle for electrophilic methane oxidation using stoichiometric oxidant YO and redox potentials of methane oxidation catalysts. (b) Proposed mechanism for electrochemical methane functionalization by a putative $Pd_2^{III,III}$ intermediate. Source: Xie et al. [1]. © 2018, Elsevier.

the reoxidation catalyst (Figure 12.2). In fact, the status of the reoxidation catalyst in almost all CH_4 functional cycles is usually the pathogen. Electrochemically regenerated electrophilic high-price catalysts can overcome this oxidation problem [5].

The driving force for reoxidation is controlled systematically by changing the electrode potential to remain abundant, highly active high-valance substances. Based on this, Zimmermann and colleagues [5] reported that oxidation of $PdSO_4$ electrolyte condensed sulfuric acid generates a hypothesized Pd_2^{III}, and CH_4 could be fast activated by these species to form methyl hydrogen sulfate (CH_3OSO_3H) and methyl sulfonic acid (CH_3 3H), the methanol precursors (Figure 12.2). This simple system catalyzes CH_4 by highly functionalized high-temperature reactions with no preconditions, reaching a 2300 H-1 flip frequency at 140 °C. Therefore, the use of electrochemical methods to generate and maintain the high activity of high-priced metals is useful. Functionalizing other inert substrates can be achieved by this strategy via accessing catalytic intermediates, thus accelerating a variety of challenging reactions [6].

12.1.3 Alkaline or Neutral Media

Electrochemical CH_4 conversion can occur in basic or neutral media, and many anodic materials, including metal hydroxides and oxides, can be used for such conversion. Nickel hydroxide-modified electrode was developed by Jafarian et al. [7] to show CH_4 electro-oxidation activity in an alkaline solution, while CO_2 remained the main product. Dong and coworkers [8] prepared a nickel hydroxide modified by multiwall carbon nanotubes (CNTs) to form CH_3OH by double electron transfer. Recently, Pt/C cathode, a metal oxide anode, and an alkaline electrolyte film were consisted of Lu and Zhuang [9, 10] to fabricate the alkaline fuel cell reactor

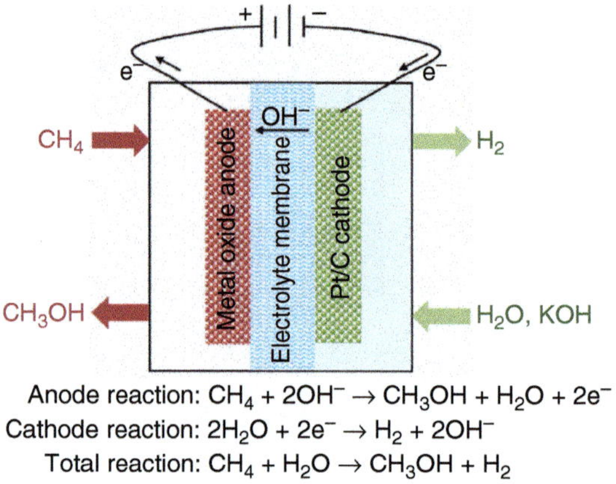

Figure 12.3 Schematic illustration of alkaline fuel cell reactor using alkaline electrolyte membrane for methane conversions. Source: Xie et al. [1]. © 2018, Elsevier.

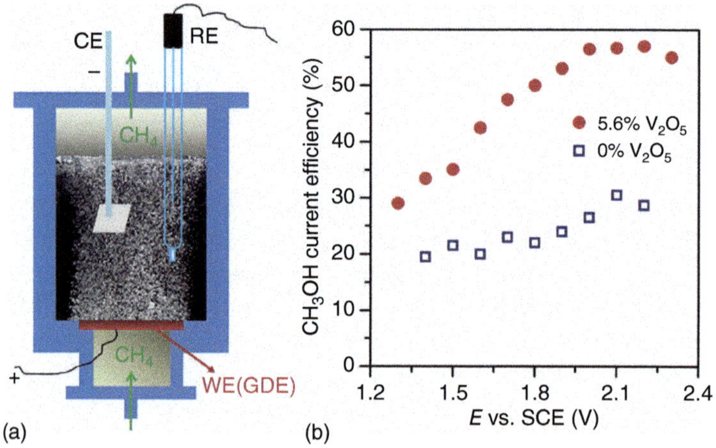

Figure 12.4 (a) Schematic illustration of the electrocatalytic cell using gas diffusion electrode (GDS) for methane conversions. (b) Methanol current efficiency of TiO_2/RuO_2/PTFE GDE with 5.6% V_2O_5 or without V_2O_5. Source: Xie et al. [1]. © 2018, Elsevier.

as shown in Figure 12.3. CH_4 electro-oxidation generates CH_3OH at 160 °C on the metal oxide catalyst, with high selectivity for CH_3OH generation obtained on the anode of $Ni(OH)_2/CoO(9:1)/Ag/C/CeO_2(85/5/10)$. It is found that the NiOOH led to the complete oxidation of CH_4 on the electrode surfaces. In addition, the cathode also generates H_2. Further basic research for the electrocatalyst is required.

Gas diffusion electrode (GDE) consists of porous structure in the electrode together with high conductivity, making its wide use in CO_2 electrochemical reduction, fuel cells, depolarized O cathode of chlor-alkali industry, and oxygen reduction electrode for electrosynthesis [11]. These applications are most based on the fact that more contact interfaces are generated in the GDE. Therefore, for the electro-oxidation of CH_4, GDE may be a promising electrode due to the solution to overcome the bottleneck of low CH_4 solubility in the most electrolyte (Figure 12.4).

12.2 Methanol Oxidation

12.2.1 Reaction Thermodynamics and Mechanism

Electrocatalytic methanol oxidation was widely received study for developing direct methanol fuel cells (DMFCs) [12–15] Due to the occurrence of carbonate generation in alkaline media, methanol oxidation is usually performed in acidic.

$$CH_3OH + H_2O \rightarrow CO_2 + 6H^+ + 6e^- \quad (E^0 = 0.04\,V) \tag{12.5}$$

The thermodynamic equilibrium potential of methanol oxidation reaction (MOR) and hydrogen reaction is close. However, 6e transfer was involved in the MOR accompanied with the generation of various intermediates. 2e is involved in the hydrogen reaction. Thus, these intermediates on the surface may cause catalyst poisoning in DMFCs. As a result, the development of excellent catalysts to avoid catalyst poisoning from surface intermediates produced partial oxidation is necessary.

The basic steps of methanol adsorption, dehydrogenation, and subsequent oxygen introduction lead to the formation of CO_2. For such processes, Pt is the best available catalyst [16]. Bimetallic systems (Pt–M) as a second substrate, the generation of oxophilic species from the water on metal surfaces (M) at relatively negative potentials, beneficial for partial oxidation intermediates to be removed away from surfaces of electrocatalysts, thus the poisoning effect would be mitigated on the catalyst surfaces [17–19]. Some of the possible reaction processes are shown in Figure 12.5.

As mentioned earlier, other related surface poisoning effects, as well as the sluggish kinetics, are the main basic limitations or disadvantages of Pt. High electrocatalyst loading and/or using high-purity fuels may benefit to overcome these limitations. Pt catalysts have the disadvantages of low efficiency, sluggish kinetics, and high cost because of its scarcity. All of these disadvantages prevent its

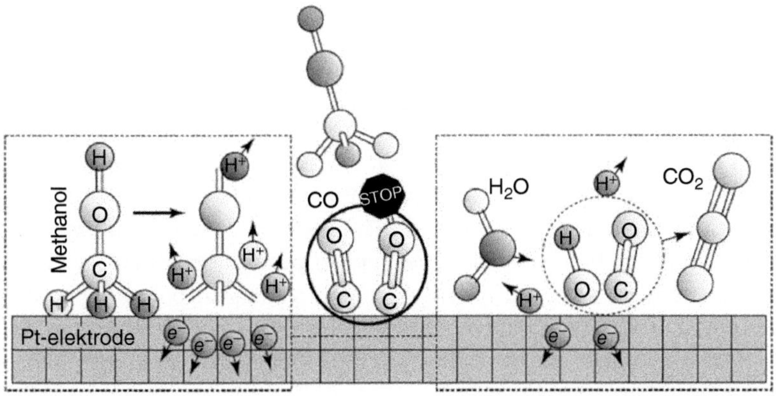

Figure 12.5 Schematic diagram of different reaction steps in the methanol oxidation process on the surface of the model catalyst. Source: Kakati et al. [16]. © 2014, American Chemical Society.

wide commercialization applications in fuel cells. Thus, developing electrocatalyst candidates, which possess high activity and stability, as well as cost-effective, is indispensable. Luckily, ultralow content of Pt electrocatalysts was used in the fuel cells by current studies [20–25]. Besides, some studies show that non-Pt electrocatalysts can be used to displace Pt [26]. To further improve the reaction kinetics of these non-Pt electrocatalysts, the studies showed that applications of alkaline media to replace the acid media are a choice [27]. However, the low chemical stability of anion exchange membranes has become a key issue in alkaline fuel cells.

Many studies were focused on the anode electrocatalysts in acidic electrolytes for DMFCs. Research shows that to achieve large-scale practical application of DMFC, the application of Pt catalysts in DMFC is inevitable. Pt has been modified by other metals [17, 27–37] or formed core–shells [38–44], metal oxides [45–71], and other nanostructured particles [72–82] in order to obtain the expected Pt-based catalysts for methanol oxidation. Carbon carriers are commonly used in fuel cells to increase the dispersion of precious metals and to facilitate charge transfer in external circuits. Recent advances in carbon nanomaterials, including layered structures, e.g. CNTs and graphene, have broadened the range of catalyst support materials. These developments provide alternative directions for the development of fuel cell electrocatalyst systems. Due to the unique characteristics of nanomaterials, especially the vector transfer of charge in the dimensional support, high efficiency can be achieved [83]. Other advantages include good chemical resistivity, high surface area, and excellent material strength.

12.2.2 Catalyst Advances

12.2.2.1 Pd-Based Catalysts

As is well known that in acidic solutions, methanol electro-oxidation on Pd electrocatalysts is difficult [84]. Interestingly, according to the mechanism proposed below in alkaline media, Pd electrocatalysts are favored for methanol electro-oxidation [27, 85–91]:

$$\mathrm{Pd} + \mathrm{CH_3OH} \leftrightarrow \mathrm{Pd} - \mathrm{CH_3OH_{ads}} \tag{12.6}$$

$$\mathrm{Pd} - \mathrm{CH_3OH_{ads}} + n\mathrm{OH^-} \rightarrow \mathrm{Pd} - \mathrm{intermediates} + n\mathrm{H_2O} + ne^- \tag{12.7}$$

In these reaction processes, intermediates of $-CHO_{ads}$, $-CH_3O_{ads}$, $-CH_2O_{ads}$, $-CO_{ads}$, –COOH, etc. may be generated from methanol oxidation, which can be further reacted on Pd catalyst surface with the OH– species generated, to ultimately produce the H_2O and CO_2. In this reaction mechanism, adsorption with CO species on the Pd surface may be included in the final step leading to poisoning of Pd catalysts, thereby MOR may be inhibited to lower catalytic activity. To remove these poisoning species and refresh Pd active sites, additional chemical reactions on the surface should be employed [86, 89].

$NaBH_4$ reduction approach was used to produce Pd–Ag/CNT and Pd–Ag/C electrocatalysts by Wang et al. [92]. These catalysts exhibit higher activity for MOR than Pd/C, and they have better CO tolerance, more negative reaction potentials, and better stability. Besides increasing kinetics, the additional Ag activates water at a lower potential under the bifunctional mechanism. Thus, the surface-adsorbed CO may be oxidized and removed, refreshing more active Pd sites.

Yin et al. investigated bimetallic PdAu particles for MOR in alkaline electrolyte [93]. The preparation of unique PdAu nanoparticles with Au-rich core and/or Pd-rich shell was realized by emulsion-assisted synthesis. As determined by the component test, $Pd_{64}Au_{36}$, $Pd_{45}Au_{55}$, $Pd_{30}Au_{70}$ with total metal content ranging from 10 to 20 wt% on C support were obtained. $Pd_{30}Au_{70}$ particles have the best performance, as depicted in Figure 12.6, which is obviously better than Pt/C electrocatalysts obtained commercial approaches. The initial maximum currents on $Pd_{30}Au_{70}$/C obtained in the chronoamperometric test (Figure 12.6) were clearly larger compared to the corresponding values on Pd/C electrocatalyst. This result suggests the $Pd_{30}Au_{70}$/C has better tolerance to poisoning intermediates,

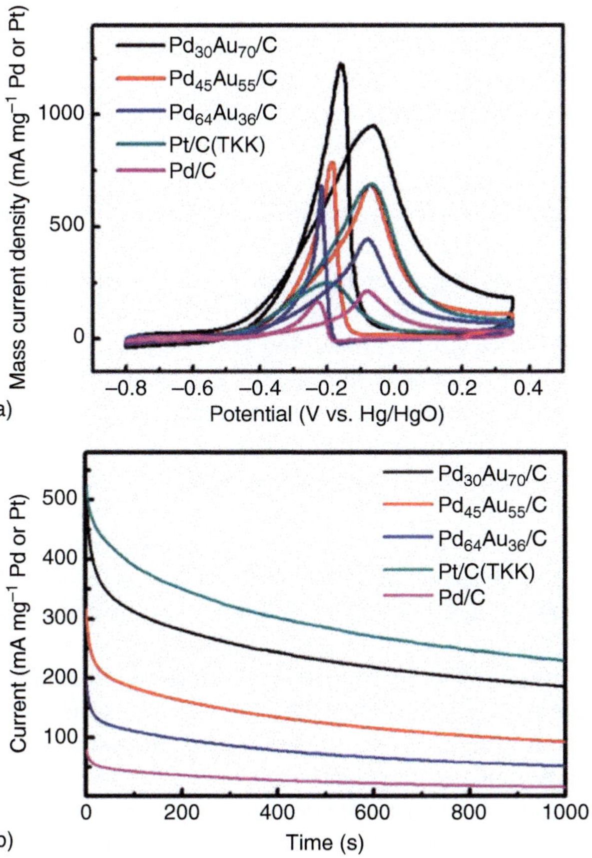

Figure 12.6 (a) Cyclic voltammograms methanol oxidation on the heat-treated Pd/C, PdAu/C, and commercial Pt/C (TKK) catalysts in 1 M KOH + 1 M CH_3OH at 50 mV s^{-1}, room temperature. (b) Chronoamperometry measurements of methanol oxidation with PdAu/C, Pd/C, and commercial Pt/C (TKK) catalysts at −0.2 V vs. Hg/HgO in 1 M CH_3OH and 1 M KOH solution. Source: Yin et al. [93]. © 2013, Royal Society of Chemistry.

which is the reason for the better MOR activity relative to Pd/C. In addition, the emulsion-assisted synthesis approach seems to lead to better activity for PdAu nanoparticles compared to the direct $NaBH_4$ reduction approach. This phenomenon also suggests the influence from structure effects toward MOR activity [93–95].

Similarly, in order to prepare Pd–Au with different structures, Zheng et al. used Good's Buffer 2-[4-(2-hydroxyethyl)-1-piperazinyl]ethane sulfonic acid (HEPES) as reduction and guide agents to prepare PD-modified core–shell AuPd materials (AuPd@Pd) [96]. Agglomeration may occur when used repeatedly for catalysis; therefore, catalyst carriers such as CNTs, carbon black, and graphene can be used to enhance dispersion. For example, to sum up, many metal catalyst enhancement effect can arise from at least one of these effects: (i) the cooperativity between the metal and the other metal oxide intermediate adsorption in the active Pd sites (for example, Pd-CO_{ads}) at a low potential, so as to improve the regeneration rate of Pd active site, (ii) expand surface area provided by the specific synthetic methods (morphological changes provide additional opportunities for an active site with atomic step, edge, and corner atoms), and (iii) strengthen the support of the material, it can provide the specific surface area, promote the fast transfer of electrons, and thus the possibility for agglomeration would be decreased.

Oxides such as Co_3O_4, CeO_2, NiO, and Mn_3O_4 can be used to modify Pd and improved the activity/stability of oxidative alcohol in alkaline media would be achieved over the commercial PtRu/C electrocatalysts. Effects of this oxide amount for MOR were investigated by Xu et al. [97], for which Pd–NiO (4 : 1, w/w)/C, Pd–CeO_2 (1.3 : 1, w/w)/C, Pd–Mn_3O_4/C (2 : 1, w/w), and Pd–Co_3O_4 (2 : 1, w/w)/C were prepared to uncover some of the relationships between content and activity Pd electrocatalysts. It has been shown that electrocatalytic activity and electrode stability were influenced by the content of these metal oxides; specifically, electron conductivity would be reduced at a high amount of oxides owing to the nature of the semiconductor of these oxides [27]. Ye et al. [98] displayed a similar work to further confirm that the oxidation activity of methanol by adding oxide was higher than Pd/C. In addition, such as the linear sweep voltammetry (LSV) curve is shown in (Figure 12.7), –NiO/C and Pd–Co_3O_4/C electrode on the activity of methanol oxidation is clearly higher than Mn_3O_4–Pd/C and Pd electrode and the activity of CeO_2/C electrode was Ye et al. [98] by cyclic voltammetry observation (Figure 12.7), the intermediate substance was adsorbed on the electrode surface and poisoned the electrocatalyst during the forward scanning process. In the process of reverse scan, CO and other adsorption material with palladium oxide reoxidation, compared with palladium oxide resistance enhancement of toxic substances. This is considered to be the lattice oxygen in oxide; as a result, it is considered to be reactive oxygen species; it removes methanol oxidation of CO intermediates [27, 84, 97].

By reducing the battery voltage and increase the required oxygen measurement than that occur in the cathodic reduction of methanol, dramatically reducing the overall performance [99, 100]. In order to address the issue of transboundary of

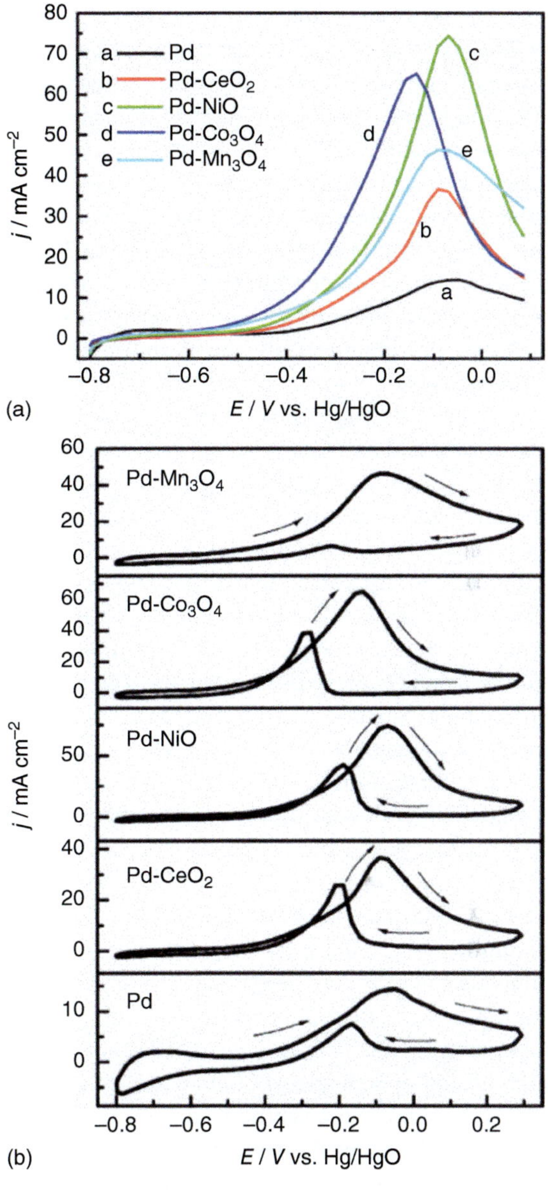

Figure 12.7 (a) LSV curves of methanol oxidation on the Pd/C, Pd–CeO_2/C, Pd–NiO/C, Pd–Co_3O_4/C, and Pd–Mn_3O_4/C electrodes. (b) Cyclic voltammogram of methanol oxidation on the Pd/C, Pd–CeO_2/C, Pd–NiO/C, Pd–Co_3O_4/C, and Pd–Mn_3O_4/C electrodes in 1.0 M KOH solution containing 1.0 M methanol with a sweep rate of 50 mV s^{-1} and Pd loading of 0.30 mg cm^{-2} at 303 K. Source: Ye et al. [98]. © 2013, Elsevier.

methanol, extensive studies were conducted on nafion-based membrane modifications, with the focus on maintaining the proton conductivity while blocking the transmission of methanol. Nafion membrane was combined with Pd to form a filler for decreasing permeability of methanol and also to keep a high permeability for proton [99–104].

Methanol crossover can be reduced by clamping a layer of Pd film between Nafion films, Pd nanoparticle deposition via ion exchange, and/or of Pd-based alloy film deposition.

12.2.2.2 Pt-Based Catalysts

To increase the area of the active component, some traditional methods, such as dispersing active metals on supports and giving priority to obtain necessary active site, and specific exposure surface of dispersion on electrocatalyst surfaces are used. To produce active metal components on support systems, a lot of synthetic strategies have been proposed. As wide accepted, the size and shape of the metal supported and the properties of the supporting metal have significant effects on the surface properties of the supporting metal system. In addition, since the loaded metal is in the nanometer state in electron transfer processes, the dimension would become a variable to utilize. According to the shape, size, dimension, and properties of the supporting materials, the platinum nanostructures exhibit various catalytic properties [13].

Although the dispersion of metal nanocrystals on the carrier has these advantages, the use of nanocatalysts has its limitations. First, the effective electrochemical surface area (ECSA) decreases [72] due to the accumulation of nanoparticles, and effective charge transfer and mass diffusion would be prevented by the more interfaces between nanoparticles [73]. However, it was demonstrated that electrocatalysts with 1D nanostructures, such as nanotubes (NTs) or nanowires (NWs), were more efficient compared 3D nanoparticles, thus offering a large surface area to make it easier to overcome these limitations without the need to use support with large surface area [72–74] In brief, many ways were proposed to make nanoPT structures. For example, wires can be electrodeposited into polycarbonate membranes. [73, 75] In addition, metal NWs are also serving as templates to generate Pt NTs via simple galvanic substitution reaction [72, 76]. Since the dimension of Pt nanostates has a considerable influence on catalytic behavior, many types of Pt nanoparticles, especially active types, would be briefly discussed below.

12.2.2.3 Platinum-Based Nanowires

Platinum NW compounds for methanol oxidation were synthesized by polymer template and improved electrodeposition [73]. A polymer template was used for the successful separation of Pt NWs via washing with ethanol and CH_2Cl_2. The higher mass activity was obtained on these unloaded Pt NWs for MOR compared with high loaded or unloaded Pt nanoparticles typically required by DMFC. Platinum NW compounds were synthesized from platinum and tungsten by polyol method. These Pt NWs showed excellent MOR activity on the Pt gauze owing to the higher Pt(110) surface exposure based on Pt NWs structure [74]. Pt and W gauzes on Pt NWs were characterized by scanning electron microscopy (SEM), as depicted in Figure 12.8. Pt nanoparticle catalysts-modified electro-spinning anatase nano-fibers were studied by Formo et al. [79]. Higher electrochemical activity/durability was found on the supported nanostructures compared to commercial Pt/C catalysts. Pt can also be deposited on titanium substrate for application in DMFC. Pt nanoparticles loaded on TiO_2 support, which reduces CO adsorption energy intermediate and increases CO mobility would be advantageous. In anodic MOR catalysis, the specific role of metal oxides would be specified later. Pt NWs were prepared by electrospinning polyvinylpyridinone (PVP)-PT composite fiber in air at high

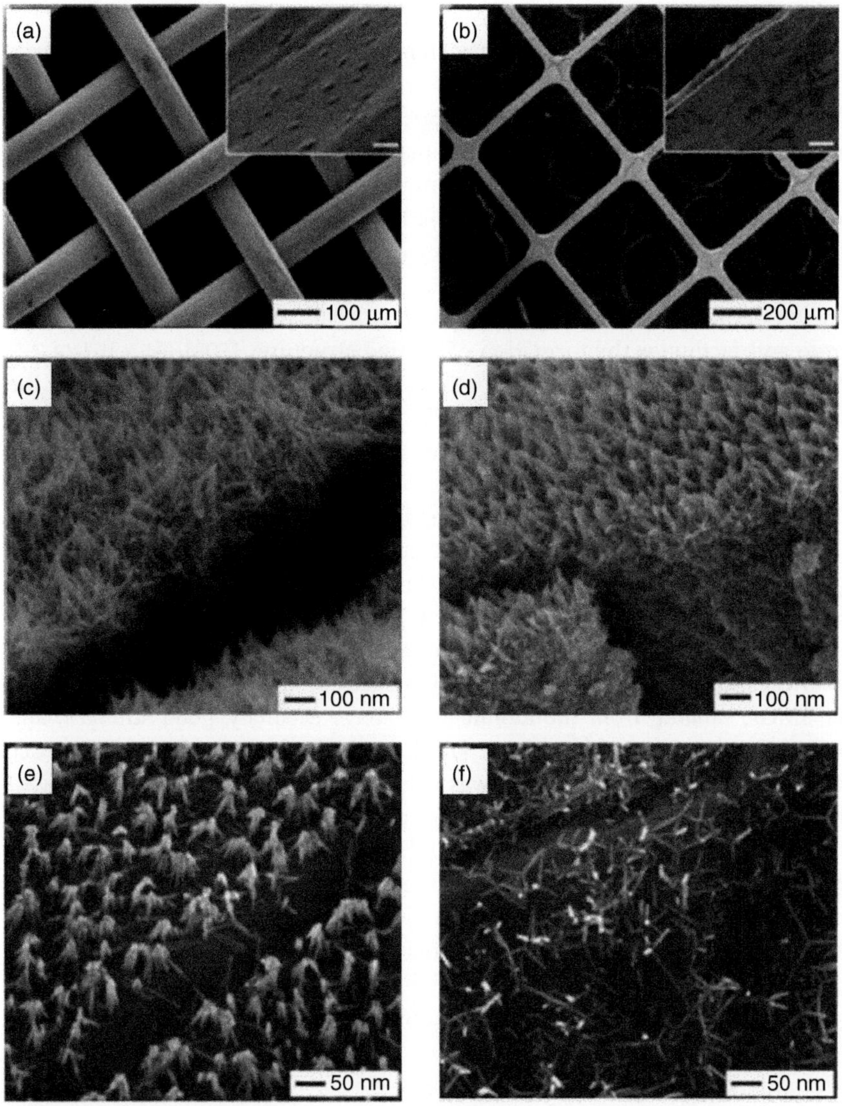

Figure 12.8 SEM images of (a) Pt and (b) W gauzes prior to Pt NW growth, with insets showing the detailed surface morphology. (c, d) SEM images of Pt NWs on Pt and W gauzes, respectively, were obtained as the final products of an iron-mediated polyol process. (e, f) SEM images of Pt NWs grown on the surface of Pt and W gauzes. Source: Lee et al. [74]. Reproduced with permission of the American Chemical Society.

temperature using PVP-Pt degradation and platinum precursor reduction [105]. For improving the catalytic activity of one-step synthesis NWs, a heat treatment at two-step processes was adopted. The ECSA of the NWs in the two-step heat treatment was approximately doubled. Although commercial Pt catalysts still have higher activity compared to these Pt NWs, Pt NWs catalysts have higher specific activity for MOR. Pt nanoparticles are better than spherical candidates in regard to electron conductivity, micro-degree, and specific activity. Thus, tuning the roughness and diameter of Pt NWs may significantly improve the electrocatalytic performance for MOR.

Pt NW array catalyst electrode was fabricated by Zhao et al. on the Ti–Si substrate by employing porous anodic alumina templates (AAO) via a direct Pt deposition process. The performance of the electrode in methanol oxidation is better than that of the platinum film on titanium substrate [106]. They reported that Pt–Ru NW was synthesized by the same method and compared with the MOR activity of Pt–Ru NW. Pt–Ru NW was formed on Vulcan XC-72 carbon surface as a MOR catalyst, with better performance than Platinum-NW [107]. The pH value controls the reaction rate and facilitates the formation of desired structures [108]. At a pH of 2.5, the NWs can be generated, and the optimal load of toner was 40 wt%. Since the formation of internal channels in THE NW network improves the mass transfer effect of methanol oxidation, the loose NW network is preferable to the compact NW network.

Other Pt-based polymetallic NWs were prepared beyond Pt–Ru and Pt NWs [109], taking into consideration of the electron effects from the other metals components toward Pt. Recently, Pt-based metallic glass (BMG) NWs, which are composed of Pt, Cu, Ni, and P as electrocatalysts, were demonstrated to be active for electro-oxidation of CO, methanol, and ethanol. Due to the absence of grain boundaries and dislocations in the body structure, the amorphous properties of the BMGs, ranging from the atomic scale, have elasticity, good strength, and excellent tolerance against corrosion [110]. Excellent catalytic activity for CO, methanol, ethanol electro-oxidation was achieved on these PT-BMG NWs based on the intrinsic strain actions and electron properties of these systems. Thermal decomposition of $Fe(CO)_5$ combined with $Pt(ACAC)_2$ and $Pd(ACAC)_2$ was used for the synthesis of Fe–Pt–Pd NWs for continuous MOR [111], and a ratio of 28 : 38 : 34 was realized as the optimal composition ratio for Fe–Pt–Pd NWs. A porous AAO membrane was used for the synthesis of Pt-co NWs via electrodeposition processes [112]. The methanol oxidation current of these NWs was significantly increased compared to the corresponding membrane structure (Figure 12.9).

12.2.2.4 Platinum-Based Nanotubes

Using the AAO template, bimetallic NTs, which is consisted of Sn-PT, were synthesized by combining evaporation with electrodeposition (Figure 12.10) [113]. Pt presents on the inner wall of Sn–Pt NTs, as shown in the HRTEM image of Sn-Pt NTs in Figure 12.10. Sn–Pt bimetallic NTs have high catalytic activity, which are twice higher than that of the Sn–Pt membrane. Tin NTs had a nanochannel structure, which facilitates the dispersion of Pt particles and the reduction of particle sizes. The generated small particles offer a high surface area for MOR. Shin et al. used AAO and polyaniline nanorod templates to synthesize smooth, porous Au NTs as templates for the deposition of thin platinum layers [114]. The electrocatalytic activity of nanopore gold-bearing NTs on methanol oxidation was found to be better compared to smooth gold NTs obtained from commercial approaches. Thus, 1D structures seem to be more suitable for electrode

Figure 12.9 (a) SEM image of Ag NWs, (b) Transmission electron microscope (TEM) image of Ag NWs, (c) SEM image of Pt NTs, (d) TEM image of Pt NTs, (e) high resolution transmission electron microscope (HRTEM) image of Pt NTs, and (f) selected area diffraction pattern of Pt NTs. Source: Kakati et al. [16]. Reproduced with permission of the American Chemical Society.

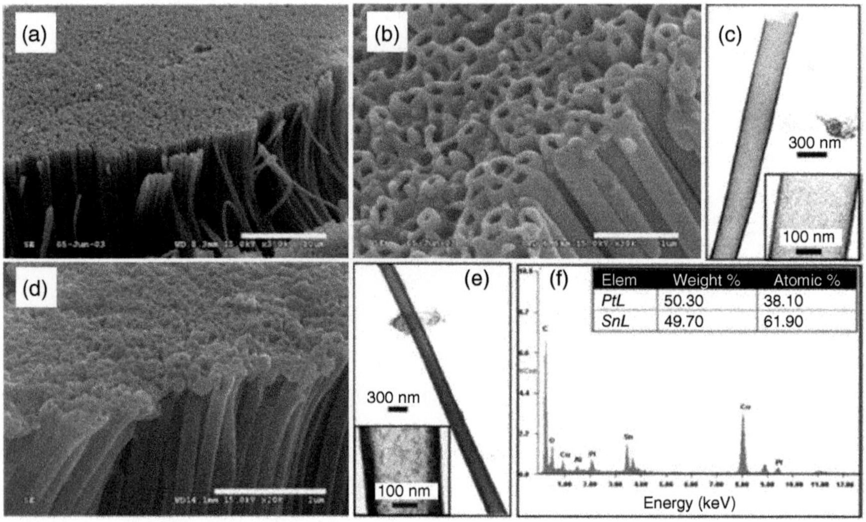

Figure 12.10 Typical SEM images of Sn NT (a, b), Sn–Pt bimetallic NT (d). Typical TEM image of Sn NT (c), Sn–Pt bimetallic NT (e). (f) Energy-dispersive X-ray profile of the Sn–Pt NT. Source: Guo et al. [113]. © 2005, John Wiley & Sons.

applications because of the potential for electron vector transfer from these structural materials.

12.2.2.5 Platinum-Based Nanoflowers

A simple electrochemical approach was proposed by He et al. to prepare a large-scale assembly of Pt flower-like nanorods to form 3D superstructures at 25 °C without using any templates. The total number of these Pt nanocrystals showed a highly enhanced electrochemical activity on methanol oxidation while Pt nanoparticles aggregated and commercial Pt. By platinum nanoflow-decorated multilayer CNTs, (thermal) wet hydrogen reduction routes by chemical synthesis demonstrated significant electrocatalytic activity to reduced oxygen and oxidized methanol [80]. Three-dimensional flowered Pt nanoparticle clusters can be electrodeposited on polythermal-bonded CNTs by a simple process that involves a critical process (potential pulse train) and is a totally electrochemical process [81]. This structure results in better MOR activity due to an order of magnitude higher than the 2D dispersed Pt nanoparticle in ECSA. In recent years, and polyoxometalates (POMs)-consisted multilayer films have been prepared by the stratification method. The films were applied as the substrate to electrodeposited novel flower-like platinum clusters [82]. Based on the special morphology of Pt micro-nano clusters, the catalyst showed excellent activity for MOR. Excellent transport characteristics, as well as the ability toward CO tolerance, have both make POMs to have unique "pseudo-liquid phase" electrochemical reaction conditions. These results indicate the possible application of POMs for electrochemical alcohol oxidation. POM may be reactive for MOR among these structures. Thus, the found effects could be synergistic and/or cumulative.

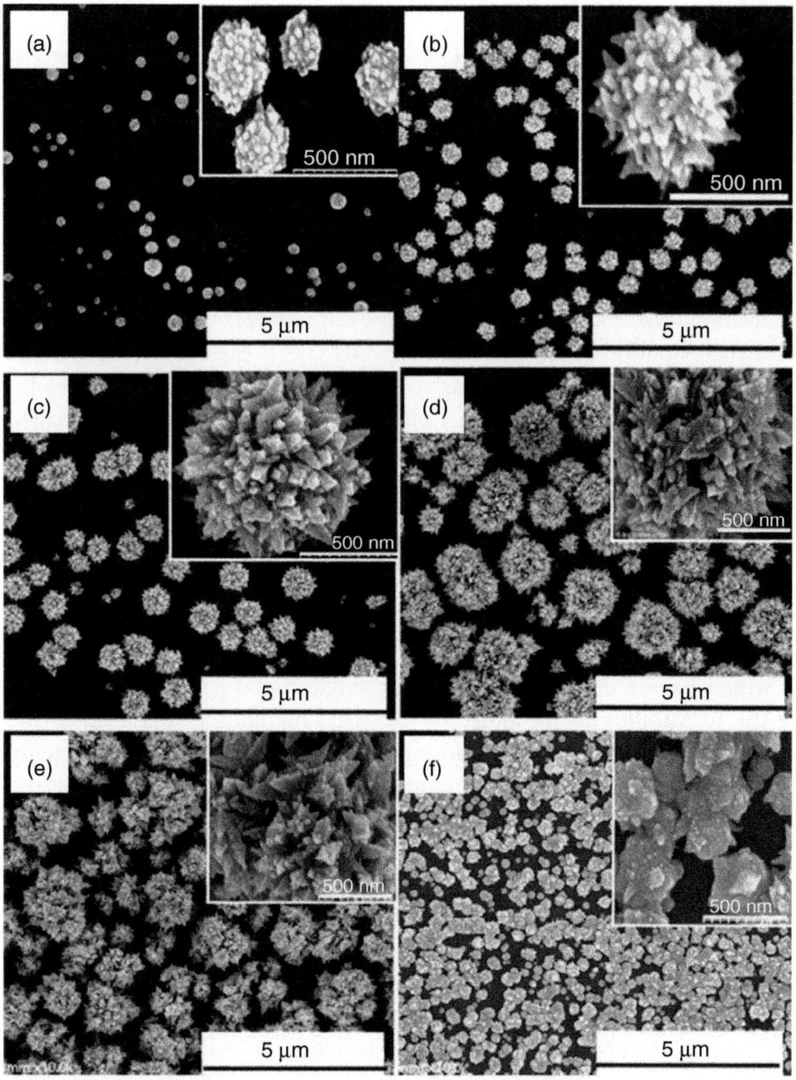

Figure 12.11 SEM images of Pt nanostructures (a) 25 seconds, (b) 400 seconds, (c) 1000 seconds, (d) 4000 seconds, and (e) 10 000 seconds electrodeposited on ITO at −0.2 V under various conditions. Source: Zhang et al. [115]. Reproduced with permission of Elsevier.

Platinum nanoflowers were prepared on indium tin oxide (ITO) substrates by simple electrodeposition in a solution containing H_2PtCl_6 and H_2SO_4 [115]. Figure 12.11 exhibits platinum nanostructures under different electrodeposition cases. The catalytic activity of platinum nanoparticles was enhanced, and the peak current density was about 4.4 times that of platinum nanoparticles. The Pt and Pt–Ag mesoflower [116] were synthesized using bovine serum albumin as a template. The sacrificial templates employed silver mesoporous catalysts for the preparation of porous Pt–Ag alloy and Pt catalysts. Methanol transport to the Pt surface could be accelerated by the 3D expansion channels; the electrical conductivity could also be improved by these, both resulting in the enhanced electrocatalytic activity of the mesoporous substructures. As expected, poisoning resistance and catalytic activity were improved on the electrocatalyst, both better than traditional Pt black electrocatalyst. The electron effect of Ag in platinum-silver alloy makes the electrocatalyst better resistant to CO poisoning. Besides, dealumination can reduce the content of silver in such alloy to form a catalyst with porosity, thus improving the MOR activity.

Recently, Yao et al. designed Pt nanoflowers deposited on rGO-modified carbon cloth using a simple electrochemical method. The nanoflowers were used as anodic materials for the oxidation of HCOOH and CH_3OH [117]. The carbon cloth fibers fully coat the graphene sheets, generating lots of folds which become the sites for nucleating of the platinum nanoflowers. Compared with bare carbon cloth-loaded Pt nanoflowers and carbon cloth-supported Pt nanoparticles commercially obtained, the activity of graphene-carbon cloth-supported Pt nanoparticles was improved for the oxidation of HCOOH and CH_3OH.

12.2.2.6 Platinum-Based Nanorods

Generally, Pt and oxyphilic metals interface provides a key site for electrooxidation MOR. This has stimulated the preparation of Pt–Ni, Pt–Ru, and Pt–Ru–Ni nanorods, which have a customizable alloy or single metal length. Sequentially, depositing the metal/alloy in the pores of the AAO membrane was used in the preparation. The obtained Pt–Ru–Ni multisegmented nanorods have higher activity than Pt–Ni and Pt–Ru multisegmented nanorods [118–121]. It was found that the best Ru : Ni ratio of Pt–Ru–Ni nanorod was 4.24 : 1, which showed the highest MOR activity. When increasing the number of interfaces and the nanorod constant length, the MOR catalytic activity, which clearly indicated the existence of a bimetal mechanism in the DMFC anodic reaction. Multilayer Pt–Ru nanorods that have controllable bimetallic sites were prepared by Yoo et al., and the results were demonstrated to be similar [122]. It is obvious from their study that Pt–Ru nanorods with Ar heat treatment results in an optimal effect on MOR. This is because of the bifunctional mechanism and the electronic effect of the ruthenium atom. The intermetallic compounds Pt–In, Pt–Pb, and Pt–Sn were demonstrated to be effective for MOR [123]. The oxidation currents for MOR are 44 times higher than pure Pt. In particular, Pt–Pb exhibited extremely high current, which was 40 times higher compared to Pt. Intermetallic Pt–Pb compound nanorods were prepared by Maksimuk et al. via coreducing Pt^{2+} and Pb^{2+} salts [124]. The formation of platinum-lead nanorods was performed by adding adhesive to the growing surface with a good binding effect. To evaluate MOR activity, carbon black was used to deposit the as prepared Pt–Pb nanorods then the composite was treated by plasma in air. The obtained nanorods were further heat treated in H_2–Ar atmosphere at 600 °C for two hours, resulting in the change of nanorod morphology. It was found that the peak current density of Pt–Pb–C was higher than that of Pt–Ru–C electrocatalyst. TEM images and loop voltammogram of PtPb nanorods are depicted in Figure 12.11.

12.2.2.7 Platinum-Based Nanocubes

Another Pt and Pt-based alloy nanostructure widely used in MOR is nanocube (NC). It has been well accepted that with the increasing surface area, the electrocatalytic activity of nanomaterials increases. Besides, the shape and structure of platinum and platinum-based catalysts have attracted much attention due to the importance of the atom arrangement on the surface of nanocatalysts to the activity/selectivity. For instance, the Pt(100) surface exhibits higher oxygen reduction catalytic activity in acidic solution compared to the Pt(111) surface [125, 126].

Han et al. synthesized Pt NCs by using PVP/Fe^{3+} ion aided polyol process [127]. Compared with polycrystalline Pt nanocatalysts, Pt nanocatalysts with the surface (100) showed better catalytic performance for ethanol and methanol electrooxidation with respect to the initial current and potential density. With the existence of PVP, the authors also prepared Pt NC by heat decomposition processes [128]. Compared with the spherical Pt nanoparticles, these Pt nanoparticles also showed advantages (100), with improved electrooxidation activities for ethanol, methanol, and HCOOH. Without using PVP, scattered Pt particles may induce little aggregation, which means the formation of PVP micelle resulting in monodispersed particles dominated with planes (100). The electrochemical recombination approach was developed by Peng et al. using truncated octahedron Ag nanotemplate to form Pt nanoboxes or nanospheres [129]. The catalytic performance and TEM images of these materials are depicted in Figure 12.12. The formation of the nanobox can be understood as (i) in H_2SO_4 solution, the low index Pt surface is relatively stable; (ii) when a linear potential cycle is used, the reconstruction of Pt(100) facet is harder compared to other low index surfaces [130, 131]. Since the Pt(100) surface of MORs

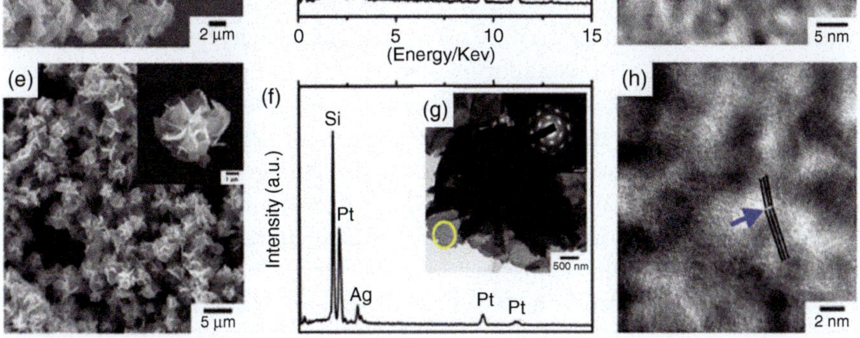

Figure 12.12 Representative (a) SEM, (b, c) TEM and elemental analysis, and (d) HRTEM images of $Pt_{45}Ag_{55}$ and (e) SEM, (f, g) TEM and elemental analysis, and (h) HRTEM images of $Pt_{72}Ag_{28}$. Source: Zhuang et al. [116]. © 2012, Elsevier.

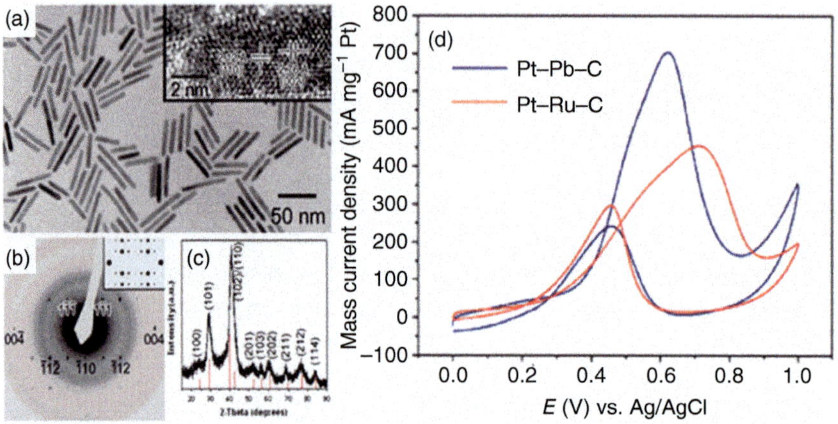

Figure 12.13 (a) TEM image and selected area electron diffraction pattern and (b) of a single Pt–Pb nanorod. (c) Powder X-ray diffraction pattern of an ensemble of Pt–Pb nanorods. (d) Cyclic voltammograms of electrocatalytic oxidation of methanol with Pt–Pb–C and a commercial Pt–Ru–C catalyst. Source: Maksimuk et al. [124]. © 2007, American Chemical Society.

is more active than the Pt(111) surface [132–135]. In comparison with commercial Pt particle, Pt nanoparticle with the surface (100) can improve the activity of MOR (Figure 12.13).

There are only a few reports on the use of platinum with bimetallic and trimetallic NCs structures for MOR. But there are also some reports on PtPd, PtCu, PtCo, PtPdCu, PtFeCo, etc. Platinum iron–cobalt particles with different compositions and shapes were investigated for MOR [136] (Figure 12.14).

Using OLA-oleic acid as an agent for shape control, Pt3Co NTs with (100) orientation were obtained [137]. The methanol oxidation activity of these NC was obviously better compared to the activity of Pt NC. There was no bifunctional effect in the electrocatalyst, and the electron effect of Co was completely responsible for the enhanced electrocatalytic activity. PtCo spherical nanoparticles, as well as their NCs toward the MOR activity, were studied. The precursors of a diamond monocarboxylic acid, acetylamino, and Co were used as coating agents, heterogeneous and supporting metals, and platinum alloys. Cubic nanoparticle growth was facilitated along the [111] direction. It is found that spherical cobalt platinum exhibited higher MOR activity compared to the cubic ones. Taking into consideration of the selectivity of capping agents and small ions, synthesis of Pt–Pd nano-tetrahedrons (NTHs), and NCs via a hydrothermal process would be possible (Figure 12.15).

Br–I$^-$ and PVP systems were used for the synthesis of Pt–Pd NCs by Yin et al. [138] and Pt–Pd NTHs were synthesized using formaldehyde, $Na_2C_2O_4$, and PVP. The Pt–Pd NC is surrounded by a (100) surface, while Pt–Pd NTHs is surrounded by a [111] surface. This is because Br–I – ions have (100) – surface selectivity, while HCHO-$N_2C_2O_4$ has (111)-surface selectivity. The electrocatalytic activity sequence for MOR was decreasing from Pt–Pd NCs, Pt–Pd NTHs to Pt–C (commercial), as shown in Figure 12.16. Nevertheless, Pt–Pd NCs showed poorer durability than

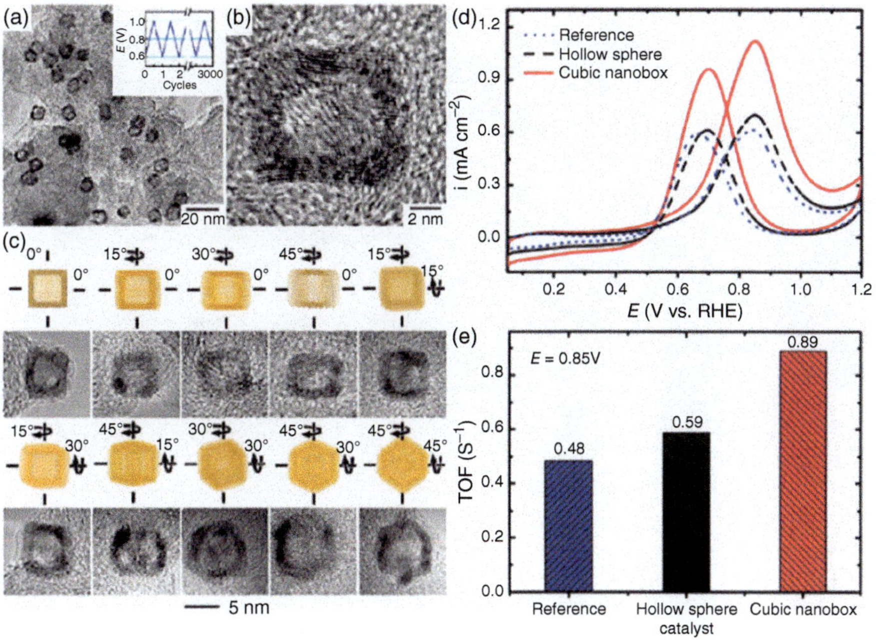

Figure 12.14 TEM images of Pt hollow nanocubes at (a) low and (b) high magnifications and (c) individual cubes imaged under various tilting angles with respect to the direction of the imaging beam. (d) *I* vs. *E* for MOR catalyzed by Pt cubic nanobox, hollow nanosphere, and the reference catalysts, and (e) their turnover frequencies (TOF) at the peak potentials. Source: Peng et al. [129]. © 2010, American Chemical Society.

Pt–Pd NTHs because of the high durability based on the Pt catalyst (111), such as the example of oxidation of formic acid on the Pt catalyst [139]. Different adoption of reaction approach via facets of (111) and (100) is the main reason for the difference in electrocatalytic activity of Pt–Pd NCs and Pt–Pd NTHs (Scheme 12.1) [140–142].

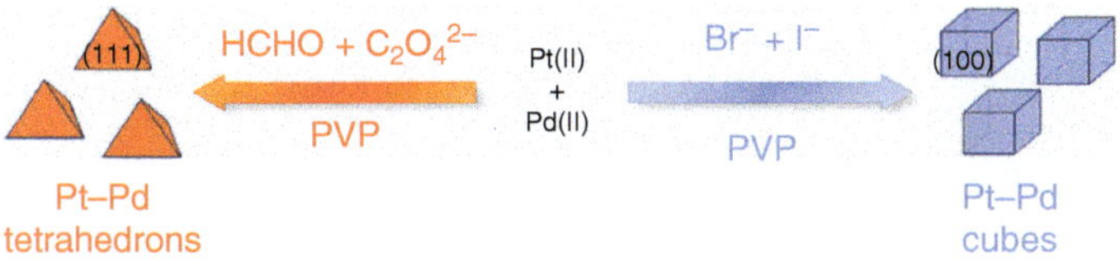

Scheme 12.1 Shape-selective synthesis of Pt–Pd NTHs and NCs. Source: Yin et al. [138]. © 2011, American Chemical Society.

12.2.3 Pt–Ru System

In the bimetallic catalysts studied, Pt–Ru has a unique position because it can greatly mitigate CO toxicity to the surface-active sites. This is done by generating hydroxyl groups on the Ru sites. An excellent activity of the catalysts was achieved when Ru is maintained in the solid solution in platinum [143]. More substances like $Ru(Pt)_2COH$ were also proposed for continuous dehydrogenation through

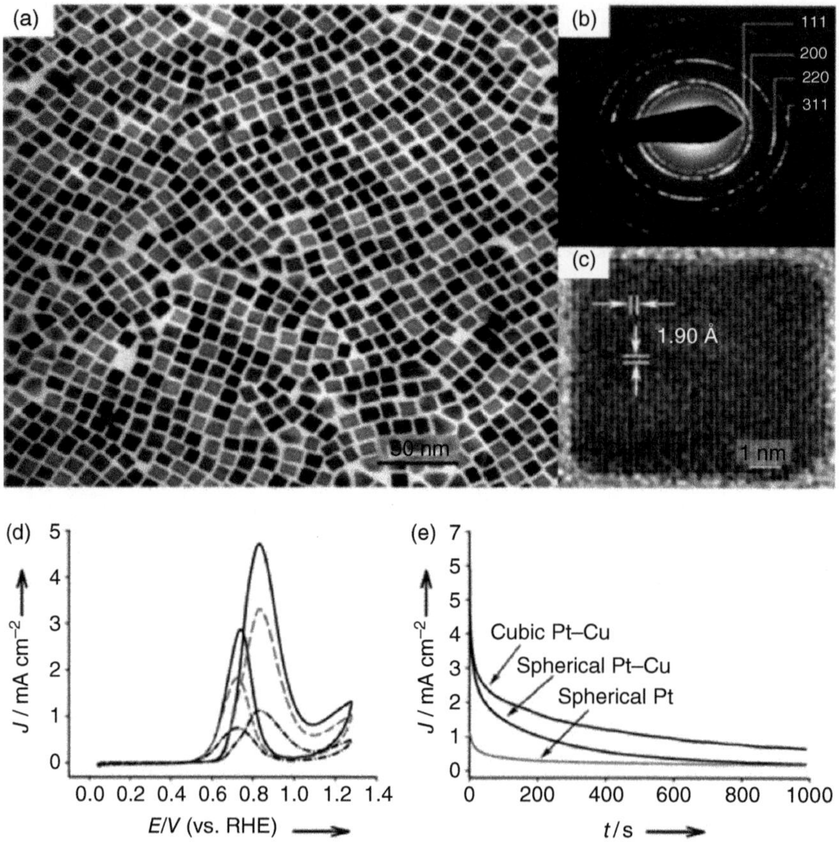

Figure 12.15 (a) Low-magnification TEM image of the overall morphology of Pt–Cu NCs. (b) TEM selected area electron diffraction pattern of the Pt–Cu NCs. (c) High-resolution TEM image of a selected Pt–Cu NC. (d) Cyclic voltammograms of MeOH oxidation on Pt–Cu NCs, Pt–Cu nanospheres (---), and Pt nanospheres (-·-) in 0.1 M $HClO_4^{-1}$ M MeOH (scan rate 0.02 V s^{-1}). The electrode potentials are reported vs. a reversible hydrogen electrode (RHE). (e) Chronoamperometric results of MeOH oxidation at 0.8 V on Pt–Cu NCs, Pt–Cu nanospheres, and Pt nanospheres in 0.1 M $HClO_4^{-1}$ M MeOH. Source: Kakati et al. [16]. © 2014, American Chemical Society.

methanol. The study on the stripping voltammetry of CO gave the evidence to prove the key role of the Pt–Ru domain for high activity in the system [144]. As mentioned previously, oxidation of the ruthenium surface is one of the important reasons for the great activity, as observed [145]. The mixed electron and proton conductivity cause increased activity on the Pt–Ru electrocatalyst [146]. No matter which reason, the platinum–ruthenium bimetallic electrocatalysts are widely proved to be a good choice as electrocatalytic material for DMFC applications.

12.2.4 Pt–Sn Catalysts

The activity of Pt–Sn catalysts can be divided into good and bad [147]. The reason for the controversy may be that tin can be adsorbed to platinum, which shows

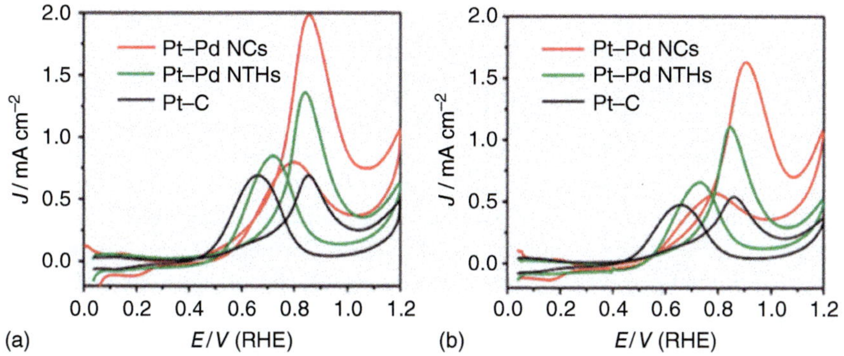

Figure 12.16 (a) Stable cyclic voltammograms obtained for the Pt–Pd NCs and NTHs and Pt–C. (b) Cyclic voltammograms obtained after 4000 additional cycles. Source: Yin et al. [138]. © 2011, American Chemical Society.

excellent activity for MOR, while the alloy with good morphology shows poor activity. Shukla et al. [148] studied the effects of tin and ruthenium additives on platinum carbon by X-ray photoelectron spectroscopy (XPS) to elucidate the differences in their mechanical behavior. Sn is believed to modify the electron structure of Pt (via electron transfer within the alloy); however, the Ru site in the Pt–Ru alloy promotes the generation of lattice-bound oxides near methane residues adsorbed at Pt. Studies of the system formed by electrodeposited tin on platinum have shown that surface Sn concentration has a certain effect on the observed activity [149, 150], and this activity has a relationship with potential. Besides this binary system, more systems containing Ru and W [151, 152] are also used for MOR. Overall, other metals, including Mo, Sn, Cr, Os, Ti, Fe, Re, and Ta, can be used to combined with Pt to enhance the activity for MOR.

12.3 Formic Acid Oxidation

12.3.1 Reaction Mechanism

With increasing interest in other fuels, formic acid was used in polymer electrolyte film fuel cells (PEMFCs), known as direct formic acid fuel cells (DFAFCs). Compared with DMFC, the advantages of formic acid oxidation (FAO) over MOR are very obvious: which involve increasing reaction rate, low toxicity, avoiding or reducing cross-effects through the membranes, and the possibility for operation at low temperatures. Low volume energy density is the drawback of DFAFCs. However, increasing formic acid concentration would alleviate such limitations. The FAO mechanism is usually concluded as dual or parallel pathways. Direct FAO goes through the dehydrogenation process, which did not involve the generation of CO in the reaction:

$$HCOOH \rightarrow CO_2 + 2H^+ + 2e^- \quad (12.8)$$

Indirect oxidation of formic acid refers to the dehydration approach, with the formation of CO adsorbed on the electrocatalyst surface:

$$HCOOH \rightarrow -CO_{ads} + H_2O \rightarrow CO_2 + 2H^+ + 2e^- \quad (12.9)$$

Since formic acid is completely oxidized to CO_2, the pathway of dehydrogenation in Eq. (12.8) is a favorable process. During the dehydration reaction of Eq. (12.9), the CO species formation on the surface is undesirable, resulting in the oxidation to CO_2 requiring a higher electrode potential. The ideal DFAFC catalyst should be operated in a dehydrogenation pathway to improve the overall efficiency of the cell and avoid catalyst poisoning.

12.3.2 Catalyst Advances

12.3.2.1 Pd-Based Catalysts

It is generally believed that a direct pathway is the main pathway for the formic acid electrooxidation on the Pd surface. It avoids the CO formation on the catalyst. However, relative to Pt for formic acid electrooxidation, less studies were paid to the FAO mechanism on Pd. Generally, multistep reaction processes were involved in the reaction mechanism on Pd for formic acid electrooxidation involves, which is similar with Pt cases. Beyond the aforesaid direct/indirect oxidation mechanism as depicted earlier, electrochemical oxidation of adsorbed formate was also studied [153–156].

Instead, HCOOH is considered to be a transient active intermediate produced by the FAO. This can only be characterized as the formic acid decomposition to produce carbon dioxide that is inhibited.

$$HCOOH \rightarrow -COOH_{ads} + H^+ + e^- \rightarrow CO_2 + 2H^+ \quad (12.10)$$

Pd-based catalysts have been proved to have better performance than those composed of Pt. However, DFAFC using the Pd catalyst still showed significant attenuation several hours after discharge. It was originally thought that adsorptive co-like species could explain the decay of Pd, similar to that of Pt. However, during formic acid electrooxidation, CO species were not found [156, 157]. The mechanism of decay is still particularly unclear, so it is proposed to adsorb OH^- ions, formic acid ($HCOO^-$) ions, or other anions on the surface of Pd as toxic substances (Figure 12.17) [155, 159–161].

Although the properties of the poisoning species type remain controversial, methods to recover the toxic palladium catalyst have been explored. Anodic polarization techniques can restore the activity of the catalyst by oxidizing adsorbed toxic substances. However, one of the concerns in the anodes refers to Pd oxidation/dissolution [162]. Zhou et al. [156] reported another approach that used a non-electrochemical approach to regenerate Pd using pure water, but extending the regeneration time by more than an hour hindered proficiency. Another approach is cathodic polarization that utilizes low potential's ability for palladium corrosion and the shorter regeneration time. Jeon et al. [158] activated the Pd/C catalyst every 60 seconds by using a pulse cathode potential at −0.2 V during formic acid

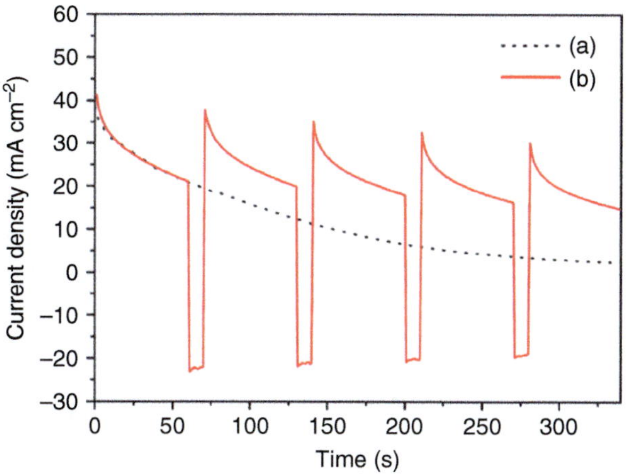

Figure 12.17 Current density vs. time profiles of Pd (a) under a constant potential of 0.5 V and (b) in the pulsed potential method (0.5 V for 60 seconds, −0.2 V for 10 seconds) in 2 M HCOOH and 0.5 M $HClO_4$. Source: Jeon et al. [158]. © 2014, Elsevier.

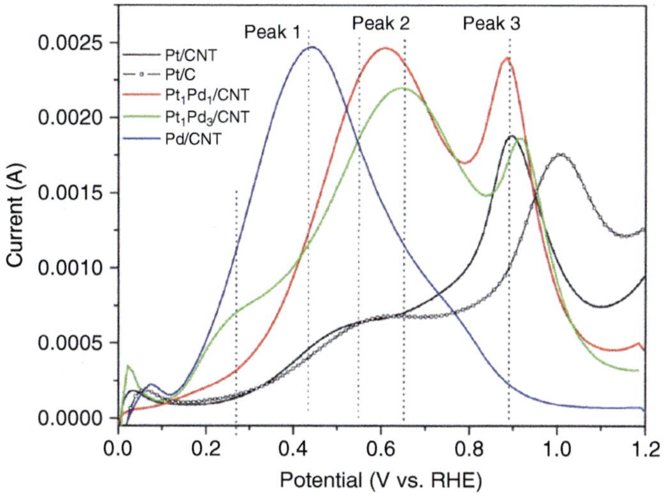

Figure 12.18 Cyclic voltammograms of formic acid oxidation on Pt/CNT, Pt_1Pd_1/CNT, Pt_1Pd_3/CNT, Pd/CNT, and Pt/C (E-TEK). Source: Winjobi et al. [163]. © 2010, Elsevier.

electro-oxidation. Figure 12.18 compares the current density of Pd at constant potential 0.5 V with that of pulsed cathode potential. It is found that periodic cathode polarization postpones the deactivation of Pd electrocatalyst to a great extent.

For an understanding of the electro-oxidation mechanism of formic acid on the palladium surface, catalysts having high electrocatalytic activity, excellent durability, and low cost should be produced. Controlling size is known to change the electron properties of metal particles. Generally speaking, with the decreasing nanoparticle, the D-band center would be negatively shifted, resulting in a decrease of its adsorption energy. For the formic acid electro-oxidation, the negative shift of the D-band center decreases the binding energy of the assumed intermediate, thus increasing the reaction rate [155, 164]. Size effects of catalysts for formic acid electro-oxidation were studied by Zhou and Lee [165] using carbon-supported Pd particles with different sizes from 2.7 to 9.0 nm. It was found that formic acid electro-oxidation is a reaction with high surface structure-sensitive natures. Strong size effects were revealed to be related to the geometric requirements of active sites on Pd nanoparticles.

Both the size and the surface could influence the electrocatalytic activity. The surface orientation of formic acid has a great influence on the current when it is oxidized on palladium [165, 166]. Structure effects for formic acid electro-oxidation were investigated by Hoshi et al. [167, 168]. For structure-defined Pd single crystals with low and high index planes, according to the maximum current density of forwarding scanning, the low exponential plane presents the following reaction activity series: Pd(100) > Pd(111) > Pd(110). As to the planes with high index, the maximum current density increased in the order Pd(S)-[n(111) × (111)] < Pd(S)-[n(111) × (100)] < Pd(S)[n(100) × (110)] < Pd(S)-[n(100) × (111)]. Therefore, Pd(511) had the best electrocatalytic activity.

In order to improve the electrocatalytic performance of formic acid, various preferred surfaces were prepared via nanocrystal shape controlling. Lee et al. [169] synthesized a shape-controlled polygon Pd nanostructure catalyst with the main surface (111). Compared with Pd/C, polygonal Pd/C produced higher current densities in formic acid electro-oxidation, which were 1.908 and 0.478 mA^{-2}, respectively. In addition, the dendritic structure was used for the bifurcated branch; it offers many facets with a high index [170, 171]. The topography, such as palladium nanocrystals, provides similar improvements for the dendrites. The Pd catalyst with flower-like nanostructures showed significantly higher activity in electro-oxidizing formic acid than traditional Pd nanoparticles [172, 173].

A variety of binary palladium-based catalysts were synthesized to improve ethanol catalytic activity. Many metals can be included, e.g. Pt, Cd, Cu, Pb, Sn, Au, Ag, Fe, Ni, and Ir [153, 163, 174–183]. The activities of Pd, Pt, and PtPd alloyed particles supported on CNTs were studied [163]. The CV shown in Figure 12.19 indicates the increase of Pd content in the electrocatalyst is more in favor of the direct oxidation pathway of formic acid. Due to the possible inhabitation from the palladium dissolution in the high oxidation potential conditions, Au is applied in the PdAu-based electrocatalysts. Low and high alloying Pd–Au/C electrocatalysts with such property were prepared [180]. It was found that a high alloying degree in PdAu catalysts results in high catalytic activity/stability compared to the counterparts at a low alloying degree. The reason for such a phenomenon is arisen from increased tolerance against CO during FAO and possible inhibition against the dehydration pathway.

Recently, the electrocatalytic activity of porous palladium and four palladium-M binary catalysts (M = Pb, Cd, Pt, and Ir) with the general atomic ratio of 90 : 10 was studied for formic acid electrooxidation [153]. The initial activity of the as-prepared Pd-M catalysts for formic acid electro-oxidation was nearly independent from alloy material. However, adding the second metals largely affect the stability of these Pd catalysts. To understand the degradation of specific catalysts and the FAO mechanism on these Pd catalysts, the in situ electrochemical attenuating total reflection Fourier transform infrared (ATR-FTIR) spectra were used. Infrared spectroscopy

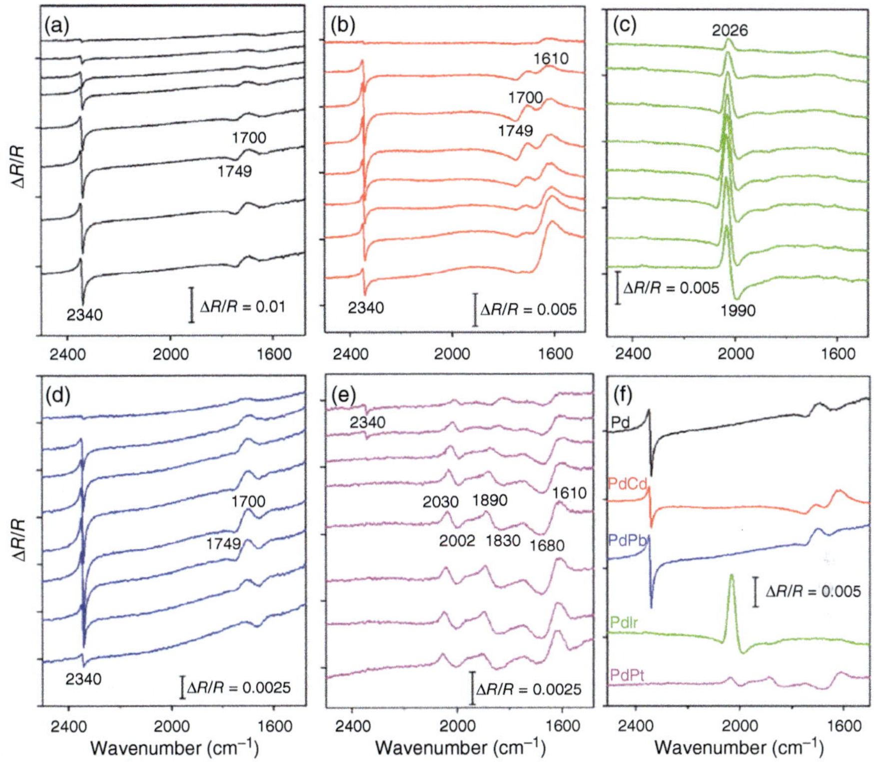

Figure 12.19 FTIR spectra of the formic acid oxidation (a) Pd, (b) PdCd, (c) PdIr, (d) PdPb, (e) PdPt from −0.1 V (top) to +0.6 V (bottom) taken at 0.1 V intervals, and (f) of all catalysts at +0.3 V. Source: Adams et al. [153]. © 2014, American Chemical Society.

(Figure 12.18) uncovered that the oxidation of PdCd, PdPb, and Pd electrocatalysts was carried out mainly through direct mechanisms to generate CO_2 peak at $2349\,cm^{-1}$, demonstrating complete electro-oxidation of formic acid, but other indirect mechanisms were associated with major poisoning by CO (CO_L) extended $2030–1990\,cm^{-1}$; PdIr and PdPt electrodes were found with bipolar brigded bondede (CO_b) peaks between 1890–1830 and $1680–1610\,cm^{-1}$. A small addition amount of Pt and Ir was found to inhibit the FAO and produce a large amount of CO, thus triggering the quick attenuation in the electrocatalytic activity. Additionally, adding cheap Pb and Cd accelerate the formation of the direct mechanism. Thus, it further makes the Pd-based catalyst cost effective. In summary, Pd has a high electrocatalytic activity for FAO but also has the ability to overcome the toxic effects of CO. Therefore, the development of advanced palladium-based electrocatalyst used in FAO will be a necessary condition for promoting DFAFCs' development.

12.3.2.2 Pt-Based Catalysts

Most of the studies on Pt or Pt-based alloys for electrocatalytic oxidation of FAO mainly focus on the activity of the catalyst, and their stability/durability during long-term tests received seldom study. Although CO poisoning is usually related to the stability/durability of Pt-based catalysts for FAO, studies have recently proved the fact that the performance of the catalyst is related to Pt dissolution.

A comprehensive study of FAO Pt dual-pathway was proposed several decades ago as an implementation mechanism [184–186] including (i) formic acid dehydrogenation, which produces carbon dioxide through more than one active intermediates (direct pathway) and (ii) CO_{ad} formed by dehydration that acts as a "toxic" or active intermediate (indirect pathway) [151, 184, 186–188]. Since then, developing highly active/durable Pt-based bimetallic electrocatalysts for fast FAO reaction is the main goal in this field [189–191], and various Pt electrocatalysts such as PtRu, PtPd [176], PtPb [192–194], PtCu [195, 196], PtBi [194, 197, 198] were reported, these catalysts could either minimize poisoning effects related to the indirect pathway or tune the reaction to the direct approach. These electrocatalytic materials were designed to form ultra-thin films or simple nanoparticles to solve the current bottleneck of that high cost of Pt. For the past few years, researchers have been focused on the optimal activity of Pt-based bimetallic electrocatalysts for enhancing the activity by the well-known bifunctional effects, electronic and geometric effects [199]. The action mechanism of Pt on FAO has been studied for several decades. However, some of the essential aspects, including the detailed mechanisms, especially at atomic levels, as well as the properties of the resulting intermediates, are still not clear and in a debate [185, 200–206]. Particularly, the nature of FAO intermediates produced via the direct pathway was controversial; the work from Behm and coworkers [201, 203, 206], Osawa and coworkers [205, 207], and Cuesta et al. [204] was trying to make the issue clear. The complexity of FAO was illustrated from these controversies; besides, strong dependence on applied potential and reaction intermediates or CO adsorbate species was revealed.

The focus on FAO mechanisms, together with the development of more active Pt-based catalysts, has led to little attention being paid to the durability of catalysts and Pt dissolution aspects during the FAO period. To date, the durability of Pt and Pt alloys has been examined in part by a limited number of FAO studies using potentiometric methods [193, 196] (for short periods of time, 10–30 minutes) [208] or potentiometric methods (pulsed or cyclic) [194, 195, 209]. Potentiometric tests generally focus on the poisoning of CO_{ad} at low potentials, which largely results in lowering electrocatalytic activity but are limited to a certain extent. Potential tests reflect well the complexity during fuel cell operations [210], while its ability to control carbon monoxide poisoning/reaction effects depends greatly on the potential values. Therefore, surface oxidation and/or oxygen adsorption would be promoted by higher positive potentials, leading to the oxidation and removal of adsorbed CO species from the electrocatalyst surfaces [184, 211]. With supporting the latter case, a study suggests that the selection of positive potential limits ranging 1.3–1.4 V (new humane) marginalizing the effects of adsorbed CO poisoning during durability measurements on bulk Pt catalyst. The durability testing shows another important result that Pt electrocatalysts (Pt and PtCu alloys) have been strongly influenced by the dissolution of Pt, and that the life of the catalyst is limited by the Pt amount (nanoparticles and thin films) [195, 209]. It can be concluded from the aforesaid results that catalyst durability was largely influenced by Pt dissolution, the balance between (i) Pt dissolution and (ii) the interaction between CO_{ad} poisoning, intermediate adsorption, and Pt oxidation was shown during the potential cycle. FAO does not have more basic insights into the durability of catalysts, as most Pt dissolution studies have been conducted through the development of PEMFCs, with attention paid to oxygen reduction reaction (ORR) in sulfate acid or perchlorate electrolyte. More detailed, areas effects were highlighted by recent studies on Pt-based catalysts [210, 212] in PEMFCs: (i) solubility of Pt together with its dissolution mechanism at equilibrium/nonequilibrium cases, [210] (ii) Pt nanoparticle degradation in support of carbon and dissolution [212, 213], and (iii) degradation of nanoparticles at different cycling conditions and polarization of Pt and Pt-alloy [213–215]. However, despite recent advances, the mechanism by which platinum dissolves on flat surfaces in background electrolytes is still not fully understood. It is not clear whether the dissolution of platinum is an anodic process with the oxidation of platinum or a cathodic process with the chemical dissolution of oxidized Pt [216–220]. For these two cases, an important contributor to dissolution processes is platinum oxides. Mayrhofer and coworkers proven this view by using inductively coupled plasma mass spectromertry (ICP-MS) studies [220], and they offer significant evidence to know the strong effects at different operating conditions (e.g. polarization, cycling, pH, etc.) on the dissolution of platinum in $HClO_4$ electrolyte. More basic studies on the dissolution of Pt in background electrolytes have also been carried out using potential cycle (often called electrochemical activation) [216, 217, 220–224] and potential statically determined conditions [225, 226]. Some higher rates for dissolution have been achieved in potential cyclic experiments. [221, 225] Rand and Woods [221], Johnson et al. [216], and Kinoshita et al. [222] compared different influence from different acidic media toward the dissolution of Pt, for which sulfuric acid (1 M H_2SO_4) and hyperchlorine (0.1 M $HClO_4$) electrolyte were used. It is interested to know from the reports about the influence from negative potential limits, the specific surface area evolution of platinum during dissolution, and the properties of platinum complexes as dissolved. In the negative potential limit of 1.2–1.5 V (NHE) [216, 221–223] potential cycle, Pt's dissolution rate was 2.0–5.5 ng cm^{-2}/cycle. So far, in the $HClO_4$ electrolyte, it was found that the main dissolved products are Pt(II) [217] complexes, while in sulfuric acid, the main oxidized substances are Pt(IV) complexes [216, 221]. By identifying that surface (111) is more soluble than surface (100), Komanicky et al. recently demonstrated in situ effects about Pt dissolution dependent on crystal orientation on in $HClO_4$ at a constant potential [226].

Platinum dissolves during FAO. Control the dissolution of Pt thin film with thickness. In order to better understand the dissolution process during FAO and evaluate the quantitative dissolution rate, we detected Pt films with different thicknesses. The surface limited redox replacement (SLRR) method is used for strict thickness control of platinum deposition to ensure that one platinum monolayer grows during each time of replacement. To further avoid crystal orientation and ambiguity effects on the metal dissolution, Pt faces control were studied previously (for example, surface (111) is less durable than the surface (100)), the authors have study face structures that grow Au-poly(polycrystalline) matrix uniformly and reproductively in Pt films [190].

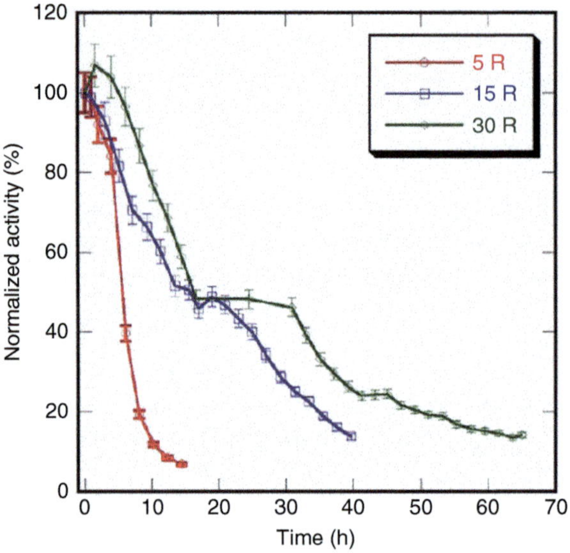

Figure 12.20 Normalized activity decay for Pt films deposited on Au-poly after 5 (red), 15 (blue), and 30 (green) SLRR events. Solution 2 M HCOOH + 0.1 M $HClO_4$. Scan rate 0.050 V s^{-1} [227]. Source: Fayette, M. et al. © 2013, American Chemical Society.

During the FAO cycle, the activity gradually decreased at a stable rate of 0.050 V s^{-1}, as depicted in Figure 12.20. The sample's activity was expressed as a standardized value relative to the initial value. The direct relationship between catalyst durability and Pt dosage is illustrated in Figure 12.20. Taking into consideration of the catalyst's "total loss" of time activity being reduced to only 20%, it was found that the puea Thai films required approximately eight hours increased by the five SLRR events, approximately 30 hours 15 SLRR events, and then ultimately approximately 60 hours 30 SLRR events. It indicates the relationship between the durability of catalyst and the platinum film thickness is close to linear. A close examination on the curves in Figure 12.20 indicates the decay phenomenon, characterized by plains periods of approximately 50–60% activity (evident for films with 15 and 30 SLRR events growing). It should be noted that reduction of the limit of positive cycling on Pt film results in a similar qualitative decline in the shape of the activity, although it is 1.5–2 times slower at a 20% mark. In general, the general activity attenuation trend of platinum-alloyed electrocatalysts with limiting thickness surely determines that the main factor of catalytic performance loss in the potential cycling process is arisen from the dissolution of platinum. This is further confirmed by the fact that in the large number of Pt samples reported in our earlier work, activity did not decline due to the "unlimited supply" of Pt.

References

1 Xie, S.J., Lin, S.Q., Zhang, Q.H. et al. (2018). *J. Energy Chem.* 27: 1629–1636.
2 Periana, R.A., Taube, D.J., Evitt, E.R. et al. (1993). *Science* 259: 340–343.
3 Lee, K.J. and Dempsey, J.L. (2017). *ACS Cent. Sci.* 3: 1137–1139.
4 O'Reilly, M.E., Kim, R.S., Oh, S., and Surendranath, Y. (2017). *ACS Cent. Sci.* 3: 1174–1179.

5 Zimmermann, T., Soorholtz, M., Bilke, M., and Schuth, F. (2016). *J. Am. Chem. Soc.* 138: 12395–12400.

6 Periana, R.A., Taube, D.J., Gamble, S. et al. (1998). *Science* 280: 560–564.

7 Jafarian, M., Mahjani, M.G., Heli, H. et al. (2003). *Electrochem. Commun.* 5: 184–188.

8 Qiao, J., Tang, S.N., Tian, Y.N. et al. (2009). *Sens. Actuators, B* 138: 402–407.

9 Lu, S.Q. and Zhuang, Z.B. (2016). *Sci. China Mater.* 59: 217–238.

10 Wang, H. and Abruna, H.D. (2017). *J. Am. Chem. Soc.* 139: 6807–6810.

11 Forti, J.C., Rocha, R.S., Lanza, M.R.V., and Bertazzoli, R. (2007). *J. Electroanal. Chem.* 601: 63–67.

12 Wasmus, S. and Kuver, A. (1999). *J. Electroanal. Chem.* 461: 14–31.

13 Arico, A.S., Srinivasan, S., and Antonucci, V. (2001). *Fuel Cells* 1: 133–161.

14 Iwasita, T. (2002). *Electrochim. Acta* 47: 3663–3674.

15 Li, X.L. and Faghri, A. (2013). *J. Power Sources* 226: 223–240.

16 Kakati, N., Maiti, J., Lee, S.H. et al. (2014). *Chem. Rev.* 114: 12397–12429.

17 Watanabe, M. and Motoo, S. (1975). *J. Electroanal. Chem.* 60: 259–266.

18 Arico, A.S., Creti, P., Giordano, N. et al. (1996). *J. Appl. Electrochem.* 26: 959–967.

19 Wang, K., Gasteiger, H.A., Markovic, N.M., and Ross, P.N. (1996). *Electrochim. Acta* 41: 2587–2593.

20 Sasaki, K., Wang, J.X., Balasubramanian, M. et al. (2004). *Electrochim. Acta* 49: 3873–3877.

21 Chen, T.Y., Lin, T.L., Luo, T.J. et al. (2010). *ChemPhysChem* 11: 2383–2392.

22 Brankovic, S.R., Wang, J.X., and Adzic, R.R. (2001). *Electrochem. Solid-State Lett.* 4: A217–A220.

23 Alayoglu, S., Nilekar, A.U., Mavrikakis, M., and Eichhorn, B. (2008). *Nat. Mater.* 7: 333–338.

24 Luo, J., Wang, L., Mott, D. et al. (2008). *Adv. Mater.* 20: 4342–4347.

25 Zeng, J., Yang, J., Lee, J.Y., and Zhou, W. (2006). *J. Phys. Chem. B* 110: 24606–24611.

26 Serov, A. and Kwak, C. (2009). *Appl. Catal., B* 90: 313–320.

27 Bianchini, C. and Shen, P.K. (2009). *Chem. Rev.* 109: 4183–4206.

28 Kakati, N., Maiti, J., Oh, J.Y., and Yoon, Y.S. (2011). *Appl. Surf. Sci.* 257: 8433–8437.

29 Oh, J.Y., Jee, S.H., Kakati, N. et al. (2010). *Jpn. J. Appl. Phys.* 49: 259–266.

30 Ishikawa, Y., Liao, M.-S., and Cabrera, C.R. (2000). *Surf. Sci.* 463: 66–80.

31 Wang, J.G. and Hammer, B. (2006). *J. Catal.* 243: 192–198.

32 Zhu, Y.M. and Cabrera, C.R. (2001). *Electrochem. Solid-State Lett.* 4: A45–A48.

33 Choi, J.H., Park, K.W., Park, I.S. et al. (2004). *Electrochim. Acta* 50: 787–790.

34 Choi, J.H., Park, K.W., Kwon, B.K., and Sung, Y.E. (2003). *J. Electrochem. Soc.* 150: A973–A978.

35 Alcaide, F., Alvarez, G., Cabot, P.L. et al. (2011). *Int. J. Hydrogen Energy* 36: 4432–4439.

36 Feng, L., Gao, G., Huang, P. et al. (2011). *Nanoscale Res. Lett.* 6: 551.

37 Gasteiger, H.A., Markovic, N., Ross, P.N., and Cairns, E.J. (1994). *Electrochim. Acta* 39: 1825–1832.
38 Wang, R.F., Wang, H., Wei, B.X. et al. (2010). *Int. J. Hydrogen Energy* 35: 10081–10086.
39 Zhang, H., Yin, Y.J., Hu, Y.J. et al. (2010). *J. Phys. Chem. C* 114: 11861–11867.
40 Kristian, N. and Wang, X. (2008). *Electrochem. Commun.* 10: 12–15.
41 Wu, Y.N., Liao, S.J., Liang, Z.X. et al. (2009). *J. Power Sources* 194: 805–810.
42 Wang, S.Y., Kristian, N., Jiang, S.P., and Wang, X. (2009). *Nanotechnology* 20: 0957–4484.
43 Zhao, H.B., Li, L., Yang, J., and Zhang, Y.M. (2008). *Electrochem. Commun.* 10: 1527–1529.
44 Wang, R.F., Li, H., Feng, H.Q. et al. (2010). *J. Power Sources* 195: 1099–1102.
45 Zhou, C., Wang, H., Peng, F. et al. (2009). *Langmuir* 25: 7711–7717.
46 Kakati, N., Maiti, J., Jee, S.H. et al. (2011). *J. Alloys Compd.* 509: 5617–5622.
47 Wang, G.X., Takeguchi, T., Zhang, Y. et al. (2009). *J. Electrochem. Soc.* 156: B862–B869.
48 Hirakawa, K., Inoue, M., and Abe, T. (2010). *Electrochim. Acta* 55: 5874–5880.
49 Tian, J.A., Sun, G.Q., Jiang, L.H. et al. (2007). *Electrochem. Commun.* 9: 563–568.
50 Yang, S., Zhao, C., Ge, C. et al. (2012). *J. Mater. Chem.* 22: 7104–7107.
51 Jusys, Z., Schmidt, T.J., Dubau, L. et al. (2002). *J. Power Sources* 105: 297–304.
52 Lei, Y., Zhao, G., Tong, X. et al. (2010). *ChemPhysChem* 11: 276–284.
53 Guo, J.W., Zhao, T.S., Prabhuram, J. et al. (2006). *J. Power Sources* 156: 345–354.
54 Sun, Z., Wang, X., Liu, Z. et al. (2010). *Langmuir* 26: 12383–12389.
55 Saha, M.S., Li, R.Y., and Sun, X.L. (2007). *Electrochem. Commun.* 9: 2229–2234.
56 Zhang, H.L., Hu, C.G., He, X.S. et al. (2011). *J. Power Sources* 196: 4499–4505.
57 Tsang, K.Y., Lee, T.C., Ren, J.W. et al. (2006). *J. Exp. Nanosci.* 1: 113–123.
58 Gu, Y.J. and Wong, W.T. (2006). *J. Electrochem. Soc.* 153: A1714–A1718.
59 Maiyalagan, T. and Viswanathan, B. (2008). *J. Power Sources* 175: 789–793.
60 Ke, K. and Waki, K. (2007). *J. Electrochem. Soc.* 154: A207–A212.
61 Hu, C.G., He, X.S., and Xia, C.H. (2010). *J. Power Sources* 195: 1594–1598.
62 Xu, C.W., Zeng, R., Shen, P.K., and Wei, Z.D. (2005). *Electrochim. Acta* 51: 1031–1035.
63 Lee, J.M., Han, S.B., Kim, J.Y. et al. (2010). *Carbon* 48: 2290–2296.
64 Saha, M.S., Li, R.Y., Cai, M., and Sun, X.L. (2007). *Electrochem. Solid-State Lett.* 10: B130–B133.
65 Huang, S.H., Susanti, D., Tsai, D.S. et al. (2008). *Langmuir* 24: 2785–2791.
66 Huang, S.Y., Chang, C.M., Wang, K.W., and Yeh, C.T. (2007). *ChemPhysChem* 8: 1774–1777.
67 Zhao, J., Chen, W.X., and Zheng, Y.F. (2009). *Mater. Chem. Phys.* 113: 591–595.
68 Hayden, B.E., Malevich, D.V., and Pletcher, D. (2001). *Electrochem. Commun.* 3: 395–399.
69 Tamizhmani, G. and Capuano, G.A. (1994). *J. Electrochem. Soc.* 141: 968–975.
70 Jian, X.H., Tsai, D.S., Chung, W.H. et al. (2009). *J. Mater. Chem.* 19: 1601–1607.
71 Shen, P.K. and Tseung, A.C.C. (1994). *J. Electrochem. Soc.* 141: 3082–3090.

72 Chen, Z., Waje, M., Li, W., and Yan, Y. (2007). *Angew. Chem. Int. Ed.* 46: 4060–4063.

73 Choi, S.M., Kim, J.H., Jung, J.Y. et al. (2008). *Electrochim. Acta* 53: 5804–5811.

74 Lee, E.P., Peng, Z., Chen, W. et al. (2008). *ACS Nano* 2: 2167–2173.

75 Walton, A.S., Allen, C.S., Critchley, K. et al. (2007). *Nanotechnology* 18: 0957–4484.

76 Mayers, B., Jiang, X., Sunderland, D. et al. (2003). *J. Am. Chem. Soc.* 125: 13364–13365.

77 Gorzny, M.L., Walton, A.S., and Evans, S.D. (2010). *Adv. Funct. Mater.* 20: 1295–1300.

78 He, Y.B., Li, G.R., Wang, Z.L. et al. (2010). *J. Phys. Chem. C* 114: 19175–19181.

79 Formo, E., Peng, Z.M., Lee, E. et al. (2008). *J. Phys. Chem. C* 112: 9970–9975.

80 Ghosh, S. and Raj, C.R. (2010). *J. Phys. Chem. C* 114: 10843–10849.

81 Zhao, Y., Fan, L.Z., Zhong, H.Z. et al. (2007). *Adv. Funct. Mater.* 17: 1537–1541.

82 Li, Z.S., Lin, S., Chen, Z.L. et al. (2012). *J. Colloid Interface Sci.* 368: 413–419.

83 Hsin, Y.L., Hwang, K.C., and Yeh, C.T. (2007). *J. Am. Chem. Soc.* 129: 9999–10010.

84 Antolini, E. (2009). *Energy Environ. Sci.* 2: 915–931.

85 Li, Z.Y., Liang, Y.J., Jiang, S.P. et al. (2012). *Fuel Cells* 12: 677–682.

86 Hsieh, C.T., Gu, J.L., Chen, Y.C., and Tzou, D.Y. (2013). *Electrochim. Acta* 98: 39–47.

87 Miao, F.J., Tao, B.R., Sun, L. et al. (2010). *J. Power Sources* 195: 146–150.

88 Prabhuram, J., Manoharan, R., and Vasan, H.N. (1998). *J. Appl. Electrochem.* 28: 935–941.

89 Tao, B.R., Zhang, J., Hui, S.C. et al. (2010). *Electrochim. Acta* 55: 5019–5023.

90 Borasio, M., Rodriguez de la Fuente, O., Rupprechter, G., and Freund, H.J. (2005). *J. Phys. Chem. B* 109: 17791–17794.

91 Baumer, M., Libuda, J., Neyman, K.M. et al. (2007). *Phys. Chem. Chem. Phys.* 9: 3541–3558.

92 Wang, Y., Sheng, Z.M., Yang, H. et al. (2010). *Int. J. Hydrogen Energy* 35: 10087–10093.

93 Yin, Z., Chi, M.F., Zhu, Q.J. et al. (2013). *J. Mater. Chem. A* 1: 9157–9163.

94 Su, Y.Z., Zhang, M.Z., Liu, X.B. et al. (2012). *Int. J. Electrochem. Sc.* 7: 4158–4170.

95 Yin, Z., Zheng, H.J., Ma, D., and Bao, X.H. (2009). *J. Phys. Chem. C* 113: 1001–1005.

96 Zheng, J.N., Li, S.S., Ma, X. et al. (2014). *J. Power Sources* 262: 270–278.

97 Xu, C., Tian, Z., Shen, P., and Jiang, S.P. (2008). *Electrochim. Acta* 53: 2610–2618.

98 Ye, K.-H., Zhou, S.-A., Zhu, X.-C. et al. (2013). *Electrochim. Acta* 90: 108–111.

99 Iwai, Y., Ikemoto, S., Haramaki, K. et al. (2014). *J. Supercrit. Fluids* 94: 48–58.

100 Thiam, H.S., Daud, W.R.W., Kamarudin, S.K. et al. (2013). *Energy Convers. Manage.* 75: 718–726.

101 Yoon, S.R., Hwang, G.H., Cho, W.I. et al. (2002). *J. Power Sources* 106: 215–223.

102 Choi, W.C., Kim, J.D., and Woo, S.I. (2001). *J. Power Sources* 96: 411–414.

103 Hejze, T., Gollas, B.R., Sauerbrey, R.K. et al. (2005). *J. Power Sources* 140: 21–27.

104 Brandao, L., Rodrigues, J., Madeira, L.M., and Mendes, A. (2010). *Int. J. Hydrogen Energy* 35: 11561–11567.

105 Kim, J.M., Joh, H.I., Jo, S.M. et al. (2010). *Electrochim. Acta* 55: 4827–4835.

106 Zhao, G.Y., Xu, C.L., Guo, D.J. et al. (2007). *Appl. Surf. Sci.* 253: 3242–3246.

107 Zhao, G.Y., Xu, C.L., Guo, D.J. et al. (2006). *J. Power Sources* 162: 492–496.

108 Meng, H., Xie, F., Chen, J. et al. (2011). *Nanoscale* 3: 5041–5048.

109 Carmo, M., Sekol, R.C., Ding, S. et al. (2011). *ACS Nano* 5: 2979–2983.

110 Schroers, J. (2010). *Adv. Mater.* 22: 1566–1597.

111 Guo, S., Zhang, S., Sun, X., and Sun, S. (2011). *J. Am. Chem. Soc.* 133: 15354–15357.

112 Bertin, E., Garbarino, S., Ponrouch, A., and Guay, D. (2012). *J. Power Sources* 206: 20–28.

113 Guo, Y.G., Hu, J.S., Zhang, H.M. et al. (2005). *Adv. Mater.* 17: 746–750.

114 Shin, T.Y., Yoo, S.H., and Park, S. (2008). *Chem. Mater.* 20: 5682–5686.

115 Zhang, H.M., Zhou, W.Q., Du, Y.K. et al. (2010). *Electrochem. Commun.* 12: 882–885.

116 Zhuang, L.N., Wang, W.J., Hong, F. et al. (2012). *J. Solid State Chem.* 191: 239–245.

117 Yao, Z.Q., Zhu, M.S., Jiang, F.X. et al. (2012). *J. Mater. Chem.* 22: 13707–13713.

118 Liu, F., Lee, J.Y., and Zhou, W.J. (2005). *Adv. Funct. Mater.* 15: 1459–1464.

119 Liu, F., Lee, J.Y., and Zhou, W.J. (2004). *J. Phys. Chem. B* 108: 17959–17963.

120 Liu, F., Lee, J.Y., and Zhou, W.J. (2006). *J. Electrochem. Soc.* 153: A2133–A2138.

121 Liu, F., Lee, J.Y., and Zhou, W.J. (2006). *Small* 2: 121–128.

122 Yoo, S.J., Jeon, T.Y., Kim, K.S. et al. (2010). *Phys. Chem. Chem. Phys.* 12: 15240–15246.

123 Casado-Rivera, E., Volpe, D.J., Alden, L. et al. (2004). *J. Am. Chem. Soc.* 126: 4043–4049.

124 Maksimuk, S., Yang, S., Peng, Z., and Yang, H. (2007). *J. Am. Chem. Soc.* 129: 8684–8685.

125 Markovic, N.M., Gasteiger, H.A., and Ross, P.N. (1995). *J. Phys. Chem.* 99: 3411–3415.

126 Kinoshita, K. (1990). *J. Electrochem. Soc.* 137: 845–848.

127 Han, S.B., Song, Y.J., Lee, J.M. et al. (2008). *Electrochem. Commun.* 10: 1044–1047.

128 Lee, Y.W., Han, S.B., Kim, D.Y., and Park, K.W. (2011). *Chem. Commun. (London)* 47: 6296–6298.

129 Peng, Z., You, H., Wu, J., and Yang, H. (2010). *Nano Lett.* 10: 1492–1496.

130 Stamenkovic, V.R., Fowler, B., Mun, B.S. et al. (2007). *Science* 315: 493–497.

131 Clavilier, J., Orts, J.M., and Feliu, J.M. (1994). *J. Phys. IV* 4: 303–308.

132 Hamnett, A. (1997). *Catal. Today* 38: 445–457.

133 Herrero, E., Franaszczuk, K., and Wieckowski, A. (1994). *J. Phys. Chem.* 98: 5074–5083.

134 Kita, H., Gao, Y.Z., Nakato, T., and Hattori, H. (1994). *J. Electroanal. Chem.* 373: 177–183.

135 Sun, S.G. and Clavilier, J. (1987). *J. Electroanal. Chem.* 236: 95–112.

136 Kim, S., Kim, C., and Lee, H. (2010). *Top. Catal.* 53: 686–693.

137 Yang, H., Zhang, J., Sun, K. et al. (2010). *Angew. Chem. Int. Ed.* 49: 6848–6851.

138 Yin, A.X., Min, X.Q., Zhang, Y.W., and Yan, C.H. (2011). *J. Am. Chem. Soc.* 133: 3816–3819.

139 Markovic, N.M. and Ross, P.N. (2002). *Surf. Sci. Rep.* 45: 121–229.

140 Ferrin, P. and Mavrikakis, M. (2009). *J. Am. Chem. Soc.* 131: 14381–14389.

141 Housmans, T.H., Wonders, A.H., and Koper, M.T. (2006). *J. Phys. Chem. B* 110: 10021–10031.

142 Jarvi, T.D., Sriramulu, S., and Stuve, E.M. (1998). *Colloids Surf., A* 134: 145–153.

143 Tseung, A.C.C. and Chen, K.Y. (1997). *Catal. Today* 38: 439–443.

144 Dinh, H.N., Ren, X.M., Garzon, F.H. et al. (2000). *J. Electroanal. Chem.* 491: 222–233.

145 Arico, A.S., Creti, P., Modica, E. et al. (2000). *Electrochim. Acta* 45: 4319–4328.

146 Rolison, D.R., Hagans, P.L., Swider, K.E., and Long, J.W. (1999). *Langmuir* 15: 774–779.

147 Wei, Z.D., Guo, H.T., and Tang, Z.Y. (1996). *J. Power Sources* 58: 239–242.

148 Shukla, A.K., Arico, A.S., El-Khatib, K.M. et al. (1999). *Appl. Surf. Sci.* 137: 20–29.

149 Markovic, N.M., Widelov, A., Ross, P.N. et al. (1997). *Catal. Lett.* 43: 161–166.

150 Rahim, M.A.A., Khalil, M.W., and Hassan, H.B. (2000). *J. Appl. Electrochem.* 30: 1151–1155.

151 Parsons, R. and Vandernoot, T. (1988). *J. Electroanal. Chem.* 257: 9–45.

152 Arico, A.S., Shukla, A.K., El-Khatib, K.M. et al. (1999). *J. Appl. Electrochem.* 29: 671–676.

153 Adams, B.D., Asmussen, R.M., Ostrom, C.K., and Chen, A.C. (2014). *J. Phys. Chem. C* 118: 29903–29910.

154 Mikolajczuk, A., Borodzinski, A., Kedzierzawski, P. et al. (2011). *Appl. Surf. Sci.* 257: 8211–8214.

155 Zhou, W.P., Lewera, A., Larsen, R. et al. (2006). *J. Phys. Chem. B* 110: 13393–13398.

156 Zhou, Y., Liu, J.G., Ye, J.L. et al. (2010). *Electrochim. Acta* 55: 5024–5027.

157 Baik, S.M., Han, J., Kim, J., and Kwon, Y. (2011). *Int. J. Hydrogen Energy* 36: 14719–14724.

158 Jeon, H., Jeong, B., Choun, M., and Lee, J. (2014). *Electrochim. Acta* 140: 525–528.

159 Miyake, H., Okada, T., Samjeske, G., and Osawa, M. (2008). *Phys. Chem. Chem. Phys.* 10: 3662–3669.

160 Sennik, E., Soysal, U., and Ozturk, Z.Z. (2014). *Sens. Actuators, B* 199: 424–432.

161 Meng, D., Yamazaki, T., and Kikuta, T. (2014). *Sens. Actuators, B* 190: 838–843.

162 Pan, Y.H., Zhang, R.M., and Blair, S.L. (2009). *Electrochem. Solid-State Lett.* 12: B23–B26.

163 Winjobi, O., Zhang, Z.Y., Liang, C.H., and Li, W.Z. (2010). *Electrochim. Acta* 55: 4217–4221.

164 Corcoran, C.J., Tavassol, H., Rigsby, M.A. et al. (2010). *J. Power Sources* 195: 7856–7879.

165 Zhou, W.J. and Lee, J.Y. (2008). *J. Phys. Chem. C* 112: 3789–3793.

166 Jiang, K., Zhang, H.X., Zou, S., and Cai, W.B. (2014). *Phys. Chem. Chem. Phys.* 16: 20360–20376.

167 Hoshi, N., Nakamura, M., and Kida, K. (2007). *Electrochem. Commun.* 9: 279–282.

168 Hoshi, N., Kida, K., Nakamura, M. et al. (2006). *J. Phys. Chem. B* 110: 12480–12484.

169 Lee, Y.W., Oh, J.K., Kim, H.S. et al. (2010). *J. Power Sources* 195: 5896–5901.

170 Zhao, Y., Qin, S.J., Li, Y. et al. (2014). *Electrochim. Acta* 145: 148–153.

171 Hong, W., Fang, Y.X., Wang, J., and Wang, E.K. (2014). *J. Power Sources* 248: 553–559.

172 Zhang, B.Q., Peng, H.L., Yang, L.J. et al. (2015). *J. Mater. Chem. A* 3: 973–977.

173 Ren, M.J., Zou, L.L., Yuan, T. et al. (2014). *J. Power Sources* 267: 527–532.

174 Yi, Q.F., Huang, W., Liu, X.P. et al. (2008). *J. Electroanal. Chem.* 619: 197–205.

175 Yang, L., Hu, C., Wang, J. et al. (2011). *Chem. Commun. (London)* 47: 8581–8583.

176 Arenz, M., Stamenkovic, V., Schmidt, T.J. et al. (2003). *Phys. Chem. Chem. Phys.* 5: 4242–4251.

177 Lu, Y.Z. and Chen, W. (2010). *J. Phys. Chem. C* 114: 21190–21200.

178 Ojani, R., Abkar, Z., Hasheminejad, E., and Raoof, J.B. (2014). *Int. J. Hydrogen Energy* 39: 7788–7797.

179 Wang, X., Tang, Y., Gao, Y., and Lu, T.H. (2008). *J. Power Sources* 175: 784–788.

180 Zhang, G.J., Wang, Y.E., Wang, X. et al. (2011). *Appl. Catal., B* 102: 614–619.

181 Tian, M., Malig, M., Chen, S.A., and Chen, A.C. (2011). *Electrochem. Commun.* 13: 370–373.

182 Dai, L. and Zou, S.Z. (2011). *J. Power Sources* 196: 9369–9372.

183 Chen, D., Cui, P.L., He, H.Y. et al. (2014). *J. Power Sources* 272: 152–159.

184 Capon, A. and Parsons, R. (1973). *J. Electroanal. Chem.* 45: 205–231.

185 Capon, A. and Parsons, R. (1973). *J. Electroanal. Chem.* 44: 1–7.

186 Capon, A. and Parsons, R. (1973). *J. Electroanal. Chem.* 44: 239–254.

187 Sun, S.G., Clavilier, J., and Bewick, A. (1988). *J. Electroanal. Chem.* 240: 147–159.

188 Chang, S.C., Leung, L.W.H., and Weaver, M.J. (1990). *J. Phys. Chem.* 94: 6013–6021.

189 Obradovic, M.D., Tripkovic, A.V., and Gojkovic, S.L. (2009). *Electrochim. Acta* 55: 204–209.

190 Fayette, M., Liu, Y., Bertrand, D. et al. (2011). *Langmuir* 27: 5650–5658.

191 Brankovic, S.R., McBreen, J., and Adzic, R.R. (2001). *J. Electroanal. Chem.* 503: 99–104.

192 Zhang, L.J., Wang, Z.Y., and Xia, D.G. (2006). *J. Alloys Compd.* 426: 268–271.

193 Hwang, S.M., Bonevich, J.E., Kim, J.J., and Moffat, T.P. (2011). *J. Electrochem. Soc.* 158: B1019–B1028.

194 Liu, Y., Abe, H., Edvenson, H.M. et al. (2010). *Phys. Chem. Chem. Phys.* 12: 12978–12986.

195 Xu, D., Bliznakov, S., Liu, Z. et al. (2010). *Angew. Chem. Int. Ed.* 49: 1282–1285.

196 Yang, H.Z., Dai, L., Xu, D. et al. (2010). *Electrochim. Acta* 55: 8000–8004.

197 Tripkovic, A.V., Popovic, K.D., Stevanovic, R.M. et al. (2006). *Electrochem. Commun.* 8: 1492–1498.

198 de-los-Santos-Alvarez, N., Alden, L.R., Rus, E. et al. (2009). *J. Electroanal. Chem.* 626: 14–22.

199 Neurock, M., Janik, M., and Wieckowski, A. (2008). *Faraday Discuss.* 140: 363–378; discussion 417–337.

200 Samjeske, G., Miki, A., Ye, S. et al. (2005). *J. Phys. Chem. B* 109: 23509–23516.

201 Chen, Y.X., Ye, S., Heinen, M. et al. (2006). *J. Phys. Chem. B* 110: 9534–9544.

202 Chen, Y.X., Heinen, M., Jusys, Z., and Behm, R.J. (2007). *ChemPhysChem* 8: 380–385.

203 Gao, W., Keith, J.A., Anton, J., and Jacob, T. (2010). *J. Am. Chem. Soc.* 132: 18377–18385.

204 Cuesta, A., Cabello, G., Gutierrez, C., and Osawa, M. (2011). *Phys. Chem. Chem. Phys.* 13: 20091–20095.

205 Mukouyama, Y., Kikuchi, M., Samjeske, G. et al. (2006). *J. Phys. Chem. B* 110: 11912–11917.

206 Chen, Y.X., Heinen, M., Jusys, Z., and Behm, R.J. (2006). *Langmuir* 22: 10399–10408.

207 Chen, Y.X., Heinen, M., Jusys, Z., and Behm, R.J. (2006). *Angew. Chem. Int. Ed.* 45: 981–985.

208 Grozovski, V., Climent, V., Herrero, E., and Feliu, J.M. (2010). *Phys. Chem. Chem. Phys.* 12: 8822–8831.

209 McCurry, D.A., Kamundi, M., Fayette, M. et al. (2011). *ACS Appl. Mater. Interfaces* 3: 4459–4468.

210 Borup, R., Meyers, J., Pivovar, B. et al. (2007). *Chem. Rev.* 107: 3904–3951.

211 Tian, M. and Conway, B.E. (2008). *J. Electroanal. Chem.* 616: 45–56.

212 Shao, Y.Y., Yin, G.P., and Gao, Y.Z. (2007). *J. Power Sources* 171: 558–566.

213 Shao-Horn, Y., Sheng, W.C., Chen, S. et al. (2007). *Top. Catal.* 46: 285–305.

214 Hiraoka, F., Matsuzawa, K., and Mitsushima, S. (2013). *Electrocatalysis* 4: 10–16.

215 Carlton, C.E., Chen, S., Ferreira, P.J. et al. (2012). *J. Phys. Chem. Lett.* 3: 161–166.

216 Johnson, D.C., Napp, D.T., and Buckenst, S. (1970). *Electrochim. Acta* 15: 1493–1509.

217 Mitsushima, S., Koizumi, Y., Uzuka, S., and Ota, K.I. (2008). *Electrochim. Acta* 54: 455–460.

218 Inzelt, G., Berkes, B., and Kriston, A. (2010). *Electrochim. Acta* 55: 4742–4749.

219 Rinaldo, S.G., Stumper, J., and Eikerling, M. (2010). *J. Phys. Chem. C* 114: 5773–5785.

220 Topalov, A.A., Katsounaros, I., Auinger, M. et al. (2012). *Angew. Chem. Int. Ed.* 51: 12613–12615.

221 Rand, D.A.J. and Woods, R. (1972). *J. Electroanal. Chem.* 35: 209–218.

222 Kinoshita, K., Lundquist, J.T., and Stonehart, P. (1973). *J. Electroanal. Chem.* 48: 157–166.

223 Wang, X.P., Kumar, R., and Myers, D.J. (2006). *Electrochem. Solid-State Lett.* 9: A225–A227.

224 Ofstad, A.B., Thomassen, M.S., de la Fuente, J.L.G. et al. (2010). *J. Electrochem. Soc.* 157: B621–B627.

225 Ota, K.I., Nishigori, S., and Kamiya, N. (1988). *J. Electroanal. Chem.* 257: 205–215.

226 Komanicky, V., Chang, K.C., Menzel, A. et al. (2006). *J. Electrochem. Soc.* 153: B446–B451.

227 Fayette, M., Nutariya, J., Vasiljevic, N., and Dimitrov, N. (2013). *ACS Catal.* 3: 1709–1718.

13

C_{2+} Molecule Oxidation

13.1 Ethanol Oxidation

13.1.1 Reaction Mechanism

The issues arisen from toxicity and crossover of methanol have prevented the commercial applications for direct methanol fuel cells (DMFCs). This case has motivated more study on the alternative fuel candidates to replace methanol; among various fuels available, ethanol has the advantages of low toxicity, and more importantly, the ideal energy density as high as 8.1 kW h kg^{-1}, which is obviously higher than methanol of 6.1 kW h kg^{-1}, and thus ethanol and its use in direct basic alcohol fuel cells (DAFCs) are promising to replace the utilization of fossil fuels [1, 2].

A rather complex multi-electron transfer pathway was involved in the ethanol electro-oxidation in alkaline media, which could result in the production different products via 2-, 4-, or 12-electron transfer. In a typical direct ethanol fuel cell (DEFC), the anode Pd-based electrocatalysts without Pt could electrochemically convert ethanol into acetate in high pH conditions (Figure 13.1) [2, 3]. In particular, a four-electron transfer process may occur for the production of acetate as the final product in such high OH^- concentration. When the pH is increased for restraining effectively, OH^- concentration, complete oxidation may occur to produce CO_2 ultimately [2, 4]. Mechanically, the oxidation degree of ethanol depends on the coverage of OH_{ads} and CH_3CO_{ads}; this suggests that the amount of OH^- ions available in the electrolyte together with the coverage level of OH^- on catalyst surfaces are the key factors to improve the ethanol electro-oxidation kinetics [5].

13.1.2 Catalyst Advances

13.1.2.1 Pd-Based Catalysts

Toward substitution of platinum-based catalysts for palladium-based catalysts, it is emphasized that the palladium-based catalyst has high catalytic activity and can improve the steady-state oxidation behavior of ethanol on palladium, while platinum is the opposite [5–13]. Fang et al. [13] studied the ethanol electro-oxidation

Electrocatalysis in Balancing the Natural Carbon Cycle, First Edition. Yaobing Wang.
© 2021 WILEY-VCH GmbH. Published 2021 by WILEY-VCH GmbH.

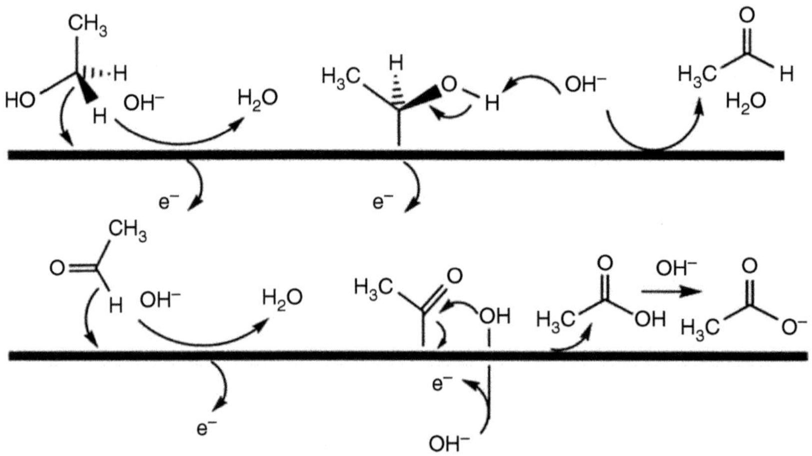

Figure 13.1 Proposed mechanism for the selective conversion of ethanol into acetate on Pd-based electrocatalysts in alkaline media. Source: Bianchini et al. [3]. © 2009, Elsevier.

mechanism of Pd at different pH values using in situ Fourier-transform infrared spectroscopy (FTIR) spectroscopy (Figure 13.2). When 0.5 M NaOH or a higher concentration was used, the selectivity of ethanol for Pd oxidation in a strongly alkaline medium was observed, only with characteristic bands for acetic acid ions found (1558 and 1415 cm^{-1}). This is different from platinum-based catalysts in acidic media, which always causes ethanol oxidation into a mixture of aldehyde, acetic acid, and carbon dioxide. The FTIR spectra were used for mechanism analysis. When the pH value was lower or equal to CO_2 (2343 cm^{-1} represents CO_2 formation band), it was found that the catalyst efficiency decreased along with the increase in pH value.

Compared with Pt-based materials, Pd-based materials have obvious advantages over ethanol electro-oxidation in terms of toxicity tolerance reactivity and cost. Nevertheless, C—C bonds cleavage is relatively not easy to occur on Pd. In order to enhance the catalytic performance of Pd for ethanol oxidation, surface modification, alloying, and oxidation promotion were used. Palladium-based electrocatalysts were studied in an alkaline medium. Nguyen et al. [14] indicated that when Pd and Ag were allowed to generate Pd-Ag/C catalyst with enhanced alkaline ethanol electro-oxidation activity, in comparison with Pd/C and Pt/C, the D-band center changed. At the same time, they found that the Ag addition promoted the oxidization of poisonous CO species on the Pd-Ag/C surface, resulting obvious improvement in the electrocatalytic activity of Pd. Recently, the catalytic activity of Pd/GO and Pd-Ag/GO was compared during ethanol electro-oxidation [15]. Similarly, the Ag addition also improved catalytic activity for ethanol electro-oxidation, which was further depicted by D-band center theory. Many researchers used Pd–Au to promote the electrocatalytic performance of Pd [16–20]. Studies have shown that gold can catalyze partial electro-oxidation of hydrocarbons and also CO oxidation. Pd–Au catalyst has unique electron-absorbing properties and can stabilize nearby metal catalysts. It has a broad application prospect in the ethanol oxidation reaction

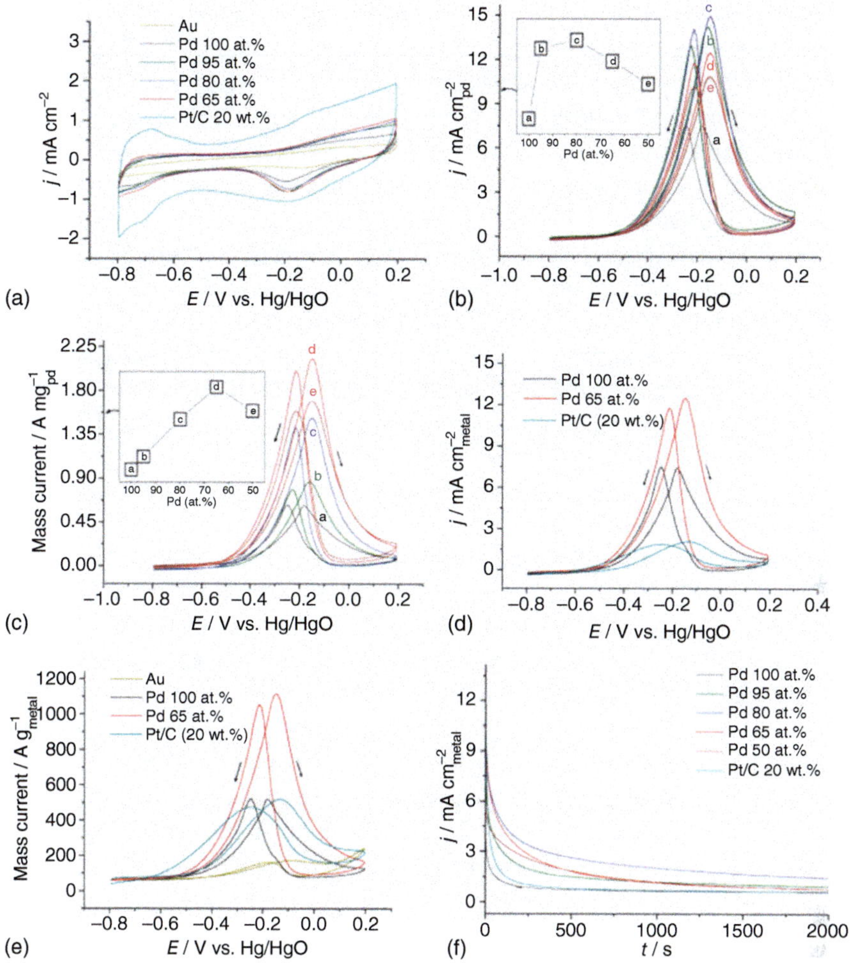

Figure 13.2 In situ FTIR spectra obtained under potential step polarization in 1 M ethanol with different NaOH concentrations of (a) 5 M, (b) 2 M, (c) 1 M (pH 14), (d) 0.5 M NaOH + 0.5 M $NaClO_4$ (pH 13.3), (e) 0.1 M NaOH + 0.9 M $NaClO_4$ (pH 13), and (f) 0.05 M NaOH + 0.95 M $NaClO_4$ (pH 12.7). E_r = 24 mV, scan number 128. The arrows indicate the potential polarization direction. Source: Fang et al. [13]. © 2010, Elsevier.

(EOR) [19]. Cui et al. [16] reported the unloaded Pd–Au nanotubes with adjustable interfaces and proved the enhancement effect of Au and Pd particles during ethanol oxidation. Similar work was reported by Qin et al. [19], with focusing on the influence from alloying degree between Au and Pd particles toward the ethanol oxidation performance. Studies have shown that, compared with high-level alloyed PdAu/C catalysts, PdAu/C catalysts with low-level alloy have higher catalytic activity.

Some transition metal oxides have promoted the effects to Pd catalysts, which involves Co_3O_4, SnO_2, ZrO_2, Fe_3O_4, NiO, and MnO_2 [2, 4, 7, 21]. Based on this, the electrocatalytic activity of palladium on EOR was studied. There are three aspects for the promoted catalytic performance of Pd-based metal oxides catalysts: (i) the electronic structure of Pd would be changed by the strong Pd-oxide support interactions, with the formation of electron transfer or interface bonding, (ii) the improved stability of Pd components during sintering, and (iii) the increased content of OH^- species on the catalyst surfaces which is beneficial for the proceeding electrochemical ethanol oxidation by oxidation/remove of poisonous intermediates adsorbed on Pd surfaces.

Nickel oxides have been proved as an effective co-catalyst to promote Pd electrocatalytic activity during ethanol oxidation [3, 21]. In combination with cobalt oxides, a hybrid $NiCoO_x$ was synthesized on the near sites of Pd electrocatalyst (Pd-$NiCoO_x$/C) by growing Pd epitaxy into NiCo oxides alloys [4]. The close proximity of Pd nanoparticles and $NiCoO_x$ resulted in strong metal–support interaction, showing high conductivity and the resulting enhanced catalytic activity of ethanol electro-oxidation in alkaline electrolyte. Martinez et al. [2] experimentally understood the difference between intrinsic alloy effects and the synergies arisen from addition oxides. SnO_2 phase was prepared by the sacrificial support method (SSM) as well as spray pyrolysis (SP), and Pd was chemically reduced to the oxide phase. X-ray diffraction (XRD) analysis shows that Pd–SnO_2 (SSM) and Pd–SnO_2 (SP) samples have palladium and SnO_2 phases. To understand the ethanol electro-oxidation mechanism, Pd–SnO_2 (SP) catalyst was studied by in situ infrared reflection and absorption spectra with recording in 1.0 M KOH + 1.0 M ethanol solution. The obvious peaks concentrated in 1045, 1410, 1086, and 1549 cm^{-1} (Figure 13.3), which confirmed the adsorption of ethanol on the surface accompanied with acetate generation. During the forward scan of cyclic voltammetry (CV) diagram representing ethanol oxidation, ethanol characteristic peak integration field (ν = 1045 cm^{-1}) and acetic acid (ν = 1410 cm^{-1}) (Figure 13.3) are proposed. It was revealed that acetic acid species showed complete potential for ethanol oxidation and increased potential for acetic acid production by increasing the number, reaching a maximum peak potential (−0.10 V). A similar experiment was recorded using 0.1 M KOH ethanol electrolyte; the results show that the produced products and intermediates were distinct with those with high pH (1.0 M KOH). The results show that SnO_2, as the co-catalyst, provides OH^- to the layer in the interface at high hydroxide ion concentration, which improves the turnover rate and restricts the product effectively to acetate. On the contrary, at a limited OH^- concentration, the OH^- species adjacent to reaction surfaces are partially dehydrogenated by ethanol, thus limiting the formation of acetic acid and allowing the adsorbed ethanol to be further oxidized completely.

The approaches of environmental adaptation via tuning the pH and concentration dependence, forming alloy, adding modification, and promoting oxidation, would also be used for enhancing oxidation of ethanol on Pd-based catalysts. The design of advanced catalysts for oxidation of other organic fuels can take into consideration of these factors. In addition, these Pd characteristics also highlight the great potential of these Pd-based catalysts for DEFCs as a power source for motor vehicles and portable applications in the future.

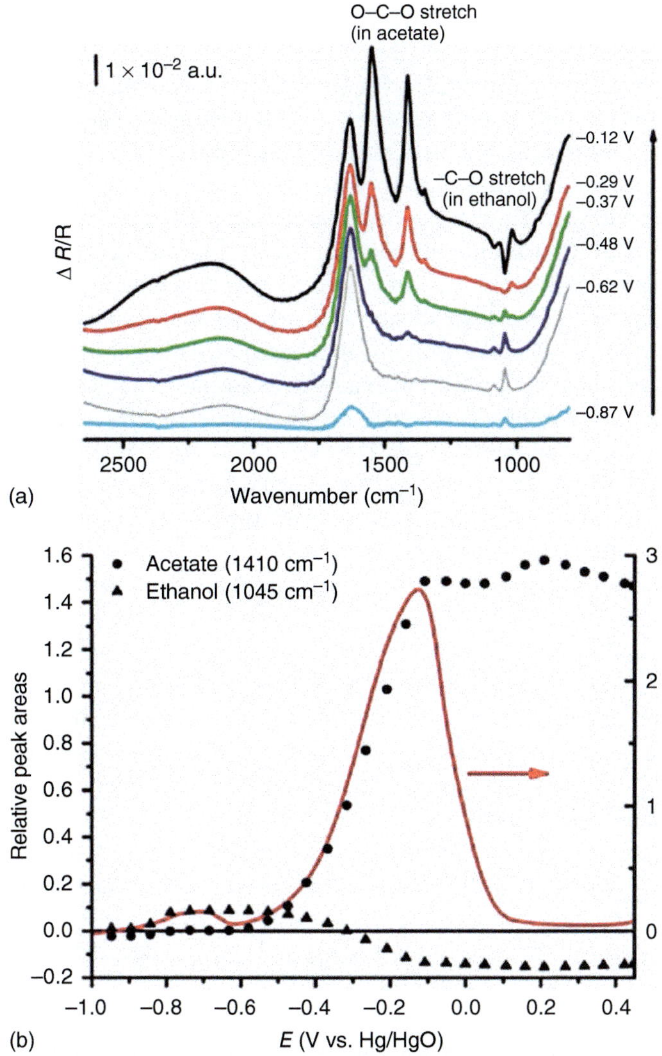

Figure 13.3 (a) In situ infrared reflection-absorption spectroscopy (IRRAS) spectra of the ethanol electro-oxidation reaction by Pd–SnO_2 (SP) catalyst in a 1 M KOH electrolyte. (b) Forward sweep voltammogram of the oxidation reaction. Conditions: 1 M ethanol, 1 mV s^{-1}, 228 °C, 0 rpm. Source: Martinez et al. [2]. © 2014, John Wiley & Sons.

13.1.2.2 Pt-Based Catalysts

Pt–Ru system. Face-centered cubic (FCC) is the typical crystal structure of Pt, on the other hand, while hexagonal close packed (HCP) is the typical structure of Ru. Pt and Ru can form a solid solution in which Pt atoms were replaced by Ru atoms at its FCC structure as the Ru content increase to 70%. The typical lattice size decrease from 3.923°A (pure Pt) to 3.83°A when Ru content is 67.5%. On the other hand, Ru with the HCP structure would be replaced by Pt atoms when the Ru content is above

0.7. Although Pt–Ru alloys are easy to be obtained, acquiring carbon-loaded Pt–Ru is relatively hard [22].

On the oxidation mechanism of ethanol, with regard to the ethanol electro-oxidation mechanism, it was found that the generation of adsorbed substances from dissolved ethanol would be partially inhibited with the existence of Ru [23]. Such a case promotes the formation of weakly adsorbed substances and it is beneficial to proceed the oxidation pathway. Thus, it has been found that ethanol production would be easier than pure Pt. Consistent with this result, Camara et al. [24] found that dissociation adsorption of ethanol was prevented by Ru. This effect may be arisen from the reduction of the adjacent Pt site, which is required to break the C—C bond. Additionally, Ru was found to promote the electro-oxidation of strongly binding adsorption intermediate [25], resulting in a relative excellent CO_2 production compared to Pt. Temperature effects were studied by Jiang et al. [26], that is, the electro-oxidation of ethanol on the Pt/C and Pt–Ru/C by CV. For temperatures increasing from 25 to 80 °C, the CV data displayed that Pt/C increased by only four times the current, while Pt–Ru/C increased by eight times the temperature range. The authors suggest that the addition of ruthenium significantly improves the oxidation performance of ethanol, possibly because of its strong adsorption for OH^-. Compared with Pt/C, Pt–Ru/C displayed greater apparent activation energy of oxidation, indicating that Ru has a higher dependence on temperature for OH^- adsorption.

Ru and Ir oxides were combined, or carbon-supported platinum was combined by Calegaro et al. via a sol–gel method [27], keeping the total Pt mass percentage of 25%. Pt–RuO_2/C has a higher activity for ethanol electro-oxidation, as depicted by the polarization curve, compared to the Pt/C and Pt–Ru/C catalysts. In addition, the recovery factor would be enhanced by adding IrO_2 into Pt–RuO_2/C composites. Boron-doped diamond (BDD) material was used for the fixation of Pt–RuO_2/C composite by the researchers [28]. The oxidation of ethanol in H_2SO_4 solution over BDD-supported Pt–RuO_2/C catalyst was tested by CV, Tafel, and time amperometry.

Ethanol electro-oxidation performance would be influenced by ruthenium content in ruthenium electrocatalysts. Pt–Ru catalyst with the atomic ratio of 4 : 1 prepared by Lamy et al. [29] had a relatively poor electro-oxidation activity to ethanol at 25 °C. Camara et al. [24] reported that the composition range of Pt–Ru with high ethanol oxidation rate was narrow: when the Ru content was below 20 at%, the adsorbed intermediate electro-oxidation cannot effectively by the catalysts due to the lack of Ru locus, and the ethanol electro-oxidation current would keep almost the same as pure Pt. Therefore, the poor electrocatalytic activity of this Pt–Ru (4 : 1) electrocatalyst was explained by the too low Ru content. By the CV measurements, Spinace et al. [30] reported that with increasing Ru content in the Pt-based catalysts, e.g. Pt:Ru with the ratio of 1 : 3, the ethanol oxidation activity on this Pt–Ru/C would be promoted. Similarly, with the increasing content of the metal Ru up to Pt:Ru with the ratio of 3 : 2, ethanol oxidation activity was observed to be increased on the Pt–Ru/C electrocatalyst, as reported by Oliveira Neto et al. [31].

The experiment of DEFC shows that the performance of Pt–Ru/C electrocatalysts is generally better than the activity of Pt/C anode material [29, 32, 33]. At all temperatures studied, Pt–Ru/C-based DEFC has the highest maximum power density (MPD) compared to that of Pt/C cells.

Zhou et al. [32] tested the ethanol electro-oxidation activity on binary/ternary Pt–Ru-based catalysts supported on carbon by CV and in a single DEFC. The addition of Mo and W in Pt–Ru/C had a negligible influence on the XRD spectrum, indicating the same crystal structure of Pt–Ru/C after adding these additives. Pt–Ru–Mo (1 : 1 : 1) and Pt–Ru–W (1 : 1 : 1) performed well. The single DEFC test using the above ternary catalyst results are shown in Table 13.1, where the working output voltage at current densities of 30 and 60 mA cm^{-2}, the open-circuit voltage of each monomer, and the maximum power density are shown. It was found that the open-circuit voltage was increased by Mo addition, while the maximum power density was promoted by the W addition. Both Pt–Ru–Mo and Pt–Ru–W cells showed higher performance compared to that of Pt–Ru (1 : 1) cell, but these performances were still lower than those of Pt–Sn (1 : 1) cell. Similarly, ethanol electro-oxidation was studied by Zhou et al. in a single DEFC using Pt–Ru–Sn (1 : 1 : 1) catalyst supported on carbon as anodic material [34]. For Pt–Ru–Sn, it can be seen from the XRD pattern that its lattice parameters are between those of Pt–Ru/C and Pt–Sn/C. The same excellent performance was obtained on the cells employed with Pt–Ru–Sn–C and Pt–Ru–W/C, and the cells employed with Pt–Ru–Sn–C outperformed those employed with Pt–Ru–W/C in both the internal resistance control area and the activation control area. However, the performance of DEFC using Pt–Ru–Sn was higher than that using Pt–Ru (1 : 1), but it is still lower than that using Pt–Sn (1 : 1).

Au substrates were used to deposit binary/ternary platinum-based catalysts by Tanaka et al. [35]. Contrary to the research from Spinace et al. [30], the activity of Pt–Ru–W was superior to binary catalysts like Pt–Sn, Pt–W, and Pt–Ru, and its electrocatalytic performance was higher than that of a single Pt layer electrode. In the Pt–Ru–W alloy system, the electro-oxidation of ethanol showed greater cathode potential displacement and current than that of the binary alloy electrode.

Table 13.1 Summary of performance of single direct ethanol fuel cell tests adopting different anode catalysts (90 °C).

Catalyst	Open-circuit voltage (mV)	Voltage at 30 mA cm^{-2} (mV)	Voltage at 60 mA cm^{-2} (mV)	Maximum power density (mW cm^{-2})
Pt	547	275	177	10.85
Pt–Ru (1 : 1)	677	461	368	28.54
Pt–Sn (1 : 1)	811	662	576	52.22
Pt–Ru–W (1 : 1 : 1)	698	503	425	38.54
Pt–Ru–Mo (1 : 1 : 1)	720	486	389	31.19

Source: Data from Zhou et al. [32]. 2003 Elsevier.

A direct chemical reduction approach was used to prepare Pt–Ru/C (1 : 1) and Pt–Ru–Ni/C (6 : 3 : 1) catalysts for ethanol electro-oxidation in 0.5M CH_3CH_2OH and 0.5M H_2SO_4 solutions by CV and time amperometry. Considering that the ternary electrocatalyst had more Pt atoms relative to the binary electrocatalyst, the atom ratio constant of Pt–Ru/C is 1 : 1, and that for Pt–Ru–Ni/C is 6 : 3 : 1, indicating ternary Pt–Ru–Ni alloy was formed. Cyclic voltammetry did not find any significant difference in the electro-oxidation of ethanol at the two electrodes, especially regarding the initial potential of the electro-oxidation of ethanol. The electrochemical activity-specific surface areas of Pt–Ru/C and Pt–Ru–Ni/C catalysts were basically the same. However, Pt–Ru–Ni/C catalyst had an obviously higher electrocatalytic activity for ethanol electro-oxidation compared to Pt–Ru/C electrocatalyst. Promoting effects attributed to the increase of nickel company – oxidation platinum ruthenium – ethanol has special important significance in low potential, can see clearly in Figure 13.4, which shows the ratio of the current densities, measured under quasisteady-state conditions, with and without the addition of nickel, as a function of the oxidation potential function. The graph shows that adding nickel as the dopant leads to the highest peak at around 0.3 V, with the current density for ethanol electro-oxidation increasing by 5.4 times.

Oliveira Neto et al. [37] used CV and linear sweep voltammetry (LSV) to analyze the electro-oxidation of methanol and ethanol over Pt–Ru–Mo and Pt–Ru catalysts supported on carbon synthesized via the modified Bonnemann method. PtO_2, metal Pt, Mo oxides, and RuO_2 were found in the X-ray photoelectron spectroscopy (XPS) measurements. They reported that the oxidation performance of Pt–Ru–Mo/C (1 : 1 : 1) catalyst was superior to that of methanol. Therefore, the author speculated that the addition of Mo not only promoted the removal of CO species but also promoted the fracture of C—C bonds.

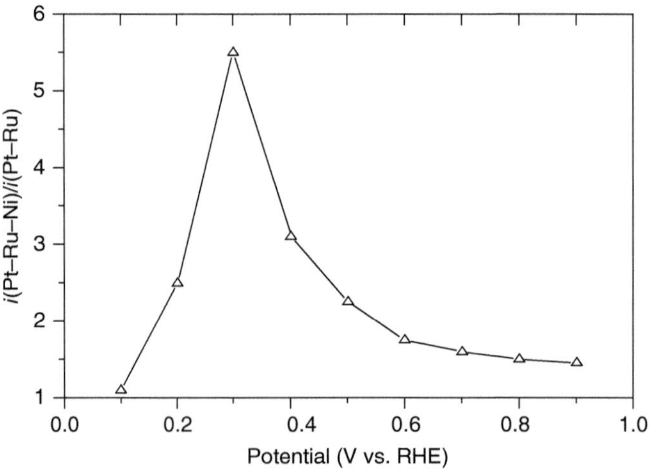

Figure 13.4 Ratio of the current densities of ethanol oxidation recorded on the Pt–Ru–Ni/C and Pt–Ru/C electrodes as a function of the potential. Source: Wang et al. [36]. © 2006, Elsevier.

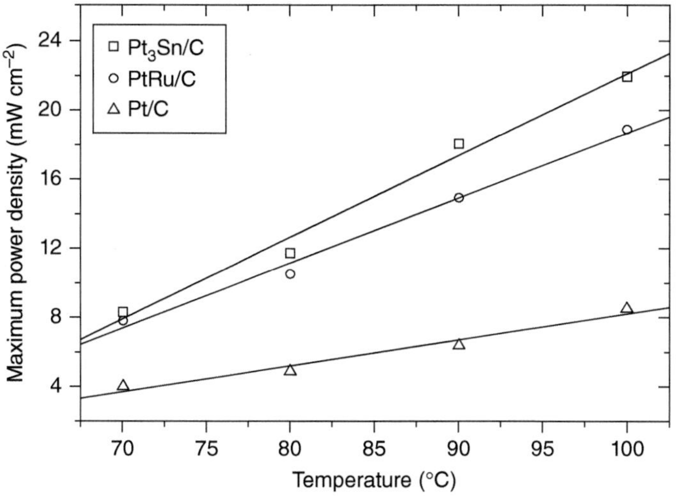

Figure 13.5 Dependence of the maximum power density of DEFCs with Pt/C, Pt–Ru (1 : 1), and Pt–Sn/C (3 : 1) as anode catalysts on cell temperature. O_2 pressure: 3 atm; ethanol solution: 1 mol l^{-1}. Anode metal loading 1 mg cm^{-2}. Cathode 20 wt% Pt/C, Pt loading 1 mg cm^{-2}. Source: Colmati et al. [33]. © 2006, Elsevier.

Pt–Ru–Pb and Pt–Pb were synthesized by Liu et al. [38] by depositing Pb on Pt–Ru or Pt materials. No Pt–Ru–Pb and Pt–Pb catalysts were found. Adding Pb to Pt–Ru and Pt can enhance ethanol oxidation activity, particularly at high potentials.

The BDD surface modified with Pt–RuO_2–RhO_2, Pt–RuO_2, and Pt was studied by Salazar-Banda et al. during ethanol oxidation [39]. The results from the CV test indicated that the CO species formation was clearly prevented during ethanol electro-oxidation on Pt–RuO_2–RhO_2/BDD catalyst. As analyzed from the polarization curve and the derived Tafel slope, it indicated that adding RhO_2 and RuO_2 to Pt/BDD could produce stronger catalyst activity and change the initial reaction time to the direction of low potential by 180 mV. In addition, at a fixed potential, the obtained current for ethanol oxidation on Pt–RuO_2–RhO_2/BDD catalyst was over 10 times higher compared to Pt/BDD catalysts (Figure 13.5).

13.1.2.3 Pt–Sn System

Pt and tin can form five intermetallic compounds, namely Pt_2Sn_3, Pt–Sn, $PtSn_2$, Pt_3Sn, and $PtSn_4$; among these compounds, Pt–Sn and Pt_3Sn are confused. The XRD patterns and the derived unique crystal structures indicated that these intermetallic compounds are distinguished. Kuznetsov et al. [40] asserted that platinum could be combined with tin to form almost any alloy possible. Therefore, a lower angle exists in the FCC peak of Pt–Sn catalyst compared to the metal Pt, but a higher angle than the FCC Pt_3Sn compound. This result indicates that a solid solution is formed in the Pt–Sn material, in which Sn is added into the Pt lattice with FCC structure. Instead, Radmilovic and Richardson [41] assigned the lattice constant of 0.3965 nm of the Pt–Sn/C catalyst with Pt:Sn with the ratio of 1.23 : 1 to a mixture of Pt_3Sn catalysts, and Pt_9Sn (0.3934 nm) [42]. The catalyst was prepared using a conventional precipitation pathway for the decomposition of Pt–Sn precursors at 500 °C. A similar diffraction pattern was generated by the mixture of stoichiometric Pt_3Sn and Pt_9Sn to that of non-stoichiometric Pt_3Sn compound, taking into account that the reflections of Pt_9Sn and Pt_3Sn are nearly the same as the widening of particle size.

The preparation of carbon-supported Pt–Sn catalysts usually does not require heat treatment, so the catalysts consisted of a mixture of Pt_9Sn and Pt_3Sn species, or a Pt–Sn alloy with FCC structure, and tin and platinum oxides. Relative amounts of platinum–tin alloy and SnO_2 influence the electrocatalytic performance of the catalysts. In contrast to DMFC, the performance of DEFCs using Pt–Sn/C for anode catalyst is better than that using Pt–Ru/C [29, 32, 33], as depicted in Figure 13.6

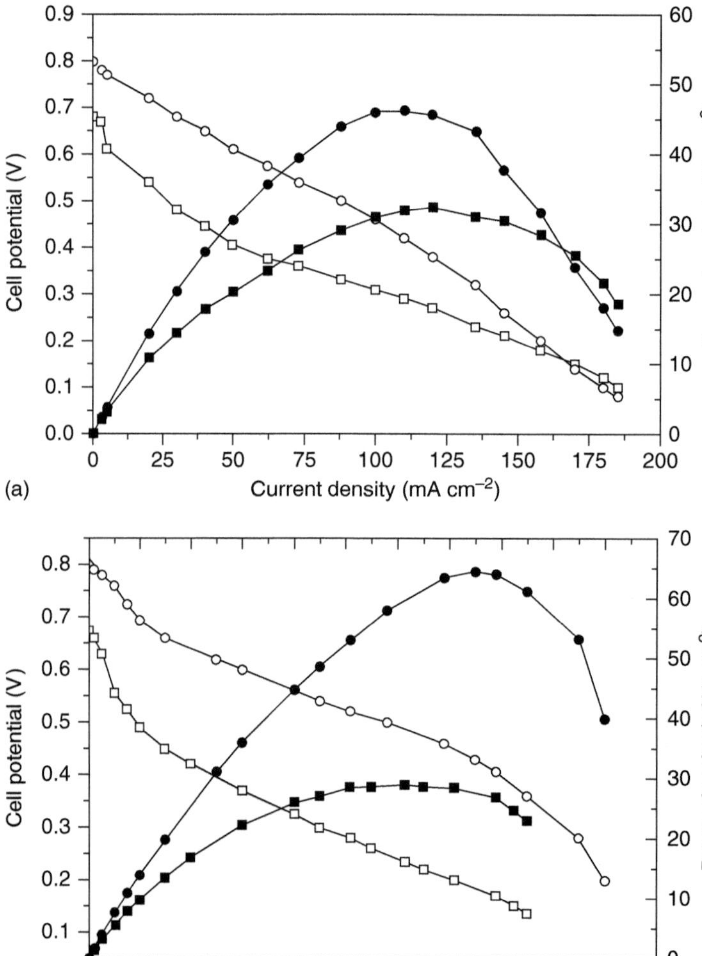

Figure 13.6 Polarization curves and power density curves in single DEFC with Pt–Sn/C (2 : 1) and Pt–Ru/C (1 : 1) catalysts. (a) Pt–Sn/C-a and commercial Pt–Ru/C, (b) Pt–Sn/C-b and in-house Pt–Ru/C. Cell temperature: 90 °C [40]. Source: Song et al. [40]. © 2005, Elsevier.

and Table 13.1. The DEFC performance with using different Pt–Ru/C (1 : 1) and Pt–Sn/C (2 : 1) electrocatalysts as anode materials were studied by Song et al. [43]. It was found that DEFC employing Pt–Sn/C catalyst synthesized via different methods performed better compared to those using the internally prepared and commercial Pt–Ru/C electrocatalysts.

Lamy et al. [29, 44, 45] and Xin and coworkers [32, 46, 47] studied ethanol electro-oxidation on Pt–Sn catalysts [44] supported on carbon, which were synthesized via co-dipping reduction, Bonnemann [29], and modified polyols. In all cases, most tin is in a non-alloy oxidation state. The influence of the content of tin in Pt–Sn catalyst on the performance of batteries has been investigated. Pt–Sn/C catalysts were synthesized by Lamy et al. [29], and the sample has the atomic ratio of Pt:Sn ranging between 90 : 10 and 50 : 50. It was found that the optimal composition of tin was ranging from 10 to 20 at%. On the contrary, Zhou et al. [32] analyzed that at Pt:Sn atom ratios of 66 : 33, 60 : 40, and 50 : 50, the Pt–Sn/C catalysts exhibited obviously better activity than the ratios of 75 : 25 and 80 : 20. Among these ratios, the optimal ratio was between 33 and 40 at%, which also relied highly on the DEFC operating temperatures. When the ethylene glycol was used as a solvent and reducing agent, Pt–Sn/C catalysts at different tin contents were synthesized via the direct alcohol reduction method. Spinace et al. [48] investigated the electro-oxidation activity of ethanol with using the CV measurement, and the results are depicted in Figure 13.7. Ethanol electro-oxidation started at a very lower potential of c. 0.25 V on the Pt–Sn/C catalysts with a Pt:Sn atom ratio of 25 : 75 and 50 : 50. At the potential higher than 0.4 V, the Pt–Sn electrocatalyst with at a ratio of 50 : 50 showed higher performance. As to the Pt–Sn catalyst at the atom ratio of 75 : 25, the ethanol electro-oxidation starts at c. 0.35 V, while the current

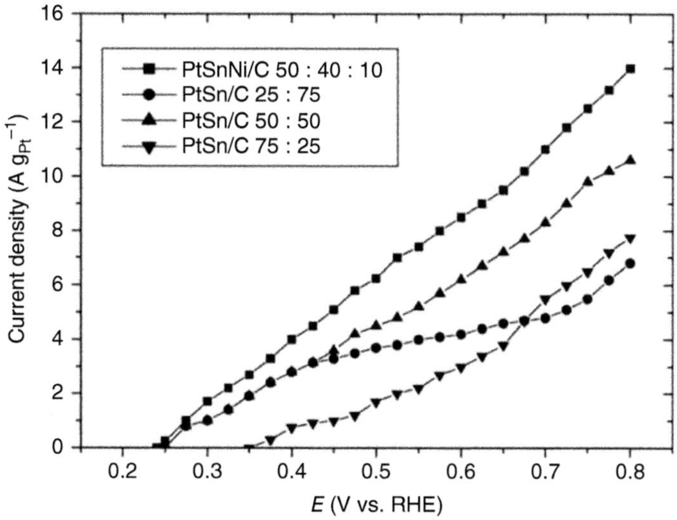

Figure 13.7 Cyclic voltammetry of Pt–Sn/C and Pt–Sn–Ni/C electrocatalysts in 0.5 M H_2SO_4 containing 1.0 M of ethanol with a sweep rate of 10 mV s^{-1}, considering only the anodic sweep. Source: Antolini [49]. © 2007, Elsevier.

density is clearly smaller compared to the catalysts with higher Sn content. Similar results were found for the Pt–Sn catalyst synthesized by Zhou et al. [32] using a similar method, while different from Lamy et al.'s [29] results using co-dipping reduction approaches. These phenomena suggested that the Pt–Sn catalysts' performance relied highly on the synthesis. The influence of the synthesis methods was also proven by Song et al. [43]. The Pt–Sn/C electrocatalyst was synthesized via a polyalcohol method. Sn and Pt were both deposited on the carbon carrier (Pt–Sn/C-b), while tin was deposited on a pre-formed Pt/C (Pt–Sn/C-a). From the obtained results, different Pt–Sn/C-based fuels showed similar performance at low current densities. However, the cell performance employing Pt–Sn/C-a is obviously lower compared to the cell employing Pt–Sn/C-b at a higher current. The author suggested the reason to be that the content of Pt–Sn/C-a in the cell is greater compared to the use of Pt–Sn/C-b, thus the resistance would be increased in the cell and reduce the performance of the cell. Additionally, the partial coverage of Pt-active site would occur by the direct deposition of Sn on Pt/C, which prevent the effective adsorption/oxidation of ethanol, resulting in poor activity especially in terms of high currents when the anode employing Pt–Sn/C-a electrocatalyst.

Bimetal Pt–Sn nanoparticles supported on BDD substrates were synthesized by Sine et al. [50] , which leads to highly effective ethanol electro-oxidation. It was found that the diamond substrate can increase the turnover rate for producing products by promoting ethanol activation.

The alloying degree affects the ethanol oxidation activity, which was studied by Jiang et al. [51]. They had analyzed the electrocatalytic activity of the synthesized Pt–Sn catalysts and quasi-alloy $PtSnO_x$ catalysts. The results show that the catalytic activity of $PtSnO_x$ is better compared to Pt–Sn alloy. The authors concluded that lattice parameters of unchanged platinum in the catalyst $PtSnO_x$ are conducive to the ethanol adsorption and that Sn oxide near platinum nanoparticles can provide oxygen clusters, which can conveniently remove Co from ethanol residue and free the active site of platinum. On the converse, polyol-type Pt–Ru/C (1 : 1), Pt/C, and Pt–Sn/C (3 : 1) electrocatalysts were analyzed by Colmenares et al. [52] with a comparison with commercial electrocatalysts correspondingly. Incomplete ethanol oxidation occurs on these electrocatalysts, and they found that the ratio for CO_2 production is less than 1%. Therefore, polyalcohol catalysts have low activity for ethanol oxidation to produce acetaldehyde; these EOR activities are obviously lower compared to those of industrial electrocatalysts. The reason for this phenomenon is due to the low alloy degree of the catalysts as prepared by the polymerized methods.

EOR activity was studied on Pt–Sn/C electrocatalysts, which were prepared by formic acid reduction method (Pt:Sn ratio of 90 : 10, 75 : 25, and 66 : 33) [53]. It was observed that the amount between alloyed and non-alloyed tin would significantly influence the electrocatalytic activity of Pt–Sn catalysts during ethanol oxidation. Reaction temperature would affect the rate-determining step for ethanol electro-oxidation that controls the optimized distribution between non-alloy and alloy tin. At low temperature as well as low current density, the ethanol electro-oxidation rate is not fast, and thus the electrochemical oxidation rate of adsorptive CO and CH_3CO intermediate controls the process rates. In this case, the existence of SnO_x would promote the oxidation of ethanol. Oppositely, the increase in lattice parameters at high temperature increases the current density ethanol electro-oxidation: assuming that Pt and Sn forms a solid solution, the cleavage C—C bond requires a larger lattice parameter. On the contrary, the increase in lattice parameters is to increase the amount of Pt–Sn pair; it is needed to complete proceeding ethanol oxidation to acetaldehyde; that is, the C—C bond was not broken. Under the hypothesis of two-phase formation, it is deducted that the EOR activity of Pt_3Sn catalyst is better compared to that of the Pt_9Sn_1 catalyst.

With regard to the cleavage of C—C bond, it must be noteworthy that although Gursahani et al. [54] have shown that EOR on pure Pt may undergo cleavage of C—C bond, [55] recently working with the same research group report, increased Sn in the Pt electrocatalyst may inhibit the decomposition of ethanol, because of the inhibition of C—C bond cleavage. The addition of tin to the platinum catalyst inhibited the cracking reaction of the C—C bond because of the changes in the surface geometry and electronic properties. Similarly, experiments by DEFC Rousseau's research group [56] observed the adding tin in Pt catalysts could not only enhance ethanol electro-oxidation of the active catalyst and thus promote DEFC performance but also would change the distribution of different products. It was found that the amount of acetic acid increased, while the amount of acetaldehyde and carbon dioxide was decreased. The Pt/C electrocatalyst produced twice as much carbon dioxide as the Pt–Sn/C electrocatalyst. The result should be interpreted by the requisite condition for several Pt sites adjacent with each other to dissociate the adsorbed ethanol molecules and to make a breakage of the C—C bond. The latter reaction is undesirable when the Sn component was incorporated into Pt catalysts. Once acetaldehyde forms, however, it may further bond on the Pt-active sites, which results in the formation of Pt–CH_3 company moieties: afterward, since tin is well known with having lower energy for the activation of water at a lower potential compared to Pt, abundant hydroxide radical can be generated from Sn sites at a low potential, resulting in the further reaction of acetaldehyde intermediate with these OH^- to ultimately generate acetic acid.

Colmati's research group [57] prepared a C-loaded Pt–Sn catalyst, using the Pt–Sn precursors by reducing Pt and Sn with a ratio for Pt:Sn of 75 : 25 in 200 °C. For the preparation of the catalyst, the formed Pt–Sn solid solution and the existence of tin oxide was assumed to be FCC based on the lattice parameters smaller compared to the pure Pt phase and higher compared to the Pt_3Sn material. Alternatively, a mixture of Pt_9Sn, Pt_3Sn, and SnO_2 phases is used. The Pt–Sn (3 : 1) catalyst formed a cubic Pt_3Sn phase when treated at 200 °C. The catalyst treated at 200 °C was pretreated with greater ethanol oxidation activity and precursors to single DEFC. The reason for this result should be ascribed to the outstanding electrocatalytic activity of Pt_3Sn phase over Pt–Sn solid solution/tin oxide mixture with an FCC structure or Pt_9Sn/Pt_3Sn/tin oxide mixture. On the contrary, Pt–Sn catalyst showed a lower ethanol oxidation activity compared to the Pt_3Sn catalyst as well as Pt–Sn solid solution/tin oxide with FCC structure (Figures 13.8 and 13.9).

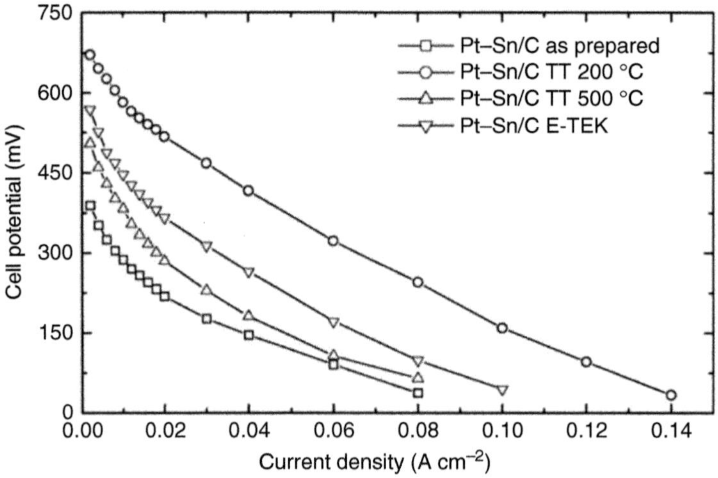

Figure 13.8 Polarization curves and power density curves in single DEFC. Source: Antolini [49]. © 2007, Elsevier.

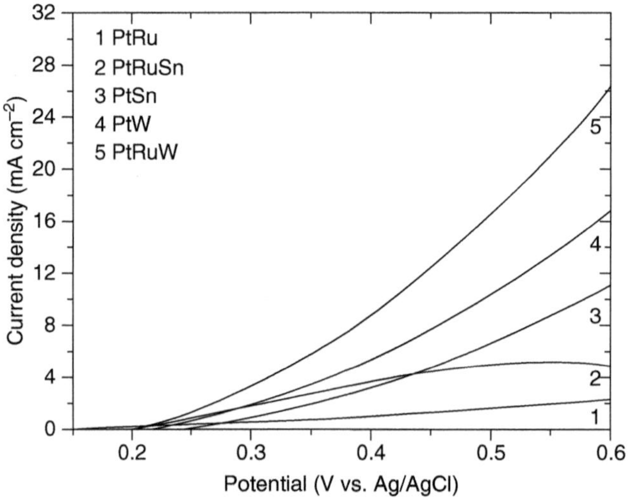

Figure 13.9 Voltammograms of Pt–metal alloy electrodes in 0.5 mol dm^{-3} H_2SO_4 + 1 mol dm^{-3} ethanol solution compared to Pt–Ru. The substrate, Au; heating temperature 300 °C; scan rate 10 mV s^{-1}. Source: Antolini [49]. © 2007, Elsevier.

Ethanol oxidation on Pt-Sn-Ni/C catalyst (50 : 40 : 10) was investigated by Spinace et al. [48]. This ternary electrocatalyst was synthesized via a process of alcohol reduction by CV and time amperometry method. The XRD patterns of the Pt–Sn–Ni/C (50 : 50) catalyst indicated the presence of some characteristic peaks and the cassiterite phase in the platinum catalytic cracking structure. Through CV measurements, they found that small amounts of tin alternatives to nickel have little chance of change (~0.25 V), while the current value is higher than the potential of those obtaining Pt–Sn/C catalyst with the ratio of 50 : 50, in the entire range. The superior Pt–Sn–Ni/C catalyst has been found for the timing current test. In the test, a current drop was found initially within the first five minutes, then the current decay was slowed down; however, the current density of the electro-catalytic Pt–Sn–Ni/C catalyst is greater compared to that of the electro-catalytic Pt–Sn/C at the atom ratio of 50 : 50.

Distribution of products from ethanol oxidation in the DEFC employed with Pt–Sn–Ru (86 : 10 : 4), Pt–Sn (90 : 10), and Pt was investigated by Rousseau et al. [56]. The incorporation of Ru into Pt–Sn catalyst largely improves the overall performance of DEFC based on the high activity of the used electrocatalyst. However, the product distribution was not altered. In the XRD patterns of the SnO_x-containing materials, the platinum diffraction peak did not shift, indicating that the co-reduction process did not induce the formation of alloy material. Based on this case, the ratio between non-alloyed Ru to non-alloyed Sn in the Pt–Sn–Ru material (86 : 10 : 4) was ≤0.4.

C-supported Pt–Ru–Sn (1 : 1 : 3 and 1 : 1 : 1) alloy and Pt–Sn (1 : 1) ternary electrocatalysts were synthesized by the Antolini research group [58] via reducing metal precursors using formic acid, and their ethanol oxidation activity was analyzed in comparison with commercial Pt–Ru/C and Pt/C catalysts. The two as-prepared electrocatalysts are composed of catalytic cracking of Sn and Pt–Sn alloy, Ru oxides. Pt–Sn–Ru (1 : 1 : 1) and Pt–Sn–Ru (1 : 1 : 0.3) have the ratio for non-alloy Ru to non-alloy Sn of 0.86 and 0.43, respectively. The electrocatalytic activity of catalysts would be obviously influenced by the ratio. The ethanol electro-oxidation activity of Pt–Sn–Ru (1 : 1 : 3.3) catalyst was better compared to the binary Pt–Ru, Pt–Sn, and ternal Pt–Sn–Ru (1 : 1 : 1 : 1.1) catalysts as determined from by LSV measurement and DEFC test (Figure 13.10). The active role of Ru in Pt–Sn–Ru (1 : 1 : 3.3) ternary electrocatalyst is attributed to the interaction between Ru oxide and Sn. In contrast, the Pt–Sn–Ru (1 : 1 : 1) catalyst had a lower activity compared to that of Pt–Sn

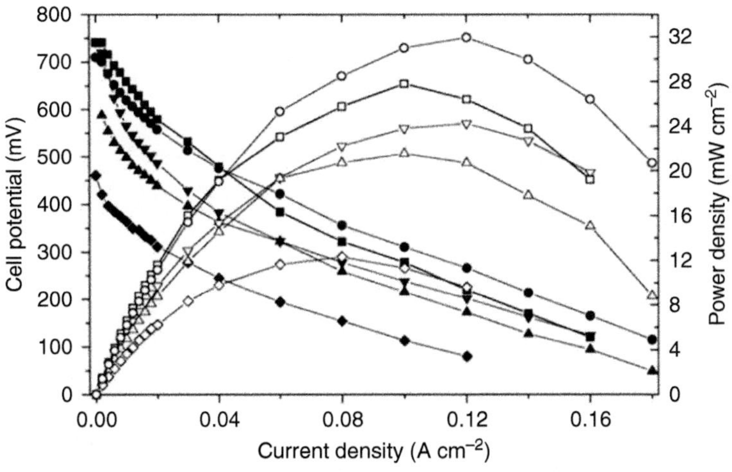

Figure 13.10 Polarization curves and power density curves in single DEFC. Source: Antolini et al. [58]. © 2007, Elsevier.

and Pt–Ru binary electrocatalysts with the same ratio of 1 : 1. With the increasing Ru content in the ternary catalysts, the interaction of Pt–RuO_2 substituted the interaction of Pt–SnO_2, so their electrochemical behaviors are similar.

Sine et al. [59] studied the electro-oxidation reaction of ethanol and methanol on Pt–Ru–Sn ternary nanocatalysts prepared via a microemulsion method, followed by its deposition on the B-doped diamond support electrode. The content of ruthenium and tin oxides has not been reported. For the test of methanol electro-oxidation, ternary catalysts showed lower potentials for ethanol oxidation compared to bimetal Pt–Sn and Pt–Ru electrocatalysts or even Pt electrocatalyst. Besides, methanol electro-oxidation on ternary Pt–Ru–Sn electrocatalyst was demonstrated to be more effective compared to ethanol electro-oxidation. This is due to the termination with the formation of acetaldehyde and/or acetic acid during ethanol electro-oxidation, suggesting the C—C bond was hard to be activated and disconnected on the ternary Pt–Ru–Sn electrocatalyst.

The method using metal precursors reduced by formic acid was further employed for preparing C-supported ternary Pt–Sn–Rh (1 : 1 : 1, 1 : 1 : 3) electrocatalyst, and ethanol electrocatalytic tests were analyzed compared to binary Pt–Rh/C and Pt–Sn/C, which were synthesized via the same processes. XRD pattern shows that the grain size and lattice parameters of Pt–Sn–Rh material are larger compared to that of Pt–Rh/C catalyst and smaller compared to that of Pt–Sn/C material. Ternary Pt–Sn–Rh alloys may be formed in Pt–Sn–Rh/C electrocatalysts. LSV test show that at the potentials more than 0.45 V and flow value, Pt–Sn–Rh material ethanol electro-oxidation alloy catalyst has the best performance, while the electrochemical activity of ternary Pt–Sn catalyst with potential lower than 0.45 V and flow value is less compared to Pt–Sn catalyst.

13.2 Glucose Oxidase

For the applications in biological fuel cells, wastewater treatment, medical fields, and food analytical fields, the electrochemical catalysis of sugars have received significant attention.

$$\beta - \text{Glucose} + \text{FAD} - \text{oxide} \rightarrow \text{glucolactone} + \text{FADH}_2\ \text{oxidase} \tag{13.1}$$

$$\text{Glucolactone} + H_2O \rightarrow \text{gluconic acid} \tag{13.2}$$

$$\text{FADH}_2\ \text{oxidase} \rightarrow \text{FAD} - \text{oxidase} + 2H^+ + 2e^- \tag{13.3}$$

and at the Pt counter electrode:

$$2H^+ + 2e^- + O_2 \rightarrow H_2O_2 \tag{13.4}$$

where flavin adenine dinucleotide (FAD) is the associated co-enzyme required for the action of glucose oxidase, and $FADH_2$ is the reduced form of FAD.

Since the 1960s, glucose electro-oxidation has received an investigation, and it has kept an active research field until now. Glucose electro-oxidation was first reported by Vassilyev and Bagotzky in acidic media. Meanwhile, the conditions of high temperatures and alkaline media were investigated by Bockris et al. Recently, the glucose adsorption and subsequent electro-oxidation on a Pt catalyst electrode were studied using infrared (IR) spectrometry. The Fourier transform IR spectroscopy was used by the Bae research group to firstly reveal that the intermediate CO is the primary species absorbed on the electrode during glucose electro-oxidation in both alkaline and acidic electrolyte, which is similar to electro-oxidation of small organic molecules like HCHO, HCOOH, and CH_3OH. The study also reveals that the crystallographic orientation of the Pt single-crystal catalyst surface had a significant influence on such a reaction.

13.3 Ethylene Glycol Oxidation

For the past few years, ethylene glycol was studied by many researchers as an alternative fuel for use in the fuel cells [60, 61]. Ethylene glycol has the advantages of low crossover based on the large size of the molecule, as well as the high energy density compared to other fuels. Particularly, ethylene glycol can be produced at a large scale, which is beneficial for its cost advantages and wide applied potential in the fuel cells. Significantly, ethylene glycol did not have the C—C bond, and its complete oxidation is rather complex compared to methanol electro-oxidation. Two reaction pathways were concluded for ethylene glycol electro-oxidation, that is, the non-poisoning and poisoning pathways [62–64]. In the latter non-poisoning pathway, oxalate would be obtained from complete electro-oxidation of ethylene glycol, while the intermediate format would be obtained from the partial oxidation of ethylene glycol in the former poisoning pathway, leading to the formation of poisoning species after the subsequent oxidation.

13.4 Glycerol Oxidation

As an important by-product of soap, fatty acids, biodiesel, and bioethanol industries, glycerol electro-oxidation has also received wide attention. Although a surplus of glycerol has caused a decrease in the value of glycerol on the global market nowadays, which is because of the fast expansion of biodiesel producers all over the world, a lot of researchers have made their efforts to figure out novel methods for treating, managing, and converting glycerol to value-added chemicals. By amounting the alternative approaches, the electrochemical approach is beneficial because of its simplicity. In addition, including the pH value of electrolyte, current density, electrode materials, application potential, temperature, and additives, these factors would affect the process mechanism and product distribution during glycerol electro-oxidation. The value-added compounds produced by the electrochemical transformation of glycerol include glyceraldehyde, dihydroxyacetone, glycolic acid, lactic acid, glycerol, 1,2-propanediol, tartaric acid, 1,3-propanediol, and mesoxalic acid. These chemicals were widely used in medicine, cosmetics, polymer industries, and food (Figure 13.11).

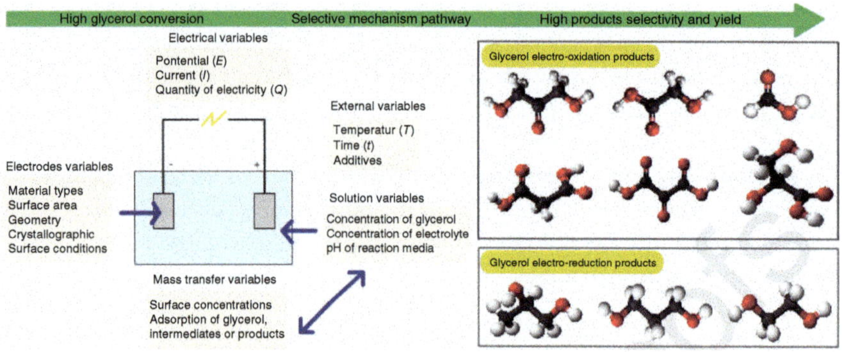

Figure 13.11 The glycerol oxidation. Source: Rahim et al. [65]. © 2020, Elsevier.

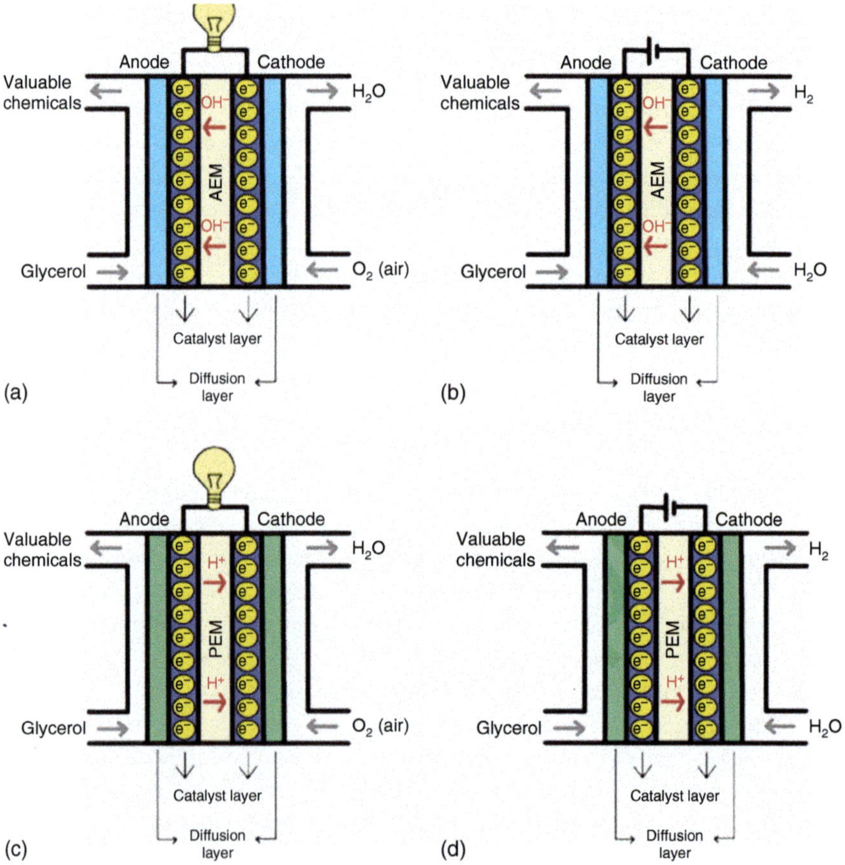

Figure 13.12 Working principles of (a) AEM fuel cell, (b) AEM electrolysis, (c) PEM fuel cell, and (d) PEM electrolysis. Source: Rahim et al. [65]. © 2020, Elsevier.

Since most of the aforesaid compounds offer potential in various industry applications, basically understanding the electrochemical and chemistry processes as well as their engineering aspects is requisite. Recently, electrolytic cells and fuel cells have been used for electrochemical glycerine conversion. Electrical energy, together with some value-added compounds, would be provided by the fuel cell operation via oxidizing glycerol at the anode and oxygen reduction reaction (ORR) at the cathode (Figure 13.12). Electrolysis, on the other hand, combines glycerol oxidation at the anode with ORR at the cathode to produce valuable compounds and H_2 (Figure 13.12).

In addition to single cells, bicompartmental cells are usually carried out by anion exchange membrane (AEM), in which the ion transport direction is in reversion with that of proton exchange membrane (PEM). Theoretically, when large currents are used, more and faster ion transport must be achieved through these membranes to keep the high ion conductivity. Meanwhile, the membrane should have a low permeability, which suggests the solvent has the least exchange with neutral molecules. The difference between the PEM and AEM is that hydrogen ions (H^+) are the charge carrier in PEM, and it would be transferred through PEM from anode to cathode; on the contrary, OH^- is the charge carrier in AEM, and they would be transferred from cathode to anode through AEM. Both cases will allow redox catalysis to be completed (Figure 13.12). The electronic release EOR will transfer via the external circuit and make a complete electrochemical loop. The generated electrons can be used to combine with the glycerol and convert it into corresponding oxidation (such as glycerol and glyceraldehyde) and reduction (e.g. 1,3-propylene glycol and 1,2-propylene glycol) products without the presence of any chemical oxidants (such as H_2O_2 and O_2) or reductases (such as H_2) (Scheme 13.1).

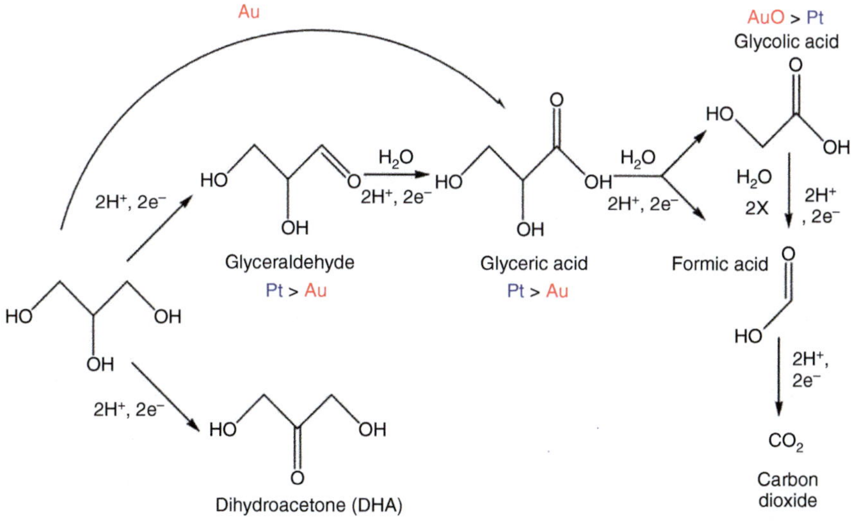

Scheme 13.1 Mechanism of glycerol oxidation on platinum and gold electrodes in alkaline medium. Source: Rahim et al. [65]. © 2020, Elsevier.

References

1 Bianchini, C. and Shen, P.K. (2009). *Chem. Rev.* 109: 4183–4206.
2 Martinez, U., Serov, A., Padilla, M., and Atanassov, P. (2014). *ChemSusChem* 7: 2351–2357.
3 Bianchini, C., Bambagioni, V., Filippi, J. et al. (2009). *Electrochem. Commun.* 11: 1077–1080.
4 Wang, W., Yang, Y., Liu, Y.Q. et al. (2015). *J. Power Sources* 273: 631–637.
5 Liu, H.P., Ye, J.P., Xu, C.W. et al. (2007). *Electrochem. Commun.* 9: 2334–2339.
6 Wang, H., Xu, C.W., Cheng, F.L., and Jiang, S.P. (2007). *Electrochem. Commun.* 9: 1212–1216.
7 Xu, C.W., Shen, P.K., and Liu, Y.L. (2007). *J. Power Sources* 164: 527–531.
8 Carrera-Cerritos, R., Fuentes-Ramirez, R., Cuevas-Muniz, F.M. et al. (2014). *J. Power Sources* 269: 370–378.
9 Liang, Z.X., Zhao, T.S., Xu, J.B., and Zhu, L.D. (2009). *Electrochim. Acta* 54: 2203–2208.
10 Hong, W., Fang, Y.X., Wang, J., and Wang, E.K. (2014). *J. Power Sources* 248: 553–559.
11 Li, Y.H., Xu, Q.Z., Li, Q.Y. et al. (2014). *Electrochim. Acta* 147: 151–156.
12 Hu, F.P., Ding, F.W., Song, S.Q., and Shen, P.K. (2006). *J. Power Sources* 163: 415–419.
13 Fang, X., Wang, L.Q., Shen, P.K. et al. (2010). *J. Power Sources* 195: 1375–1378.
14 Nguyen, S.T., Law, H.M., Nguyen, H.T. et al. (2009). *Appl. Catal. B* 91: 507–515.
15 Kakaei, K. and Dorraji, M. (2014). *Electrochim. Acta* 143: 207–215.
16 Cui, C.H., Yu, J.W., Li, H.H. et al. (2011). *ACS Nano* 5: 4211–4218.
17 He, Q., Chen, W., Mukerjee, S. et al. (2009). *J. Power Sources* 187: 298–304.
18 Zhu, L.D., Zhao, T.S., Xu, J.B., and Liang, Z.X. (2009). *J. Power Sources* 187: 80–84.
19 Qin, Y.H., Li, Y.F., Lv, R.L. et al. (2014). *Electrochim. Acta* 144: 50–55.
20 Xu, J.B., Zhao, T.S., Shen, S.Y., and Li, Y.S. (2010). *Int. J. Hydrog. Energy* 35: 6490–6500.
21 Xu, C., Tian, Z., Shen, P., and Jiang, S.P. (2008). *Electrochim. Acta* 53: 2610–2618.
22 Antolini, E. (2003). *Mater. Chem. Phys.* 78: 563–573.
23 Schmidt, V.M., Ianniello, R., Pastor, E., and Gonzalez, S. (1996). *J. Phys. Chem.* 100: 17901–17908.
24 Camara, G.A., de Lima, R.B., and Iwasita, T. (2004). *Electrochem. Commun.* 6: 812–815.
25 Fujiwara, N., Friedrich, K.A., and Stimming, U. (1999). *J. Electroanal. Chem.* 472: 120–125.
26 Jiang, L.H., Zhou, Z.H., Li, W.Z. et al. (2004). *Energy Fuels* 18: 866–871.
27 Calegaro, M.L., Suffredini, H.B., Machado, S.A.S., and Avaca, L.A. (2006). *J. Power Sources* 156: 300–305.
28 Suffredini, H.B., Tricoli, V., Vatistas, N., and Avaca, L.A. (2006). *J. Power Sources* 158: 124–128.

29 Lamy, C., Rousseau, S., Belgsir, E.M. et al. (2004). *Electrochim. Acta* 49: 3901–3908.
30 Spinace, E.V., Neto, A.O., Vasconcelos, T.R.R., and Linardi, M. (2004). *J. Power Sources* 137: 17–23.
31 Neto, A.O., Giz, M.J., Perez, J. et al. (2002). *J. Electrochem. Soc.* 149: A272–A279.
32 Zhou, W.J., Zhou, Z.H., Song, S.Q. et al. (2003). *Appl. Catal. B* 46: 273–285.
33 Colmati, F., Antolini, E., and Gonzalez, E.R. (2006). *J. Power Sources* 157: 98–103.
34 Zhou, W.J., Li, W.Z., Song, S.Q. et al. (2004). *J. Power Sources* 131: 217–223.
35 Tanaka, S., Umeda, M., Ojima, H. et al. (2005). *J. Power Sources* 152: 34–39.
36 Wang, Z.B., Yin, G.P., Zhang, J. et al. (2006). *Electrochim. Acta* 51: 5691–5697.
37 Neto, A.O., Franco, E.G., Arico, E. et al. (2003). *J. Eur. Ceram. Soc.* 23: 2987–2992.
38 Liu, F., Lee, J.Y., and Zhou, W.J. (2006). *Small* 2: 121–128.
39 Salazar-Banda, G.R., Suffredini, H.B., Calegaro, M.L. et al. (2006). *J. Power Sources* 162: 9–20.
40 Kuznetsov, V.I., Belyi, A.S., Yurchenko, E.N. et al. (1986). *J. Catal.* 99: 159–170.
41 Radmilovic, V., Richardson, T.J., Chen, S.J., and Ross, P.N. (2005). *J. Catal.* 232: 199–209.
42 Harris, I.R., Norman, M., and Bryant, A.W. (1968). *J. Less-Common Met.* 16: 427.
43 Song, S.Q., Zhou, W.J., Zhou, Z.H. et al. (2005). *Int. J. Hydrog. Energy* 30: 995–1001.
44 Vigier, F., Coutanceau, C., Perrard, A. et al. (2004). *J. Appl. Electrochem.* 34: 439–446.
45 Vigier, F., Coutanceau, C., Hahn, F. et al. (2004). *J. Electroanal. Chem.* 563: 81–89.
46 Jiang, L.H., Sun, G.Q., Zhou, Z.H., and Xin, Q. (2004). *Catal. Today* 93–95: 665–670.
47 Zhou, W. (2004). *Solid State Ionics* 175: 797–803.
48 Spinace, E.V., Linardi, M., and Neto, A.O. (2005). *Electrochem. Commun.* 7: 365–369.
49 Antolini, E. (2007). *J. Power Sources* 170: 1–12.
50 Sine, G., Foti, G., and Comninellis, C. (2006). *J. Electroanal. Chem.* 595: 115–124.
51 Jiang, L.H., Sun, G.Q., Sun, S.G. et al. (2005). *Electrochim. Acta* 50: 5384–5389.
52 Colmenares, L., Wang, H., Jusys, Z. et al. (2006). *Electrochim. Acta* 52: 221–233.
53 Colmati, F., Antolini, E., and Gonalez, E.R. (2007). *J. Electrochem. Soc.* 154: B39–B47.
54 Gursahani, K.I., Alcala, R., Cortright, R.D., and Dumesic, J.A. (2001). *Appl. Catal. A* 222: 369–392.
55 Alcala, R., Shabaker, J.W., Huber, G.W. et al. (2005). *J. Phys. Chem. B* 109: 2074–2085.
56 Rousseau, S., Coutanceau, C., Lamy, C., and Leger, J.M. (2006). *J. Power Sources* 158: 18–24.
57 Colmati, F., Antolini, E., and Gonzalez, E.R. (2007). *Appl. Catal. B* 73: 106–115.

58 Antolini, E., Colmati, F., and Gonzalez, E.R. (2007). *Electrochem. Commun.* 9: 398–404.
59 Sine, G., Smida, D., Limat, M. et al. (2007). *J. Electrochem. Soc.* 154: B170–B174.
60 Serov, A. and Kwak, C. (2010). *Appl. Catal. B* 97: 1–12.
61 An, L., Zhao, T.S., Shen, S.Y. et al. (2010). *Int. J. Hydrog. Energy* 35: 4329–4335.
62 Matsuoka, K., Iriyama, Y., Abe, T. et al. (2005). *J. Electrochem. Soc.* 152: A729–A731.
63 An, L. and Chen, R. (2016). *J. Power Sources* 329: 484–501.
64 Wang, H., Jusys, Z., and Behm, R.J. (2006). *J. Electroanal. Chem.* 595: 23–36.
65 Rahim, S., Lee, C.S., Abnisa, F. et al. (2020). *Sci. Total Environ.* 705: 135137.

14

Fuel Cell Devices

14.1 Introduction

As one typical fuel cell type with high promise, direct liquid fuel cell (DLFC) has attracted wide interest based on its advantages: small fuel box, simple structure, high energy density, fast charging, and easy storage and transportation. Ethanol and methanol are the most frequently used alcohol types; in addition, formic acid and ethylene glycol can also be used in DLFC. The high loading of electrocatalysts, especially using precious metals, and the resulting high cost are the main issues in DLFC. To improve DLFCs' performance, problems including generation of various by-products, cathode flooding, fuel crossover, poor durability in long-term tests, and fuel safety should be fully addressed. More research is needed to improve DLFC's overall performance and promote commercialization. Up to date, some DLFCs such as direct ethanol fuel cells (DEFCs) and direct methanol fuel cells (DMFCs) are developed on the way to commercial applications; however, other DLFC types are still in their initial developing stage.

Nowadays, energy supplies with high power density have been intensively pursuing for satisfying the increasing demand for portable electronic devices such as laptops and mobile phones. DLFC is considered to be an ideal device for medium power applications that require a wide voltage supply from several hundreds of watts to about 3 kW, as well as a very specific military application [1, 2]. In addition, devices powered by lithium-ion batteries have movement restrictions due to the inadaptability of them for application of remote areas without electricity [3], while DLFCs can operate without electricity. In addition, higher energy density could be easily achieved by using liquid fuels compared to lithium-ion polymers, about 600 $W\,h\,kg^{-1}$. That suggests the DLFC when using mobile phones can provide a longer time, when using a laptop can provide a longer time, can provide more energy to meet the needs of consumers, and compared with rechargeable batteries, rechargeable batteries after use need several hours of charging, can easily charge moment [4]. In DLFC, the anode was directly imported with the liquid fuels, while for the cathode, the oxygen from the air was imported to be a further reduction. During the operation, electrons go through the external circuit while ions pass across the membrane/electrolyte. Based on the possible incomplete oxidation of various fuels, there will always be some by-products produced at the anode, while

Electrocatalysis in Balancing the Natural Carbon Cycle, First Edition. Yaobing Wang.
© 2021 WILEY-VCH GmbH. Published 2021 by WILEY-VCH GmbH.

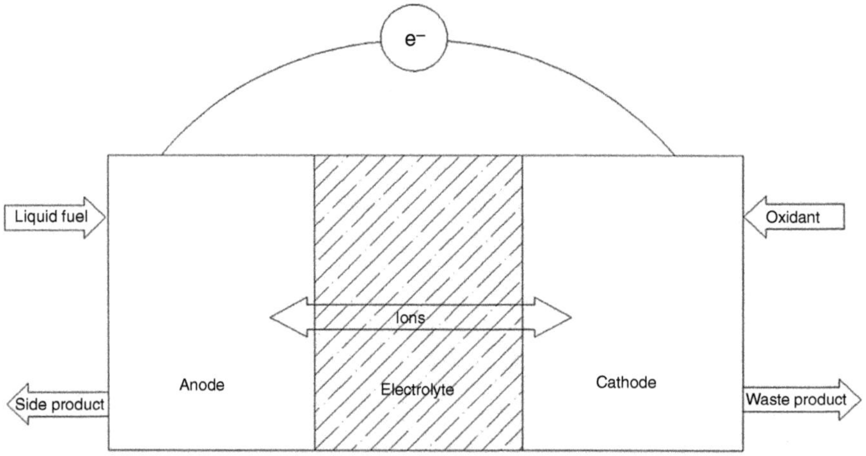

Figure 14.1 General operating principle of DLFCs. Source: Ref. Ong et al. [5]. © 2017, Elsevier.

the waste is mainly water at the cathode. The operation principle of DLFC is shown in Figure 14.1.

Different types of DLFC were introduced by de León et al. with focusing on toxicological, ecological hazards, and thermodynamic-energetic data of the reactants of liquid fuels [6]. Different types of inorganic and organic fuels were also focused on by Soloveichik [7] with considering their performance of DLFC. Herein, a general review on DLFC considering alkaline and acid types of fuel cells will be discussed. Different types of fuels in the DLFC will be discussed, respectively.

14.2 Types of Direct Liquid Fuel Cells

Many fuel types can be used in DLFC, and every fuel has intrinsic disadvantages and advantages. Ethanol and methanol are among the most commonly used alcohols. In addition, formic acid and amino borane dimethyl are the types of nonalcoholic fuels, which were also used in DLFC. The energy densities and theoretical cell potentials of each fuel used in the DLFCs are summarized in Table 14.1, in which purity of the fuel with 100% is used for the evaluation of the energy densities.

14.2.1 Acid and Alkaline Fuel Cells

Acidic and alkaline fuel cells are the basic types for DLFCs. The primary difference for these two DLFCs is based on the ions transfer types across the membrane/electrolyte and the electrolyte types. Acidic fuel cells use polymer electrolyte membrane (PEM) as the electrolyte, and Nafion membranes are the best-used type up to date. At the same time, the electrolyte in alkaline DLFC would be either liquid or solid. In the early days, alkaline fuel cells used only alkaline aqueous solutions as electrolytes; however, the corrosive nature of the liquid electrolyte complicated

Table 14.1 General reactions of DLFC with different fuels.

Type of fuel	Reaction		Standard theoretical potential, E^0 (V)	Energy density (Wh l^{-1})
Methanol	Anode	$CH_3OH + H_2O \rightarrow CO_2 + 6H^+ + 6e^-$		
	Cathode	$6H^+ + 6e^- + (3/2)O_2 \rightarrow 3H_2O$		
	Overall	$CH_3OH + (3/2)O_2 \rightarrow CO_2 + 2H_2O$	1.213	4820
Ethanol	Anode	$C_2H_5OH + 3H_2O \rightarrow 2CO_2 + 12H^+ + 12e^-$		
	Cathode	$12H^+ + 12e^- + 3O_2 \rightarrow 6H_2O$		
	Overall	$C_2H_5OH + 3O_2 \rightarrow 2CO_2 + 3H_2O$	1.145	6280
Propanol	Anode	$C_3H_7OH + 5H_2O \rightarrow 3CO_2 + 18H^+ + 18e^-$		
	Cathode	$18H^+ + 18e^- + (9/2)O_2 \rightarrow 9H_2O$		
	Overall	$C_3H_7OH + (9/2)O_2 \rightarrow 3CO_2 + 4H_2O$	1.122	7080
Ethylene glycol	Anode	$C_2H_6O_2 + 2H_2O \rightarrow 2CO_2 + 10H^+ + 10e^-$		
	Cathode	$10H^+ + 10e^- + (5/2)O_2 \rightarrow 5H_2O$		
	Overall	$C_2H_6O_2 + (5/2)O_2 \rightarrow 2CO_2 + 3H_2O$	1.220	5800
Glycerol	Anode	$C_3H_8O_3 + 3H_2O \rightarrow 3CO_2 + 14H^+ + 14e^-$		
	Cathode	$14H^+ + 14e^- + (7/2)O_2 \rightarrow 7H_2O$		
	Overall	$C_3H_8O_3 + (7/2)O_2 \rightarrow 3CO_2 + 7H_2O$	1.210	6400
Formic acid	Anode	$HCOOH \rightarrow CO_2 + 2H^+ + 2e^-$		
	Cathode	$2H^+ + 2e^- + (1/2)O_2 \rightarrow H_2O$		
	Overall	$HCOOH + (1/2)O_2 \rightarrow CO_2 + H_2O$	1.400	1750
Dimethyl ether	Anode	$(CH_3)_2O + 3H_2O \rightarrow 2CO_2 + 12H^+ + 12e^-$		
	Cathode	$12H^+ + 12e^- + 3O_2 \rightarrow 6H_2O$		
	Overall	$(CH_3)_2O + 3O_2 \rightarrow 2CO_2 + 3H_2O$	1.198	5610
Hydrazine	Anode	$N_2H_4 \rightarrow N_2 + 4H^+ + 4e^-$		
	Cathode	$4H^+ + 4e^- + O_2 \rightarrow 2H_2O$		
	Overall	$N_2H_4 + O_2 \rightarrow N_2 + 2H_2O$	1.615	5400
Ammonia borane	Anode	$BH_2^- + 6OH^- \rightarrow BO_2^- + 4H_2O + 6e^-$		
	Cathode	$3H_2O + 6e^- + (3/2)O_2 \rightarrow 6OH^-$		
	Overall	$BH_2^- + (3/2)O_2 \rightarrow BO_2^- + H_2O$	1.620	6100

Source: Ref. Ong et al. [5]. © 2017, Elsevier.

their design. In order to address this issue, an anion exchange membrane (AEM) used as a solid electrolyte in alkaline DLFC was proposed [8]. DLFCs, including DMFCs, have been measured under two conditions. For example, in Oh et al.'s study [9], DMFCs were operated under acid conditions with Nafion membrane as the electrolyte, while Kanninen et al. [10] used an anion exchange membrane as the electrolyte in DMFCs, which were known as alkaline DMFCs. When DLFCs operate in alkaline environments, ultra-alkaline solutions such as KOH can be incorporated into the anode to form the alkaline condition. The ionic categories of electrolytes through acidic and alkaline fuel cells are cation and anion, respectively. Acidic fuel cells operate at low temperatures, are easy to start, and have high power densities; however, the main problem is that they require expensive precious metal catalysts to operate [11]. Alternatively, alkaline fuel cells could solve the problem of expensive precious metal catalysts that plague acid fuel cells by using cheap catalysts, including nickel, iron, and cobalt, which could reduce manufacturing costs [12]. In addition, the reverse ions transport from cathode to anode would cause a lower fuel cross rate of the alkaline fuel cell, which reduces the permeability resistance effect of the acid fuel cell. However, with the exception of hydrazine and sodium borohydride, most of the fuels used for DLFCs are composed of carbon atoms, and carbonate is the main disadvantage faced by alkaline DLFC when fuels are oxidized to form carbon dioxide. Abdullah et al. [13] suggested that the AEM conductivity decreases after the formation of carbonate, causing the decrease of overall efficiency of cells. In the current research, acidic DLFCs should be a better choice due to the higher power density. This could change, however, if the problem of carbonate formation could be addressed properly. In general, both acidic and alkaline environments can be adapted in DLFCs, to satisfy various requirements in practical applications.

14.2.2 Direct Methanol Fuel Cells (DMFCs)

DMFC is the first type of direct alcohol fuel cell (DAFC) and also the most commonly used DLFC. Due to methanol's excellent performance, DMFC has been widely studied. When using methanol in the DMFC, it enters the anode directly in its liquid form, and no performing processes are needed. Due to most catalysts having higher activity for methanol oxidation compared to other fuel oxidation, and the intrinsic properties of methanol such as biodegradable, continuously available, and ease of transport, handling, and storage, low cost, and a high energy–carbon ratio, thus it has been intensively studied in DAFC [14–16]. An impressive energy density of $4820\,W\,h\,l^{-1}$ was achieved with methanol, which is obviously higher than hydrogen with the energy density at $180\,W\,h\,l^{-1}$. However, the disadvantage of methanol is its slight toxicity, so attention should be paid when dealing with it, especially its use in a large-scale commercial device. The anodic catalyst for DMFCs is usually composed of Pt- or Pd-based binary catalysts, which are highly effective compared to a single metal catalyst. Among them, Pt–Ru material is the most frequently used material type with excellent performance. When ruthenium is incorporated into platinum on carbon-based support, it inhibits the toxic effect of CO on the DMFC anode. The catalytic activity and stability of Pt–Ru were also better than those of pure Pt. This improves the whole performance of the DMFC [17, 18]. For the cathode of DMFCs, a variety of electrocatalysts are being studied, and Pt is still the best electrocatalyst for oxygen reduction reaction (ORR), especially in acidic media [19]. An open-circuit voltage as high as 0.59 V was acquired by Santasalo-Aarnio et al. [20] in the tested DMFC employing Pt–Ru electrocatalyst, which also exports a current density as high as 299 mA cm^{-2}. Ahmad et al. [21] indicated that in passive DMFCs, Pt–Ru was used as the anodic electrocatalyst and the Pt as the cathode electrocatalyst, and 3 M methanol could obtain the peak power density of 23 mW cm^2.

14.2.3 Direct Ethanol Fuel Cells (DEFCs)

Compared to methanol, ethanol is more renewable due to the ease of production from sugary feedstocks or biomass in agricultural processes. The advantages of non-toxicity and the higher energy at 8030 W h kg^{-1} make ethanol to have more potential in the applications of DLFC [22, 23]. However, the C—C bond is stable and difficult to be broken during ethanol electro-oxidation; thus, the complete oxidation of ethanol is difficult, and the overall oxidation efficiency is low in most cases. The increase of ethanol oxidation temperature may address this issue; however, it is not the best way. This is because polymer electrolyte membranes dehydrate as observed by Yang et al. [24], resulting in deterioration of their performance. Song et al. [25] indicated that the membrane permeability for ethanol was lower than that for methanol because ethanol has a large molecular size; thus, the influence of ethanol on the cathode during operation is ignorable. Due to the low electrochemical activity of ethanol, its anodic oxidation process is slow, making its output power density c. 1/7 of the power density in DMFC [23]. During DEFC operation, by-products such as formaldehyde and acetic acid are produced. It reduces the activity of the product. In addition, it is a dead end to generate acetic acid during electro-oxidation because this product is very hard to be completely oxidized. Therefore, acetic acid should be removed from the system, adding to the design complexity of the defensive device [26].

14.2.4 Direct Ethylene Glycol Fuel Cells (DEGFCs)

There are two hydroxyl groups in the glycol. Glycol is a popular product from the automotive industry as a raw material for antifreeze and polyethylene terephthalate, and its supply chain is stable, with annual production exceeding 7 million tons. Compared with monohydroxy ethanol fuel, it has superior performance because its vapor pressure is much higher, reducing the loss of fuel based on evaporation. A good performance in direct ethylene glycol fuel cells (DEGFCs) was obtained by using Pt–Ru electrocatalyst for the anode reaction, and the oxalic acid, carbon dioxide, and glycolic acid were found as the by-products. de Lima et al. [27] found that the anode activities were enhanced by Ru catalyst; however, the anode was not completely oxidized because of the release of excess by-products. Under the catalyst of Pt, ethylene glycol could be completely oxidized, so the composite Pt–Ru catalyst is regarded as one of the best choices of DEGFCs. By using Pt–Ru as an electrocatalyst in the DEGFCs, a maximum power density as high as 320 mW cm^2 was obtained. [28] Beyond the aforesaid electrocatalysts, other electrocatalysts were also developed for DEGFCs, which include Pt–Sn–Ni [29], Pt–Sn [30], Pt–Ru–W [31], Pd, and Pd–Ag [32]. A C-supported Pd–Ag electrocatalyst was prepared in 2015 by Yang et al. [32], which had obviously enhanced catalytic activity for electro-oxidation of ethylene glycol compared to Pd catalyst. Correspondingly, the current reached 4.85 in the DEGFCs employing this bimetallic Pd–Ag electrocatalyst in the anode, which is obviously larger compared to the devices employing pure Pd-based electrocatalyst. Anyhow, the best electrocatalysts for DEGFCs are still needed to be developed to further their activity.

14.2.5 Direct Glycerol Fuel Cells (DGFCs)

As one of the main by-products from biodiesel production, glycerol applications in DLFC are also attractive. Production of glycerol is expected to exceed its demand by six times by 2020. The use of waste glycerin for appropriate purposes, including fuel cells, is desirable. Glycerol has a low price because of being as a waste product in the biodiesel industry. The price of crude glycerol is approximately 0.74–0.89 USD gal^{-1}, which is much cheaper than methanol 1.34 USD gal^{-1} and ethanol 3.15 USD gal^{-1}. The cost of fuel cells could be lowered by the low price of glycerol; thus, the higher feasibility for their commercial applications could be achieved [33]. Beyond this, glycerol has nonflammability, nonvolatility, biorenewability, and no toxicity, making it a potential fuel in direct glycerol fuel cell (DGFC) systems. DGFCs have a theoretical energy density as high as 6.4 kW h l^{-1} [34]. As for the types in DAFCs, DGFCs may be the latest type to be developed, and thus far, its study is still at the early stages. Han et al. [35] developed a DGFC, which had a high power density of about 45 MW cm^2 at a current density of 200 mA cm^2. Different DGFC catalysts, such as Au, Pd, and Pt, were introduced. The DGFC employing Pt-based electrocatalyst had an outstanding performance, and the peak power density was 154.8 mW cm^2. The reaction rate of the Pd based catalyst was 80.8 mW cm^2, while the reaction rate of the Au based catalyst was the lowest, 43.1 MW cm^{-2} [33].

14.2.6 Direct Formic Acid Fuel Cells (DFAFCs)

Compared with methanol, formic acid has the advantages of low toxicity and high safety; formic acid can be used even as a food additive. Thus the concerns about safety can be reduced when the fuel cell is using formic acid as the fuel [36]. Beyond these aspects, the unique properties of formic acid also include its incombustibility, rapid electro-oxidation, and accessibility [37]. The electro-oxidation of direct formic acid fuel cells (DFAFCs) is fast and efficient. The open-circuit voltage of fuel cell using formic acid is as high as 1.45 V, clearly higher compared to that of DMFC at 1.18 V and that of polymer electrolyte membrane fuel cell (PEMFC) at 1.23 V [38]. Therefore, improved DFAFCs efficiency in some lower loads and a high-power density would be achieved by these high open-circuit potentials. A maximum current density as high as 134 mA cm^2 was achieved by using 12 M formic acid in the cell, as indicated by Rice et al. [36]. However, one of the main drawbacks of using formic acid as the fuel is its rather high market price; another drawback is the relatively low volume energy density of only 1750 W h l^{-1} [39]. DFAFCs react with Pt in two ways, including further oxidation of the reaction intermediates of formic acid to CO_2 (via direct hydrogenation pathway) and CO intermediates generation (via indirect dehydration pathway) [40]. Because of the poisoning effect, CO is undesirable because it would cause a decrease in overall battery performance, including durability. It was found that Pd addition into Pt would promote the formic acid oxidation through the direct path without CO formation at 0.5 V [36], and a higher current density of 62 mA cm^2 could be obtained through the use of Pt–Pd electrocatalyst compared with 33 mA cm^2 through pure Pt catalyst.

14.2.7 Direct Dimethyl Ether Fuel Cells (DDEFCs)

As a simple ether, dimethyl ether has a vapor pressure close to that of butane and propane, which are the basic components for liquefied petroleum gas (LPG). This similar nature suggests that the technology for fuel storage and treatment is mature. Dimethyl ether is volatile but nontoxic, not carcinogenic, teratogenic, and nonmutagenic [41], and has less toxicity compared to methanol. It occurs in nature as a gas, and some studies have used it directly in fuel cells [42, 43]. However, it can also be used as a DLFCs [44] in a liquid state. Direct dimethyl ether fuel cells (DDEFCs) yields a standard cell voltage of about 1.2 V, with 12 electron transferred during dimethyl ether electro-oxidation, more than 6 electron from methanol, thus requiring less fuel to produce the same amount of electricity, reducing fuel intake [3, 45] by 3424. The Pt anode based on carbon support was employed as the anode by Cai et al. [44] in DDEFCs [14]. A maximum power density as high as 67 MW cm^2 was acquired. Methanol and formaldehyde were found to be the primary by-products during the operation of DDEFCs, which were possibly oxidized further to a degree similar to, but twice as great as, DMFCs. The electro-oxidation activity for dimethyl ether is less compared to the methanol oxidation activity. Therefore, mixing methanol with dimethyl ether was tried by Yoo et al. [46] to get a higher overall oxidation activity. At 80 °C, the power density of DDEFCs was 40 mW cm^{-2}, but after adding methanol, the DDEFC's performance doubled to 850 MW cm^2. The result indicates its high application potential in the hybrid fuel cells.

14.2.8 Other DLFCs

The electrochemical energy density of propanol is 7080 W h l^{-1}, which is one of the fuels with the highest electrochemical energy density in DLFCs [22]. The main advantages of propanol are safe operation, obvious odor, immediate detection in case of leakage or failure, and high energy density. The disadvantage is that many unnecessary by-products, such as carbon monoxide, propanol, acetone, acetone, and carbon dioxide, were produced via this device; thus, the performance of direct propanol fuel cells (DPFCs) would be influenced [14]. Yamada et al. [47] found that propanol oxidation produced various intermediates and poisoned the anodic Pt–Ru electrocatalyst. Thus, the electrodes should be recleaned from time to time, being time and labor-consuming. Kobayashi et al. [48] indicated that the open-circuit voltage of Ni catalyst to DPFCs was 350 MV, indicating that the nickel catalyst had high potential. Habibi and Dadashpour [49] pointed out that the existence of Ni rendered the Pt–Ni complex/ceramic carbon to show better performance compared to the smooth Pt electrode, thus makes it to be a promising electrocatalyst for DPFCs. However, there was a problem in the DPFC anode as many harmful products would be generated to poison the electrocatalyst. It makes the difficulty of DPFC toward commercialization. A complicated system that facilitates the removal of by-products is required in this fuel cell. Ammonia-borane can be utilized as a hydrogen source of PEMFCs through hydrolysis and pyrolysis. However, as the energy density of liquid ammonia-borane is as high as $8.4\,\mathrm{Wh\,g^{-1}}$, the current research has proposed the direct application of liquid ammonia-borane as the reactant in DLFCs. In addition, it has the advantages: being safe for the environment and biology, high availability, low cost, nontoxicity, and chemical stability [50]. Pt electrocatalyst was utilized by Zhang et al. [50] on either electrode side of direct formic acid fuel cells (DABFCs). A high-power density higher than $110\,\mathrm{mW\,cm^2}$ is obtained at 45 °C with a current density of $185\,\mathrm{mA\,cm^2}$ as well as a battery potential of 0.6 V. The hydrolysis of ammonia-borane was the primary issue in DABFCs, which reduces the efficiency of the cell and increases its complexity since additional devices would be necessary to remove the generated hydrogen. Replacing Pt catalyst with Au catalyst can address this problem. This is due to the elimination of the unnecessary hydrolysis reaction by Au, resulting in an enhanced performance of the device without production of hydrogen [51].

14.2.9 Challenges of DLFCs

DLFCs have the advantage of the steam feed system and have a broad application prospect. However, DLFC still has some restrictions, which hinder its commercialization. To replace current energy sources, including fossil fuels and batteries, all these problems should be solved properly. In what follows, the general issues from DLFCs will be discussed.

The catalyst cost is relatively high, and the amount of catalyst is large. One of DLFCs main challenges is the high cost and high load of catalysts. The high cost makes the difficulty for rendering DLFCs with high price competitive compared to the most advanced power supplies, including lithium-ion batteries. Currently, most of the DLFCs still need to use expensive Noble catalysts to promote cellular reactions, and their loads are higher than those of hydrogen-filled PEMFCs. For example, for DMFCs, Pt–Ru is regarded as one of the best electrocatalyst for applications in the anode.

14.2.10 Fuel Conversion and Cathode Flooding

The transport of the used fuel from anode to cathode across the membrane is another challenge for DLFCs. For gaseous fuels, this problem is insignificant, but for liquid, this problem is more serious. The reason for this phenomenon has arisen from the difference in fuel concentration between anode and cathode regions, which drives the diffusion of fuel molecules through the membrane, and the electroosmosis resistance of proton migration through the membrane is the main reason for the diffusion of fuel in DLFCs. This liquid fuel crossover will result in lower fuel efficiency, lower cathode activity, additional heat generation, lower battery potential, and lower overall battery performance. The factors that the operating conditions of the fuel cell system, as well as the types and concentration of fuels supplied, would influence the crossover. In general, the higher the fuel concentration, the higher the crossover rate of liquid fuel. Since the too low concentration of liquid fuel will significantly decrease the overall performance of the fuel cell, and too high concentration would intensity the fuel crossover, thus it is a significant problem for the DLFC.

14.2.11 Chemical Safety and By-product Production

In fact, some organic fuels for DLFCs are not safe, which particularly relates to the toxicity and corrosion of chemical fuel, and it is also a big risk to hinder the commercialization applications of DLFCs. For example, dimethyl ether and methanol have some toxicity, although low levels of their toxicity are reported. Another example is the corrosion of formic acid, which is harmful when the eyes and skin are contacted. Therefore, a tight and integrated system for the storage of fuel is requisite to achieve this goal. In addition, the flammable nature of most fuels used in DLFC should be noted, especially the alcohols. Due to the ease of ignition of the fuels, they should be kept away from any combustion source. Therefore, it is better to design a fireproof system, which can prevent the contact of fuels with combustion source toward their commercialization with high safety [6].

In addition, the oxidation reaction of the fuel in DLFCs will produce many by-products. Some of these byproducts are toxic anodic catalysts that reduce the battery's durability and performance. In DEFCs, DMFCs, and DFAFCs, the Pt anodic electrocatalyst is toxic because of the existence of trace CO in the incomplete reaction. It should be continuously oxidized and removed on anodes to prevent poisoning the catalyst. Electrocatalysts for the inactivity of by-products need to be developed [36, 52, 53]. In addition to intermediate poisoning, some by-products are harmful. For example, formaldehyde is a typical by-product from DEFC, and this by-product damages the optic nerve and can lead to blindness [6]. DEGFCs may produce glycolic acid, which is dangerous if ingested [34]. In addition, DLFCs produce environmentally harmful by-products. These harmful or dangerous by-products should be treated by various methods before they can be let out into the environment, adding to the device cost of running a fuel cell. However, it is necessary to study how to reduce the formation of unnecessary by-products of single DLFCs.

14.2.12 Unproven Long-term Durability

It is indispensable to study the durability of fuel cells before their commercialization. The stability of DLFC in long-term working should be tested. Generally, the performance of all fuel cells tends to decay over the course of operation. The performance decay is usually associated with dissolution inactivation and/or agglomeration of electrocatalyst, change of electrode structure and porosity, and electrolyte degradation. The Pt–Ru catalyst is the most general type for the anodic electrocatalyst in DLFCs. With the prolonged operation, the catalyst may dissolve, and meanwhile, the Ru may transport across the membrane to the cathode side. This case would reduce the efficiency of cathode ORR and also reduce the tolerance of cathode against fuel. Cathode hydrophobicity may decrease due to the degradation of electrocatalyst, causing flooding of the cathode, and inhibit oxygen transfer to the electrocatalyst layers. All these factors would lead to an overall decline in cell performance over operation [54].

Rapid advances in fuel cell technology over the past few decades have made them promising for transportation, portable electronic devices, and remote areas. For different fuel reactants, catalyst and cell design show different characteristics. In this chapter, we discuss key electrochemical parameters for evaluating catalyst/cell performance, progress in reactant types for various state-of-the-art Pt, Pd, and other metal catalysts and outline innovative system designs for various fuel cell devices. Controlling the composition and nanometer/microstructure of the electro-oxidation catalyst for a fuel cell is the key factor to improve its electrocatalytic activity and stability. In addition, combined with optimization, the hybrid catalyst shows great potential in fuel cell applications. Although great progress has been made, there are still some challenges and problems in the development of anodic electrocatalysts and the overall design of batteries. The challenges and prospects are given below.

So far, the challenges of developing high-performance fuel cells include: (i) the activity and stability of anodic electrocatalysts used for fuel electro-oxidation are still low, and the toxic effects of fuel electro-oxidation have not been adequately addressed. (ii) For multicarbon fuels, most catalysts cannot completely convert them into CO_2, which also reduces their electrocatalytic efficiency. (iii) Due to the rare use of in situ technology and other advanced characterization technologies, the research on the mechanism of electrocatalytic stability and durability is still insufficient, which also limits the effective design of high-durability electrocatalysts. In addition, postanalysis of this morphology and composition change is also required to help understand the stability/durability of electrocatalysts in fuel cells. (iv) There is a gap between the performance of the electrocatalyst characterized well in the three-electrode electrochemical test and the practical application. For example, frequent starting and stopping operation, high operating temperature and humidity, and long-period operation under actual working conditions, electrochemical evaluation should be considered. (v) The total cost of fuel cells must be reduced, especially the cost of anodic catalysts and membranes. (vi) Excessive water migration from anode to cathode causes water flooding, resulting in an extremely low utilization ratio of the cathode to electrocatalyst and high resistance to oxygen transport. (vii) Lack of systematic consideration in battery design and optimization of various components. For example, the use of nonconventional metal electrocatalysts, including M–N–C materials in the ORR cathode, may give less consideration to fuel crossings.

Despite these challenges, there is still much room for improvement in fuel cell performance by developing high-performance, low-cost electrode materials and membranes and optimizing device structure and operating parameters. To be specific, we need to (i) design, develop, and synthesize new, highly active, and durable small organic molecular oxidation catalysts for nonnoble metals to fundamentally reduce the cost of cell devices; (ii) use advanced characterization techniques to reveal degradation mechanism and promotion mechanism of active sites; (iii) further verify the good performance of the catalyst in the electrochemical test, membrane electrode assembly (MEA), and single-cell test; (iv) improve the quality of transportation by optimizing the mesoscale structures of the electrocatalyst and the structures of fuel cells; (v) develop acid and alkali dielectric membranes with high conductivity, high durability, and low cost; (vi) optimize operation parameters of specific fuel cells to maximize system output power; and (vii) optimize fuel cell performance by building hybrid power units, such as coupling supercapacitors with fuel cells to maximize their energy and power.

In summary, recent breakthroughs in electrocatalysts, membranes, electrode structures, and stack assembly have facilitated the application of different fuel cells from a variety of fundamental and engineering aspects.

References

1 Wee, J.H. (2007). *J. Power Sources* 173: 424–436.

2 Winter, M. and Brodd, R.J. (2004). *Chem. Rev.* 104: 4245–4269.

3 Kamarudin, S.K., Achmad, F., and Daud, W.R.W. (2009). *Int. J. Hydrogen Energy* 34: 6902–6916.

4 Dillon, R., Srinivasan, S., Arico, A.S., and Antonucci, V. (2004). *J. Power Sources* 127: 112–126.

5 Ong, B.C., Kamarudin, S.K., and Basri, S. (2017). *Int. J. Hydrogen Energy* 42: 10142–10157.

6 de León, C.P., Walsh, F.C., Rose, A. et al. (2007). *J. Power Sources* 164: 441–448.

7 Soloveichik, G.L. (2014). *Beilstein J. Nanotechnol.* 5: 1399–1418.

8 Li, G.W., Wang, Y., Pan, J. et al. (2015). *Int. J. Hydrogen Energy* 40: 6655–6660.

9 Oh, Y., Kim, S.K., Peck, D.H. et al. (2014). *Int. J. Hydrogen Energy* 39: 15760–15765.

10 Kanninen, P., Borghei, M., Sorsa, O. et al. (2014). *Appl. Catal., B* 156: 341–349.

11 Kirubakaran, A., Jain, S., and Nema, R.K. (2009). *Renewable Sustainable Energy Rev.* 13: 2430–2440.

12 Yu, E.H., Krewer, U., and Scott, K. (2010). *Energies (Basel, Switz.)* 3: 1499–1528.

13 Abdullah, N. and Kamarudin, S.K. (2015). *J. Power Sources* 278: 109–118.

14 Demirci, U.B. (2009). *Environ. Int.* 35: 626–631.

15 Lamy, C., Lima, A., LeRhun, V. et al. (2002). *J. Power Sources* 105: 283–296.

16 Zainoodin, A.M., Kamarudin, S.K., Masdar, M.S. et al. (2014). *Appl. Energy* 135: 364–372.

17 Huang, S.Y. and Yeh, C.T. (2010). *J. Power Sources* 195: 2638–2643.

18 Sahin, O. and Kivrak, H. (2013). *Int. J. Hydrogen Energy* 38: 901–909.

19 Bruno, M.M., Viva, F.A., Petruccelli, M.A., and Corti, H.R. (2015). *J. Power Sources* 278: 458–463.

20 Santasalo-Aarnio, A., Borghei, M., Anoshkin, I.V. et al. (2012). *Int. J. Hydrogen Energy* 37: 3415–3424.
21 Ahmad, M.M., Kamarudin, S.K., Daud, W.R.W., and Yaakub, Z. (2010). *Energy Convers. Manag.* 51: 821–825.
22 Qian, W.M., Wilkinson, D.P., Shen, J. et al. (2006). *J. Power Sources* 154: 202–213.
23 Fujiwara, N., Siroma, Z., Yamazaki, S.I. et al. (2008). *J. Power Sources* 185: 621–626.
24 Yang, C., Costamagna, P., Srinivasan, S. et al. (2001). *J. Power Sources* 103: 1–9.
25 Song, S.Q., Zhou, W.J., Liang, Z.X. et al. (2005). *Appl. Catal. B* 55: 65–72.
26 Jablonski, A. and Lewera, A. (2012). *Appl. Catal. B* 115: 25–30.
27 de Lima, R.B., Paganin, V., Iwasita, T., and Vielstich, W. (2003). *Electrochim. Acta* 49: 85–91.
28 Livshits, V. and Peled, E. (2006). *J. Power Sources* 161: 1187–1191.
29 Neto, A.O., Linardi, M., and Spinace, E.V. (2006). *Ionics* 12: 309–313.
30 Neto, A.O., Vasconcelos, T.R.R., Da Silva, R.W.R.V. et al. (2005). *J. Appl. Electrochem.* 35: 193–198.
31 Chetty, R. and Scott, K. (2007). *J. Appl. Electrochem.* 37: 1077–1084.
32 Yang, Y., Wang, W., Liu, Y.Q. et al. (2015). *Int. J. Hydrogen Energy* 40: 2225–2230.
33 Zhang, Z.Y., Xin, L., Qi, J. et al. (2013). *Appl. Catal. B* 136: 29–39.
34 Zhang, Z.Y., Xin, L., and Li, W.Z. (2012). *Int. J. Hydrogen Energy* 37: 9393–9401.
35 Han, X.T., Chadderdon, D.J., Qi, J. et al. (2014). *Int. J. Hydrogen Energy* 39: 19767–19779.
36 Rice, C., Ha, S., Masel, R.I., and Wieckowski, A. (2003). *J. Power Sources* 115: 229–235.
37 Miesse, C.M., Jung, W.S., Jeong, K.J. et al. (2006). *J. Power Sources* 162: 532–540.
38 Aslam, N.M., Masdar, M.S., Kamarudin, S.K., and Daud, W.R.W. (2012). *APCBEE Proc.* 3: 33–39.
39 Yu, X.W. and Pickup, P.G. (2008). *J. Power Sources* 182: 124–132.
40 Lovic, J.D., Tripkovic, A.V., Gojkovic, S.L.J. et al. (2005). *J. Electroanal. Chem.* 581: 294–302.
41 Semelsberger, T.A., Borup, R.L., and Greene, H.L. (2006). *J. Power Sources* 156: 497–511.
42 Ueda, S., Eguchi, M., Uno, K. et al. (2006). *Solid State Ionics* 177: 2175–2178.
43 Jensen, J.O., Vassiliev, A., Olsen, M.I. et al. (2012). *J. Power Sources* 211: 173–176.
44 Cai, K., Wang, C., Pu, W.H. et al. (2012). *Int. J. Hydrogen Energy* 37: 12605–12608.
45 Xing, L.H., Gao, Y.Z., Wang, Z.B. et al. (2011). *Int. J. Hydrogen Energy* 36: 11102–11107.
46 Yoo, J.H., Choi, H.G., Chung, C.H., and Cho, S.M. (2006). *J. Power Sources* 163: 103–106.
47 Yamada, K., Yasuda, K., Tanaka, H. et al. (2003). *J. Power Sources* 122: 132–137.
48 Kobayashi, T., Otomo, J., Wen, C.J., and Takahashi, H. (2003). *J. Power Sources* 124: 34–39.
49 Habibi, B. and Dadashpour, E. (2013). *Electrochim. Acta* 88: 157–164.
50 Zhang, X.B., Yan, J.M., Han, S. et al. (2008). *J. Power Sources* 182: 515–519.

51 Zhang, X.B., Han, S., Yan, J.M. et al. (2007). *J. Power Sources* 168: 167–171.
52 Patel, P.P., Datta, M.K., Jampani, P.H. et al. (2015). *J. Power Sources* 293: 437–446.
53 Zainoodin, A.M., Kamarudin, S.K., and Daud, W.R.W. (2010). *Int. J. Hydrogen Energy* 35: 4606–4621.
54 Wang, J.Y. (2015). *Energy* 80: 509–521.

Part VII

CO_2 Reduction and Device

The electrochemical CO_2 reduction reaction (ECDRR) has attracted wide attention because of not only providing sustainable and clean renewable energy at ambient conditions but also addressing both energy and environmental issues in one process by reducing the CO_2 level in the atmosphere. In this part, the experimentation fundamentals and the electrocatalytic advances of the ECDRR on various catalytic materials are presented in addition to the available fabricated devices and their aspects.

Electrocatalysis in Balancing the Natural Carbon Cycle, First Edition. Yaobing Wang.
© 2021 WILEY-VCH GmbH. Published 2021 by WILEY-VCH GmbH.

15

Introduction

In the twenty-first century, energy and environment are the most severe issues due to rapid industrialization and deforestation, respectively. For several decades, carbon-based fossil fuels like coal, natural gas, and petroleum have been the dominating energy sources [1], which provide a vast majority (~80%) of global energy [2]. Unfortunately, these fossil fuels are not renewable and rapidly depleted. As per International Energy Agency, the combustion of these fossil fuels enhanced the global energy demand is projected from 18 TW in 2013 to 24 or 26 TW in 2040 with a corresponding rise in carbon dioxide (CO_2) emissions from 32 Gt per year in 2013 to 37 or 44 Gt per year in 2040, respectively [2]. To addressing this issue of the increasing amount of CO_2 (greenhouse gas) emission in the atmosphere, extensive technologies have been proposed, such as carbon capture and sequestration (CCS) [3], carbon capture and utilization (CCU) [4], and carbon capture and conversion. Among them, carbon capture and conversion showed the most acceptable route to produce a variety of value-added chemicals and fuels ranging from carbon monoxide, formate, alcohol, methane, and C_{2+} (e.g. ethane and ethylene) to higher hydrocarbon and oxygenate (e.g. aldehyde and so on). Recently, various strategies, such as photochemical [5–8], thermochemical [9], photo-electro-chemical [6, 10], electrochemical [11–13], bio-electrocatalysis [14], and other procedures have been proposed and extensively studied for CO_2 conversion. Among them, the electrochemical CO_2 reduction reaction (ECDRR) strategy could provide an attractive method for the carbon-neutral production of essential chemicals and value-added fuels [15, 16].

It is well understood that the carbon dioxide (molecular weight 44.0) is a colorless and odorless, fully oxidized gaseous molecule having linear and center of symmetrical geometry. Carbon dioxide molecule has two oxygen (O) atoms that are covalently doubly bonded with single carbon (C) atom (i.e. O=C=O) where C=O bond length is 116.3 pm. Although CO_2 is mainly in gaseous form in the air, it has a solid form at −78 °C called "dry ice" and a liquid form when it partially dissolves and is even dissociated in water (1.45 g l^{-1} at 25 °C, 100 kPa). The bond enthalpy of C=O (~806 kJ mol^{-1}), which is significantly higher than that of the C—C bond (336 kJ mol^{-1}), C—H bond (411 kJ mol^{-1}), or C—O bond (327 kJ mol^{-1}) [17, 18]. Due to its highly thermodynamically stable and kinetically inertness of CO_2 structure, CO_2 molecules activated hardly to participate in chemical reaction unless treated under chemically, photochemically, and electrochemically. Besides,

Electrocatalysis in Balancing the Natural Carbon Cycle, First Edition. Yaobing Wang.
© 2021 WILEY-VCH GmbH. Published 2021 by WILEY-VCH GmbH.

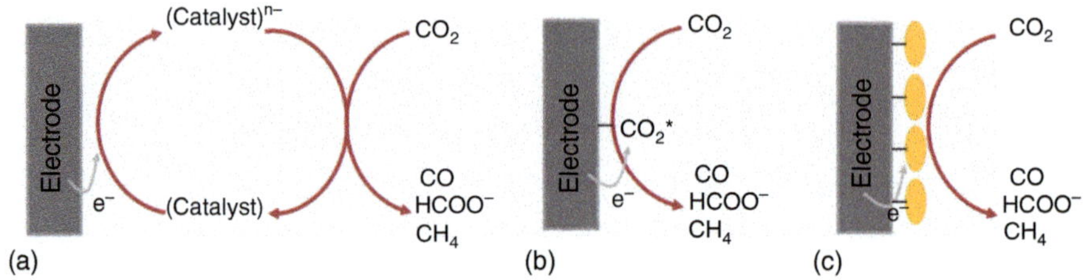

Figure 15.1 Schematic of different mechanisms for ECDRR. (a) Homogeneous electrocatalyst. (b) Heterogeneous electrocatalyst. (c) An immobilized electrocatalyst. Source: Varela et al. [21]. © 2018, John Wiley & Sons.

the linear molecule CO_2 is stable and chemically inert with a low electron affinity and a large energy gap (13.7 eV) between its lowest unoccupied molecular orbital and highest occupied molecular orbital [19].

Over several decades, homogeneous and heterogeneous metal complex-based catalysts for the CO_2 reduction have been reported that explored include homogeneous catalysts in solution, catalysts immobilized on surfaces, and heterogeneous catalysts [20]. In homogeneous catalysis, (as shown in Figure 15.1a) reduction occurred following diffusion of the catalyst to an electrode where multiple proton-coupled electron transfer (PCET) reduction occurs [20], in which the transition metal complexes such as Ir/Ru are widely studied with organic ligands (bipyridine derivatives) and electron transfer properties that utilized combinations of metal and ligand redox levels [20].

In a homogeneous catalysis mechanism, the metal complex acts as a redox shuttle between the electrode and CO_2 [20]. Where, in most cases, the electrode could supply electrons to metal complex, giving a more highly reduced state of the catalyst [20]. Following reduction, the catalyst donates electrons to the substrate (CO_2) in solution which can be reduced into value-added chemicals (HCOOH, CO, CH_4, or other higher hydrocarbons), and simultaneously reduced form of metal complex returned to its original state as shown in Figure 15.1a with the overall reaction called indirect electrolysis [21]. Recently in 2018, Ryoichi Kanega et al. reported the electrochemical reduction of CO_2 into formate by homogeneous Ir catalysts in water. The authors revealed that the metal hydride species, as shown in Figure 15.2, are proposed as intermediates that could avoid one-electron transfer to CO_2, which is the cause of the increased overpotential [22]. However, for a large-scale synthesis, a significant drawback arises during homogeneous electrolysis, such as the difficulty of catalyst recovery at the end of catalytic cycles with considerable energy demands from distillation, filtration, or crystallization to recover the catalyst [20].

To overcome these issues, in surface or electrochemical reactions, catalyst immobilization combines the advantages of homogeneous catalysis and selectivity with the ability to recycle the catalysts [20]. In 1974, Meshitsuka et al. first reported the catalytic activity of immobilized complexes, such as cobalt and nickel phthalocyanines, for the electrochemical CO_2 conversion. Though their pioneer works are not clear of which product is formed [23], this work has to disclose the fundamental probability and an understanding of the crucial role of transition metal complexes for

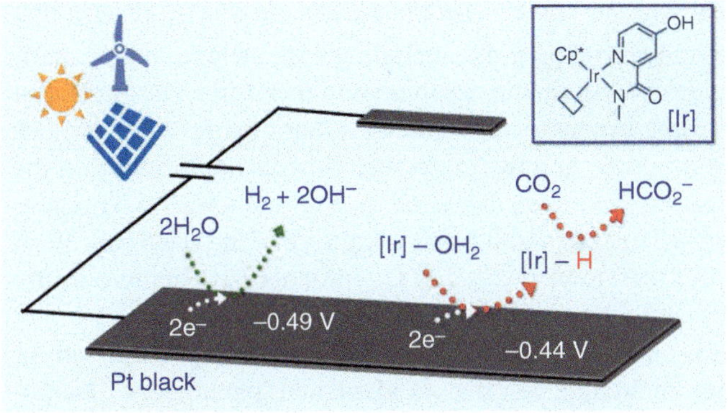

Figure 15.2 Homogeneous Ir catalysts mechanism for the conversion of CO_2 into formate in water. Source: Kanega et al. [22]. © 2018, American Chemical Society.

CO_2 reduction by electrocatalysis. Later, in 1980, Fisher and Eisenberg reported the tetraazomacrocyclic cobalt and nickel complexes, which could act as homogeneous catalysts for the reduction of CO_2-to-CO at potentials between −1.3 and −1.6 V vs. saturated calomel electrode (SCE) [24]. However, these complexes showed very low turnover frequencies (TOFs) in the range of 2–9 per hour. Therefore, the well-tuned and functionalized ligand could overcome the kinetic challenges of electrochemical CO_2 reduction. In the last decades, several types of research had been done in this field. Jing Shen et al. reported the cobalt protoporphyrin immobilized complex for the reduction of CO_2-to-CO as the main product and CH_4 as a by-product at −0.6 V vs. reference hydrogen electrode (RHE) in an aqueous acidic solution [25]. Their study showed that the binding could facilitate electron transfer from the electrode to metal complex catalysts on the surface and the further for the conversion of CO_2 to value-added chemicals, as shown in Figure 15.1c. However, the immobilized, homogeneous catalysts could be suffered from the performance losses and even decomposition during long-term CO_2 reduction cycles, but they are amenable to detailed mechanistic investigations [20].

On the contrary, heterogeneous electrocatalysis has been extensively studied in the last seven decades, which could overcome the above issues. The heterogeneous catalyst strategy is the promising and an attractive way to carry out ECDRR with the electrode (material have a high surface area) which could facilitate the electron transfer to the substrate (CO_2) for further reduction of CO_2 into various value-added products, with the overall reaction called direct electrolysis as shown in Figure 15.1b. Compared to conventional electrodes, the use of nanostructure catalysts significantly enhanced reactive surfaces with a sizable fraction of the reactive sites available for catalysis [20]. Therefore, nanostructured metal, bimetal/alloy, oxide metal, porous materials, and carbon-based metal-free materials have been investigated as an electrocatalyst for CO_2 reduction. The results of a series of studies have revealed that control morphologies and compositions of supported nanocatalyst structures can lead to high reactivity and selectivity [20].

In the last two decades, tremendous work has been done, and several review papers have already appeared, which contain the history and classical works in this field. In this chapter, we will mainly provide an overview to emphasize the pioneering work to recent advanced progress, including bulk metal to nanomaterials, then single-atom catalysts (SACs), and carbon-based metal-free electrocatalyst for ECDRR. Along with this, fundamental challenges, basic parameters, various structural factors, device structure, and the practical/industrial applications on the heterogeneous ECDRR aspects have been discussed. In addition to ECDRR, we also highlight the recent progress, understanding, and advantage of the electrochemical CO reduction reaction (ECORR) toward C_{2+} and C_{3+} production. Finally, a future perspective on the research into various electrocatalysts is presented.

Thus, it is crucial to outline basic parameters of the CO_2 reduction in ECDRR/ECORR process, including: (i) five fundamental parameters: overpotential (η), Faradaic efficiency (FE), current density (j), energy efficiency (EE), and Tafel slop that are usually measured to evaluate the cell performance of the electrocatalysts; (ii) factors affecting electrolytes, pH, concentration, temperature, and pressure; (iii) electrode: loading method and preparation; and (iv) experimental process and analysis methods. All these parameters will be cover in Section 15.1.

15.1 Basic Parameters of the CO_2 Reduction Reaction

15.1.1 The Fundamental Parameters to Evaluate the Catalytic Activity

15.1.1.1 Overpotential (η)

Practically, the electrochemical reduction of CO_2 over metal catalysts requires an operation potential more negative than its equilibrium potential (E_{eq}), resulting in an overpotential (η) [26]. As the reduction of CO_2 has to overcome the kinetic energy barrier, the onset potential is always lower than the standard reduction potential of CO_2 [27]. The difference between the onset potential and the standard reduction potential is defined as the overpotential [27]. The onset potential could directly show the overpotential for the reduction of CO_2 to products. This parameter is highly dependent on the performance of solid-state catalysts as the working electrode [27].

15.1.1.2 Faradaic Efficiency (FE)

The FE toward various products (i.e. CO, $HCOO^-$, EtOH, MeOH, and so on) describes the proportion of charge that is being utilized to generate the specific product [28] during the reduction and it is calculated by Eq. (15.1).

Without CO_2 gas bubbling during electrolysis:

$$FE = \frac{z \cdot n \cdot F}{Q} \tag{15.1}$$

where

- z Number of electrons transferred for a given product (e.g. for CH_4 formation from CO_2, $z = 4$)
- n Number of moles produced for a given product
- F Faraday's constant (96 485 C mol^{-1})
- Q Input charge passed (C)

From the above relationship, the FE is dependent on the electrocatalyst product selectivity, and this is reliant on the type of active sites present within electrocatalysts. In addition to the kind of active sites, the reaction pathways and the adsorption energy of the reaction intermediates also play a key role in determining the FE for ECDRR [28].

The FE can be calculated by Eq. (15.2):

$$\text{FE} = \frac{\text{mole of product}}{\text{mole of electron}} \times \frac{1}{\begin{array}{c}\text{mole equivalents of electron needed}\\ \text{for the conversion of electron}\end{array}} \tag{15.2}$$

where the number of mole of products is obtained from the gas chromatography, liquid chromatography, or mass spectrometry [27].

15.1.1.3 Current Density (j)

The current density (j) is defined as the total measured current (i) flowing at a given cell potential normalized using the electrochemical surface area (ECSA) instead of geometric surface area, especially for very rough electrodes, to reflect the intrinsic performance of a catalyst. This metric provides a measure of the rate of CO_2 conversion into reduced products and therefore is an indication of the production rate [28]. In other words, catalysts with similar FE but higher current density would have a higher absolute yield of the product. Unlike FE, j is also dependent on the number of active sites, improved mass transport (how fast reactants and products are removed from electrolyte), and system impedance (relates to how fast electrons are transferred to the reactants on electrode surface) [28].

Current density is the total current (I, in Amps) per unit area of the cathode (A, m^2, or commonly cm^2) calculated by Eq. (15.3) and describes the total rate of reaction, so it is a valuable input to estimate electrolyzer size and capital cost for ECDRR process [29]:

$$j = \frac{I}{A} \tag{15.3}$$

Partial current density (j_{Product}) for a specific product can be calculated using the following Eq. (15.4):

$$j_{\text{Product}} = \text{FE}_{\text{Product}} \times j \tag{15.4}$$

15.1.1.4 Energy Efficiency (EE)

Energy efficiency (EE) is the most important parameter to determine the overall energy utilization in the cathodic and anodic reactions. EE is a measure of net energy consumption toward a specific product expressed in Eq. (15.5) as a ratio of the amount of energy used to produce the specific product to the net electrical energy supplied to the system [29]:

$$\text{EE}_{\text{Product}} = \frac{E^0 \times \text{FE}_{\text{Product}}}{E^0 + \eta} \tag{15.5}$$

where E^0 is the equilibrium cell potential for the desired product ($E^0{}_{cathode} - E^0{}_{anode}$) and η is the sum of overpotentials on the cathode and anode [29].

15.1.1.5 Tafel Slope

The Tafel slope is a plot of overpotential vs. the logarithm of the partial current density [19, 27]. The calculated number is an indicator of the reaction pathways and the rate-determining step (RDS) [27]. A Tafel slope of ~118 mV dec^{-1} implies that the RDS for CO_2 reduction is the initial step of the $CO_2^{\cdot-}$ generation. In contrast, the RDS is the chemical step following the fast one-electron pre-equilibrium when the Tafel slope is −59 mV dec^{-1} [19, 30].

15.1.2 Factors Affecting ECDRR

In this section, we will mainly discuss the solvent/electrolytes effects (aqueous, non-aqueous, and ionic liquid [IL]), the major variables within the electrolyte (pH [31–33] cations, anions [33, 34], and concentration [34, 35]), and electrochemical conditions (temperature and pressure) [33, 36–39] and their impacts on ECDRR performance.

15.1.2.1 Solvent/Electrolyte

In ECDRR, electrolytes/solvent act as a medium to transfer coupled electrons and protons (e^-/H^+). The three main components, including an inert electrolyte or salt, the electroactive species, and solvent (e.g. water), are commonly present in the electrolyte. However, solvent quality plays a vital role in ECDRR performance. Therefore, the solution should (i) have a high solubility for the reactant (CO_2) and desired electrolyte to provide conduction, (ii) be electrochemically stable, (iii) be chemically compatible with the electrode materials, and (iv) be easy to handle and store, and safe [29]. The electrocatalyst's performance during the ECDRR process also depends on various electrolytes and their concentration, which will discuss in Section 15.1.2.1–15.1.2.4.

Aqueous Electrolyte Most of the studies in ECDRR examined on weakly acidic or alkaline CO_2-saturated aqueous electrolytes containing inorganic salts with HCO_3^-, SO_4^{2-}, or Cl^- anions and alkali metal cations (e.g. Na^+ and K^+) [19]. Particularly in acidic solutions, H_2 evolution (hydrogen evolution reaction [HER]) is prevalent, whereas CO_2 molecules do not exist in a basic solution [40, 41]. Therefore, the majority of the ECDRR was reported in neutral electrolyte solutions. Hori and co-worker study observed that a small cation size (Li^+) electrolyte was responsible for H_2 evolution, while a large cation size (Na^+, K^+, and Cs^+) containing electrolytes for the ECDRR selectivity [40]. They also studied the various types of electrolytes, including K_2SO_4, KCl, $KClO_4$, phosphate buffer, and dilute and concentrated $KHCO_3$ solutions and their selectivity in ECDRR. Nevertheless, the performance of electrocatalysts (product selectivity and current density) can be influenced by electrolyte compositions, pH values, ion species, and concentrations in the ECDRR process (see Sections 15.1.2.1, 15.1.2.2 and 15.1.2.4 for more detail).

Non-aqueous Electrolyte Aqueous electrolytes being used predominantly in the electrochemical process in which large amounts of solvents are applied and consumed because water has apparent advantages such as low price, wide availability, and high sustainability [42]. However, for the low solubility of the aqueous solvent in the ECDRR process, there are plenty of studies on ECDRR in a non-aqueous solution [42] (does not contain water except a small amount of residual one) have been reported. The application of organic solvents in ECDRR is intriguing for multiple reasons though their price, toxicity, and safety hazards. Generally, organic solvent (*N,N′*-dimethylformamide [DMF], dimethyl sulfoxide [DMSO], MeOH, and acetonitrile [AN]) showed higher solubility of CO_2 compared with water and enabled different reaction products, including value-added C_{2+} products.

CO_2 reduction in MeOH (protic solvent, five-time high solubility than water) has extensively been investigated, showing promising results toward the ECDRR process. Because, protic solvent with a pK_a value only slightly higher than that of water and reaction products observed in the ECDRR with MeOH are like those in aqueous solvents. Some studies investigated the methanol-based addition of various supporting salts (LiI, NaI, KI, NaOH, KOH, and CsOH) and various anionic species (chloride, bromide, iodide, thiocyanate, and acetate) which could affect the selectivity of final products [43, 44]. They found that the FE of hydrocarbons were reached over 80% on Cu electrodes when sodium hydroxide (NaOH) used in methanol [44]. Furthermore, the influence of anionic species in methanol also studied, which showed that the anions sequence bromide > iodide > chloride > thiocyanate > acetate enhanced the selective C_2H_4 formation over CH_4 [27, 43].

Ionic Liquid ILs have been proposed to serve as potential co-catalysts in the CO_2 reduction [42, 45, 46] which could enhance the CO_2 solubility and conductivity. Due to their polar nature, it can be tuned to a specific application relatively easily by changing the cation or anion species [42]. The ILs that are commonly used in ECDRR include 1-ethyl-3-methylimidazolium tetrafluoroborate ($EMIM\text{-}BF_4$), 1-butyl-3-methylimidazolium hexafluorophosphate ($BMIM\text{-}PF_6$), 1-butyl-3-methylimidazolium triflate (BMIM-OTf), 1-ethyl-3-methylimidazolium trifluoromethanesulfonate (EMIM-TFO), and 1-ethyl-3-methyl-imidazolium triflate (EMIM-OTf) [19, 27]. Some studies consider certain ILs ($BMIM\text{-}PF_6$) showed as an electrolyte and a promotor for CO_2 activation resulting in lowering the free energy for the formation of $CO_2^{\cdot -}$ intermediate [45] which means lowering CO_2 reduction overpotential. Besides the capability of ILs (both act as a solvent or as a supporting electrolyte) to absorb CO_2 physically and chemically, it can open a variety of options to convert CO_2 into valuable chemical products efficiently. Moreover, $[EMIm]^+$-based IL improves the partial current density for CO production by a 14-fold in comparison to CO_2 reduction on Ag_2S nanowires in $KHCO_3$ aqueous electrolyte. This is because initially $CO_2^{\cdot -}$ coordinates with ILs resulting in the $[EMIm\text{-}CO_2]^+$ complexes as intermediates, and these complexes further adsorbed on the electrocatalytic surface, thereby increasing the local CO_2 concentration near the surface, which can potentially reduce energy barriers for the reduction of CO_2. Several recent studies summarized the ILs solvent effects and co-catalytic properties of imidazolium-based supporting electrolytes [42, 47, 48]. Nevertheless, to overcome the relatively high cost and viscosity of ILs, they are usually dissolved in organic solvents or water for use as electrolytes [19]. Some reports studied that 75 mol% of water and 25 mol% of EMIM-BF_4 resulted in approximately five times higher CO_2 reduction current density compared to pure EMIM-BF_4. Despite IL-based electrolytes benefits of high CO_2 solubility and low overpotential, their high cost and poor stability in the presence of H_2O need to be addressed for practical CO_2 electrolyzer [19].

15.1.2.2 pH

It is well understood that the reaction's selectivity of Cu-based catalysts can be influenced by the pH of the electrolytes for ECDRR [49]. The pioneering work by Hori et al. [50, 51] studied the pH toward selectivity between CH_4 and C_2H_4. Experimental studies showed that weak buffering capabilities, such as KCl, $KClO_4$, and $K_2SO_4C_2$, might favor C_2H_5OH and C_2H_4 product formation in electrolytes allowing increased local pH at the electrode–electrolyte interface, while stronger buffers like $KHCO_3$ and $K_2H_2PO_4$ favor CH_4 and H_2 due to a local pH close to neutral [41, 50, 52]. Hori et al. [53] study demonstrated that the rate-limiting step (RLS) for the formation of methane is dependent on the pH value. In contrast, the RLS for the $C_{2+}(C_2H_4)$ products is pH-independent on pH on the standard hydrogen electrode (SHE) scales (while it is dependent on pH on the RHE scale), suggesting that the RDS toward C_2H_4 does not involve a PCET. Similarly, a recent study by Chan and co-workers demonstrated the effect of pH on C_1 vs. C_2 product activity and selectivity on Cu(111) electrode for ECDRR via a pH-dependent microkinetic model [32]. According to their study, the rate-determining proton–electron transfer steps could play a deciding factor for the C_1 vs. C_2 pathways at differences in pH with water media act as the proton source. Furthermore, density functional theory (DFT) calculations with explicit solvation investigated the C_1 vs. $C_2(C_3)$ product pathways of ECORR at different pH values on Cu(111) electrode [54]. They found that at first, at acidic conditions, i.e. at low pH (pH = 1), the formation of C_2 products was found to be kinetically blocked, and the C_1 pathways proceed through the COH to CHOH intermediate resulting to form CH_4 [52]. Second, the C_1 and $C_2(C_3)$ pathways proceed through COH common intermediate at neutral pH, while CO–COH pathway for C–C coupling. Third, at high pH (pH = 12), the first C–C coupling through adsorbed CO dimerization was dominated by suppressing the C_1 pathway [55]. In general, an increased ECDRR activity and selectivity over the HER can be observed with an increase in pH. Therefore, a pH range to work with a CO_2 saturated solution is limited to neutral and acidic electrolytes [49]. Moreover, several kinds of literature studies and review revealed that the use of glass diffusion electrodes (GDEs) could help to overcome the pH limitations by having the gas stream and electrolyte stream separated by the electrode [49, 56].

15.1.2.3 Cations and Anions

The impacts of the cation in the electrolyte on electrochemical ECDRR activity and selectivity have been studied in various literature. Pioneering work by Murata et al. [40] first reported the influence of cation size on polycrystalline Cu in the electrolyte for selectivity of CO_2 and CO reduction. Their study showed that the cation size (Li^+-to-Cs^+) is directly proportional to the selective formation of CO and C_{2+} species and inversely proportional to the selectivity for the HER. Furthermore, several reports have confirmed and expanded on these experimental observations [33, 57–62]. Besides, a theoretical study suggested that due to the very negative reduction potentials of alkali ions calculated on transition metal electrodes, the specific adsorption of alkali cations does not apply to the conditions of ECDRR in part [33, 60]. Resasco et al. [58] and Perez-Gallent et al. [59] studies proposed an essential role of a cation such as a promoter and stabilizing ECDRR/ECORR intermediates on the surface that have favorable electrostatic interactions with solvated cations. Also, the adsorption energy $^*CO_2$ and *OCCO or *OCCHO decreased for the precursor to $2e^-$ products and precursors to C_{2+} products, respectively. This is due to the cation-created electrostatic field stabilization [58]. Furthermore, the DFT study confirmed the larger hydrated cations are more energetically favored at the outer Helmholtz plane than smaller ones, which suggests that a higher concentration of cations will accumulate with increasing cation size leading to a larger local electric field [33, 58].

The majority of CO_2 reduction catalysis has been studied in $KHCO_3$ electrolytes due to the CO_2-(bi)carbonate–water equilibrium, which helps to maintain a neutral bulk pH [33]. Because of this, very few reports have been studied for the influence of anions in electrolyte for CO_2 reduction. However, initial work by Hori and his co-workers studied the various electrolyte solutions at constant current electrolysis, in which they found that non-buffering anions (Cl^-, ClO_4^-, and SO_4^{2-}) are most favored for the formation of C_2H_4 and alcohols and dilute $KHCO_3$ solution, whereas concentrated bicarbonate and phosphate solutions are preferentially produced CH_4 and H_2 [50, 63]. This is because the availability of H/protons on the surface, which is controlled by pH at the electrode [31], could play a crucial role in the product selectivity. During the reduction process, OH^- ions are released in the electrolyte, which can be neutralized by HCO_3^- or $H_2PO_4^-$ [33]. This study showed that whether it is the non-buffering electrolytes or dilute $KHCO_3$, there is likely an insufficient neutralization leading to a rise in pH, which favors ECDRR over HER and C_{2+} over C_1 products [33]. Furthermore, the impact of $KHCO_3$ concentration has also been studied by Varela et al. [35] and Kas et al. [64] research group (see Section 15.1.2.4 for more details). Recently, Bell and co-workers have studied that the changes in the pH near the electrode surface were insufficient to explain differences in activity and selectivity observed with changes in anion buffering capacity [34]. The experimental and computational studies confirmed that the buffering anion pK_a is inversely proportional to the H_2 and CH_4 product activity ($\leq$1 order of magnitude) and inactive for non-buffering anions. While, anion identity was inactive toward the product formation of CO, $HCOO^-$, and C_{2+}. This study revealed that buffering anions could play a crucial role as a competent proton donor with their effectiveness increases with decreasing pK_a of buffering anion. Moreover, the influence of halide ions (Cl^-, Br^-, and I^-) on Cu electrochemistry has also been reported [65–68]. Some studies suggest that the specific adsorption of these anions on the Cu surface could alter the charge density and selectivity for ECDRR products, with the suppression of the HER [33, 65, 68]. To be specific, adsorbed anions are linked to an increased adsorbed CO_{ads} coverage on the catalyst surface stabilizing the intermediate which may accelerate the activity of the electrocatalyst by favoring the protonation of CO_{ads} and increasing FE to hydrocarbon products like methane [35, 42, 68] and C_{2+} products [69]. However, the Cu surface structure and morphology could alter by halide ions which play an essential role in enhancing the selectivity of C_{2+} products (C_2H_4) [66, 67]. The exact dominating effect(s) of employing halide interactions during ECDRR catalysis is still unclear, and it is an urgent need for future studies.

15.1.2.4 Concentration

As we discussed in Section 15.1.2.2, how CO_2 reduction is usually carried out at close to neutral pH. Under such conditions, a small change in proton concentration could have a dramatic effect on the pH. Therefore, to understand the exact role of electrolyte concentration for ECDRR, Hori and co-workers first studied CO_2 and CO reduction at one constant current in different $KHCO_3$ electrolyte concentrations. They claimed that methane was favored in high concentrations of bicarbonate, which are attributed to a difference in the buffer capacity of an electrolyte. Furthermore, the potential dependence study of the ECDRR on polycrystalline copper in different $KHCO_3$ concentrations have also been studied. Their study observed that high concentrations of bicarbonate electrolyte improve the H_2 and CH_4 formation rates, while low concentration of bicarbonate could enhance the ethylene formation rate [35], because of diluted $KHCO_3$ solutions allow for more alkaline local pH values during ECDRR. Recently, experimental and computational studies proposed that the composition and concentration of electrolyte anions have relatively little effect on the formation of CO, $HCOO^-$, C_2H_4, and CH_3CH_2OH but has a significant impact on the formation of H_2 and CH_4 [34, 55]. This suggests that the buffering anions can serve as competent proton donors with their effectiveness increasing with decreasing pK_a [33]. In addition to the electrolyte concentration effect, very few numbers of the study also have been studied the effect of local CO_2 concentration on C–C coupling in ECDRR.

15.1.2.5 Temperature and Pressure Effect

To date, most of the ECDRR/ECORR studies have been conducted under ambient temperature and pressure, which have also been explored to manipulate the local reaction conditions [33]. Herein, we will discuss the nonambient environment trends toward ECDRR product selectivity. According to Henry's law and the acid–base buffer equilibria ($CO_2/HCO_3^-/CO_3^{2-}$), it is understood that increasing the concentration of dissolved CO_2 in the electrolyte can be achieved by decreasing temperature or increasing pressure [33, 37, 38, 64, 70]. Therefore, several studies have been reported at varying temperature [36, 37, 70, 71] and pressure [38, 39, 64, 70] during ECDRR/ECORR. The study revealed that the CH_4 product selectivity increased as the temperature decreases [37, 70, 71], and increased product selectivity of C_2H_4 [36, 39] and/or CO [36, 37] as the temperature increases over Cu foil at 5 mA cm^{-2}. Ahn et al. [37] also confirmed this trend with a temperature range between 2 and 22 °C. The proximity of *CO intermediates is important for the formation of the C—C bond, meaning that a high *CO coverage should make it easier for the C_2H_4 evolution [29]. However, low *CO coverage could suppress the C_2H_4 selectivity at high temperature, and it suggests that *CO coverage may not play a dominating role. Furthermore, Hussain et al. [72] theoretical studies revealed that *COH and *CHO intermediates are responsible for the CH_4 or C_2H_4 production, respectively, on Cu electrode. Besides, several recent studies demonstrate that selectivity of HER was suppressed at lower temperatures, while selectivity increased at higher temperatures [37, 39, 70]. However, some studies show that at below room temperature, the total partial current density for ECDRR has minimal changes [71], while the reaction rate increases at above room temperature [37, 71]. The reasons for these temperature effects are not entirely clear; changes in temperature affect the CO_2 solubility, thermal diffusion rates, pH, and solvent resistance, and they may also alter adsorption equilibria [33]. This is an urgent need to understand the exact temperature role on ECDRR product activity and selectivity over a broader range of potentials.

Wang et al. [39] reported that, below 1 atm CO_2 or CO pressure, minimal ECDRR/ECORR is detected, and HER is dominant, however, as the total ECDRR current generally increases with increasing pressures (with the exception of Hara et al. [38] study of the highest current at pressures 40–60 atm). Elevation of CO_2 pressure enhances the solubility of CO_2 in H_2O. The solubility at 25 °C is 0.034 mol l^{-1} at 1 atm and magnified by 36 times to 1.22 mol l^{-1} at 50 atm. Hara et al. [38] studied the pressure effect of a Cu wire (0.16 cm^2) for ECDRR products, resulting in the H_2 formation at 1 atm that further converted to hydrocarbons and then to HCOOH and/or CO at higher pressure 30 atm in 0.1 M $KHCO_3$. Furthermore, Kas et al. [64] study found that the selectivity of C_2H_4 production increased with increase in pressure because of an increase in local CO concentration and surface coverage. Recently, Hashiba et al. [70] reported that the effect of CO_2 pressure and stirring rate hardly affected the CH_4 production; however, the varying reaction temperature has been a significant contribution to CH_4 selectivity. Despite that, more systematic studies are needed for a better understanding of the impact of pressure over a broader range of potentials and various conditions.

15.1.3 Electrode

15.1.3.1 Loading Method

The various loading methods of as-synthesized electrocatalysts onto a prepared electrode have been reported, including drop-casting [73], dip coating [74, 75], airbrushing [76], and spraying [77]. In the case of the drop-casting method, the electrocatalyst ink was prepared by dispersing 3 mg of catalysts in a mixture solution of 200 μl of deionize (DI)-water, 370 μl of ethanol, and 30 μl of Nafion (5 wt%, Dupont) solution via sonication for 3 hours to get a homogeneous catalyst ink [73]. The as-prepared electrocatalyst ink (120 μl) was deposited onto a carbon paper (1 cm^2) by dropwise using a pipette until catalyst loading was achieved 0.5 mg cm^{-2} [73]. The deposited carbon paper was further dried at room temperature in a vacuum for 24 hours. In the case of the airbrushing method, catalysts were coated on carbon cloths with a micropore layer on the microporous layer (MPL) face of the GDE by airbrush. The catalyst ink was prepared by mixing 50 mg of catalyst, 10 ml of deionized water, 2.5 ml of isopropanol, and 100 μl of Nafion® perfluorinated resin solution that further sonicated for 30 minutes. In the case of spraying method, the ink was prepared by dispersing 1.0 mg of as-synthesized catalysts into a mixture solution of 1.0 ml of isopropyl alcohol and 30 μl of Nafion with sonication for 30 minutes and then further loaded on glassy carbon electrode (GCE) or carbon paper with an airbrush gun.

15.1.3.2 Preparation

The various types of electrodes such as (i) simple porous electrodes (e.g. carbon paper, carbon cloth, or mesh) [78, 79], (ii) planar electrodes (e.g. metal foil, glassy carbon plat, [30, 80, 81] or GCE), and (iii) GDE [82–85] have been reported. Among them, planar electrodes or simple porous electrodes have been widely studied in ECDRR because these types of electrodes are relatively simple to construct and immerse in a CO_2-saturated electrolyte [29]. However, before loading the synthesized electrocatalyst onto the electrode, the electrode preparation plays an important role in ECDRR performance. For example, the carbon paper was prepared by cutting in a circle with a diameter of 1.25 cm and immersed for 15 minutes in 65% HNO_3 to remove iron impurities possibly present in the fibers [74]. In the case of a glassy carbon plate, the plat was first to cut into 2.2 cm × 2.2 cm dimensions and cleaned by sonicating in acetone, followed by isopropanol and with deionized water, and then finally mechanically polished using an alumina suspension of up to 0.05 μm on a polishing cloth [86]. The glassy carbon plates were soaked in 1 M of high purity nitric acid for 2 hours and rinsed with deionized water to remove any possible metallic impurities [86]. Before every experiment, only the nitric acid treatment and mechanical polishing were carried out on the glassy carbon [86]. Similarly, a GDE with different sizes of diameter has also been used and prepared in the same procedure. A flow cell equipped with a GDE device developed to overcome the mass transport limitation of CO_2 in aqueous solution, resulting in the advance toward industrial current densities. The electrodes were cut into 1.5 cm × 1.5 cm pieces when assembling the cell.

15.1.3.3 Experimental Process and Analysis Methods

The most crucial parameter for newcomers is to understand the fundamental of the ECDRR process in aqueous electrolytes. Here, we introduce several aspects for experimenting ECDRR, including the electrolyte solution, CO_2 gas flow rate, catholyte stirring rate, electrolysis cell types (for more information, see Sections 20.1 and 20.2), and product analysis.

In general, before the electrolysis experiments, first, the electrolyte should be saturated with CO_2 gas by bubbling for ~0.5 hours. Meanwhile, newcomers must pay attention to the pH value change before and after CO_2 bubbling as well as cation size (e.g. $NaHCO_3$, $KHCO_3$, etc.), since both parameters have effects on the selectivity of products as we discussed previously [87]. The pH of 0.1, 0.2, 0.3, and 0.5 M $KHCO_3$ saturated with CO_2 is 6.8, 6.9, 7.0, and 7.2, respectively [87]. Moreover, various research groups have been studied the influence of flow rate (standard cubic centimeters per minute [sccm] or ml min^{-1}) of CO_2, and the catholyte stirring rate also plays an important role for catalyst activity during the ECDRR process [38, 88–90]. Their study observed that the rate of H_2 formation increased as the stirring rate increased while the product selectivity switched from CH_4-to-CO, though there was the increased availability of CO_2 at the electrode surface. Because, the higher stirring rate of the catholyte could cause the enhanced mass transfer of dissolved CO away from the electrode surface and then less adsorbed CO left for further reduction [87, 90].

In the last few decades, a variety of cells have been reported in the literature, such as H-type cells (see Section 20.1) and flow cell (see Section 20.2). Among them, H-type cells are widely used for ECDRR. In a typical H-type cell, two compartments referred as the cathode (working electrode and reference electrode having CO_2 gas inlet and outlet) and anode (counter electrode with or without gas inlet and outlet) are separated by activated ion exchange membrane such as Nafion membrane (e.g. Nafion@117 with 0.180 mm thickness and >0.90 mequiv g^{-1} exchange capacity). The nafion membrane should be activated first, usually by boiling in 3–5 wt% H_2O_2, DI water, 0.5 M H_2SO_4 and DI water at 80 °C respectively for 0.5–1 hours in the cathode compartment. Furthermore, the detection of the different products will be confirmed by a universal method such as gas product analysis, in which flame ionization detector (FID) with methanizer is usually used to quantify CO, CH_4, and C_2H_6, and thermal conductivity detector (TCD) is used to quantify H_2 [87]. However, TCD detector also able to detect a mixture of 100 ppm CO and 100 ppm H_2, and FID detector for 50 ppm CH_4, 50 ppm C_2H_4, and 50 ppm C_2H_6. However, the high-performance liquid chromatography (HPLC) and nuclear magnetic resonance (NMR) are used for liquid products (alcohol, formate, and acetic acid) detection [87].

References

1 Zang, W.J., Kou, Z.K., Pennycook, S.J., and Wang, J. (2020). *Adv. Energy Mater.* 10, 1903181.
2 Seh, Z.W., Kibsgaard, J., Dickens, C.F. et al. (2017). *Science*: 355.
3 Siegelman, R.L., Milner, P.J., Kim, E.J. et al. (2019). *Energy Environ. Sci.* 12: 2161–2173.
4 Joos, L., Huck, J.M., Van Speybroeck, V., and Smit, B. (2016). *Faraday Discuss.* 192: 391–414.
5 Wu, J.C.S. (2009). *Catal. Surv. Asia* 13: 30–40.
6 Kumar, B., Llorente, M., Froehlich, J. et al. (2012). *Annu. Rev. Phys. Chem.* 63: 541–569.
7 Lehn, J.M. and Ziessel, R. (1982). *Proc. Natl. Acad. Sci. U.S.A.* 79: 701–704.

8 Yu, H., Haviv, E., and Neumann, R. (2020). *Angew. Chem. Int. Ed.* 59: 6219–6223.
9 Chueh, W.C., Falter, C., Abbott, M. et al. (2010). *Science* 330: 1797–1801.
10 Deng, X., Li, R., Wu, S. et al. (2019). *J. Am. Chem. Soc.* 141: 10924–10929.
11 Hernandez, R.M., Marquez, J., Marquez, O.P. et al. (1999). *J. Electrochem. Soc.* 146: 4131–4136.
12 Gao, S., Lin, Y., Jiao, X. et al. (2016). *Nature* 529: 68–71.
13 Liu, M., Liu, M.X., Wang, X.M. et al. (2019). *Joule* 3: 1703–1718.
14 Cai, R., Milton, R.D., Abdellaoui, S. et al. (2018). *J. Am. Chem. Soc.* 140: 5041–5044.
15 Kondratenko, E.V., Mul, G., Baltrusaitis, J. et al. (2013). *Energy Environ. Sci.* 6: 3112–3135.
16 Ganji, P., Borse, R.A., Xie, J.F. et al. (2020). *Adv. Mater. Interfaces* 4, 2000096.
17 Wu, J., Huang, Y., Ye, W., and Li, Y. (2017). *Adv. Sci. (Weinh.)* 4, 1700194.
18 Jia, C., Dastafkan, K., Ren, W.H. et al. (2019). *Sustainable Energy Fuels* 3: 2890–2906.
19 Sun, Z.Y., Ma, T., Tao, H.C. et al. (2017). *Chem* 3: 560–587.
20 Zhang, S., Fan, Q., Xia, R., and Meyer, T.J. (2020). *Acc. Chem. Res.* 53: 255–264.
21 Varela, A.S., Ju, W., and Strasser, P. (2018). *Adv. Energy Mater.* 8, 1802905.
22 Kanega, R., Onishi, N., Wang, L., and Himeda, Y. (2018). *ACS Catal.* 8: 11296–11301.
23 Meshitsuka, S., Ichikawa, M., and Tamaru, K. (1974). *J. Chem. Soc. Chem. Commun.*: 158–159.
24 Fisher, B. and Eisenberg, R. (1980). *J. Am. Chem. Soc.* 102: 7361–7363.
25 Shen, J., Kortlever, R., Kas, R. et al. (2015). *Nat. Commun.* 6: 8177.
26 Wu, J., Sharifi, T., Gao, Y. et al. (2019). *Adv. Mater.* 31, e1804257.
27 Zhang, L., Zhao, Z.J., and Gong, J. (2017). *Angew. Chem. Int. Ed.* 56: 11326–11353.
28 Daiyan, R., Saputera, W.H., Masood, H. et al. (2020). *Adv. Energy Mater.* 10, 1902106.
29 Garg, S., Li, M.R., Weber, A.Z. et al. (2020). *J. Mater. Chem. A* 8: 1511–1544.
30 Lu, Q., Rosen, J., Zhou, Y. et al. (2014). *Nat. Commun.* 5: 3242.
31 Varela, A.S., Kroschel, M., Leonard, N.D. et al. (2018). *ACS Energy Lett.* 3: 812–817.
32 Liu, X., Schlexer, P., Xiao, J. et al. (2019). *Nat. Commun.* 10: 32.
33 Nitopi, S., Bertheussen, E., Scott, S.B. et al. (2019). *Chem. Rev.* 119: 7610–7672.
34 Resasco, J., Lum, Y., Clark, E. et al. (2018). *Chemelectrochem* 5: 1064–1072.
35 Varela, A.S., Kroschel, M., Reier, T., and Strasser, P. (2016). *Catal. Today* 260: 8–13.
36 Hori, Y., Kikuchi, K., Murata, A., and Suzuki, S. (1986). *Chem. Lett.*: 897–898.
37 Ahn, S.T., Abu-Baker, I., and Palmore, G.T.R. (2017). *Catal. Today* 288: 24–29.
38 Hara, K., Tsuneto, A., Kudo, A., and Sakata, T. (1994). *J. Electrochem. Soc.* 141: 2097–2103.
39 Wang, L., Nitopi, S.A., Bertheussen, E. et al. (2018). *ACS Catal.* 8: 7445–7454.
40 Murata, A. and Hori, Y. (1991). *Bull. Chem. Soc. Jpn.* 64: 123–127.

41 Hori, Y., Vayenas, C.G., White, R.E., and Gamboa-Aldeco, M.E. (2008). Electrochemical Co_2 reduction on metal electrodes. In: *Modern Aspects of Electrochemistry*, vol. 42 (eds. C.G. Vayenas, R.E. White and M.E. Gamboa-Aldeco), 89–189. Chiba University, Japan: Springer.

42 Konig, M., Vaes, J., Klemm, E., and Pant, D. (2019). *iScience* 19: 135–160.

43 Kaneco, S., Iiba, K., Yabuuchi, M. et al. (2002). *Ind. Eng. Chem. Res.* 41: 5165–5170.

44 Kaneco, S., Katsumata, H., Suzuki, T., and Ohta, K. (2006). *Energy Fuels* 20: 409–414.

45 Rosen, B.A., Salehi-Khojin, A., Thorson, M.R. et al. (2011). *Science* 334: 643–644.

46 Sun, L., Ramesha, G.K., Kamat, P.V., and Brennecke, J.F. (2014). *Langmuir* 30: 6302–6308.

47 Faggion, D. Jr.,, Goncalves, W.D.G., and Dupont, J. (2019). *Front. Chem.* 7: 102.

48 Sharma, P.P. and Zhou, X.-D. (2017). *Wiley Interdiscip. Rev. Energy Environ.* 6, e239.

49 Varela, A.S., Ju, W., Bagger, A. et al. (2019). *ACS Catal.* 9: 7270–7284.

50 Hori, Y., Murata, A., and Takahashi, R. (1989). *J. Chem. Soc. Faraday Trans.* 1 (85): 2309–2326.

51 Hori, Y., Murata, A., Takahashi, R., and Suzuki, S. (1987). *J. Am. Chem. Soc.* 109: 5022–5023.

52 Todorova, T.K., Schreiber, M.W., and Fontecave, M. (2019). *ACS Catal.* 10: 1754–1768.

53 Hori, Y., Takahashi, R., Yoshinami, Y., and Murata, A. (1997). *J. Phys. Chem. B* 101: 7075–7081.

54 Xiao, H., Cheng, T., Goddard, W.A. 3rd,, and Sundararaman, R. (2016). *J. Am. Chem. Soc.* 138: 483–486.

55 Sundararaman, R., Figueiredo, M.C., Koper, M.T.M., and Schwarz, K.A. (2017). *J. Phys. Chem. Lett.* 8: 5344–5348.

56 Dinh, C.T., Burdyny, T., Kibria, M.G. et al. (2018). *Science* 360: 783–787.

57 Kyriacou, G.Z. and Anagnostopoulos, A.K. (1993). *J. Appl. Electrochem.* 23: 483–486.

58 Resasco, J., Chen, L.D., Clark, E. et al. (2017). *J. Am. Chem. Soc.* 139: 11277–11287.

59 Perez-Gallent, E., Marcandalli, G., Figueiredo, M.C. et al. (2017). *J. Am. Chem. Soc.* 139: 16412–16419.

60 Singh, M.R., Kwon, Y., Lum, Y. et al. (2016). *J. Am. Chem. Soc.* 138: 13006–13012.

61 Lum, Y.W., Yue, B.B., Lobaccaro, P. et al. (2017). *J. Phys. Chem. C* 121: 14191–14203.

62 Birdja, Y.Y., Perez-Gallent, E., Figueiredo, M.C. et al. (2019). *Nat. Energy* 4: 732–745.

63 Hori, Y., Murata, A., Takahashi, R., and Suzuki, S. (1988). *J. Chem. Soc. Chem. Commun.*: 17–19.

64 Kas, R., Kortlever, R., Yılmaz, H. et al. (2015). *Chemelectrochem* 2: 354–358.

65 Ogura, K., Ferrell, J.R., Cugini, A.V. et al. (2010). *Electrochim. Acta* 56: 381–386.

66 Roberts, F.S., Kuhl, K.P., and Nilsson, A. (2015). *Angew. Chem. Int. Ed.* 54: 5179–5182.

67 Lee, S., Kim, D., and Lee, J. (2015). *Angew. Chem. Int. Ed.* 54: 14701–14705.

68 Varela, A.S., Ju, W., Reier, T., and Strasser, P. (2016). *ACS Catal.* 6: 2136–2144.

69 Gao, D.F., McCrum, I.T., Deo, S. et al. (2018). *ACS Catal.* 8: 10012. 10020.

70 Hashiba, H., Yotsuhashi, S., Deguchi, M., and Yamada, Y. (2016). *ACS Comb. Sci.* 18: 203–208.

71 Dewulf, D.W., Jin, T., and Bard, A.J. (1989). *J. Electrochem. Soc.* 136: 1686–1691.

72 Hussain, J., Jonsson, H., and Skulason, E. (2018). *ACS Catal.* 8: 5240–5249.

73 Pan, F.P., Zhang, H.G., Liu, K.X. et al. (2018). *ACS Catal.* 8: 3116–3122.

74 De Riccardis, A., Lee, M., Kazantsev, R.V. et al. (2020). *ACS Appl. Mater. Interfaces* 12: 5251–5258.

75 Reske, R., Mistry, H., Behafarid, F. et al. (2014). *J. Am. Chem. Soc.* 136: 6978–6986.

76 Gu, J., Hsu, C.S., Bai, L. et al. (2019). *Science* 364: 1091–1094.

77 Jung, H., Lee, S.Y., Lee, C.W. et al. (2019). *J. Am. Chem. Soc.* 141: 4624–4633.

78 Li, Q., Fu, J., Zhu, W. et al. (2017). *J. Am. Chem. Soc.* 139: 4290–4293.

79 Dai, L., Qin, Q., Wang, P. et al. (2017). *Sci. Adv.* 3, e1701069.

80 Liu, S., Xiao, J., Lu, X.F. et al. (2019). *Angew. Chem. Int. Ed.* 58: 8499–8503.

81 Hori, Y., Wakebe, H., Tsukamoto, T., and Koga, O. (1994). *Electrochim. Acta* 39: 1833–1839.

82 Kopljar, D., Inan, A., Vindayer, P. et al. (2014). *J. Appl. Electrochem.* 44: 1107–1116.

83 Whipple, D.T., Finke, E.C., and Kenis, P.J.A. (2010). *Electrochem. Solid State Lett.* 13: D109–D111.

84 Ma, S.C., Sadakiyo, M., Luo, R. et al. (2016). *J. Power Sources* 301: 219–228.

85 Park, S., Lee, J.W., and Popov, B.N. (2006). *J. Power Sources* 163: 357–363.

86 Lum, Y., Kwon, Y., Lobaccaro, P. et al. (2015). *ACS Catal.* 6: 202–209.

87 Zhao, J., Xue, S., Barber, J. et al. (2020). *J. Mater. Chem. A* 8: 4700–4734.

88 Rasul, S., Anjum, D.H., Jedidi, A. et al. (2015). *Angew. Chem. Int. Ed.* 54: 2146–2150.

89 Kas, R., Hummadi, K.K., Kortlever, R. et al. (2016). *Nat. Commun.* 7: 10748.

90 Lim, C.F.C., Harrington, D.A., and Marshall, A.T. (2017). *Electrochim. Acta* 238: 56–63.

16

Electrocatalysts-1

16.1 Heterogeneous Electrochemical CO_2 Reduction Reaction

CO_2 transformation is dominated by nucleophilic attacks at carbon, which is an uphill process requiring substantial input of energy (~750 kJ mol^{-1} required for dissociation of the C=O bond) [1]. However, among various methods for CO_2 conversion, electrochemical CO_2 reduction reaction (ECDRR) has the most widely studied and attractive route, which produced different value-added chemicals and fuels from electrochemical CO_2 reduction. Because, the electricity required to drive the reduction reaction can be driven by clean energy resources (such as wind and solar) to balance the carbon-neutral cycle and regenerating fuels. However, the heterogeneous catalysts considered the most promising way to control the selectivity and accelerate the slow kinetic reduction for the conversion of CO_2 into more than two electron products like CH_4, CH_3OH, and C_{2+} products. In this regard, this chapter will mainly be focused on heterogeneous electrocatalyst for ECDRR.

16.2 Thermodynamic and Kinetic Parameters of Heterogeneous CO_2 Reduction in Liquid Phase

The reduction of CO_2 involves (i) a proton-assisted multiple-electron transfer process taking place at cathode and (ii) a one-electron transfer process; however, proton-assisted multiple-electron transfer steps are generally more favorable than one-electron reductions [2]. Because the direct reduction of CO_2 involved one-electron transfer to form CO_2 anion radical ($CO_2{\cdot}^-$) which is highly unfavorable, having a negative formal redox potential −1.97 V (vs. standard hydrogen electrode [SHE]) in an aprotic solvent (*N*,*N*′-dimethylformamide [DMF]) and −1.90 V (Eq. (16.7)) in water (pH 7) [1]. Catalytic strategies have thus been developed to bypass the formation of $CO_2{\cdot}^-$ through proton-assisted multiple-electron transfer to reduce CO_2 at lower energetic costs [1]. Depending on the number of electrons and protons transferred, CO_2 can be reduced into 16 different products [1, 3–6], as shown in Figure 16.1. However, the different standard redox potentials for commonly reported products are shown in Table 16.1 (Eqs. (16.1)–(16.7); CO_2 reduction

Electrocatalysis in Balancing the Natural Carbon Cycle, First Edition. Yaobing Wang.
© 2021 WILEY-VCH GmbH. Published 2021 by WILEY-VCH GmbH.

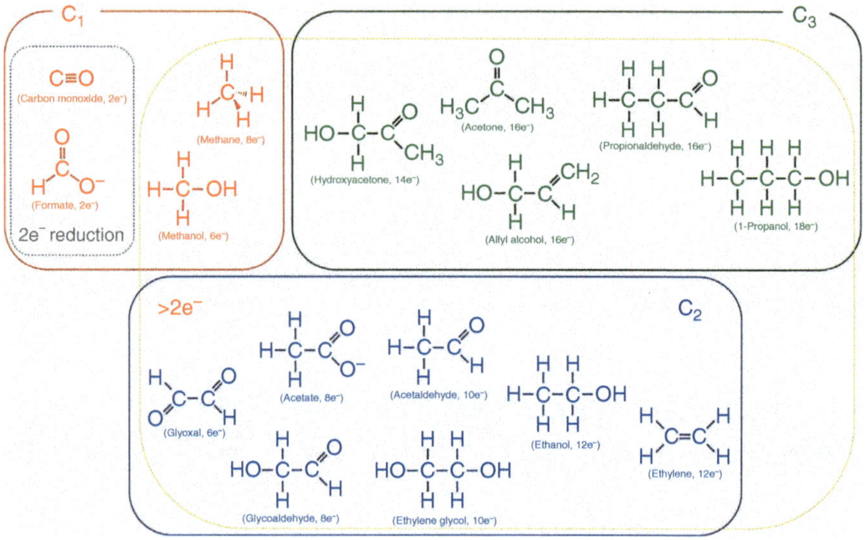

Figure 16.1 Various products formed by electrochemical reduction of carbon dioxide in aqueous media. Electron numbers in the parentheses indicate the number of electrons required to generate the products from CO_2. Source: Xie et al. [6]. © 2019, American Chemical Society.

Table 16.1 CO_2 electrolysis products at different standard redox potentials.

CO_2 reduction halfelectrochemical reaction	E^0 (V vs. SHE)	
$CO_2(g) + 2H^+ + 2e^- \rightarrow CO(g) + H_2O$	−0.53	Eq. (16.1)
$CO_2(g) + 2H^+ + 2e^- \rightarrow HCO_2H(l)$	−0.61	Eq. (16.2)
$CO_2(g) + 4H^+ + 4e^- \rightarrow HCHO(l) + H_2O$	−0.48	Eq. (16.3)
$CO_2(g) + 6H^+ + 6e^- \rightarrow CH_3OH(l) + H_2O$	−0.38	Eq. (16.4)
$CO_2(g) + 8H^+ + 8e^- \rightarrow CH_4(g) + 2H_2O$	−0.24	Eq. (16.5)
$2CO_2(g) + 12H^+ + 12e^- \rightarrow C_2H_4(g) + 4H_2O$	−0.34	Eq. (16.6)
$CO_2(g) + e^- \rightarrow CO_2{\cdot}^-$	−1.90	Eq. (16.7)
H_2O reduction half reaction		
$2H_2O + 2e^- \rightarrow H_2(g) + 2OH^-$	−0.42	Eq. (16.8)

Note: [1]pH = 7 in aqueous solution vs. standard hydrogen electrode (SHE), 25 °C, 1-atm gas pressure, and 1.0 M for other solutes.

potentials vs. SHE at pH 7). However, Eq. (16.8) shows a competitive two-electron process called hydrogen evolution reaction (HER) from H_2O reduction.

According to the relationship $\Delta G = -nFE^0$, the reduction reaction thermodynamically more feasible at more positive E^0, where F is the Faraday constant, and n is the number of electrons transferred during the redox reaction. Based on E^0, the reductions of CO_2 toward hydrocarbon or alcohol products should be thermodynamically more favorable than CO, HCOOH, HCHO, and H_2 production [1]. In addition to a thermodynamic barrier, CO_2 reduction also has a kinetic dependence on the concentration of available protons in solution [1]. Besides, the kinetics of electrochemical CO_2 reduction involves a complicated reaction mechanism, typically generating a mixture of the products as mentioned earlier [7]. This study implies that a preferred catalyst likely entails catalytic sites transferring electrons that are close to sites providing protons [1]. The hydrogenation of adsorbed C_1 intermediates is kinetically easier than the formation of C—C bonds, limiting the rate and selectivity of C_2 and higher hydrocarbon production [1], the overall mechanistic pathways in ECDRR shown in Figure 16.2. In the multi-electron transfer reduction of CO_2, the adsorption energies of reaction intermediates seem to follow linear scaling relationships, thus limiting catalytic efficiency, which can be realized by (i) reducing coordination numbers, (ii) doping with P-block elements, (iii) introducing oxophilic sites, and (iv) coating the catalyst surface with active ligands [1].

In advancing CO_2 reduction, increasing energy efficiency (EE) and low product selectivity are the main challenges that have been reported, which could primarily enhance the overpotentials for CO_2 reduction and competitive HER, respectively [8, 9]. Theoretically, the conversion of CO_2 into hydrocarbon and other products takes place at −0.3 to +0.3 V vs. reversible hydrogen electrode (RHE); however, due to the high stability of the CO_2 molecule, experimentally very large overpotential (~1 V) should be applied to reduce CO_2. To solve these problems, a catalyst (Figure 16.3) is typically required to minimize the energy input and improve reaction rates and energy efficiency needed for device-level commercialization [10].

In general, heterogeneous ECDRR takes place at the interface of an electrolyte and electrode, mainly involved in three steps [11] as follows.

Chemical adsorption of the substrate (CO_2) on the surface of a catalyst (cathode) act. (Note: In this step, the concentration of CO_2 in the electrolyte is the most direct influencing factor)

Electron transfer and/or proton migration between CO_2 and cathode catalysts cause the breaking of the C—O bond and/or the form of the C—H bond. It has been referred to as the rate-determining step in the ECDRR because enormous energy is always needed to break the C=O bond to form an intermediate $CO_2^{\cdot -}$ and to convert into different valuable materials, including C_1, C_2, and other product [11] as shown in Table 16.1 the CO_2 electrolysis products at different standard redox potentials.

Finally, the product desorption takes place at the electrode surface (cathode) and diffusion into the electrolyte.

Therefore, various types of electrocatalyst are reported, and categories, including transition metal noble catalyst (Au, Ag, and Pd) [5, 12], and non-noble (Cu, Ni, and Zn) [4, 10], bimetallic/alloy, single-atom catalyst (SAC), porous organic framework catalyst (i.e. molecular/covalent, i.e. metal-organic frameworks [MOFs] and covalent organic frameworks [COFs]), and metal-free catalysts [13] have been explored for ECDRR, and more detailed study will be discussed later.

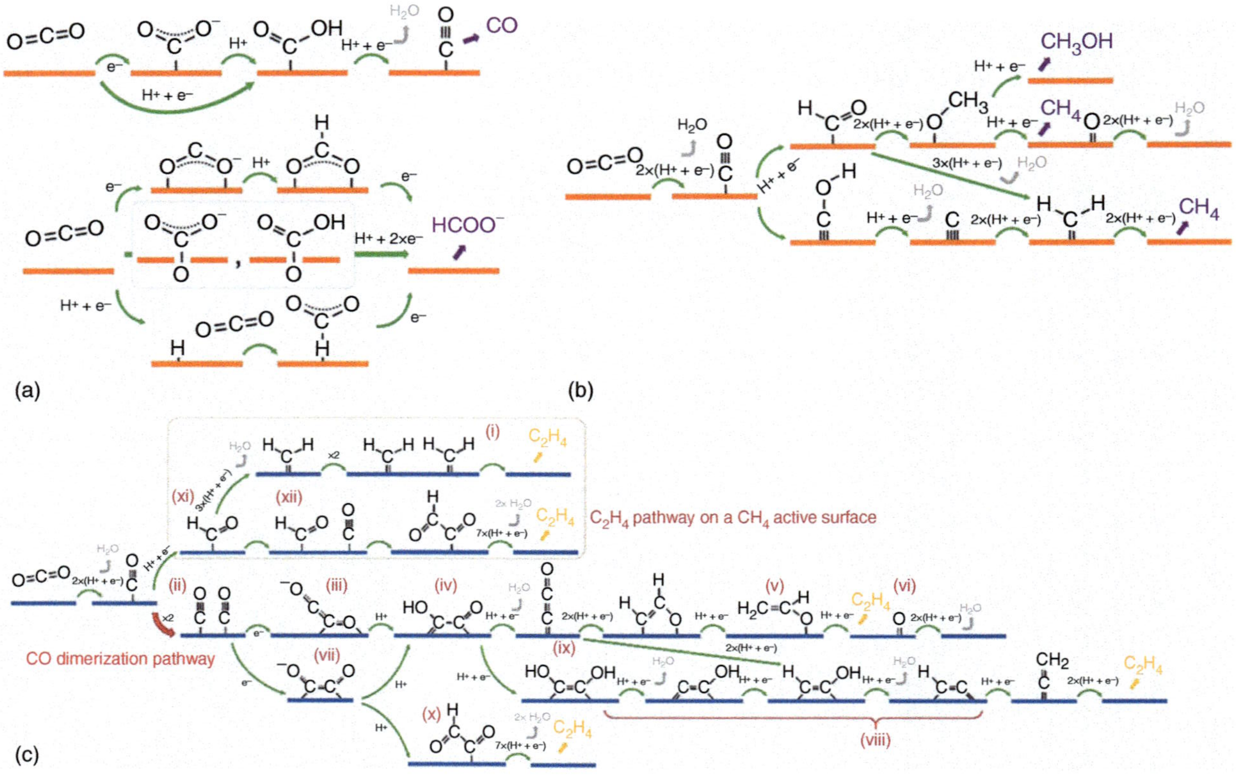

Figure 16.2 Overall mechanistic pathways in ECDRR. (a) $2e^-$ (C_1) product pathway (CO and $HCOO^-$), (b) $>2e^-$ (C_1) product pathway (CH_3OH and CH_4), and (c) (C_{2+}) product pathway (C_2H_4). Source: Xie et al. [6]. © 2019, American Chemical Society.

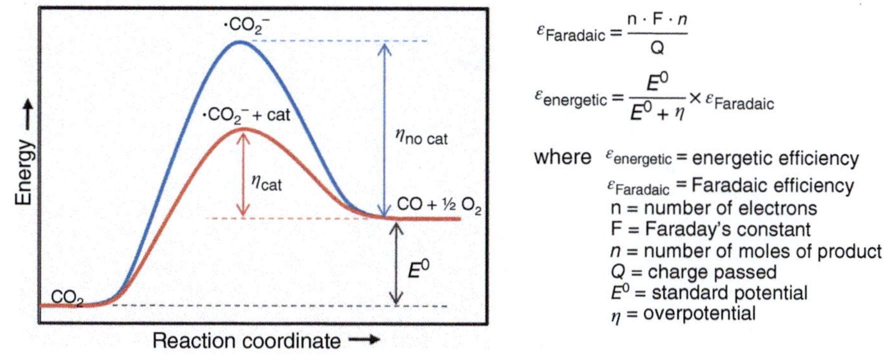

Figure 16.3 Catalysts and electrolytes acting as cocatalysts can lower the energy of the intermediate and thus improve the energetic efficiency of the conversion. ε as the symbol for the faradaic and energetic efficiencies to avoid confusion with the η symbol for the overpotential. Source: Whipple et al. [8]. © 2010, American Chemical Society.

16.2.1 Bulk Metals

In 1985, Hori et al. reported various bulk metals with the most excellent catalytic activity toward electrochemical CO_2 conversion into the different value-added products (formate, CO, and CH_4) [5]. Metal electrodes, typically in their bulk states, are classified into several subgroups by the primary product selectivity of ECDRR. First formate ($HCOO^-$)-selective metals Pb, Hg, In, Sn, Cd, Tl, and Bi, second CO-selective metals Au, Ag, Zn, Pd, and Ga, third H_2-(HER)-selective metals Ti, Pt, Fe, and Ni do not practically give products from CO_2 reduction continuously in aqueous solutions under 1 atm, and fourth hydrocarbon-selective (CH_4, C_2H_4, aldehyde, and alcohols) metal Cu showed uniquely produces in quantitative and reproducible amounts [12, 14]. Cu is thus the only pure metal that reduces CO_2 to high-value products that required more than two electrons transfers (collectively referred to as "further reduced products" or ">2e$^-$ products") with substantial faradaic efficiencies [15]. In addition to Cu, Au, and Au-based materials were considered to be the promising ECDRR catalysis for selective CO formation, though large overpotentials are required due to a wide range of products and competing for HER [5, 10].

To date, researchers have identified several bulk metallic catalysts, such as Ag, Au, Pt, Ni, Cu, Fe, and Zn, for the conversion of CO_2 (electrochemically) into more reactive molecules in aqueous electrolytes [4, 14]. However, the high cost (Au, Ag, and Pt), limited supply, and scarcity of these catalysts hamper the widespread use of electrochemical energy conversion (EEC) technology. Recently, significant effort has been made for developing and designing electrocatalysts by adopting a different strategy. Nanostructured materials tend to exhibit enhanced catalytic activity because of the presence of unique active sites on their surfaces, which are absent in bulk materials [6, 9]. It is well understood that reducing the particle size from bulk catalyst to nano-atom catalyst and further SACs as shown in Figure 16.4 (see more details in Section 19.1) could be the most efficient method for enhancing the electrocatalytic activity, total atom utilization with increase in surface free energy, and specific surface area [16]. Because, the atomic ratio at the corner, edge, and

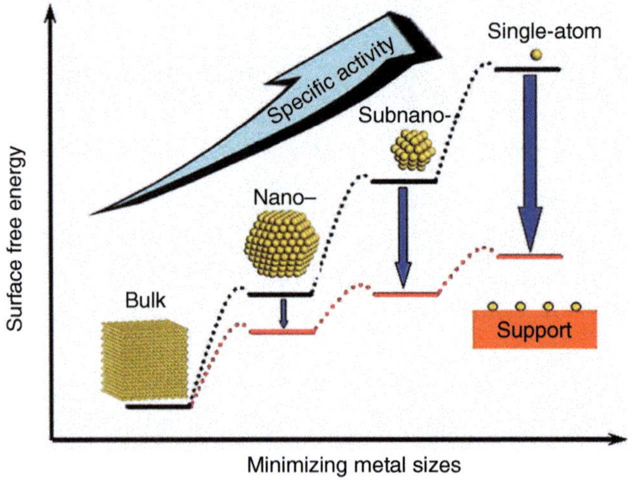

Figure 16.4 Schematic correlation between metal size and specific activity. [Note: Nano: nanoparticles and sub-nano: nanoclusters]. Source: Yang et al. [16]. © 2013, American Chemical Society.

surface positions are strongly affected by the catalyst particle size. To be specific, atoms at the corner or edge positions have high surface energy and low coordinated metal atoms often function as the catalytically active sites [17, 18]. Besides, decreasing the particle size also enhances the exposed surface atoms, modifies the surface electronic and atomic structure, and creates the defects [19]. However, reducing particle size, promoting aggregation of small clusters could address an appropriate support material (i.e. carbon paper, glassy carbon, metal oxide, metal foil, metal plate, and graphene electrode, etc.,) resulting in the prevention of this aggregation. Due to the strong interaction between supporting materials and the metal species, it further creates a stable electrocatalyst for ECDRR. More details about nanomaterial electrocatalyst for ECDRR will be discussed in Section 16.2.2.

16.2.2 Nanoscale Metal and Oxidant Metal Catalysts

In general, the catalyst should bind *COOH strongly enough and weakly enough *CO to facilitate the reduction of CO_2 where * indicates the surface-adsorbed species. The binding energies of *COOH (EB[*COOH]) and *CO (EB[*CO]) are typically proportionally correlated so-called scaling relations, making them challenging to control individually [20]. Therefore, optimum CO binding is required for that various transition metal nanomaterials (noble and non-noble), and their oxides derived (OD) showed high selectivity for conversion of CO_2-to-CO due to the stabilization of the first reduction intermediate (COOH*) [9, 21–23]. As we discussed in Section 16.2.1, the sizes of metal nanomaterial play a crucial role in the catalytic performances of ECDRR, but the optimal size varies with different metals [24]. Herein, we will discuss widely used electrocatalysts, including (i) the noble metal-based: Au, Ag, and Pd transition metal as the most effective catalysts for ECDRR toward CO formation, because of the strong binding between these metals and intermediates *COOH, and weak bind with *CO species, (ii) transition non-noble metals: Ni, Co, Zn, and Cu, which are low cost and highly abundant in the earth, have been widely explored [11], and (iii) nanomaterial along with their oxide could form different value-added products from the reduction of CO_2 molecule in aqueous media. Recent advances in the synthesis of nanoparticles (NPs), nanorods (NRs), nanoclusters (NCs), nanowires (NWs), and other nanostructure materials [25] have been widely studied. Because of the controlled surface area and surface morphology, the reaction kinetics potentially increased for ECDRR.

16.2.2.1 Gold (Au)

Gold (Au) is one of the few metals that can produce CO from the reduction of CO_2 with considerable selectivity, even in its bulk and polycrystal state [5, 26]. Several recent studies demonstrated that gold electrodes with modified surface morphologies have superior catalytic properties to traditional polycrystalline gold electrodes [21, 22, 26, 27]. Au NPs in the different nanometer size (5~200 nm) altered the electronic and geometric structural effect, resulting in the 10-fold increase in catalytic activity as the NPs size decreased to 5 nm [28]. Mistry and his co-workers studied the monodispersed Au NPs using the inverse micelle encapsulation method for CO formation with ~45% of Faraday efficiency (FE) at −1.2 V vs. RHE in 0.1 M $KHCO_3$ aq. solution [26]. Chen et al. [27] reported the Au NPs ("oxide derived Au," i.e. OD Au) synthesized by the reduction of thick Au oxide film exhibited high FE over 90% for CO formation at −0.35 V vs. RHE in 0.5 M $NaHCO_3$ and retained the activity for eight hours. In the case of traditional polycrystalline Au catalyst, the first step of CO_2 reduction is commonly a rate-determining e^- transfer to CO_2 resulting in an adsorbed $CO_2{\cdot}^-$ intermediate. Unfortunately, large overpotential is required due to poor stabilization of $CO_2{\cdot}^-$ intermediate on polycrystalline metal electrodes. In contrast, CO_2 reduction on OD Au support involved a reversible e^- transfer to CO_2 and formed adsorbed $CO_2{\cdot}^-$ intermediate, followed by a rate-determining H^+ transfer with HCO_3^- serving as the H^+ donor. The rate-determining $1e^-$ transfer suggested that active sites on the surfaces of OD Au NPs provided better stabilization for $CO_2{\cdot}^-$ than the sites on polycrystalline Au [27, 29] as illustrated in Figure 16.5a. The overall FEs for CO and HCO_2^- production on oxide-derived Au and polycrystalline Au electrodes at various potentials are shown in Figure 16.5b [27]. Moreover, an earlier study reported that the use of metal oxides in E CDRR is advocated considering the acidic (Lewis acid) nature of CO_2. Since metal oxides are basic in nature, their use as electrocatalysts can enhance the adsorption of CO_2, thereby promotes the CO_2 reduction process [30].

Kauffman et al. [22] studied experimental and computational investigation on the Au_{25} cluster, which promoted the CO_2-to-CO conversion FE ~ 100% at −1.0 V vs. RHE that was 7–700 times higher compared to larger Au catalysts. Zhu et al. [21] reported that the monodisperse polyhedral Au NPs (4, 6, 8, and 10 nm) were prepared via burst nucleation and investigated for ECDRR in 0.5 M $KHCO_3$. Among them, Au NPs (8 nm) showed the highest selectivity for CO with the FE reached 90% at −0.67 V vs. RHE, although the Au NPs (4 nm) displayed the best mass activity. The

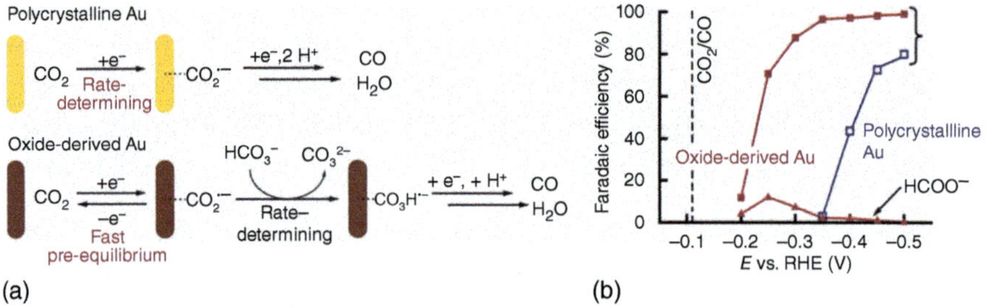

Figure 16.5 (a) Proposed mechanisms for CO_2 reduction to CO and (b) FEs for CO and HCO_2^- production on oxide-derived Au and polycrystalline Au electrodes at various potentials between −0.2 and −0.5 V in 0.5 M $NaHCO_3$, pH 7.2. The dashed line indicates the CO equilibrium potential. Source: Chen et al. [27]. © 2012, American Chemical Society.

author also studied density functional theory (DFT) calculations, which suggested that the presence of dominant edge sites (Au has a coordination number [CN] 7) has a reactive site for selective CO formation. However, the Au NPs with the size of 8 nm showed the highest selectivity for the CO product, which is due to the most optimized edge-to-corner ratio.

In 2014, Zhu and co-workers reported a new class of Au catalysts with ultra-thin (2-nm diameter) Au nanowires (NWs) morphology synthesized via facile seed-mediated growth method, which showed considerably enhanced CO_2-to-CO conversion with FE 94% at only −0.35 V vs. RHE in 0.5 M $KHCO_3$ [31]. According to their DFT calculations, the excellent catalytic performance of these Au NWs is attributed both to their high mass of reactive edge sites (≥16%) and to the weak CO binding on these sites [31]. Similarly, Liu et al. [32] successfully synthesized the Au needles, rods, and NPs via field-induced reagent concentration (FIRC) for ECDRR. Among them, Au needles exhibited excellent catalytic selectivity CO_2-to-CO formation around 95% FE at only −0.35 V vs. RHE, pH of 7.2 in 0.5 M $KHCO_3$, which obtained a high current density of ~15 mA cm^{-2} and retained the high activity for at least eight hours. The authors revealed that the superior performance toward CO formation in ECDRR is due to the locally concentrated cations (K^+). Because of adsorbed K^+, a greater electron density was found on the carbon of the COOH* intermediate. This suggested a stronger C—Au bond and further indicating that adsorbed cations could also modulate the CO_2 reduction reaction [32].

In 2019, Mezzavilla et al. [33] reported the structure sensitivity for the ECDRR with Au single crystals in which coordinated sites are responsible for catalytic activity and selectivity studied with Au(110), Au(211), and Au(100). Among this, Au(110) and Au(211) showed 90% CO FE at -0.6 V vs. RHE in 0.1 M $KHCO_3$ electrolyte and have at least 20-fold higher electrocatalytic activity than that of Au(100) for ECDRR.

16.2.2.2 Silver (Ag)

Silver (Ag) is an attractive alternative for the Au noble metal electrode, which has been considerable interest because of its lower cost compared with Au and the ability to selectively reduction of CO_2 into CO. However, most of the study revealed that the Ag-based catalyst requires higher overpotential ($\eta > 0.9$ V) compared to the Au-based catalyst for the selective conversion of CO_2 into value-added chemicals. Despite that, nanostructured Ag-based catalysts (Ag_{nano}) like Ag NP, Ag nanoplate, and Ag nanoporous (Ag_{npo}) catalysts were developed to enhance the ECDRR owing to the significantly increased active sites on Ag-based nanostructure catalyst [34]. Salehi-Khojin et al. [28] studied the effect of various sizes of 70-, 40-, 5-, and 1-nm Ag NPs on its catalytic performance. Their study revealed the conversion rate of CO_2-to-CO increases with decreasing particle size until a specific NPs particle size, i.e. 5 nm (in <5 nm NPs case, the intermediates bind too strongly) showed 10 times higher activity compared to bulk silver electrode in an ionic liquid (1-ethyl-3-methylimidazolium tetrafluoroborate EMIM-BF_4) electrolyte solution.

Mistry et al. [35] synthesized the oxidized and highly defective nanostructured Ag catalysts (i.e. by treating Ag foils in low-pressure plasmas of H_2, Ar, or O_2 gas) via plasma treatment for the selective CO_2-to-CO conversion with FE 90% at −0.6 V vs. RHE in 0.1 M $KHCO_3$ aqueous electrolyte. Liu et al. [36] successfully synthesized a new class of triangular silver nanoplates (Tri-Ag-NPs) electrocatalyst, which exhibited high CO selectivity FE 96.8% (1.5-fold higher value than bulk Ag) and energy efficiency (EE) 61.7%, at −0.74 V vs. RHE in 0.1 M $KHCO_3$ with high durability around seven days. DFT study shows that the high selectivity of Tri-Ag-NPs toward CO production at low overprotection attributed to the predominant Ag(100) facet and the optimum edge-to-corner ratio in Tri-Ag-NPs.

Ma et al. [37] synthesized oxide-derived nanostructured Ag catalysts (OD-Ag) could able to reduce CO_2-to-CO with FE 80% at −0.49 V vs. RHE in 0.1 M K_2HPO_4 and CO_2-saturated 0.1 M $KHCO_3$. Dutta et al. [38] have successfully synthesized Ag nano-foam catalysts and these Ag foams showed high activity and selectivity toward CO formation, with FE 99% at very low overpotential of −0.3 V vs. RHE, and also can produce hydrocarbons like FE CH_4 51% and FE C_2H_4 8.6% at −1.5 V vs. RHE in 0.5 M $KHCO_3$. Their study showed that the meso-porosity with pore side walls is a key structural feature of the novel Ag foam catalysts that are composed of highly anisotropic and needle-shaped Ag features having dimensions in the nm range. Recently, Huang and Wu [34] synthesized the contaminate-free nanostructured Ag electrocatalyst from electrochemical-treated Ag by two different potential pulses such as double potential pulse (DP) and repetitive DP (r-DP) in a NaBr aqueous solution. Among them, Ag_{DP} with a nanoporous structure exhibited a promising candidate for the electrochemical conversion of CO_2-to-CO with FE of 97.8% at a very low overpotential of −0.25 V vs. RHE in 0.1 M $NaHCO_3$ aqueous solution. It was demonstrated that $HCO_3^{-}{}_{ads}$ assists the CO_2 from the solution phase near the Ag surface through a possible interaction, e.g. H-bonding.

16.2.2.3 Palladium (Pd)

Palladium (Pd) tends to be a unique element capable of producing both CO and formate. In contrast, most of the other elements known to be active for the two-electron reduced products favor the formation of just one of the two.[6] Because during electrolytic conditions, Pd will convert into Pd hydrides (Pd–H) and then further promote formate production by the interaction of CO_2 and *H [6, 39].

The product selectivity of Pd-containing nanostructures for ECDRR and related reaction mechanisms strongly depends on the structure and composition, and the dynamic evolution of active phases induced by the applied potential and reaction intermediate of ECDRR [12]. The general study on Pd NPs showed that the NPs are generally more active for formate and CO formation as compared to bulk Pd substrates because it is more difficult to fully convert the bulk Pd to PdHx from its low surface/volume ratio [6, 39]. Therefore, the catalysts structural factors play an important role, which may vary depending on the desired product. Rahaman et al. [40] synthesized different-sized Pd NPs (3.8–10.7 nm) by a polyvinylpyrrolidone (PVP)-assisted reduction of Pd(II) precursors in a borohydride synthesis approach. Significantly particle size of 6.5 nm showed high conversion of CO_2 reduction into formate ($HCOO^-$) around 98% FE at −0.1 V vs. RHE in 0.5 M $NaHCO_3$, whereas FE values of 89.1% and 59.6% were obtained with the 3.8 and 10.7 nm Pd NP samples, respectively due to the Pd NPs involved in the formation of a β-PdH phase as the actual active species for ECDRR.

Similarly, Gao et al. [41] reported various types of Pd NPs, ranging from 2.4 to 10.3 nm for ECDRR. Among them, 3.7-nm size Pd NPs catalyst showed selective CO formation from CO_2 reduction with FE 91.2% at −0.89 V vs. RHE in 0.1 M $KHCO_3$ and a high current density compared to 10.3 nm size NPs. This result is due to the corner and edge sites of Pd NPs showed more activity than terrace sites toward CO_2 reduction, confirmed by DFT calculations. Recently, Zhu et al. [42] synthesized nanosized Pd cubes and octahedra particles dominated by Pd(100) and Pd(111) facets. Among them, the Pd octahedra particle showed higher CO FE 95% at −0.77 V vs. RHE, and higher activity than Pd cubes and commercial particles. According to DFT calculations, the high current density of 220 mA cm^{-2} and faradaic efficiency of CO are due to the lower binding energies of CO and HOCO intermediates on Pd–H(111).

16.2.2.4 Zinc (Zn)

In Sections 16.2.2.1–16.2.2.3, we have discussed about the noble metal catalyst (Au, Ag, and Pd) for the electrochemical reduction of CO_2 into selective product formation in ECDRR. Herein, we will discuss the non-noble-based metal such as zinc (Zn) not only because of showing selective CO formation but also considering its high abundance, low cost, and low scarcity compared to noble metal electrocatalyst. In the 1980s, Hori et al. [5] has reported bulk metallic Zn non-precious metal-promising properties as an active electrocatalyst for CO formation with FE of 40–80% at −1.14 V vs. RHE for ECDRR. In addition to bulk Zn metal, nanostructure Zn electrocatalyst has also been reported, which could enhance the electrocatalyst activity and selectivity than a bulk catalyst for ECDRR. However, in air, moisture, and through prolonged contact with aqueous electrolytes, Zn can be easily oxidized and may have a significant effect on catalytic activity. Jiao and coworkers reported a nanostructured Zn dendrite electrocatalyst, which was synthesized using an electrodeposition method to minimize the surface oxide layer formation and to create a highly active catalyst with a dendritic structure [10]. The result showed the selective CO formation with FE ~80% at −1.1 V vs. RHE in 0.5 M $NaHCO_3$, which was threefold higher than bulk Zn metal. Because at potentials less negative than −0.7 V, the Zn catalysts oxidized to Zn^{2+}, while at more negative potentials, they are structurally stable under working conditions. This electrodeposition approach is not only to overcome the sensitivity of the Zn electrode but also to enhance the CO selectivity and current density (~13 mA cm^{-2}) due to a very high surface area.

Recently, Luo et al. [43] successfully synthesized a low-cost porous Zn electrocatalyst using a facile electrodeposition method to enhance ECDRR toward selective CO formation. The porous Zn catalyst showed remarkable CO formation with FE of ~95% and current density of ~27 mA cm^{-2} at −0.95 V in the H-cell reactor in 0.1 M KOH. Besides, a novel strategy to transform the porous Zn electrode into a gas diffusion electrode (GDE) revealed CO FE ~84%, and higher current density of ~200 mA cm^{-2} at −0.67 V vs. RHE in the flow-type cells reactor showed the best performance to date over non-noble ECDRR catalysts. The dramatically improved catalytic activity and selectivity for CO_2-to-CO are primarily attributed to the highly porous structure of P–Zn, which increases the number of active sites and strengthens the local pH effect [43]. This study will guide a new direction for the future development of other low cost and active electrocatalysts, which could fulfill the industrial requirement for ECDRR applications.

16.2.2.5 Copper (Cu)

Copper (Cu) is another non-noble metal that has been widely used for ECDRR. In the 1980s, Hori et al. [5] found that the Cu metal has complicated nature of electrocatalyst with exclusive catalytic behavior due to the formation of various (>2e) value-added hydrocarbon products such as methane, methanol, ethane, ethylene, and others. Since then, researchers have been designed and synthesized various morphological Cu-based catalysts, i.e. NPs, nanowires (NWs), nanopores, nanocubes (NCs), nanosheets (NSs), overlayers, and nanofoams (NFs) for ECDRR. Reske et al. [44] revealed the very first study on the influence of catalytic particle size in ECDRR using Cu NPs with sizes ranging from 2 to15 nm and compared with bulk Cu catalyst, particularly on product selectivity, as illustrated in Figure 16.6c,d. In addition, an increasing number of surface atoms with CN below eight are responsible for enhanced hydrogen and CO evolution on NPs around 2 nm and below [44]. Strong CO binding on low coordinated sites (small NPs) favors high activity and high selectivities of CO and H_2, but the production of hydrocarbons such as CH_4 and C_2H_4 dropped drastically. In short, the catalytic activity was directly depending on the size of the Cu NPs, which can be increased by decreasing particle size for the CO and H_2 formation.

Similarly to the earlier discussion, Loiudice et al. [47] synthesized copper nanocrystal Cu–NCs with spherical size (7.5 and 27 nm) and cubic shapes (24, 44, and 63 nm) by the same colloidal chemistry-based method. Among them, the cubes with 44-nm edge length exhibited a total 80% selectivity toward ECDRR and 41% FE for ethylene at −1.1 V vs. RHE in 0.5 M $KHCO_3$. The group claimed that an optimal ratio of edge sites over (100) plane sites (N_{edge}/N_{100} = 0.025 for d = 44 nm) is a crucial role for selective formation of ethylene and ECDRR product. Furthermore, O_2 plasma treatment was used to form a nanostructured oxide layer

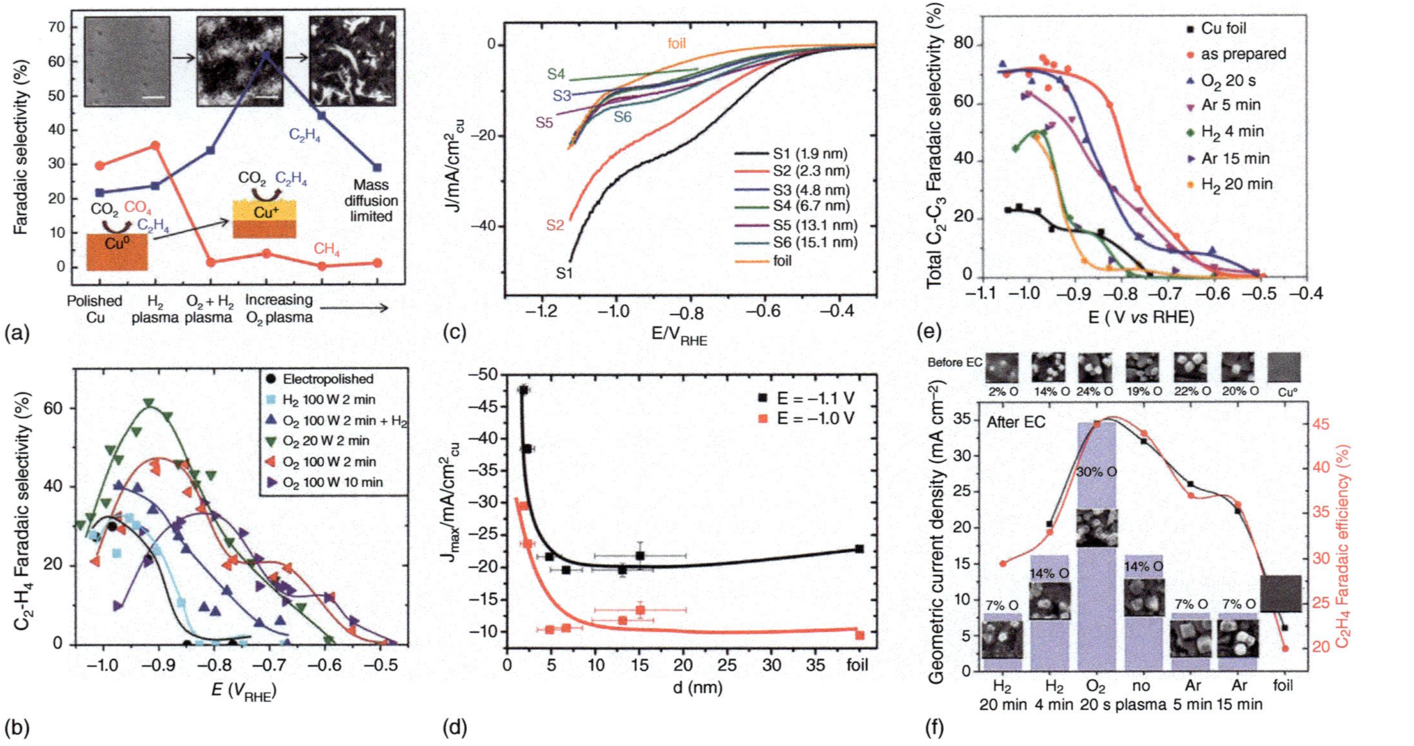

Figure 16.6 (a) Summary of hydrocarbon selectivity of plasma-treated Cu foils. From left to right, the insets show scanning electron microscope (SEM) images of the low surface area of H_2 plasma-treated metallic Cu foil, the O_2 of 20-W 2 minutes plasma-treated Cu foil with optimal ethylene selectivity, and the high surface area nanoneedles on the O_2 of 100-W 10 minutes oxidized sample after the reaction (500-nm scale bars), (b) Faradaic selectivity of the main ECDRR products. Source: (a, b) Mistry et al. [45]. Licensed under CC BY 4.0. Schemes (c and d) size-dependent ECDRR activities of Cu nanoparticles. (c) Linear sweep voltammetry of the ECDRR on Cu nanoparticle catalysts with different sizes in 0.1 M $KHCO_3$, at room temperature. (d) Particle size effect during electrocatalytic ECDRR. Source: (c, d) Reske et al. [44]. © 2014, American Chemical Society. (e) Total C_2-C_3 Faradaic efficiencies. (f) Geometrical current density (left axis) and Faradaic efficiency for C_2H_4 (right axis) after 1 h of electrochemical reaction at −1.0 V vs. RHE in CO_2-saturated 0.1 M $KHCO_3$ for different plasma treatments. Source: (e, f) Gao et al. [46]. © 2017, American Chemical Society.

on polycrystalline metal (Cu or other) that possesses a highly porous surface with controlled morphology and chemical states [48]. Mistry et al. [45] developed oxidized copper catalysts activated by facile plasma treatment, which showed the highest selectivity of ethylene formation >60% at −0.9 V vs. RHE in 0.1 M $KHCO_3$, as illustrated in Figure 16.6a,b. The surface roughness of oxide-derived copper catalysts plays a minor role in determining the catalytic performance, while the presence of copper (i.e. Cu(I)) was proposed to decrease the onset potential and to enhance the ethylene selectivity. To understand the effect of different structural (surface) parameters, such as surface roughness factor, crystal facets, and the presence of subsurface oxygen for selective ECDRR toward hydrocarbons and alcohol products, Cuenya research group conducted a well-rounded investigation using plasma-activated Cu nanocubes exposed to a Cu(100) facet [46]. The group demonstrated that the both parameters, surface roughness and specific crystal facet, played a significant role for determining the catalytic activity during the catalysis reaction, resulting in maximum ethylene FE of 45% at −1.0 V vs. RHE in 0.1 M $KHCO_3$ with current density of 35 mA cm^{-2} as illustrated in Figure 16.6e,f. Their study revealed that the presence of oxygen species in surface and subsurface regions of the nanocube catalysts is key for achieving high activity and hydrocarbon/alcohol selectivity, even more important than the presence of Cu(100) facets.

Furthermore, partially reduced copper oxide nano dendrite with rich surface oxygen vacancies (CuO_x–V_o) on carbon paper via a facile electrochemical deposition method has been developed [49]. This CuO_x–V_o act as excellent Lewis base sites for enhanced CO_2 adsorption and subsequent electrochemical reduction. This nano dendrite (CuO_x–V_o) catalyst exhibited the highest C_2H_4 production FE 63% higher than CuO (40%) and Cu (28%) at −1.4 V vs. RHE with current density of ~30 mA cm^{-2} in 0.1 M $KHCO_3$. According to their theoretical calculations, the high selectivity of C_2H_4 formation was due to the strong binding affinity to the intermediates of *CO and *COH and the weak affinity to *CH_2 on oxygen vacancy-rich CuO_x surfaces. The summary of ECDRR performances on different nanocatalysts is listed in Table 16.2.

16.2.3 Bimetallic/Alloy

In Section 16.2.2, we have discussed the different single metal (monometallic) electrocatalysts system, which have widely been studied as an electrocatalyst for ECDRR due to their high catalytic activity and selectivity enhanced by the structural modifications (i.e. morphology, size, material porosity and facet) [5, 15]. It is well understood that the product selectivity of ECDRR on metal-based catalysts is highly dependent on the binding energy of the key intermediates of ECDRR and competing HER, such as *H, *OCHO (O site binding on the surface), *COOH, *CO, *CHO, or *COH (C site binding on the surface) [50]. However, changing the composition of the single catalyst by combining it with a secondary metal might tune the selectivity and activity through optimization of the binding strength of the key intermediates

Table 16.2 Summary of ECDRR performances on different nanocatalysts in aqueous solution.

Electrocatalyst	Electrolyte	Potential (V vs. RHE)	J ($mA\ cm^{-2}$)	Device	FE (%)	Major product	Ref.
Au (NPs)	0.1 M $KHCO_3$	−1.2	NA	H-cell	~45	CO	[26]
Au (NPs)	0.5 M $KHCO_3$	-0.67	~6	H-cell	90	CO	[21]
Au_{25} (cluster)	0.1 M $KHCO_3$	-1.0	NA	H-cell	~100	CO	[22]
Au (single crystals)	0.1 M $KHCO_3$	-0.6	NA	H-cell	90	CO	[33]
Au (NWs)	0.5 M $KHCO_3$	−0.35	NA	H-cell	94	CO	[31]
Au (needles)	0.5 M $KHCO_3$	−0.35	~22	H-cell	95	CO	[32]
Au (OD)	0.5 M $NaHCO_3$	−0.35	~2-4	H-cell	>96	CO	[27]
Ag (OD)	0.1 M $KHCO_3$	−0.60	~2	H-cell	90	CO	[35]
Ag nanoplate	0.1 M $KHCO_3$	−0.85	~1.5	H-cell	~97	CO	[36]
Ag (OD)	0.1 M K_2HPO_4	−0.6	~0.4	H-cell	80	CO	[37]
Ag foams	0.5 M $KHCO_3$	−0.8	~12	H-cell	~98	CO	[38]
Ag nanopore	0.1 M $NaHCO_3$	−0.39	5	H-cell	97.8	CO	[34]
Pd (NPs) 3.7 nm	0.1 M $KHCO_3$	−0.89	~9	H-cell	91	CO	[41]
Pd (NPs) 6.5 nm	0.5 M $NaHCO_3$	−0.1	~2	H-cell	98	($CHOO^-$) Formate	[40]
Pd (NPs) octahedra PdH(111)	0.5 M $NaHCO_3$	−0.77 −3.4	~16 ~220	H-cell Flow-cell	95 >80	CO CO	[42]
Zn dendrite	0.5 M $NaHCO_3$	−1.1	~13	H-cell	80	CO	[10]
Zn porous	0.1 M $KHCO_3$ 1.0 M KOH	−0.95 −0.64	~27 ~200	H-cell Flow-cell	95 84	CO CO	[43]
Cu (NCs)	0.1 M $KHCO_3$	−1.0	~35	H-cell	45	C_2H_4	[46]
Cu (NPs)	0.1 M $KHCO_3$	−1.1	~48	H-cell	60	CH_4	[44]
Cu (NCs)	0.5 M $KHCO_3$	−1.1	~6	H-cell	41	C_2H_4	[47]
Cu (OD)	0.1 M $KHCO_3$	−0.9	~12	H-cell	60	C_2H_4	[45]
CuO_x–V_o (dendrite)	0.1 M $KHCO_3$	−1.4	~30	H-cell	63	C_2H_4	[49]

NA, not applicable/not reported.

on the catalytic surfaces. Because of this reason, the multimetallic (bimetallic/alloy) catalyst has attracted considerable attention in past decades [46, 51].

Herein, we will be discussing the bimetallic system, which could offer several distinct advantages over the single element counterparts for ECDRR [6]. Particularly, multielectron reactions involve a large number of intermediates and each intermediate required optimization of the adsorption energy, which is difficult to achieve on a surface composed of a single element [6]. By alloying, it is believed that we might be able to tune the binding strength of intermediates on a catalyst surface to enhance the reaction kinetics for CO_2 reduction [52]. In general, multimetallic catalysts have been divided into three main categories: (i) M_1 and M_2 noble metal, i.e. Pd–Au, (ii) M_1 = noble metal and M_2 = non-noble metal, i.e. Au–Cu, Ag–Cu, Ag–Zn, etc., and (iii) M_1 = noble metal and M_2 = Oxyphilic (S, Sn, or P-block elements), i.e. Ag–Sn. In the 1990s, Watanabe et al. [53] reported various Cu-based bimetallic ECDRR and the Cu alloys such as Cu–Sn, Cu–Cd, Cu–Pb, Cu–Ag, Cu–Zn, and Cu–Ni exhibited different catalytic activities and selectivity for CO, CH_3OH, and HCOOH production compared with their respective elemental metals. In this regard, multimetallic nanomaterials have been studied as CO_2 electrocatalysts with a focus on how the second element to the system affects their catalytic properties (activity and selectivity) [6, 54].

For example, Kim et al. [52] successively assembled uniform gold–copper (Au–Cu) bimetallic NPs with different stoichiometric ratios in the order (o) of Au-to-Cu (Au, Au_3Cu, Au–Cu, $AuCu_3$, and Cu monolayers were synthesized with average size all close to 10–11 nm). Their study observed that the CO FE increases and H_2 FE decreases with increasing Au content; for example, Au_3Cu has the highest catalytic activity and selectivity toward CO formation. Moreover, the high CO activity of Au–Cu bimetallic NPs observed due to the electronic effect. To be specific, the binding of intermediates can be tuned using different surface compositions and the geometric effect that has to do with the local atomic arrangement at the active site, which allows the catalyst to deviate from the scaling relation [52]. Kortlever et al. [55] reported the carbon-supported palladium–platinum ($Pd_xPt(100-x)/C$) nanomaterial for formic acid production from the CO_2 reduction, and their activity was compared with the reverse oxidation of formic acid to CO_2 formation. The $Pd_{70}Pt_{30}/C$ catalyst (4.3 nm) showed a formation of formic acid with FE of 88% at −0.4 V vs. RHE. This result revealed that the martial is capable of avoiding a reliable deactivation that was observed with Pd-based catalysts [55]. Moreover, the author hypothesized that an appropriate mixing of Pt and Pd could tune the d-band center, which may play a crucial role in the enhancement of catalytic activity.

Kim et al. [56] studied systematic atomic ordering (o) transformations of Au–Cu NPs and activates them to perform as selective electrocatalysts for ECDRR. Their study showed that disordered (d) alloy NP, which is catalytically active for hydrogen evolution and ordered AuCu NPs selectively converted CO_2-to-CO at FE, reached 80% at −0.77 V vs. RHE with 0.1 M $KHCO_3$. Some studies reported that the Au–Cu (1 : 1) alloy composition for ECDRR resulting in H_2 as a major product [52]. Besides, without altering the composition, the ordering Au–Cu transformation could enhance CO formation, with the mass activity of ~830 A/g_{Au} (2.6-fold higher than Au NPs), transitioning a CO_2 inactive catalyst to an active catalyst with high CO selectivity. Atomic structural investigation of o-AuCu NPs found the formation of a three-atom-thick gold overlayer following ordering transformation. This result suggested that surface enrichment of gold at atomic levels may be the origin of enhanced catalytic activity, which is further supported by DFT analysis [56]. Similarly, Ma et al. [57] synthesized and characterized the different elemental mixing patterns and compositions of the bimetallic Cu–Pd catalyst. A bimetallic Cu–Pd catalyst with ordered, disordered, and phase-separated atomic arrangements (Cu_{at}: Pd_{at} = 1 : 1), as well as two additional disordered arrangements (Cu_3Pd and $CuPd_3$ with Cu_{at} : Pd_{at} = 3 : 1 and 1 : 3), are studied to determine key factors needed to achieve high selectivity for C_1 or C_2 chemicals in CO_2 reduction. Their systematic study demonstrated that the ordered CuPd catalyst mostly provides C_1 products with maximum selectivity (>80%), meanwhile, compared to $CuPd_3$ and ordered CuPd, the phase-separated CuPd and Cu_3Pd achieved higher selectivity (60%) for C_2 production. This study suggested that the probability of dimerization of C_1 intermediates is higher on surfaces with neighboring Cu atoms. Based on surface valence band spectra, geometric effects rather than electronic effects seem to be key in determining the selectivity of bimetallic Cu–Pd catalysts [57].

Zhu et al. [58] studied the structural and compositional effects in Pd–Au bimetallic nanowires for high selectivity and activity toward CO production in ECDRR. These synthesized nanowires showed an excellent CO selectivity. Among them, $Pd_{0.8}Au$ nanowire achieved high selectivity for CO formation with maximum FE 94.3% at −0.6 V vs. RHE along with superior mass activity at low overpotential. This result depends on the Pd : Au ratio, to be specific, the incorporation of Au into Pd significantly increases the percentage of weakly linear bonded CO (CO_L) that is easier to desorb from the Pd shell resulting in an improved reaction rate. DFT calculations revealed that surface CO could be more facilely generated on nanowires as compared with that on particles, which is caused by the accelerated CO_2 to *COOH conversion [58]. Recently, Yuan et al. [59] synthesized ultra-thin Pd–Au alloy nanoshells with a controllable alloying degree on Pd nanocubes (NCs) via a seed-mediated growth method, as illustrated in Figure 16.7a,b. $Pd@Pd_3Au_7$ (NCs) catalyst showed superior ECDRR performance, with a 94% FE CO at −0.5 V and 100% FE CO at −0.6 to −0.9 V vs. RHE. The enhancement primarily originates from the ensemble and ligand effects, i.e. appropriately proportional Pd-Au sites, and electronic back-donation from Au-to-Pd [59]. Buonsanti and coworkers developed a novel seed-growth colloidal approach to tailor-make Ag–Cu nano diamond (NDs) with different Cu domain sizes (Ag_1–$Cu_{0.4}$, Ag_1–$Cu_{1.1}$, and Ag_1–$Cu_{3.2}$) and achieved synthetic control over the respective counterparts, Cu NPs and Ag NPs [60]. The Ag_1–$Cu_{1.1}$ NDs showed FE ~40% C_2H_4 at −1.1 V vs. RHE, which is a 3.4-fold enhancement and a twofold improvement in the overall ECDRR activity compared with the Cu NPs of similar size and shape as shown in Figure 16.7d. Their conclusion suggested that the selectivity of bimetallic catalysts in ECDRR is sensitive to the length scale at which the metals mix. Sequential catalysis and electronic effects were identified as predominant mechanisms (CO spillover), as shown in Figure 16.7c behind the increased selectivity toward multicarbon products in the Ag–Cu NDs [60]. In addition to various synthesis approaches, in the last few decades, the cost-effective electrodeposition strategy has also been extensively studied in bimetallic electrocatalyst for ECDRR. Due to this reason, Gewirth and coworkers have been successfully synthesized a nanoporous and

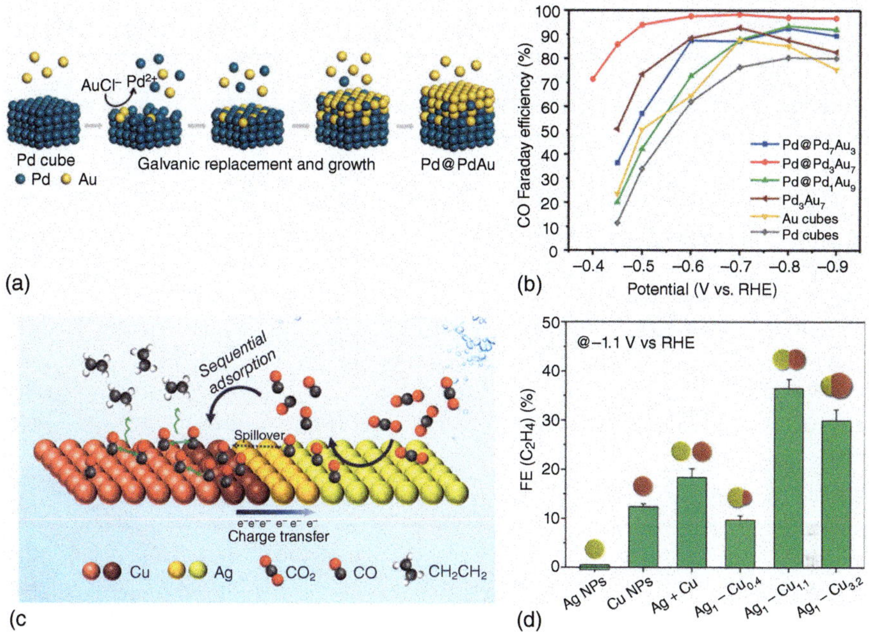

Figure 16.7 (a) Schematic and (b) CO FE. Source: (a, b) Yuan et al. [59]. © 2019, American Chemical Society. (c) Schematic and (d) C_2H_4 FE. Source: Huang et al. [60]. © 2019, American Chemical Society.

very low Ag contents of ~6% copper–silver alloy (Cu–Ag) electrocatalyst by additive-controlled electrodeposition method for ECDRR [61]. The Cu–Ag-wire electrocatalyst exhibited the highest FE C_2H_4 ~60% and C_2H_5OH ~25% at −0.7 V vs. RHE and a high current density of −300 mA cm^{-2} compared to Cu–Ag-poy and Cu-wire in an alkaline 1 M KOH flow electrolyzer for ECDRR [61]. Their study showed that the presence of Ag could help to the formation of Cu_2O on the Cu surface because Ag has high redox potential than Cu and can accept electron from Cu, resulting slight positive charge on Cu in Cu–Ag sample. To be specific, Ag plays a crucial role as a promoter for Cu_2O formation in Cu–Ag electrocatalyst which can lead to CO and C_2H_4 production, and also appropriate loading of Ag could enhance the CO production so that neighboring Cu have to participate in the C–C coupling.

Recently, Lamaison et al. [62] reported the rational design of novel and highly active Ag-alloyed Zn dendritic electrodes with remarkable CO_2-to-CO selectivity as high as 91% at −1.0 V vs. RHE over 40 hours in 0.1 M $CsHCO_3$ solution. Authors studied the effect of pressure on CO evolution current densities; their study revealed that the higher CO_2 pressure (up to 9.5 bar) resulting in an extensively enhanced CO_2 concentration allowed to achieve new record CO current densities of −286 mA cm^{-2} in neutral pH. The ECDRR performances on different bimetallic/alloy electrocatalysts are listed in Table 16.3

Table 16.3 Summary of ECDRR performances on different bimetallic/alloy catalyst in aqueous solution.

Electrocatalyst	Electrolyte	Potential (V vs. RHE)	J (mA cm^{-2})	Device	FE (%)	Major product	Ref.
Au_3Cu (NPs)	0.1 M $KHCO_3$	−0.73	~3	H-cell	>60	CO	[52]
$Pd_{70}Pt_{30}/C$	0.1 M K_2HPO_4	−0.4	~5	H-cell	88	HCOOH	[55]
AuCu (NPs) (Ordered)	0.1 M $KHCO_3$	−0.77	~1.4	H-cell	80	CO	[56]
CuPd (Ordered) (Phase separation)	1.0 M KOH 1.0 M KOH	−0.55 −0.55	~61 134	Flow-cell	~80 ~48	CO C_2H_4	[57]
$Pd_{0.8}$–Au (NWs)	0.5 M $KHCO_3$	−0.6	~2	H-cell	94	CO	[58]
Ag_1–$Cu_{1.1}$ (NDs)	0.1 M $KHCO_3$	−1.1	~2	H-cell	40	C_2H_4	[60]
CuAg alloy (Nanoporous)	1.0 M KOH	−0.7	~300	Flow-cell	60 25	C_2H_4 EtOH	[61]
Ag–Zn (Alloy)	0.1 M $CsHCO_3$	−1.0	~286 (9.5 bar)	H-cell	>90	CO	[62]

References

1 Sun, Z.Y., Ma, T., Tao, H.C. et al. (2017). *Chem-Us* 3: 560–587.

2 Benson, E.E., Kubiak, C.P., Sathrum, A.J., and Smieja, J.M. (2009). *Chem. Soc. Rev.* 38: 89–99.

3 Kuhl, K.P., Cave, E.R., Abram, D.N., and Jaramillo, T.F. (2012). *Energy Environ. Sci.* 5: 7050–7059.

4 Kuhl, K.P., Hatsukade, T., Cave, E.R. et al. (2014). *J. Am. Chem. Soc.* 136: 14107–14113.

5 Hori, Y., Kikuchi, K., and Suzuki, S. (1985). *Chem. Lett.* 14: 1695–1698.

6 Xie, C., Niu, Z., Kim, D. et al. (2020). *Chem. Rev.* 120: 1184–1249.

7 Feng, D.M., Zhu, Y.P., Chen, P., and Ma, T.Y. (2017). *Catalysts* 7: 373.

8 Whipple, D.T. and Kenis, P.J.A. (2010). *J. Phys. Chem. Lett.* 1: 3451–3458.

9 Back, S., Yeom, M.S., and Jung, Y. (2015). *ACS Catal.* 5: 5089–5096.

10 Rosen, J., Hutchings, G.S., Lu, Q. et al. (2015). *ACS Catal.* 5: 4586–4591.

11 Lü, F., Bao, H., Mi, Y. et al. (2020). *Sustain. Energy Fuels* 4: 1012–1028.

12 Gao, D.F., Zhou, H., Cai, F. et al. (2018). *ACS Catal.* 8: 1510–1519.

13 Mohamed, A.G.A., Huang, Y.Y., Xie, J.F. et al. (2020). *Nano Today* 33: 100891.

14 Hori, Y., Vayenas, C.G., White, R.E., and Gamboa-Aldeco, M.E. (2008). *Modern Aspects of Electrochemistry*, vol. 42, 89–189. New York: Springer.

15 Nitopi, S., Bertheussen, E., Scott, S.B. et al. (2019). *Chem. Rev.* 119: 7610–7672.

16 Yang, X.F., Wang, A., Qiao, B. et al. (2013). *Acc. Chem. Res.* 46: 1740–1748.
17 Sundararaman, R., Figueiredo, M.C., Koper, M.T.M., and Schwarz, K.A. (2017). *J. Phys. Chem. Lett.* 8: 5344–5348.
18 Varela, A.S., Kroschel, M., Leonard, N.D. et al. (2018). *ACS Energy Lett.* 3: 812–817.
19 Mohanty, B., Jena, B.K., and Basu, S. (2020). *ACS Omega* 5: 1287–1295.
20 Peterson, A.A. and Norskov, J.K. (2012). *J. Phys. Chem. Lett.* 3: 251–258.
21 Zhu, W., Michalsky, R., Metin, O. et al. (2013). *J. Am. Chem. Soc.* 135: 16833–16836.
22 Kauffman, D.R., Alfonso, D., Matranga, C. et al. (2012). *J. Am. Chem. Soc.* 134: 10237–10243.
23 Yin, Z., Palmore, G.T.R., and Sun, S. (2019). *Trends Chem.* 1: 739–750.
24 Yu, F., Wei, P., Yang, Y. et al. (2019). *Nano Mater. Sci.* 1: 60–69.
25 Birdja, Y.Y., Perez-Gallent, E., Figueiredo, M.C. et al. (2019). *Nat. Energy* 4: 732–745.
26 Mistry, H., Reske, R., Zeng, Z. et al. (2014). *J. Am. Chem. Soc.* 136: 16473–16476.
27 Chen, Y.H., Li, C.W., and Kanan, M.W. (2012). *J. Am. Chem. Soc.* 134: 19969–19972.
28 Salehi-Khojin, A., Jhong, H.R.M., Rosen, B.A. et al. (2013). *J. Phys. Chem. C* 117: 1627–1632.
29 Zhang, L., Zhao, Z.J., and Gong, J. (2017). *Angew. Chem. Int. Ed. Engl.* 56: 11326–11353.
30 Li, C.W. and Kanan, M.W. (2012). *J. Am. Chem. Soc.* 134: 7231–7234.
31 Zhu, W., Zhang, Y.J., Zhang, H. et al. (2014). *J. Am. Chem. Soc.* 136: 16132–16135.
32 Liu, G.Y., Pang, Y., Zhang, B. et al. (2016). *Nature* 537: 382–386. https://doi.org/10.1038/nature19060.
33 Mezzavilla, S., Horch, S., Stephens, I.E.L. et al. (2019). *Angew. Chem. Int. Ed. Engl.* 58: 3774–3778.
34 Huang, J.-F. and Wu, Y.-C. (2019). *ACS Sustain. Chem. Eng.* 7: 6352–6359.
35 Mistry, H., Choi, Y.W., Bagger, A. et al. (2017). *Angew. Chem. Int. Ed. Engl.* 56: 11394–11398.
36 Liu, S., Tao, H., Zeng, L. et al. (2017). *J. Am. Chem. Soc.* 139: 2160–2163.
37 Ma, M., Trzesniewski, B.J., Xie, J., and Smith, W.A. (2016). *Angew. Chem. Int. Ed. Engl.* 55: 9748–9752.
38 Dutta, A., Morstein, C.E., Rahaman, M. et al. (2018). *ACS Catal.* 8: 8357–8368.
39 Min, X. and Kanan, M.W. (2015). *J. Am. Chem. Soc.* 137: 4701–4708.
40 Rahaman, M., Dutta, A., and Broekmann, P. (2017). *ChemSusChem* 10: 1733–1741.
41 Gao, D., Zhou, H., Wang, J. et al. (2015). *J. Am. Chem. Soc.* 137: 4288–4291.
42 Zhu, W.L., Kattel, S., Jiao, F., and Chen, J.G.G. (2019). *Adv. Energy Mater.* 9: 1802840.
43 Luo, W., Zhang, J., Li, M., and Zuttel, A. (2019). *ACS Catal.* 9: 3783–3791.
44 Reske, R., Mistry, H., Behafarid, F. et al. (2014). *J. Am. Chem. Soc.* 136: 6978–6986.

45 Mistry, H., Varela, A.S., Bonifacio, C.S. et al. (2016). *Nat. Commun.* 7: 12123.
46 Gao, D., Zegkinoglou, I., Divins, N.J. et al. (2017). *ACS Nano* 11: 4825–4831.
47 Loiudice, A., Lobaccaro, P., Kamali, E.A. et al. (2016). *Angew. Chem. Int. Ed. Engl.* 55: 5789–5792.
48 Tomboc, G.M., Choi, S., Kwon, T. et al. (2020). *Adv. Mater.* 32: e1908398.
49 Wu, F.H., Zhang, Z.L., Zhang, F.S. et al. (2018). *Electrochim. Acta* 287: 115–123.
50 Xie, H., Wang, T.Y., Liang, J.S. et al. (2018). *Nano Today* 21: 41–54.
51 Torbensen, K., Han, C., Boudy, B. et al. (2020). *Chem* 26: 3034–3038.
52 Kim, D., Resasco, J., Yu, Y. et al. (2014). *Nat. Commun.* 5: 4948.
53 Watanabe, M., Shibata, M., Kato, A. et al. (1991). *J. Electrochem. Soc.* 138: 3382–3389.
54 Zhao, J., Xue, S., Barber, J. et al. (2020). *J. Mater. Chem. A* 8: 4700–4734.
55 Kortlever, R., Peters, I., Koper, S., and Koper, M.T.M. (2015). *ACS Catal.* 5: 3916–3923.
56 Kim, D., Xie, C., Becknell, N. et al. (2017). *J. Am. Chem. Soc.* 139: 8329–8336.
57 Ma, S., Sadakiyo, M., Heima, M. et al. (2017). *J. Am. Chem. Soc.* 139: 47–50.
58 Zhu, S.Q., Wang, Q., Qin, X.P. et al. (2018). *Adv. Energy Mater.* 8: 1802238.
59 Yuan, X., Zhang, L., Li, L. et al. (2019). *J. Am. Chem. Soc.* 141: 4791–4794.
60 Huang, J., Mensi, M., Oveisi, E. et al. (2019). *J. Am. Chem. Soc.* 141: 2490–2499.
61 Hoang, T.T.H., Verma, S., Ma, S. et al. (2018). *J. Am. Chem. Soc.* 140: 5791–5797.
62 Lamaison, S., Wakerley, D., Blanchard, J. et al. (2020). *Joule* 4: 395–406.

17

Electrocatalysts-2

17.1 Single-Atom Metal-Doped Carbon Catalysts (SACs)

Early group transition metals, such as Fe, Co, and Ni, are reported as promising electrocatalyst for H_2 evolution and rarely selected as promising electrocatalyst for CO_2-to-CO/higher hydrocarbon production due to the poor resistance to CO-poisoning (strong surface bonding) [1]. Due to this reason, recently, single-atom catalysts (SACs) have been investigated for heterogeneous electrocatalyst as they surpass the conventional catalysts in terms of having a high specific activity with a significantly reduced amount of metal usage [2]. SACs with atomically distributed active metal centers have demonstrated maximum atom efficiency and excellent selectivity in several catalytic reactions because of the unique electronic properties of a single active site [3–5]. Because of the strong bonding between the single-atom metal and subtract, the unsaturated coordination environment of the single-atom metal provides a fascinating active center. Herein, in this section, we will discuss the most promising single metal electrocatalyst, including Ni, Fe, Co, Zn, Cu, and other from lab scale to industrial application for electrochemical carbon dioxide reduction reaction (ECDRR).

17.1.1 Nickel (Ni)-SACs

Li et al. [6] successfully constructed exclusive Ni–N_4–C sites on carbon layer coating through a topo-chemical transformation strategy resulting in high activity and selectivity for CO formation from CO_2 reduction. The Ni–N_4–C structure exhibits high selectivity, with maximum FE over 99% at −0.81 V vs. RHE with an excellent current density of 28.6 mA cm^{-2} for 30 hours in 0.5 M $KHCO_3$ aqueous solution. The topo-chemical transformation strategy provided a feasible way to construct abundant and exclusive active sites, avoiding aggregation of metal atoms and the resulting loss of active sites [6]. Jiang et al. [1] reported various transition single-atom metals (Co and Ni) dispersed into graphene nanosheets (M-NG) for CO formation in ECDRR. Among them, Ni–NG nanosheet on carbon fiber paper (CFP) exhibited high catalytic selectivity and activity with maximum CO FE 95% at −0.73 V vs. RHE and ~11 mA cm^{-2} for 20 hours stability in 0.5 M $KHCO_3$ solution. The authors also studied that the CO evolution current on the Ni–NG

Electrocatalysis in Balancing the Natural Carbon Cycle, First Edition. Yaobing Wang.
© 2021 WILEY-VCH GmbH. Published 2021 by WILEY-VCH GmbH.

catalyst was limited in the H-type cell system due to the direct contact between the catalyst and water, which can overcome by introducing an anion membrane electrode assembly (MEA) to prevent the catalyst from contacting with water and also facilitate the CO_2 mass transport, greatly suppressing the competing hydrogen evolution reaction (HER) even under large overpotential. Yang et al. [3] developed atomically dispersed low-valent nickel Ni(I) anchored on nitrogen-doped graphene with the sulfur precursor (L-cysteine) Ni–NSG as an efficient electrocatalyst for ECDRR. Single-Ni-atom catalyst (Ni–NSG) exhibited high selectivity toward CO production with maximum FE 97% at −0.5 V vs. RHE in 0.5 M $KHCO_3$ solution. Moreover, the Ni–NSG catalyst exhibited high intrinsic activity with high current density of ~22 mA cm^{-2} and high turn over frequencies (TOFs) of 14 800 h^{-1} at overpotential of 0.6 V, maintaining 98% of its initial activity for CO formation after 100 hours. Furthermore, the authors studied the operando X-ray absorption and photoelectron spectroscopy measurements (out of the scope of this chapter), and the monovalent Ni(I) atomic center with a d^9 electronic configuration was identified as the catalytically active site [3].

Möller et al. [7] disclosed a nitrogen-doped nickel porous carbon catalyst (Ni–N–C), which revealed the conversion of CO_2-to-CO with high catalytic activity and selectivity of the state-of-the-art precious metal group catalysts, such as silver and gold. Ni–N–C electrodes provide maximum activity, the current density of 200 mA cm^{-2} with stable CO FE around 85% at −1.0 V vs. RHE for 20 hours, unlike their Ag benchmarks. Moreover, along with the high performance, the Ni–N–C electrocatalysts also offered the flexibility of their pH operation range, under either highly acidic or alkaline conditions. Some studies reported that the Ni-SACs could fulfill the future industrial perspective, and it may be an efficient electrocatalyst for large-scale applications [8]. Yang et al. [9] constructed a high-yield, flexible, and self-supported single-atom nickel single-atom decorated porous carbon fiber membrane catalyst (NiSA/PCFM) via electrospinning method for high efficiency in CO_2 electroreduction. The porous and interconnected carbon nanofibers (CNFs) of NiSA/PCFM provide substantial channels for CO_2 diffusion and electron transport. The authors studied the NiSA/PCFM electrocatalyst screening tests in typical H-type cells and flow-type cells equipped with a gas diffusion electrode (GDE) device, in which the H-type cell achieved maximum CO FE 96% at −0.7 V vs. RHE with partial current density of ~45 mA cm^{-2} and 15 hours stability. However, in the case of flow-type cells equipped with GDE, it not only achieved maximum CO FE 88% at −1.0 V vs. RHE but also showed high activity like a commercially relevant partial current density of 308.4 mA cm^{-2} along with 120 hours stability. Rong et al. [10] developed a facile method for control synthesis of vacancy-defect $Ni–N_3–V$ SAC on conducting substrate by using the precursors containing both oxygen and nitrogen atoms, in which the oxygen atom is initially formed $Ni–N_3O$ at 500 °C and further the oxygen can be removed under high temperature at 800 °C due to the weaker Ni–O interaction, resulting in a vacancy-defect $Ni–N_3–V$. This defective electrocatalyst showed the high selectivity with FE 90% at −0.9 V vs. RHE for 14 hours stability and achieved high current density of 65 mA cm^{-2} as well as the high TOF of 1.35×10^5 h^{-1}. The experimental and theoretical calculations revealed that the high selectivity of CO formation is because of the presence of a vacancy-defect Ni–N_3 site which can lower the energy barriers for the formation of COOH* intermediate and easy desorption of CO in ECDRR.

17.1.2 Cobalt (Co)-SACs

Wang et al. [11] prepared a series of atomically dispersed Co catalysts with different nitrogen coordination numbers (CoN_x) Co–N_4, Co–N_3, and Co–N_2 at 800, 900, and 1000 °C, respectively, for ECDRR, in which atomically dispersed Co with two-coordinate nitrogen atoms (Co–N_2) showed remarkable high catalytic selectivity toward 94% CO production at −0.68 V vs. RHE with a current density of 18.1 mA cm^{-2} and the high TOF of 23 500 h^{-1}. Both experimental and theoretical studies revealed that the lower coordination number (CN) is favorable for the activation of CO_2, which could further enhance CO_2 to the $CO_2{\cdot}^-$ intermediate and electroreduction activity. Pan et al. [12] developed an N-coordination strategy to design a robust ECDRR electrocatalyst with atomically dispersed Co–N_5 site anchored on polymer-derived hollow N-doped porous carbon spheres (Co–N_5/HNPCSs). This Co–N_5/HNPCSs catalyst reached the maximum FE CO 99% and 99.4% at −0.73 and −0.79 V vs. RHE, respectively, in 0.2 M $NaHCO_3$, which is significantly higher than CoPc at the same potential. Both the experiments and density functional theory (DFT) calculations revealed that the single-atom Co–N_5 site is the dominating active center simultaneously for CO_2 activation, for the rapid formation of key intermediate COOH* as well as for the desorption of CO. This work provided an in-depth understanding of the active catalytic centers for ECDRR.

17.1.3 Iron (Fe)-SACs

Huan et al. [13] investigated a series of iron-based catalysts synthesized by the pyrolysis of Fe-, N-, and C-containing precursors for the ECDRR to CO production in aqueous conditions. This study demonstrated that the selectivity of these materials for CO_2 reduction over proton reduction is governed by the ratio of isolated Fe–N_4 sites vs. Fe-based nanoparticles. The authors identified that isolated Fe–N_4 sites play a crucial role for the selective electrochemical reduction of CO_2-to-CO. From this study, the $Fe_{0.5}$d material proved the most effective and selective catalyst, producing CO with high FE up to 90% at −0.6 V vs. RHE in 0.1 M $NaHCO_3$ and sustaining long-term electrolysis. Besides, the selectivity of the reaction for CO formation could be further improved by using the low concentrated electrolyte. Similarly, Zhang et al. [14] synthesized the isolated Fe–N_4 structures by Fe SACs atom dispersed on nitrogen-doped graphene (Fe/NG) as an efficient electrocatalyst for CO_2 reduction to CO. Fe/NG-800 and Fe/NG-750 (calcining temperature 800 and 750 °C) electrocatalyst had maximum FE 12% and 80% toward CO at −0.60 V vs. RHE in 0.1 M $KHCO_3$ aqueous solution, respectively. This is due to the onset potentials suggested that at 750 °C more active sites are formed. Therefore, this result revealed that the annealing temperature has significant effect on the catalytic activity and should correlate with structural changes of the Fe-active site. Moreover, the DFT study showed the ECDRR mechanism on atomic Fe surrounded by four N atoms (Fe–N_4) embedded in nitrogen-doped graphene, revealing a possible promotional effect of nitrogen doping on graphene. Initially, the formation of *COOH intermediate was observed via adsorption of a CO_2 molecule on the Fe catalytic through protonation and electron transfer, then resulting Fe–N_4–COOH intermediate further reduced through proton–electron transfer which could form a CO* adduct and release one H_2O molecule. Finally, dissociation of weakly bonded CO* adduct will generate the main product CO. To be specific, introducing the trace amount of Fe atoms in nitrogen-doped graphene matrices could form the Fe–N_4 sites which could enhance CO_2 adsorption and improve CO_2 activation.

Similarly, Pan et al. [15] reported the atomically dispersed M–N_4 (M = Fe and Co) sites for the ECDRR. The (M–N–C) bulk- and edge-hosted M–N_4 coordination were prepared by nitrogen-coordinated Fe or Co site atomically dispersed into carbons precursor (i.e. metal–organic frameworks (MOFs)). However, Fe–N_4 shows intrinsically more activity than Co–N_4 for the reduction of CO_2 to CO, concerning a more significant current density and a higher CO FE 93% vs. 45% at −0.58 V vs. RHE for 20 hours. Moreover, both experimental and theoretical studies elucidated that the edge-hosted M–N_{2+2}–C_8 moieties bridging two adjacent armchair-like graphitic layers are the active sites for ECDRR compared to previously proposed bulk-hosted M–N_4–C_{10} moieties embedded compactly in a graphitic layer. During the ECDRR, when the dissociation of *COOH occurs on the M–N_{2+2}–C_8, the metal atom is the site for the adsorption of *CO and the carbon atom with a dangling bond next to an adjacent N is the other active center for the bond *OH shown in Figure 17.1. More specifically, the M centers and C atoms with dangling bonds next to N are the active sites to adsorb *CO and *OH in the cleavage of C—O bond during the ECDRR, respectively [15]. This study demonstrated that the new direction would provide guidelines on the design of advanced M–N–C cost-effective, highly efficient electrocatalyst for ECDRR into valuable fuel/chemical processes. Moreover, similar types of work have also been extended to understand the nature of real active sites in single-iron-atom (Fe–N–C) electrocatalysts that were studied by the Shao group using in situ infrared absorption spectroscopy (attenuated total reflectance-infrared [ATR-IR]) and DFT calculations [16]. Their study observed that the Fe centers of bulk- and edge-Fe–N_4 moieties in the defect-free graphitic layer were poisoned by strongly adsorbed *CO and should not be the real active sites for CO_2-to-CO conversion [16]. The high performance of Fe–N–C electrocatalyst may be originated from both Fe–N_4 moiety, and the defective graphitic layer with nanopores via synergistic interaction provides more balanced *COOH and *CO binding strength which could responsible for lower overpotential and mitigate the *CO poisoning as shown in Figure 17.1e. Furthermore, Zhang et al. [17] synthesized the robust and efficient Fe single-iron-atom electrocatalyst (Fe–N_5) by a novel synthetic approach involving prolonged thermal pyrolysis of hemin and melamine molecules on graphene (H–M–G) for ECDRR shown in Figure 17.1f. This catalyst exhibited high performance toward CO production with FE 97% at −0.46 V vs. RHE in 0.1 M $KHCO_3$ solution for 24 hours. This result is attributed to the presence of highly efficient, singly dispersed Fe–N_5 active sites supported on N-doped graphene with an additional axial

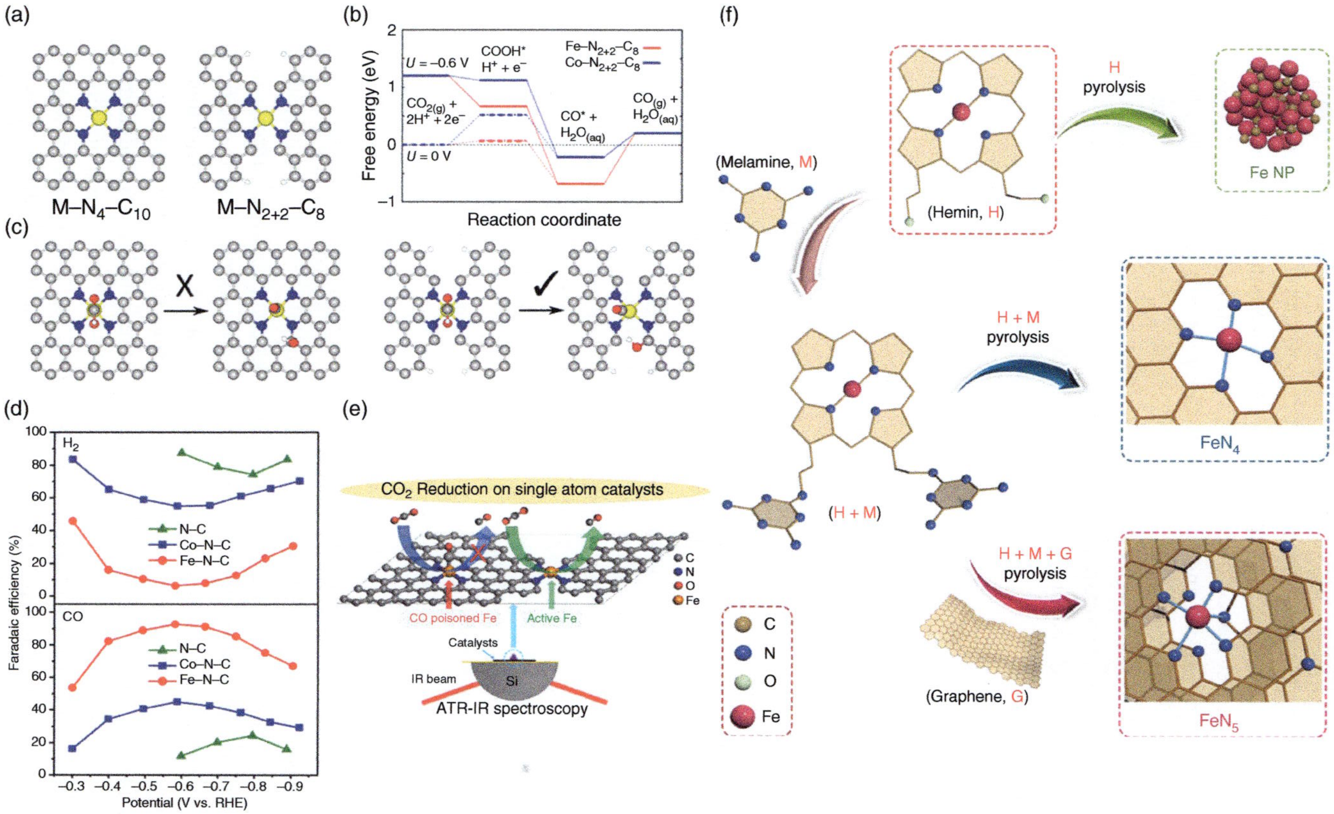

Figure 17.1 (a) Atomic structure of M−N$_4$−C$_{10}$ and M−N$_{2+2}$−C$_8$ (M = Fe or Co) active sites. (b) Calculated free energy evolution of CO_2 reduction to CO on M−N$_{2+2}$−C$_8$ sites under an applied electrode potential (U) of 0 and −0.6 V. (c) The initial and final state for the COOH dissociation reaction on M−N$_4$−C$_{10}$ and M−N$_{2+2}$−C$_8$ sites. (d) FE H_2 and CO. In the figure, the gray, blue, yellow, red, and white balls represent C, N, M, O, and H atoms, respectively. Source: Pan et al. [15]. © 2018, American Chemical Society. (e) Optimized atomic structures of Fe−N−C and proposed reaction active site paths for electrochemical reduction of CO_2. Source: Qin et al. [16]. © 2019, American Chemical Society. (f) The synthetic route toward single-atom Fe−N$_4$ and Fe−N$_5$ catalysts. Source: Zhang et al. [17]. © 2019, John Wiley & Sons.

ligand coordinated to $Fe–N_4$ [17]. DFT calculations revealed that the axial pyrrolic N ligand depletes the electron density of Fe 3d orbitals along with the reduced Fe–CO p back-donation, which play an important role for rapid CO desorption resulting the highly selective CO formation. Recently, Gu et al. [18] successfully synthesized atomically dispersed N-doped carbon support single-atom iron sites (Fe^{3+}–N–C) catalyst for CO formation in ECDRR. The catalytic activity and selectivity of Fe^{3+}–N–C catalyst was tested in both H-cell and flow-cell reactor, resulting FE CO >80% with j_{CO} reached 20 mA cm^{-2} at −0.47 V and >90% FE CO with j_{CO} reached 94 mA cm^{-2} at −0.47 V in 0.5 M $KHCO_3$ electrolyte, respectively. Their study revealed that the higher activity of Fe^{3+}–N–C compared with Fe^{2+}–N–C at high overpotentials could be rationalized by faster CO_2 adsorption and a weaker CO binding at a Fe^{3+} center than at a Fe^{2+} center.

17.1.4 Zinc (Zn)-SACs

Chen et al. [19] first reported a single-site zinc- coordinated nitrogen co-doped graphene (Zn–N–G) catalyst for CO formation in ECDRR. The Zn–N–G-800 (calcining temperature 800 °C) catalyst provides a maximum 91% CO FE at −0.5 V vs. RHE in 0.5 M $KHCO_3$ aqueous solution. DFT calculations revealed that the $Zn–N_x$ sites could facilitate the formation of COOH* intermediates and desorption of CO*, which is the origin of enhanced activity for electrochemical reduction of CO_2-to-CO. Similarly, Yang et al. [20] disclosed the cost-effective carbon-supported (i.e. Carbon Black BP-2000) nitrogen-anchored Zn SAC (ZnN_x/C with Zn 0.10 wt% and N 0.61 wt%) which showed high selectivity for CO FE up to 95% at −0.43 V vs. RHE in 0.5 M $KHCO_3$ with high TOF of 9969 h^{-1} remarkable durability of ~75 hours in ECDRR. This result was further confirmed by experimental and DFT study, which showed that the observed higher performance of ECDRR could be mainly due to the nitrogen-anchored individual Zn atoms ($Zn–N_4$) and it is the leading active site for ECDRR with low free energy barrier for the formation of *COOH as the rate-limiting step [20].

17.1.5 Copper (Cu)-SACs

At present, Cu-based catalysts are the only transition metal which not only catalyzes the formation of two-electron products but also generates complex hydrocarbons or alcohol products in ECDRR, though there is not a selective catalyst for a specific product [21]. Interestingly Cu-based catalyst selectivity challenges can be overcome by introducing the Cu SACs, which showed in higher selectivity toward different products, like CO, CH_4, and CH_3OH [21]. Jiao et al. [22] reported the two adjacent copper atom (Cu_1^{0}–Cu_1^{x}) called "atom-pair catalyst" (Cu-APC) anchored on alloy $Pd_{10}Te_3$ nanowires, which showed high activity and CO FE 92% at −0.78 V vs. RHE production (with 0.10 wt% Cu) in ECDRR. The pairs Cu_1^{0}–Cu_1^{x} stabilized by the Te surface defects of $Pd_{10}Te_3$ alloy nanowires promoted CO_2 activation because the H_2O molecule adsorbed on Cu_1^{x} stabilized the CO_2 molecule chemisorbed on the neighboring Cu_1^{0} as shown in Figure 17.2. Both experimental and theoretical results

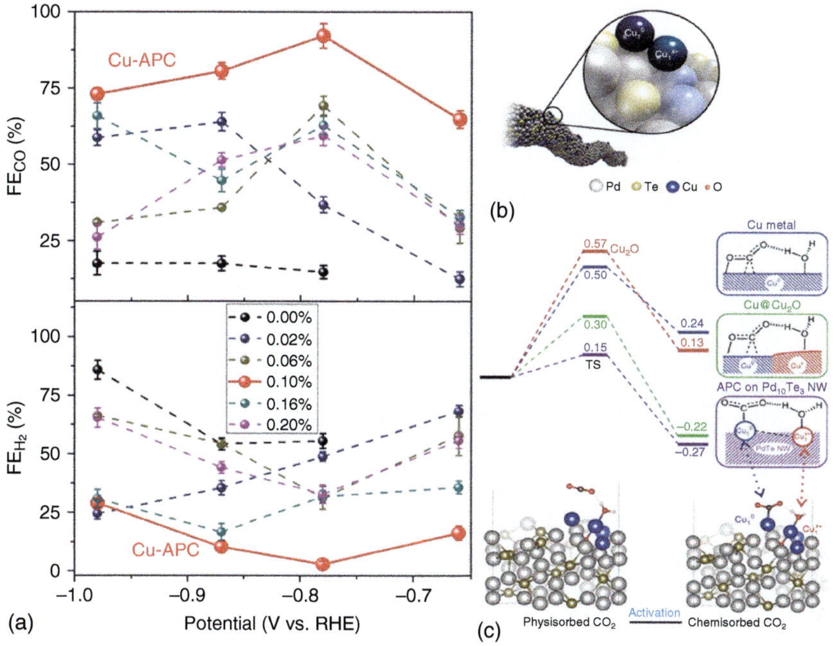

Figure 17.2 (a) FE_{CO} and FE_{H_2} on different amounts of Cu doping. (b) Cu atom pair on $Pd_{10}Te_3$ nanowires. (c) Free energy profiles for CO_2 activation mode of $Cu_1{}^0-Cu_1{}^{x+}$ on $Pd_{10}Te_3$ nanowires. Source: Jiao et al. [22]. © 2019, Springer Nature.

confirmed that the active site of $Cu_1{}^0$–$Cu_1{}^x$ promoted the CO_2 activation by improving its thermochemistry and kinetics. This method could open new opportunities for atomically dispersed catalysts to be used in more complex reactions. The Cu SACs are also able to yield more than >2e$^-$ products, which are somewhat different from other transition metal SACs.

Wang et al. [23] demonstrated an atomic design of multiple oxygen vacancy bound, single-atomic Cu-substituted $CeO_2(110)$, to optimize the ECDRR to CH_4 production. This Cu-substituted CeO_2 showed a maximum FE of ~58% at −1.8 V vs. RHE in 0.1 M $KHCO_3$. The structural characterizations indicated that the low concentration (<5%) Cu species in CeO_2 nanorods are highly dispersed at single-atomic level with an unconventionally low coordination number ~5, suggesting the direct association of three oxygen vacancies with each Cu ion on surfaces [23] which enhanced the electrocatalytic activity and selectivity. Recently, Yang et al. [24] developed and synthesized a facile strategy for the large-scale synthesis of isolated Cu decorated through-hole carbon nanofibers (CuSAs/TCNFs). The CuSAs/TCNFs can be directly used as a cathode for ECDRR due to the excellent mechanical properties of the methanol product. This CuSAs/TCNFs membrane has unique mechanical properties and can be directly used as a cathode for ECDRR, which could produce pure methanol with maximum FE of 44% at −0.9 V vs. RHE in the liquid phase for ~50 hours stability. The DFT calculations illustrated that Cu single

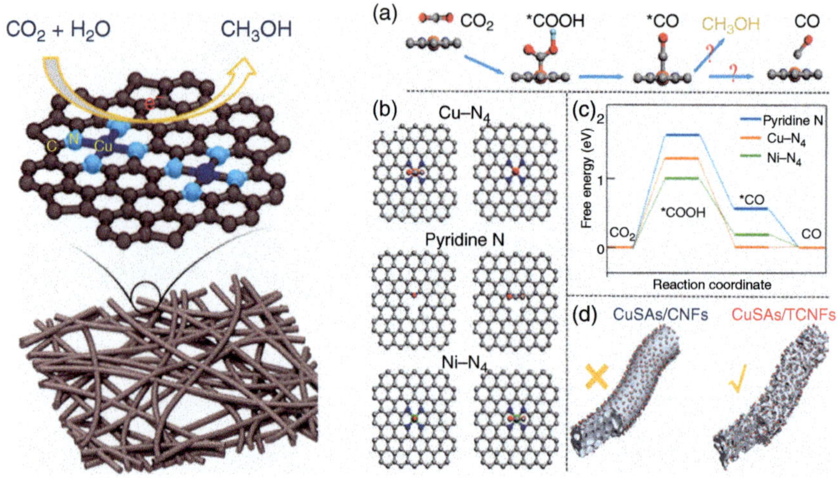

Figure 17.3 (a) Optimized atomic structures of CuSAs/TCNFs and proposed reaction paths for CO_2 electroreduction. (b, c) Free energy diagram of CO_2 to CO on pyridine N, Ni–N_4, and Cu–N_4 structure. (d) Illustration of CO_2 diffusion on two samples. Source: Yang et al. [24]. © 2019, American Chemical Society.

atoms possess a relatively higher binding energy for *CO intermediate, which could be further reduced to methanol product (Figure 17.3a), instead of being easily released from the catalyst surface as CO product [24]. Moreover, comparative study for Ni–N_4 (−0.21 eV) and pyridine N (−0.54 eV), the CO* desorption steps, both exhibited negative free energies, revealing that the CO* intermediate can be efficiently eliminated from the catalyst surface and lead to CO products [24]. In contrast, the *CO desorption on Cu–N_4 has a slightly positive free energy (0.12 eV), indicating an endergonic step. Therefore, in the case of Cu–N_4 material, *CO intermediate can be further reduced to methanol rather than being easily released as CO product [24, 25]. The various moieties such as Cu–N_4, Ni–N_4 and pyridine N-doped graphene nanosheet (Figure 17.3b), as well as free energies of all the systems are manifested (Figure 17.3c), where CuSAs/TCNFs (Figure 17.3d), shows active center for methanol product.

17.1.6 Other

In addition to the above-discussed transition metal of SACs, Mn, Sn, and Bi have also been reported as an active electrocatalyst for the conversion of CO_2-to-$HCOO^-$ and further CO products. The main advantages of these catalysts are their low cost and high abundance; therefore, metal-doped nitrogenated carbon SACs such as Fe–N–C and Mn–N–C have successfully synthesized via using pyrolysis and acid leaching for CO formation in ECDRR [26]. Among them, Mn–N–C electrocatalyst showed maximum FE CO 80% at −0.5 V vs. RHE in 0.1 M $KHCO_3$ solution. Similarly, Zhao et al. [27] also developed a low-cost tin (Sn) nanoparticle as well as an anatomically dispersed catalyst on the N-doped CNFs (Sn/N–CF) via electrospinning method along with a pyrolysis process for ECDRR. The electrocatalytic performance can be tuned by the structure of Sn species on the N-doped CNFs at 1000 °C (Sn/N-CF-1000) pyrolysis process in Ar/H_2 atmosphere. The Sn nanoparticles supported by pyridinic nitrogen-doped CNFs favor the efficient formate formation with FE of 62% −0.8 V vs. RHE and a high current density of 11 mA cm^{-2} due to energetic electron interaction between Sn nanodots and pyridinic nitrogen-doped carbon. However, atomically dispersed Sn species (Sn-SAC) on nanofibers switched to promote the selective conversion of CO_2-to-CO with a maximum FE 91% at −0.6 V vs. RHE in 0.5 M $KHCO_3$ aqueous solution due to the high activity of the Sn–N moieties. The high catalytic performance of CO formation in ECDRR is mainly due to the favorable binding strength of $CO_2\cdot^-$ intermediates improved by the interaction between Sn species and pyridinic-N. Besides, Zu et al. [28] first proposed the kilogram-scale synthesis of single-atom $Sn^{\delta+}$ on N-doped graphene ($Sn^{\delta+}$–N–G) by a quick freeze-vacuum drying-calcination method. These Sn SACs can effectively reduce CO_2 to formate with FE ∼75% at −1.6 V vs. saturated calomel electrode (SCE) in 0.25 M $KHCO_3$ aqueous solution, and a very large TOF of 11 930 h^{-1} for 200 hours stability. The theoretical study proved that the slightly positively charged $Sn^{\delta+}$ plays an important role in stabilizing $CO_2^{\cdot-}$ and $HCOO^-$, which promoted CO_2 activation and protonation to proceed spontaneously.

Zhang et al. [29] successfully developed a gas migration/NH_3-mediated strategy to prepare an atomically dispersed N-coordinated Bi electrocatalyst sites on porous carbon networks (Bi–N_4) that can be achieved through the thermal decomposition of a bismuth-based metal–organic framework (Bi-MOF) and dicyandiamide (DCD) for ECDRR. This Bi–N_4 electrocatalyst showed an excellent performance toward CO conversion, with FE CO of 97% at −0.5 V vs. RHE and high TOF of 5535 h^{-1}. Interestingly, in situ environmental transmission electron microscopy (ETEM) indicated that the introduction of DCD could effectively redisperse intermediate Bi NPs into single atoms and capture them on the defects of nitrogen-doped carbon during MOF-to-SA conversion [29]. This kind of work further confirmed the unique catalytic abilities of SACs, which would show somewhat different selectivity in electrochemical reactions compared with their counterpart bulk materials [21]. The ECDRR performances on different SACs are listed in Table 17.1.

17.2 Metal Nanoparticles-Doped Carbon Catalysts

Carbon materials such as carbon nanotubes (CNTs), activated carbon, multiwall carbon nanotubes (MCNTs), graphene, graphite, and N-doped graphene are the most popular supporting materials[30]. Because these materials have low background activity, high surface area, and high conductivity as well, it could be feasible candidates as supporting materials for ECDRR. Taking these advantages, Kim et al. [31] developed a direct one-pot synthesis of Ag nanoparticles (3-, 5-, and 10-nm size) on carbon support (Ag/C) using an anchoring agent (cysteamine). Ag NPs (5 nm) showed maximum CO FE 84.4% at −0.75 V vs. RHE in 0.5 M $KHCO_3$ solution.

Table 17.1 Summary of ECDRR performances on different SACs in aqueous solution.

Electrocatalyst	Electrolyte	Potential (V, vs. RHE)	J (mA cm^{-2})	TOF	Device	FE (%)	Major product	References
Ni–N_4–C	0.5 M $KHCO_3$	−0.81	~28	NA	H-cell	99	CO	[6]
Ni–NG	0.5 M $KHCO_3$	−0.73	~11	~59 s^{-1}	H-cell	95	CO	[1]
Ni–NSG	0.5 M $KHCO_3$	−0.5	~22	14 800 h^{-1}	H-cell	97	CO	[3]
		−1.0	~100			90		
Ni–N–C (porous carbon)	1.0 M $KHCO_3$	−1.1	~200	NA	Flow-cell	85	CO	[7]
NiSA/PCFM	0.5 M $KHCO_3$	−0.7	~45	NA	H-cell	96	CO	[9]
		−1.0	~308		Flow-cell	88	CO	
Ni–N_3–V	0.1 M $KHCO_3$	−0.9	~31	13 500 h^{-1}	H-cell	>94	CO	[10]
Co–N_2 (porous carbon)	0.5 M $KHCO_3$	−0.68	18.1	23 500 h^{-1}	H-cell	94	CO	[11]
Co–N_5/HNPCSs (hollow porous carbon spheres)	0.2 M $NaHCO_3$	−0.73	~4.5	480 h^{-1}	H-cell	99	CO	[12]
		−0.9	20			90		
Fe–N_4–C	0.1 M $NaHCO_3$	−0.6	~4.5	NA	H-cell	90	CO	[13]
Fe–N_4–C (graphene)	0.1 M $KHCO_3$	−0.6	~2	NA	H-cell	80	CO	[14]
Fe–N_{2+2}–C_8	0.1 M $KHCO_3$	−0.47	~2.8	NA	H-cell	~93	CO	[15]
Fe–N_5 (graphene)	0.1 M $KHCO_3$	−0.35	~4–5	NA	H-cell	97	CO	[17]
Fe^{3+}–N–C	0.5 M $KHCO_3$	−0.47	94	<1000	Flow-cell	>90	CO	[18]
Zn–N_4–C (graphene)	0.5 M $KHCO_3$	−0.5	~4	NA	H-cell	91	CO	[19]
Zn–N_4/C	0.5 M $KHCO_3$	−0.43	~4.8	9 969 h^{-1}	H-cell	95	CO	[20]
CuSAs/TCNFs	0.1 M $KHCO_3$	−0.9	93	NA	H-cell	56	CO	[24]
						44	CH_3OH	
Mn–N–C	0.1 M $KHCO_3$	−0.5	~5	NA	H-cell	80	CO	[26]
AD-Sn/N–C	0.1 M $KHCO_3$	−0.6	1.75	NA	H-cell	91	CO	[27]
($Sn^{\delta+}$–N–G)	0.25 M $KHCO_3$	−1.6 (SCE)	11	11 930 h^{-1}	H-cell	~75	$HCOO^-$	[28]
Bi–N_4–C	0.1 M $NaHCO_3$	−0.5	~4	5 535 h^{-1}	H-cell	97	CO	[29]

NA, not applicable/not reported

DFT studies elucidated that the Ag–S interaction induces surface localization of the unpaired electron, resulting in an enhanced intermediate stabilization and thus improved catalytic activity. Au nanoparticles (3.4 ± 1.3 nm) supported on poly(2,2′-(2,6-pyridine)-5,5′-bi-benzimidazole) polymer (PyPBI) wrapped MCNTs as the cathode catalyst (MCNT/PyPBI/Au) in an alkaline gas diffusion layer (GDL) flow electrolyzer. MCNT/PyPBI/Au electrodes obtained an exceptional CO current density of 158 mA cm^{-2} at low overpotential of −0.55 V vs. RHE for eight hours stability in 2 M KOH (13.7 pH) solution [32]. Their systematic study revealed that the use of an electrolyte with a high pH could play an important role in achieving a lower cathodic overpotential for ECDRR. Furthermore, mass- and size-dependent monodisperse Cu NPs assembled on pyridine-N-rich graphene (p-NG)-support electrocatalyst have been reported [33]. In Cu NPs (7 nm), the mass ratio of p-NG/Cu is 1 : 1, and the FE and hydrocarbon selectivity of C_2H_4 are 19% and 79% at −0.9 V vs. RHE, respectively. Their study shows that the pyridine-N acts as a CO_2 and a proton absorber and promotes hydrogenation and carbon–carbon coupling reaction on Cu to form C_2H_4. Recently, Jung et al. [34] successfully synthesized Cu_2O nanoparticles (20-nm size) directly grown on the carbon support (Cu_2O NPs/C) via a one-pot wet chemical method, using cysteamine as an immobilizing agent applied for ECDRR electrocatalysts. They observed more than five-time higher C_2H_4 56.3% than Cu foil (11.2%) at −1.1 V vs. RHE in 0.1 M $KHCO_3$ aqueous solution. The results indicate that the metallic Cu(0) state was observed under ECDRR, but the fragmented nanoparticles were more readily reoxidized at open-circuit potential inside of the electrolyte, allowing labile Cu states. It supports the assumption that the fragmented nanoparticles with a compact arrangement are favorable to have facile oxygen access on the surface.

In addition to Ag, Au, and Cu catalyst, a simple, economical, and environmentally friendly 2D metal antimony nanosheets (SbNSs) catalysts were also being reported [35]. This catalyst was designed from bulk crystal via a cathodic exfoliation method, resulting in the 2D engineering strategy that can turn inactive bulk materials into active 2D electrocatalysts. Furthermore, this composite fabricated using graphite (SbNSs–G) instead of Pt as the anode during exfoliation. SbNSs–G showed the conversion of CO_2-to-formate with maximum FE of ∼88.5% and current density of ∼8 mA cm^{-2} at −0.96 V vs. RHE in 0.5 M $NaHCO_3$ electrolyte. The enhanced catalytic ability of SbNS–G could be attributed to the synergetic effect between SbNS and graphene (increases the adsorption of CO_2 on the surface of Sb). Similarly, Chen et al. synthesized low cost, 2D–0D composite electrocatalyst composed of bismuth oxide nanosheets, and nitrogen-doped graphene quantum dots (Bi_2O_3-NGQDs) via the solvothermal method [36]. Bi_2O_3-NGQDs exhibited superior performance toward the conversion of CO_2 into formate with FE of ∼100% and current density of ∼20 mA cm^{-2} at −0.9 V vs. RHE in 0.5 M $KHCO_3$ solution. DFT calculation revealed that the origin of this superior activity attributed to the increased adsorption energy of CO_2(ads) and OCHO* intermediate after the combination with N-doped graphene quantum dots (NGQDs).

17.3 Porous Organic Material

17.3.1 Metal Organic Frameworks (MOFs)

The MOFs, also called porous coordination polymers (PCPs), are two- or three-dimensional porous crystalline materials with infinite lattices synthesized from secondary building units (SBUs), metal cations salts or clusters, and polydentate organic ligands with coordination type connections, i.e. made by linking inorganic and organic units by strong bonds (reticular synthesis) [37–39]. In the 1990s, MOFs applications and construction have been pioneered by Gardner et al. [40], Yaghi and Li [41], Munakata et al. [42], and Riou et al. [43]. According to their study, these kinds of materials showed a superior catalyst template due to their metal centers, highly ordered porous structure, uniform pore sizes, and very high surface areas (up to 10 000 $m^2\ g^{-1}$, which is much higher than zeolites and active carbons) endows superior adsorption capacity which is promising to CO_2 reduction. The most currently utilized MOFs are MIL-53, Hong Kong University of Science and Technology (HKUST-1), Fe-1,3,5-benzenetricarboxylate (BTC), zeolitic imidazolate framework (ZIF)-67, and ZIF-8. Their properties like gas storage, energy conversion, chemical sensing, drug delivery, proton conductivity, and catalysis, and their applications are in oleo-chemical and textile industries, transportation, all-electric automobile prototypes, food packaging, and respiratory systems [37, 39, 44].

Zhao et al. [45] successfully synthesized a MOF-assisted ZIF-8 strategy to generating Ni single atoms distributed in nitrogen-doped porous carbon (Ni SAs/N–C) for active CO_2 reduction by an ionic exchange of Zn nodes and adsorbed Ni salts for efficient ECDRR. Their study showed that the Ni SAs/N–C catalyst exhibited CO production FE over 71.9%, a current density of 10.48 $mA\ cm^{-2}$, with an excellent turnover frequency (5273 h^{-1}) at −1.0 V vs. RHE for 60 hours in 0.5 M $KHCO_3$ electrolyte, which was threefold compared to Ni NPs/N–C. Their study revealed that due to the low coordinated Ni SAs/N–C abundant sites on the surface, the strong bonding with $CO_2{\cdot}^-$ may account for the excellent performance for CO_2 electroreduction [45]. Cu clusters with a judiciously controlled surface CN provide active sites that simultaneously optimize selectivity, activity, and efficiency for ECDRR [46]. The ECDRR activity of Cu clusters by promoting undercoordinated Cu sites with the aid of a MOF has been optimized by Sargent and coworkers [46]. They distorted the symmetric paddle-wheel Cu dimer secondary building block of HKUST-1 to an asymmetric motif by separating adjacent benzene tricarboxylate moieties using thermal treatment [46]. This method improved MOF-derived ECDRR performance C_2H_4 FE from 10% to 45%, meanwhile decreasing HER to below 7%. Theoretical and experimental study shows that the MOF-derived Cu clusters maintained a low Cu–Cu CN control during the reaction because of local atomic structure, bond strain, and electron configuration of the asymmetric Cu dimer motif. Similarly, Wang et al. [11] prepared atomically dispersed Co catalysts with different nitrogen coordination numbers using bimetallic Co/Zn (ZIFs) synthesized at room temperature and then pyrolyzed in various temperature (800, 900, and 1000 °C). Atomically dispersed Co with two-coordinate nitrogen atoms (i.e. Co–N_2) achieved high selectivity FE of 94% CO formation and high activity current density of 18.1 mA cm^{-2} at −0.63 V vs. RHE for 60 hours in 0.5 M $KHCO_3$ solution, which is much higher than three-coordinate nitrogen atoms Co–N_3. The CO formation of TOF reached 18 200 h^{-1}. Both experimental and theoretical studies revealed that lower coordination number plays an important role in the activation of CO_2 to the $CO_2{\cdot}^-$ intermediate, resulting in the higher electrocatalyst activity and selectivity in ECDRR.

Yang et al. [47] successfully synthesized the porous carbon-supported NC (ZIF-8) surface-enriched Ni single atoms (defined as SE-Ni SAs@PNC) by a "Pac-Man"-induced thermal atomization electrocatalyst for ECDRR. The proposed "Pac-Man" mechanism is that the Ni nanoparticles could break surface C—C bonds drill into the carbon matrix, leaving abundant pores on the surface [47]. This SE-Ni SAs@PNC electrocatalyst not only transforms the nanoparticles to single atoms but also creates numerous pores to facilitate the contact of dissolved CO_2 and single Ni sites. Because of this, SE-Ni SAs@PNC could show high selectivity toward CO production with FE of 88% and a current density of about 18.3 mA cm^{-2} at −1.0 V vs. RHE. Moreover, the CO formation of TOF reached 47 805 h^{-1}, which surpasses most of the reported metal-based catalysts under comparable conditions [47]. Recently, Huang et al. [48] developed the OH^--coordinated MOF (NNU-15) to simulate the surface state of electrocatalyst in the alkaline environment of ECDRR and then further explore its activation mechanism to CO_2. NNU-15 showed high CO FE reached up to 99.2% at −0.6 V vs. RHE along with long-term stability (110 hours) in 0.5 M $KHCO_3$ electrolyte and the current density of 32.2 mA cm^{-2} at −1.1 V vs. RHE. Theoretical calculations demonstrated that the coordinated OH^- can activate the CO_2 into HCO_3^-, which will chelate to the metal center.

17.3.2 Covalent Organic Frameworks (COFs)

In addition to MOFs, covalent organic frameworks (COFs) structure has also been studied with combined advantages like molecular and heterogeneous electrocatalysts [49] for ECDRR. Lin et al. [50] successfully synthesized a model framework (COF-366-Co) by the imine condensation of 5,10,15,20-tetrakis(4-aminophenyl) porphinato cobalt [Co(TAP)] with 1,4-benzenedicarboxaldehyde (BDA) for the conversion of CO_2-to-CO which exhibited high faradaic efficiency of 90% at −0.55 V vs. RHE in aqueous solution and high turnover numbers (up to 290 000, with an initial turnover frequency of 9400 h^{-1}). Electro kinetics experiments indicated that covalently linking cobalt porphyrin active sites within a COF influences the mechanistic pathways for carbon dioxide reduction. Zhu et al. [51] designed and synthesized a series of stable and high crystalline metalloporphyrin–tetrathiafulvalene (TTF)-based COFs (i.e. M = Co or Ni). In M-TTCOFs catalyst, TTF, serving as electron donator or carrier, and metalloporphyrin, possessing conjugated π-electron system, can act as an excellent electron acceptor and electron transfer carrier. Combining TTF with metalloporphyrin might construct an intermolecular charge-transfer pathway in a structure to primarily enhance the electron transfer efficiency [51]. Remarkably, Co-TTCOF can selectively convert CO_2-to-CO with an Faraday efficiency (FF) of 91.3% at −0.7 V for >40 hours, and the maximum FE CO can reach up to almost 100% (99.7%, at −0.8 V). Furthermore, DFT calculation revealed that the Co-TTCOF has the lowest activation energy for the determining step in electrocatalytic ECDRR compared with other M-TTCOFs.

17.3.3 Metal-Free Catalyst

Carbon-based metal-free materials are referred to as carbon materials that are not composed of any metal elements, including CNFs, carbon quantum dots (CQDs), graphene, CNTs, and amorphous carbon, and these materials have extensively studied as cost-effective electrocatalysts for ECDRR. Because of the large surface area, high electrical conductivity, and tunable structure, these carbon materials could enhance the electrocatalytic activity and selectivity for ECDRR, especially by heteroatom doping (B, P, S, N, O, and F) or built into hierarchical structures [52–56]. Besides, compared to noble-metal-based systems, the carbon-based metal-free systems provide high catalytic performance and are cost-effective, and thus considered as the next-generation renewable catalysts for ECDRR.

17.4 Metal-Free Carbon-Based Catalyst

Kumar et al. [57] synthesized cost-effective metal-free, renewable, and heteroatomic CNFs for CO_2-to-CO conversion by electrospinning of polyacrylonitrile (PAN) polymer solution followed by carbonizing the resulted electrospun mats at 1050 °C in Ar environment. The catalyst showed an exceptionally high current density (~13 times higher than bulk Ag) for a selective CO formation at −0.573 V vs. standard hydrogen electrode (SHE) [57]. The catalytic proficiency of CNFs is attributed to the highly positive charges on carbon atoms due to the existence of the electronegative nitrogen atoms within the carbon lattice [57]. They found that the use of ionic liquids (ILs) (i.e. the electrolyte 1-ethyl-3-methylimidazolium tetrafluoroborate (EMIM-BF4)) enabled CO_2 to form an intermediate complex EMIM-CO_2, facilitating the product yield of CO with almost constant current density for nine hours. Sreekanth et al. [53] first time reported the electrocatalytic activity of boron-doped graphene (BG) for the reduction of CO_2 to formate. According to DFT calculations, the origin of the electrocatalytic activity of BG stems from the symmetrically distributed electron delocalization and the asymmetric charge and spin density distribution due to B-doping. The positive spin density on B and C atoms suggested that both atoms are catalytically active. In another study, Xie et al. [54] successfully synthesized a low cost-effective metal-free fluorine interlayer-doped carbon electrocatalyst (FC) by facile pyrolysis of commercial BP 2000 (carbon source) mixed with polytetrafluoroethylene (PTFE as a fluorine source). Both theoretical and practical studies showed that the fluorine interlayer doping activates the neighbor carbon atoms as active sites and facilitates the stabilization of the key intermediate

COOH*, resulting in CO_2-to-CO conversion with FE of 89.6% at −0.62 V vs. RHE in the CO_2-saturated $NaClO_4$ electrolyte.

Liu et al. [52] synthesized the dual heteroatom-doped catalyst such as boron and nitrogen co-doped nanodiamond (BND) through a hot filament chemical vapor deposition (CVD) method as an efficient and stable electrode for selective reduction of CO_2-to-ethanol with FE of 93.2% at −1.0 V vs. RHE. Theoretical and experimental calculations indicated the excellent activity and selectivity of the BND catalyst for CH_3CH_2OH production mainly because of the synergetic effect of B and N co-dopants. Besides, the B atoms in the carbon framework were able to improve the binding energy of $CO_2{}^{*-}$ intermediate, while N atom can promote *H transfer within elementary reactions. This result showed that the BND catalyst opens the new direction toward $C \geq 2$ product formation beyond the Cu benchmark. Similarly, very recently, Chen et al. [55] synthesized the N- and P-co-doped carbon aerogels (NPCA) via the gelatinization of starch for ECDRR. The NPCA of 900 °C electrocatalyst showed the maximum CO formation with FE of 99.1% at −2.5 V vs. Ag/Ag^+ in 0.5 M $[Bmim]PF_6/MeCN$ and a high partial current density of 143.6 mA cm^{-2}. Because of the higher electrochemical active area and electronic conductivity, it results in faster electron transfer to form key intermediates during ECDRR. By control experiment and theoretical calculation, they found that the pyridinic N was very active for CO_2 reduction to CO, and co-doping of P with N hinders the HER significantly, which could enhance both current density and FE simultaneously. Thereby, the synergistic effect from dual hetero-atoms doping contributes to a superior CO_2 reduction performance in ECDRR.

Wu et al. [58] reported the nanometer-size metal-free NGQDs catalyst for the production of multi-carbon hydrocarbons and oxygenated at high faradaic efficiencies, high current densities, and low overpotentials for ECDRR. These NGQDs showed superior activity with total FE of ~90% and more importantly, this material achieved ~45% selectivity for ethylene and ethanol conversions beyond Cu metal. Their study revealed that the catalytic activity of NGQDs originates from pyridinic N rather than graphitic N. Hursan et al. [59] studied uniformly sized meso or macropores (13, 27, and 90 nm) polymer-derived, interconnected, N-doped carbon structures (N–C). Catalysts were synthesized by a sacrificial support method, starting from poly(*o*-phenylenediamine) polymers and using silica nanoparticles [59] (SiO_2/PoPD), then composite was carbonized (pyrolyzed) at 900 °C in neat N_2 flow. They found that the ECDRR selectivity (vs. the HER) has been increased around three times just by introducing the porosity into the carbon structure (with an optimal pore size of 27 nm) because of the change to alterations in the wetting and CO_2 adsorption properties of the carbon catalysts. N–C-13 and N–C-27 electrocatalyst showed the formation of CO and H_2, accounting for more than 95% of the total products (trace amounts of methane and formate were also seen). The N–C-27 achieved the highest CO selectivity at all potentials, peaking at −0.6 V with FE 76%. Wang et al. [60] prepared a series of defective N-doped carbon spheres (D-NC-X and D-NC-X, where X represents the thermal treatment temperature) identified as active sites for ECDRR. The D-NC-1100 sample with the lowest N content exhibited the best FE for CO production (94.5%) at a low applied potential of −0.60 V vs. RHE in 0.1 M $KHCO_3$. Near edge X-ray absorption fine structure (NEXAFS) spectroscopy and DFT calculations revealed that octagonal and pentagonal defects involving sp^2 hybridized carbon atoms are the likely active sites for ECDRR.

17.4.1 Other Metal-Free Catalyst

In Section 17.4, we have discussed the carbon-based metal-free catalyst. Herein, we will be a focus on the non-carbon-based metal-free catalyst for CO_2 reduction.

Recently, boron phosphide (BP) carbon-free nanoparticles have been synthesized via vacuum-seal strategy to perform highly efficiently as a nonmetal electrocatalyst toward electrochemical reduction of CO_2-to-CH_3OH [61]. This BP (111) catalyst achieved a high CH_3OH FE of ~92.0% at −0.5 V vs. RHE in 0.1 M $KHCO_3$. DFT calculations revealed that B and P synergistically promote the binding and activation of CO_2 and the rate-determining step for the CO_2 reduction reaction [61]. This study not only provided us an attractive earth-abundant nanocatalyst for highly efficient and selective electrosynthesis of CH_3OH fuel but also opens an exciting new avenue to explore P as an active dopant to tune the electrocatalytic ECDRR performances for applications.

17.5 Electrochemical CO Reduction Reaction

17.5.1 The Importance of CO Reduction Study

As we have discussed in Sections 17.1–17.4, CO_2 is initially reduced to adsorbed CO and further to hydrocarbons and alcohols [62] on a various metal electrode. Among them, copper-based electrocatalyst has only one promising candidate having a unique ability to convert CO_2 into C_{2+} and C_{3+} products in significant amounts. Even after decades of studies, Cu-based catalysts are not particularly selective to any single liquid product, which leads to poor energy efficiencies, and energy-intensive product separations [63] along with low FE, high overpotential, HER side reaction, and low current density are still challenging toward the formation of C_{2+}, C_{3+}, and oxygenate in one-step electrochemical CO_2 reduction. To solve these problems, the two-step electrochemical carbon monoxide reduction (ECORR) (where CO act as reactant) at copper electrode has recently been studied, which showed superior performance compared to ECDRR including enhanced C_{2+}, C_{3+}, and oxygenates products of FE at lower overpotential by the suppression of side product (H_2), [64] enhanced product selectivity [65, 66], and achieved higher current density in aqueous solution. More importantly, alkaline conditions can increase the energy efficiency of ECORR to C–C coupling productions such as C_{2+} oxygenates (ethanol, acetate, and aldehyde) and are increasingly favored over that of C_{2+} hydrocarbons at more positive electrode potentials. Hori et al. demonstrate the molecular reaction pathway shown in Figure 17.4 for ECDRR at a Cu electrode [64]. In the case of methane and ethylene formations, the reaction paths are separated at the very early stage, where CO was formed with common intermediate species of CH_2 adsorbed

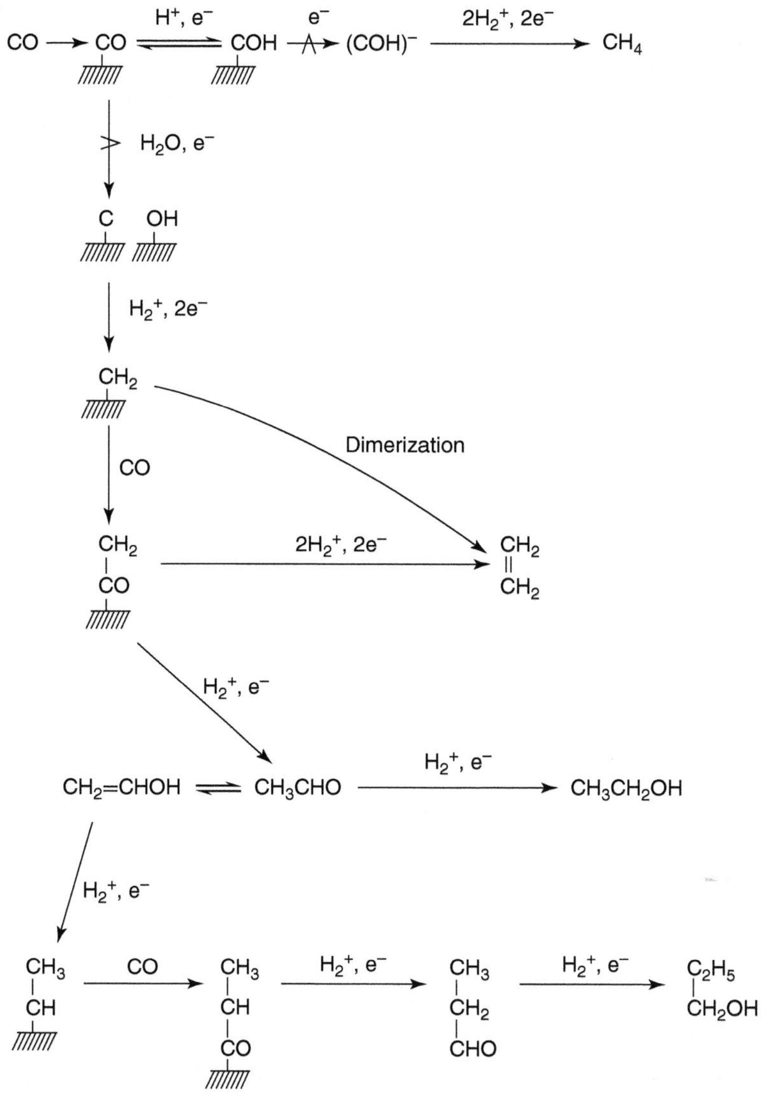

Figure 17.4 The molecular reaction pathway diagram of the ECORR. H^{2+} represents H^{2+} or a combined H(ads) + H^+. The → with a ∧ through it denote a rate-determining step. The H_2O and OH^- resulting from the reactions are excluded for brevity. Source: Hori et al. [64]. © 1997, American Chemical Society.

on the electrode surface. Authors presumed that the insertion of CO-to-CH_2(ads) leads to the formation of CH_2–CO(ads), which will further be reduced to CH_2=CH_2 via dimerization of CH_2 [64]. Moreover, CH_2–CO(ads) will also be reduced to acetaldehyde and eventually to ethanol, and a small fraction of acetaldehyde may be transformed into vinyl alcohol. In all the hydrogenation processes, H_2^+ or H(ads) + H^+ will be involved [64, 67].

17.5.2 Advances in CO Reduction

Pioneer's work by Hori et al. reported that ECDRR could serve as a sequential step in the transformation of CO_2 into multi-carbon fuels and chemicals, which is significant interest from the scientific viewpoint as well [64, 68]. It is understood that the CO is a well-known intermediate in the reduction of CO_2 [64, 69], and the reduction of CO_2 and CO has the same dependence on the atomic surface structure [70]. Moreover, the product selectivity in the EDDRR and ECORR that strongly depends on the atomic configuration of the copper electrode surface has been reported. The study revealed that the reduction of CO into ethylene could be observed via two pathways: (i) CO dimerization on Cu(100), CO reduction only by ethylene formation at low overpotential and (ii) shared intermediate on both Cu(100) and Cu(111) where CO is reduced to methane and ethylene simultaneously at higher overpotential [66, 71]. The electrochemical reduction of CO_2 and CO toward C_1, C_2, and C_3 products of ethylene, ethanol, and *n*-propanol, respectively, were investigated on GDLs as electrodes. Their study observed that the selectivity of ethylene, ethanol, and *n*-propanol on Cu nanoparticles (40–60 nm) was two times higher when CO used as the reactant (89% FE at $-300\ mA\ cm^{-2}$) in comparison with the ECDRR (34% FE cumulative at $-300\ mA\ cm^{-2}$) in a mild electrolyte at −0.97 V vs. RHE 1.0 M $KHCO_3$ solution. The results suggested that the particle size and morphology play a crucial role rather than an oxidation state of Cu in the preparation of electrodes for application at industrially relevant current densities. Besides, some previous study showed that micrometer-sized and oxide-derived Cu catalysts could reduce CO-to-C_{2+} product. Among all the C_{2+} products, acetate achieved maximum FE of ~20% in ECDRR and it is suggested that a highly alkaline environment (a high pH value) near the electrode–electrolyte interface may play an essential role for acetate production [72], but the effect of catalyst surface on acetate selectivity has not been clear.

To understand the nature of the catalytic surface toward acetate production, Luc et al. [73] reported two-dimensional (2D) Cu nanosheets (CuNSs) of ~5-nm-thick solution-phase synthesis method, which selectively exposes the (111) surface. The CuNSs exhibited high selectivity for acetate formation of maximum FE of 48% at −0.76 V vs. RHE with partial current density of up to $131\ mA\ cm^{-2}$ in 2 M KOH electrolyte for ECORR. The acetate selectivity is attributed to the suppression of ethylene and ethanol formation, probably due to the reduction of exposed (100) and (110) surfaces [73]. Moreover, computational studies suggested that the pathway for acetate formation goes through a ketene intermediate, with the incorporation of one oxygen atom from the electrolyte KOH (OH^- ions are more nucleophilic than H_2O) and the other oxygen atom from the CO reactant. Furthermore, the high roughness factor of various morphologies of Cu electrode can reduce CO with good selectivity toward multi-carbon oxygenates (C_{2+} oxygenates) products at low overpotentials range (−0.25 to −0.5 V vs. RHE) with suppressing hydrocarbon and hydrogen production[74]. Jaramillo and coworkers developed hierarchical Cu nanoflower electrodes that are highly selective for electrochemical ECORR with nearly 100% selective liquid-oxygenated products (acetaldehyde, acetate, and ethanol) at very low overpotentials (−0.23 V vs. RHE) in 0.1 M KOH saturated with

1 atm to CO at ambient temperature. However, the reduction of CO at −0.33 V vs. RHE observed ~80% liquid oxygenated with main products (ethanol ~60% and acetate ~20%) [63].

Recently, a new strategy such as Cu–Ag bimetallic electrodes has been investigated for ECORR toward acetaldehyde formation under alkaline conditions [65]. Their study showed that on planar Cu–Ag electrodes, the acetaldehyde selectivity was obtained over 50% FE and over 90% selectivity at very low overpotentials −0.536 V vs. RHE in 0.1 M KOH (pH = 13.0) saturated solution. Furthermore, the acetaldehyde FE could enhance ~70% at −0.33 V vs. RHE by increasing the roughness factor of the Cu–Ag electrode via synthesized porous of Cu–Ag (7 : 1) nanoflower electrode (Cu–Ag-NF) and showed that the surface area effect could also play an important role for product selectivity in ECORR. The DFT calculations demonstrated that the Ag ad-atoms could tune the surface binding energy of reduced aldehyde intermediates, resulting in a suppression of acetaldehyde reduction to ethanol.

References

1 Jiang, K., Siahrostami, S., Zheng, T.T. et al. (2018). *Energy Environ. Sci.* 11: 893–903.
2 Yang, X.F., Wang, A., Qiao, B. et al. (2013). *Acc. Chem. Res.* 46: 1740–1748.
3 Yang, H.B., Hung, S.F., Liu, S. et al. (2018). *Nat. Energy* 3: 140–147.
4 Pei, G.X., Liu, X.Y., Wang, A.Q. et al. (2015). *ACS Catal.* 5: 3717–3725.
5 Qiao, B., Wang, A., Yang, X. et al. (2011). *Nat. Chem.* 3: 634–641.
6 Li, X., Bi, W., Chen, M. et al. (2017). *J. Am. Chem. Soc.* 139: 14889–14892.
7 Möller, T., Ju, W., Bagger, A. et al. (2019). *Energy Environ. Sci.* 12: 640–647.
8 Ganji, P., Borse, R.A., Xie, J.F. et al. (2020). *Adv. Mater. Interfaces* 4: 2000096.
9 Yang, H., Lin, Q., Zhang, C. et al. (2020). *Nat. Commun.* 11: 593.
10 Rong, X., Wang, H.J., Lu, X.L. et al. (2020). *Angew. Chem. Int. Ed.* 59: 1961–1965.
11 Wang, X., Chen, Z., Zhao, X. et al. (2018). *Angew. Chem. Int. Ed.* 57: 1944–1948.
12 Pan, Y., Lin, R., Chen, Y. et al. (2018). *J. Am. Chem. Soc.* 140: 4218–4221.
13 Huan, T.N., Ranjbar, N., Rousse, G. et al. (2017). *ACS Catal.* 7: 1520–1525.
14 Zhang, C.H., Yang, S.Z., Wu, J.J. et al. (2018). *Adv. Energy Mater.* 8: 1703487.
15 Pan, F.P., Zhang, H.G., Liu, K.X. et al. (2018). *ACS Catal.* 8: 3116–3122.
16 Qin, X.P., Zhu, S.Q., Xiao, F. et al. (2019). *ACS Energy Lett.* 4: 1778–1783.
17 Zhang, H., Li, J., Xi, S. et al. (2019). *Angew. Chem. Int. Ed.* 58: 14871–14876.
18 Gu, J., Hsu, C.S., Bai, L. et al. (2019). *Science* 364: 1091–1094.
19 Chen, Z., Mou, K., Yao, S., and Liu, L. (2018). *ChemSusChem* 11: 2944–2952.
20 Yang, F., Song, P., Liu, X. et al. (2018). *Angew. Chem. Int. Ed.* 57: 12303–12307.
21 Lü, F., Bao, H., Mi, Y. et al. (2020). *Sustain. Energy Fuels* 4: 1012–1028.
22 Jiao, J., Lin, R., Liu, S. et al. (2019). *Nat. Chem.* 11: 222–228.
23 Wang, Y.F., Chen, Z., Han, P. et al. (2018). *ACS Catal.* 8: 7113–7119.
24 Yang, H., Wu, Y., Li, G. et al. (2019). *J. Am. Chem. Soc.* 141: 12717–12723.
25 Peterson, A.A. and Norskov, J.K. (2012). *J. Phys. Chem. Lett.* 3: 251–258.

26 Varela, A.S., Ranjbar Sahraie, N., Steinberg, J. et al. (2015). *Angew. Chem. Int. Ed.* 54: 10758–10762.
27 Zhao, Y., Liang, J.J., Wang, C.Y. et al. (2018). *Adv. Energy Mater.* 8: 1702524.
28 Zu, X., Li, X., Liu, W. et al. (2019). *Adv. Mater.* 31: e1808135.
29 Zhang, E., Wang, T., Yu, K. et al. (2019). *J. Am. Chem. Soc.* 141: 16569–16573.
30 Tomboc, G.M., Choi, S., Kwon, T. et al. (2020). *Adv. Mater.* 32: e1908398.
31 Kim, C., Jeon, H.S., Eom, T. et al. (2015). *J. Am. Chem. Soc.* 137: 13844–13850.
32 Verma, S., Hamasaki, Y., Kim, C. et al. (2017). *ACS Energy Lett.* 3: 193–198.
33 Li, Q., Zhu, W.L., Fu, J.J. et al. (2016). *Nano Energy* 24: 1–9.
34 Jung, H., Lee, S.Y., Lee, C.W. et al. (2019). *J. Am. Chem. Soc.* 141: 4624–4633.
35 Li, F., Xue, M., Li, J. et al. (2017). *Angew. Chem. Int. Ed.* 56: 14718–14722.
36 Chen, Z., Mou, K., Wang, X., and Liu, L. (2018). *Angew. Chem. Int. Ed.* 57: 12790–12794.
37 Furukawa, H., Cordova, K.E., O'Keeffe, M., and Yaghi, O.M. (2013). *Science* 341: 1230444.
38 Yaghi, O.M., O'Keeffe, M., Ockwig, N.W. et al. (2003). *Nature* 423: 705–714.
39 Dang, S., Zhu, Q.-L., and Xu, Q. (2017). *Nat. Rev. Mater.* 3: 17075.
40 Gardner, G.B., Venkataraman, D., Moore, J.S., and Lee, S. (1995). *Nature* 374: 792–795.
41 Yaghi, O.M. and Li, H.L. (1995). *J. Am. Chem. Soc.* 117: 10401–10402.
42 Munakata, M., Kurodasowa, T., Maekawa, M. et al. (1995). *Inorg. Chem.* 34: 2705–2710.
43 Riou, D., Roubeau, O., and Ferey, G. (1998). *Microporous Mesoporous Mater.* 23: 23–31.
44 Silva, P., Vilela, S.M., Tome, J.P., and Almeida Paz, F.A. (2015). *Chem. Soc. Rev.* 44: 6774–6803.
45 Zhao, C., Dai, X., Yao, T. et al. (2017). *J. Am. Chem. Soc.* 139: 8078–8081.
46 Nam, D.H., Bushuyev, O.S., Li, J. et al. (2018). *J. Am. Chem. Soc.* 140: 11378–11386.
47 Yang, J., Qiu, Z., Zhao, C. et al. (2018). *Angew. Chem. Int. Ed.* 57: 14095–14100.
48 Huang, Q., Li, Q., Liu, J. et al. (2019). *Matter* 1: 1656–1668.
49 Wu, J., Sharifi, T., Gao, Y. et al. (2019). *Adv. Mater.* 31: e1804257.
50 Lin, S., Diercks, C.S., Zhang, Y.B. et al. (2015). *Science* 349: 1208–1213.
51 Zhu, H.J., Lu, M., Wang, Y.R. et al. (2020). *Nat. Commun.* 11: 497.
52 Liu, Y., Zhang, Y., Cheng, K. et al. (2017). *Angew. Chem. Int. Ed.* 56: 15607–15611.
53 Sreekanth, N., Nazrulla, M.A., Vineesh, T.V. et al. (2015). *Chem. Commun.* 51: 16061–16064.
54 Xie, J., Zhao, X., Wu, M. et al. (2018). *Angew. Chem. Int. Ed.* 57: 9640–9644.
55 Chen, C., Sun, X., Yan, X. et al. (2020). *Angew. Chem. Int. Ed.* 59: 11123–11129.
56 Mohamed, A.G.A., Huang, Y.Y., Xie, J.F. et al. (2020). *Nano Today* 33: 100891.
57 Kumar, B., Asadi, M., Pisasale, D. et al. (2013). *Nat. Commun.* 4: 2819.
58 Wu, J., Ma, S., Sun, J. et al. (2016). *Nat. Commun.* 7: 13869.
59 Hursan, D., Samu, A.A., Janovak, L. et al. (2019). *Joule* 3: 1719–1733.
60 Wang, W., Shang, L., Chang, G. et al. (2019). *Adv. Mater.* 31: e1808276.

61 Mou, S., Wu, T., Xie, J. et al. (2019). *Adv. Mater.* 31: e1903499.
62 Hara, K., Tsuneto, A., Kudo, A., and Sakata, T. (1994). *J. Electrochem. Soc.* 141: 2097–2103.
63 Wang, L., Nitopi, S., Wong, A.B. et al. (2019). *Nat. Catal.* 2: 702–708.
64 Hori, Y., Takahashi, R., Yoshinami, Y., and Murata, A. (1997). *J. Phys. Chem. B* 101: 7075–7081.
65 Wang, L., Higgins, D.C., Ji, Y. et al. (2020). *Proc. Natl. Acad. Sci. U. S. A.* 117: 12572–12575.
66 Schouten, K.J.P., Gallent, E.P., and Koper, M.T.M. (2013). *ACS Catal.* 3: 1292–1295.
67 Sundararaman, R., Figueiredo, M.C., Koper, M.T.M., and Schwarz, K.A. (2017). *J. Phys. Chem. Lett.* 8: 5344–5348.
68 Hori, Y., Murata, A., Takahashi, R., and Suzuki, S. (1987). *J. Am. Chem. Soc.* 109: 5022–5023.
69 Hori, Y., Vayenas, C.G., White, R.E., and Gamboa-Aldeco, M.E. (2008). *Modern Aspects of Electrochemistry*, vol. 42, 89–189. New York, NY: Springer.
70 Hori, Y., Wakebe, H., Tsukamoto, T., and Koga, O. (1995). *Surf. Sci.* 335: 258–263.
71 Schouten, K.J., Qin, Z., Perez Gallent, E., and Koper, M.T. (2012). *J. Am. Chem. Soc.* 134: 9864–9867.
72 Jouny, M., Luc, W., and Jiao, F. (2018). *Nat. Catal.* 1: 748–755.
73 Luc, W., Fu, X.B., Shi, J.J. et al. (2019). *Nat. Catal.* 2: 423–430.
74 Li, C.W., Ciston, J., and Kanan, M.W. (2014). *Nature* 508: 504–507.

18

Devices

Electrochemical cell design is an indispensable facet impacting the CO_2 reduction process, which will ascertain current density, Faradaic efficiency (FE), and stability [1]. Since Hori first revealed a work of CO_2 reduction [2], various cell designs such as (i) batch or semi-batch cell (H-type cell) [3–5] which are most commonly used in laboratory electrochemical carbon dioxide reduction reaction (ECDRR) studies, (ii) continuous flow cell (i.e. polymer electrolyte membrane [PEM] [6–8] and microfluidic flow cell (MFC) [9–11]), (iii) solid oxide electrolysis cell (SOEC) [12–14], and (iv) differential electrochemical mass spectrometry (DEMS) cell [15, 16] are being developed in this field. Among them, H-type (liquid phase) and flow-type cells (gas phase) have been extensively studied in the ECDRR process, which will be discussed in Sections 18.1 and 18.2.

18.1 H-Cell

The vast majority of reported ECDRR experiments are carried out in traditional H-type cells (i.e. liquid phase) electrochemical reactor systems, considering their facile assembly for various types of catalyst electrodes, easy operation, low cost, and easy product separation. The cells that are typically encompassed into cathodic and anodic chambers are separated by a proton-conducting (e.g. Nafion 117) or an anion-exchange polymer membrane [17, 18] to prevent the reduced products from oxidizing again during the reaction [1]. Along with this, the three electrodes (e.g. working electrode, a reference electrode, and counter electrode) are placed into an electrolyte. The working electrode (as-prepared electrocatalyst) and the reference electrode (Ag/AgCl) are placed in the cathodic compartment with CO_2 gas inlet and outlet where CO_2 being reduced; meanwhile, the counter electrode (Pt) was placed in the anodic chamber where H_2O being oxidized as shown in Figure 18.1a. This set-up allows ionic conductivity, preventing the transport of cathodic products to the anode where they can be oxidized while keeping the electrodes nearby [18]. In such a system, there are several key steps involved in an ECDRR process, like (i) mass transfer of CO_2 from the gas phase to the bulk electrolyte,

Electrocatalysis in Balancing the Natural Carbon Cycle, First Edition. Yaobing Wang.
© 2021 WILEY-VCH GmbH. Published 2021 by WILEY-VCH GmbH.

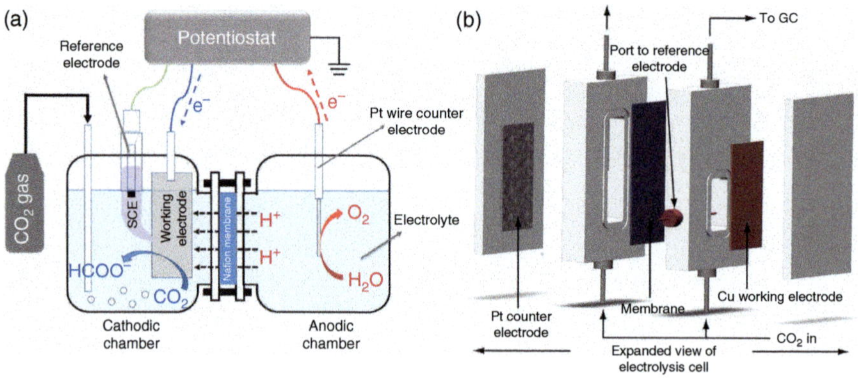

Figure 18.1 (a) A schematic of the conventional H-type electrochemical cell. Source: Zhao et al. [5]. © 2016, Elsevier. (b) Schematic of the modified electrochemical cell and experimental set-up used in electrolysis experiments. Source: Kuhl et al. [19]. © 2012, Royal Society of Chemistry.

(ii) transport of dissolved CO_2 from the bulk electrolyte to the cathode/electrolyte interface, (iii) absorption of CO_2 at the cathode surface, (iv) dissociation of adsorbed CO_2 species into adsorbed intermediates such as *COOH, *CO, *CHO, and *COH, (v) multiple electron/proton transfer from the cathode catalyst to intermediates, (vi) desorption of products from the electrode, and (vii) migration of products away from the cathode/electrolyte interface to the bulk gas or liquid phases [20]. As a result, different reaction pathways implied the formation of various products such as CO, HCOOH, CH_3OH, HCHO, CH_4, C_2H_4, and so on.

As we discussed previously, the H-type cell is widely used for gas and liquid product detection, including CO, formic acid, hydrocarbons, and so on [1]. However, the detection of liquid products in conventional H-type cells is still challenging because the FE is very low in aqueous systems. To address this issue, Jaramillo and coworkers first developed a modified H-type cell design, as shown in Figure 18.1b, where a high ratio of the electrode surface area to the electrolyte volume (S/V) > 0.5 was achieved [1]. In a modified H-type cell, the working electrode and the counter electrode were placed parallel to each other with a small distance. Meanwhile, the reference electrode was inserted into the cathodic compartment close to the working electrode to monitor the potential and reduce solution resistance between electrodes [1]. Besides, the membrane was inserted between the working and counter electrodes [1]. Their study successfully detected 16 different products compared to convention H-type cells during ECDRR, moreover, five of which were first time discovered, despite the classical H-type cells providing benefits like low cost, simple construction, commercial availability, and allows rapid screening of various types of electrocatalysts and electrolytes in ECDRR. Nevertheless, their mass transport limitation of CO_2 dissolution, limited electrode surface area, large interelectrode distance [21], high infrared (IR) losses, and low current densities (<100 mA cm^{-2}) [18] need to be addressed. Therefore, flow-type cell engineering was proposed in the last decade and will be discussed in Section 18.2.

18.2 Flow Cell

In the past decades, flow cell derived from a fuel cell or water electrolyzer has allured lots of attention due to its efficient mass transfer efficiency, which could reach over 100 mA cm^{-2} [1]. In flow cell, the higher CO_2 concentrations at the high surface area gas diffusion electrodes (i.e. GDEs) electrocatalyst could be achieved by continuously circulating the reactants and products to and away from the electrodes [1] resulting in crossover the mass transfer limitations of an H-cell. Besides, gas-phase electrochemical reduction of CO_2 is feasible within a flow cell to conquer the low solubility of CO_2 in aqueous electrolyte solutions and difficulty in recycling products from a liquid phase [1]. In short, flow-type cells are thermodynamically and kinetically different than traditional H-type cells for ECDRR, having a promising route for large-scale commercial applications.

The flow cells are generally divided into two primary architectures, including (i) membrane-based flow-cell reactor or a membrane electrode assembly (MEA) reactor and (ii) microfluidic cell reactor [17, 22] shown in Figure 18.2a,b, respectively. Among them, membrane-based flow cells were widely studied because of the choice of either a cation exchange membrane (CEM) [23] which facilitate the flow of positive ions from the anode to the cathode [17], anion exchange membrane (AEM), the flow of anions (e.g. OH^-) from the cathode to the anode, and a bipolar membrane (BPM) which affects the kinetics of ion transport pathways [17]. The typical architecture of a membrane-based flow-cell reactor is shown in Figure 18.2a, where CO_2 is delivered to the cathode dissolved in mildly basic solution (e.g. aqueous bicarbonate) [17] for an ion-exchange PEM placed between cathode and anode

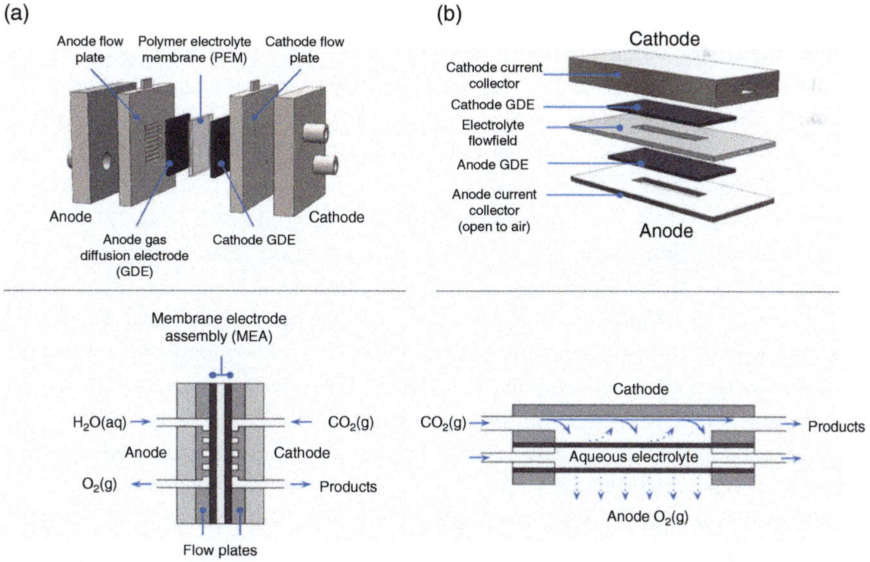

Figure 18.2 A schematic diagram of various configurations for electrochemical flow cell (a) membrane reactor and (b) microfluidic reactor. Source: Weekes et al. [17]. © 2018, American Chemical Society.

GDE. The catalyst is immobilized onto a cathode electrode such as highly porous GDE (situated between the electrode and the membrane support the catalyst), in which gaseous CO_2 is directly fed, meanwhile, another side an aqueous electrolyte supply by anode where oxygen evolution reaction takes place.

In addition to AEM and CEM flow cell, BPM flow cells have also been studied for ECDRR [8, 24]. Recently, Berlinguette research group reviewed the silver-based catalyst integrated with a carbon-based gas diffusion layer (GDL) resulting in the highest current density of 200 mA cm^{-2} with CO FE ~50 in 1 M NaOH solution for ECDRR [8]. To be specific, a BPM drives both H^+ and OH^- ions toward the anode and cathode, respectively, enabling a constant pH during electrolysis, which could play an important role in ECDRR stability [25]. Their study revealed that a configuration with only a BPM is unfavorable for robust CO_2 reduction, while modified configuration with solid-supported aqueous layer inserted between the silver-based catalyst layer (CL) and the BPM enhanced the cathode selectivity for CO_2 reduction to CO [8]. Furthermore, Cu-based cathodic catalysts supported on graphite (G), activated carbon (AC), and carbon nanofibers (CNFs) have been prepared via an impregnation technique for ECDRR [26]. Their study observed that Cu–AC-based catalysts have the highest activity toward acetaldehyde formation because of the large surface area of AC with the larger metal dispersion of the Cu particles. Furthermore, the author studied the MEA based on the Cu–AC cathodic catalyst, resulting in converting CO_2 into methanol and acetaldehyde production. In addition to CO, formate, and ethylene product, ethanol is of particular interest as it has high energy density and is used as high-octane fuel, which has a high market price and global demand as well [27]. However, electrochemical conversion of CO_2 to EtOH is still challenging because, in ECDRR, ethylene and ethanol are two main competing C_2 products which have shared a key intermediate (HOCCH*). It is well understood that ethylene is generated after C—O bond-breaking from HOCCH*. Recently, Sargent and coworkers developed an electrocatalyst with confined reaction volume by coating Cu catalysts with nitrogen-doped carbon (N-C/Cu) for ECDRR [27]. The N-C/Cu electrocatalyst achieved the highest ethanol FE 53% with a ~156 mA cm^{-2} current density and cathodic 31% energy efficiency (EE) at −0.68 V vs. reversible hydrogen electrode (RHE) in 1.0 M KOH flow cells. Density functional theory (DFT) calculations suggested that the N-C/Cu electrocatalyst promoted C–C coupling and suppressed the breaking of the C—O bond in HOCCH*. Meanwhile, their spectroscopy studies revealed that the strong electron-donating ability of the nitrogen-doped carbon layers plays an important role in improving ethanol selectivity in ECDRR. This type of research has opened new direction for higher-order value-added chemicals and fuels formation. Beside metal-based electrocatalysts, carbon-supported low-cost tin nanoparticles (NPs) (Sn/C-GDEs) have also been successfully synthesized for formate production, with 70% FE and high current density ~150 mA cm^{-2} in 0.45 M $KHCO_3$ + 0.5 M KCl saturated aqueous solution in MEA reactors [28]. This work has demonstrated the advantage of MEA reactors toward formate production in ECDRR. The summary of ECDRR performance on various catalyst materials in the gas-phase reactor is listed in Table 18.1.

Table 18.1 Summary of ECDRR performance on various catalyst materials in the gas-phase reactor.

Device[a)]	Electrocatalyst	Potential (V vs. RHE)	J (mA cm^{-2})	Electrolyte	FE (%)	Major product	References
MEA	Sn/C-GDE	NA	~150	0.45 M $KHCO_3$ 0.5 M KCl	~70	Formate	[28]
MEA (BPM)	Ag	NA	~200	1.0 M NaOH	~50	CO	[29]
MEA	Cu/AC	NA	~2.3	NA	NA	Acetaldehyde, methanol	[26]
MEA (AEM)	N–C/Cu	−0.64	~156	1.0 M KOH	53	Ethanol	[27]
MEA	Zn porous	−0.64	~200	1.0 M KOH	84	CO	[30]
MEA (AEM)	Ni–N–C (porous carbon)	−1.1	~200	1.0 M $KHCO_3$	85	CO	[31]
MEA (AEM)	Nikel single-atom/porous carbon fiber membrane catalyst (NiSA/PCFM)	−1.0	~308	0.5 M $KHCO_3$	88	CO	[32]
Microfluidic	BiOBr	−0.95	200	2.0 M $KHCO_3$	NA	Formate	[33]
Microfluidic	Au-multi-walled carbon nanotube/ poly(2,2′-(2,6-pyridine)-5,5′-bi benzimidazole) polymer (MWCNT/PyPBI) (NPs)	−0.55	158	2.0 M KOH	>90	CO	[34]
Microfluidic	Cu_2S–Cu–V	−0.92	126	1.0 M KOH	~53	Ethanol, propanol	[35]
Microfluidic	Cu–G/PTFE (NPs)	−0.54	275	5.5 M KOH (4 M KI)	~66	C_2H_4	[10]
Microfluidic	Cu nanoporous	−0.67	653	1.0 M KOH	~39	C_2H_4	[36]

a) Membrane exchange assembly: anion exchange membrane (AEM), a bipolar membrane (BPM), NA; not applicable/not reported.

On the other hand, recently, a microfluidic electrolytic cell (MEC; Figure 18.2b) electrolyzer configuration was first developed by Kenis and coworkers [37, 38]. In this device, the membrane is replaced by a thin space (<1 mm in thickness) filled with flowing electrolyte stream to separate the anode and cathode [39]. The CO_2 molecules could quickly be diffused into the electrode–electrolyte interface near the cathode, and the two fixed GDEs are flushed by the flowing electrolyte. Different from other reactors containing membrane, this configuration separates reduction and oxidation products by the diffusion of products instead of a membrane [1]. Compared to membrane-based flow-type cells, MECs not only avoid the high membrane cost but also mitigate water management, which mainly refers to the anode dry out and cathode flooding when operated at high current densities owing to the osmotic drag of water molecules along with proton transport from anode to cathode across the membrane [39, 40]. However, very few studies have reported on MECs toward C_{2+} production. Due to the caused by the "float-ing" effect that protons formed in the anode are easily drained from the cathode vicinity or washed away by the flowing electrolyte, rather than participating in the multiple protons required C_{2+} formation reaction [39]. Sargent and coworkers evaluated the ECDRR performance of the Cu_2S–Cu–V catalyst in both flow-type cells and H-type cells. Their study revealed that membrane-based flow cells showed much superior ECDRR performance compare to H-type cells [35]. To be specific, in an H-type cell, the maximum FE for C_{2+} products ~41% at −0.95 V vs. RHE with a total current density of ~30 mA cm^{-2} was achieved. However, the FE for C_{2+} products increased to 53% −0.92 V vs. RHE with an overall current density easily exceeding 126 mA cm^{-2} produced in a flow-type cell [35]. This excellent performance using flow reactor, attributed mainly originating from the triple interface interaction (gas-electrolyte-catalyst) resulting in an improved CO_2 diffusion and suppressing hydrogen evolution reaction (HER) side reaction.

Besides, a nanoporous Cu catalyst was synthesized for ECDRR toward C_{2+}, and their performance was tested using various electrolytes ($KHCO_3$, KOH, K_2SO_4, and KCl) in a membrane-based MEC [36]. Their study demonstrates that CO_2 reduction in an alkaline electrolyte (KOH) exhibited the highest reported ECDRR activity and selectivity toward C_{2+} production with FE 62% at −0.67 V vs. RHE and current density 653 mA cm^{-2} in 1.0 M KOH electrolyte [36]. Moreover, a strong correlation between the FE for C_{2+} products and current density has also been studied, which revealed that the FE of C_{2+} and current densities are directly proposal to the calculated surface pH. It is understood that the highly alkaline KOH solutions are always used in flow cells because of an excellent performance rather than H-type cells [39]. It is ascribed to the fact that KOH electrolyte could provide higher electrolyte conductivity, decrease ohmic resistance between the thin electrolyte coating on the catalyst and bulk electrolyte, and further decrease the required overpotentials for C_{2+} formation [41]. In addition to Au and Cu catalysts, bismuth oxyhalides (BiOBr) 2D catalysts were synthesized by annealing carbon paper-coated BiOBr for CO_2-to-formate production [33]. In the H-type cell, this catalyst achieved high current density ~80 mA cm^{-2} despite of mass transfer limitation; meanwhile, in the case of microfluidic flow cell, the maximum current density could be reached

~200 mA cm^{-2} in a 2.0 M $KHCO_3$. This study revealed the advantage of gas-phase reactor over liquid-phase for ECDRR process.

Since the 1980s, the working electrode (nanoporous metal foil, nanomaterial, oxide-derived material, facet defective materials, and so on) has been widely studied in the ECDRR field. However, due to the sluggish mass transfer [42], product selectivity, along with intrinsic catalyst performance, still suffers toward its practical application. To address these issues, recently, a novel electrode such as GDEs was studied for ECDRR, where the conventional GDEs were usually composed a CL and a GDL, as shown in Figure 18.3a. [43] As a result, the high cell performance is obtained than any other fuel cell because of the three phases (gas-liquid-catalyst) interface formed, which could facilitate CO_2 diffusion to catalyst particles in the GDE cell. A GDL is a porous substrate composed of carbon fibers (typically carbon paper or carbon cloth) and a microporous layer (usually CNF or pressed carbon particles) [1], where electrocatalyst is deposited onto GDL by various physical methods like drop-casting paint, air-brush, and so on. This type of engineering provided high mass transport along with high current density because of prolonged contact between the reactants and electrocatalysts on the electrode surface [1]. The GDEs are usually classified into single layer (macroporous material), where the hydrophobic layer could be a metal mesh or metal foam, and double layer (both macroporous and microporous materials) [44], wherein microporous material, a hydrophobic layer with small pores that are composed of carbon powder and polytetrafluoroethylene (PTFE) [10, 20]. Sargent and coworkers designed PTFE-based GDEs where Cu catalyst was separated by the hydrophobic PTFE layer and carbon NPs, as shown in Figure 18.3b [10]. This architecture revealed that the FE ethylene formation increased 70% in 7.0 M KOH solution and at current density ~100 mA cm^{-2} with 150 hours stability which is 300 times higher than conventional GDEs. Moreover, the various effects such as porosity, thickness, and density on GDEs have also been studied in various fuel cells and water electrolyzers in the past decade, while in the case of CO_2 electrolyzer, these parameters are in early stage. It is also well understood that in the GDL, the control wetting (appropriate hydrophilicity) of the carbon could be a main challenge due to the HER side reaction.

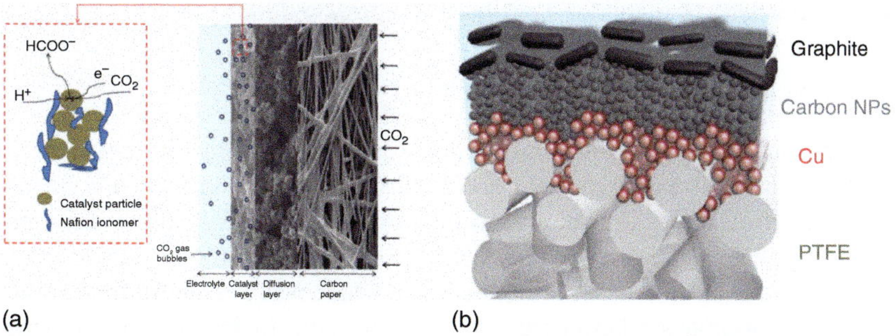

Figure 18.3 (a) A conventional schematic diagram of a gas diffusion electrode (GDE). Source: Wu et al. [43]. © 2014, Elsevier. (b) A schematic modified polymer-based GDE. Source: Dinh et al. [10]. © 2018, AAAS.

18.3 Requirements and Challenges for Next-Generation CO_2 Reduction Cell

In the ECDRR process, the electrocatalysts are not only decided on the current density and product selectivity but also largely depend on electrolytic device configuration. As we mentioned previously, a series of electrolytic cells, including H-type cells and different flow-type cells, are reviewed to provide valuable insights into cell design for ECDRR [1]. H-type cell architecture has been most widely used in laboratory-scale tests for a wide range of catalyst screening and currently commercially available to evaluate the performance of the catalyst in the liquid-phase system for ECDRR. However, the detection of liquid products in the liquid-phase system could be a challenging task which was addressed by designing a modified H-type cell in sandwich form with a high ratio of the electrode surface area to the electrolyte volume (S/V) which was also designed for better determination of liquid products [1, 19]. Nevertheless, the limitation of mass transfer efficiency derived from low solubility of CO_2 in aqueous electrolyte solutions, the flow-type cells with efficient mass transfer efficiency were proposed for further large-scale commercial application of ECDRR [1]. Despite various types of the flow cell have been studied, the status of these flow cells is still in early stage, due to the complicated cell structure, cost, and the lower electrode size 10 m^2. For the next generation, the electrochemical cell should take advantage of the development of electrolyte chemistry, catalyst, and electrode [45]. In this regard, a more profound study on gas-phase CO_2 stream, suitable PEM and microfluidic structure, GDE, SOEC, and DEMS cells needs to be considered carefully for further industrial application [1].

To summarize, in this chapter, we have discussed the progress on recent developing heterogeneous materials from pioneers to advanced nanomaterials as an electrocatalyst, fundamental influencing factors along with the device engineering for electrochemical CO_2/CO reduction toward the production of various value-added chemicals and fuels. In the last five decades, CO_2 emission in the atmosphere has been increasing due to the rapid industrialization and deforestation, causing energy and environmental issues. To solve these problems, electrochemical CO_2/CO reductions in aqueous media have widely been studied in recent years. Since the pioneering work in this field, tremendous work has been proposed including:

18.3.1 Wide Range of Electrocatalysts

Bulk metals: Initially, the bulk metals (Au, Ag, Pd, Cu, and so on) were considered as an electrocatalyst where different products (C_1 and C_{2+}) have been reported for ECDRR. Nevertheless, most of the transition metal catalysts usually required high overpotentials and provided low yield in the production of the high-order hydrocarbon due to the linear scaling reaction, which was the main challenge in transition metal catalysts. Besides, due to the scarcity, the high cost could also step down its practical application.

Nanoscale metals: Various research groups have widely studied nanoscale materials (NPs, NCs, nitrogen nano-diamonds [NDs], nanowires [NWs], and so on) due to their superior ability to improve catalytic activity and selectivity of ECDRR in contrast to bulk/polycrystalline metal. Despite these advantages, most of these class of catalysts showed a reduction of CO_2 only into $2e^-$ products (CO) with low stability and, as a result, may not meet the industrial requirements.

Advanced carbon-based electrocatalyst: In the last few years, the research on carbon-based nanostructure, i.e. single-atom catalysts (SACs) and heteroatom-doped carbon-based materials, has been extensively studied and this strategy could give new insight for an intensive direction for ECDRR. Nitrogen-doped carbon coordinate with a single metal site like $M–N_x$ in M–N–C structures (Ni, Fe, Cu, and Zn) has shown a superior performance including high FE_{CO} at moderate overpotential compared to the pure metal catalyst; this is due to breakdown of the linear scaling reaction (absence of d-band). Similar to the other catalyst selectivity and more important, the control synthesis, especially high metal loading >1% in SACs without any agglomeration of metal nanoparticles, remains challenging. The long-term durability of M–N–C catalysts in the aqueous electrolyte is another challenge due to the low energy barrier of adsorption of intermediate species on most of the metal centers, especially transition metals [45].

Two-step electrochemical reduction: In addition to one-step ECDRR, this chapter has also reviewed the two-step electrochemical reduction, such as "CO reduction," called electrochemical carbon monoxide reduction reaction (ECORR) (where CO acts as a reactant). ECORR has a superior advantage over ECDRR, such as allowing the formation of higher value-added product C_{2+} and C_{3+} along with oxygenates on the Cu electrode, as shown in Figure 18.4. This two-step strategy showed comparable performance such as high FE toward C_{2+}, C_{3+}, and oxygenate products with lower overpotential, suppression of HER, and achieved large current density in aqueous solution.

18.3.2 Fundamental Factor Influencing the Catalytic Activity for ECDRR

Electrolyte: The choice of electrolytes plays a pivotal role in determining the performance of catalysts. Two main types of aqueous electrolytes, $KHCO_3$ and KOH, were used for CO_2 reduction [45]. The HCO_3^- can reach equilibrium with CO_2, acting as a "CO_2 buffer," but whether the CO_2 source is from HCO_3^- dissolved aqueous CO_2 is not clear yet. It is well known understood that KOH solution (high concentrated) can suppress HER. But it is impossible to circulate KOH in the flow cell because KOH reacts with CO_2 to form carbonate, which in turn decreases local OH^- concentration [45]. Furthermore, the ionic liquid can coordinate with CO_2 that lowers energy barrier to form CO; however, some challenges still need to be considered, such as high cost, the solvent viscosity, and poor stability under the ECDRR condition.

The major variables within the electrolyte: In addition to electrocatalyst materials, the effect of bulk pH and local pH played deciding factors for the C_2 vs. C_1 pathways at differences in pH with water media act as the proton source in ECDRR. Several studies demonstrated that the formation of H_2 and CH_4 was pH-dependent and was consistent with a coupled elementary proton–electron transfer mechanism, whereas the CO and C_{2+} (C_2H_4) formation was pH-independent, due to the decoupled elementary proton–electron transfer mechanism on the standard hydrogen electrode (SHE) scales. In short, in highly acidic condition (at low pH = 1), the C_1 pathway was proceeding through COH_{ad} to $CHOH_{ad}$ intermediate by kinetic suppression of the multi-carbon production, at neutral condition (at pH = 6.8–7.2), C_1 and C_2 (C_3) production was proceed through a common intermediate COH_{ad}, and in highly basic conditions (at pH = 12), the selectivity of multi-carbon products was found by kinetic suppression of C_1 pathways. Furthermore, recent studies demonstrate that the ECDRR/ECORR selectivity and activity are strongly influenced by the size of cations and anions and different electrolyte concentration. Pioneering work suggested that the size of cations is directly proportional to the production of C_{2+} species and inversely proportional to the selectivity for the HER [46]. Specifically, recent studies have instead suggested that the cation acts as a promoter, stabilizing certain ECDRR/ECORR intermediates on the Cu, Ag, and Sn electrode surface. Because cation-created electrostatic field could decrease the $*CO_2$ adsorption energy, resulting in *OCCO or *OCCHO possible precursors to C_{2+} products. The impacts of the electrolyte anion have been studied to a lesser extent because the majority of ECDRR has been studied in $KHCO_3$ electrolytes. Nevertheless, according to the Resasco et al. study, it was found that formation of CO, $HCOO^-$, and C_{2+} products was insensitive to anion identity, while the formation of H_2 and CH_4 was found to be increased (≤1 order of magnitude) with decreasing pK_a of the buffering anion and was not impacted by non-buffering anions [47].

Electrochemical conditions: Several groups have also reported the impact of temperature and pressure on ECDRR selectivity and activity. The higher selectivity of ECDRR was achieved as the temperature decreases, while lower selectivity at a higher temperature where H_2 evolution becomes dominant. Moreover, as the CO_2 or CO pressure increases <1 atm, the ECDRR is dominant and HER is suppressed. It is well understood that the ECDRR selectivity and current density were enhanced by increasing CO_2 or CO pressure during ECDRR. Furthermore, Mul and coworkers study demonstrated that the C_2H_4 selectivity and activity are directly proportional to the pressure because high pressure could enhance the local CO concentration and surface coverage on the Cu electrode [48].

18.3.3 Device Engineering

H-type cells: The H-type cells' architecture is the most widely used in laboratory-scale tests to evaluate the performance of the catalyst in a liquid-phase system for ECDRR because of the wide range of catalysts screening and currently commercially available along with easy to handle and cost-effective. But, due to the low solubility of CO_2 in aqueous solution, limited electrode surface area, and large interelectrode distance, it could step down laboratory-scale to industrial-scale application.

Flow-type cells: In recent years, different types of flow cells, including MEA and MEC, have been widely studied for ECDRR in the gas phase. Despite the advanced activity, selectivity, and industrial compatible current density (>200 mA cm^{-2}) of flow-type cells compared to H-type cells, "flooding" of GDEs when liquid electrolyte flows was the major challenge in the flow-type cells.

Despite the advancement research on metal catalysts, metal-free catalysts, as well as carbon-based materials, have extensively studied in the laboratory-scale for electrocatalytic CO_2/CO reduction. Nevertheless, it is an urgent need by strengthening the laboratory-scale to the pilot-scale application; therefore, more research and developments toward practical application still need to be carefully considered in the following areas:

Next-generation electrocatalyst: Highly porous material such as metal-organic frameworks (MOFs) and covalent organic frameworks (COFs) has become newly developed in recent years, which showed high performance for ECDRR due to their structural appearance. However, the state of the art of these kinds of materials is still at an early stage. Interestingly, some studies disclosed boron and nitrogen nano-diamond (BND) and nitrogen-doped graphene quantum dots (NGQDs) as the only two most promising metal-free catalysts, which could achieve C_2H_4 and C_2H_5OH production beyond Cu metal in ECDRR [41, 49]. Unfortunately, these materials have been synthesized by costly plasma or hot filament chemical vapour deposition (CVD) methods, which could scale down the real practical application. Therefore, systematic studies are needed to understand better the impact of synthesis strategies on ECDRR product selectivity. In addition to the experimental study, recently, theoretical calculations have also been reported, including various 2D matrices such as MXenes (carbides and nitrides) for the conversion of CO_2-to-CH_4 [50] black phosphorus (BP), and boron and nitrogen (BN)-decorated SACs showed high activity toward ECDRR.

Active site and reaction mechanism: To increase the electrocatalytic performance, especially current density, more research is necessary to design more advanced catalysts containing active sites with higher intrinsic activity as well as richer density [45]. For this reason, to understand the origin of the active site of metals, carbon-based materials, metal-free carbon catalysts along with heteroatom doping would provide a new direction to design more efficient catalysts for ECDRR. To date, there is still no single established method to quantitatively measure the activity of active sites in the electrocatalysis process. The detection of the origin of the active site and understanding an appropriate reaction mechanism could be possible by reducing the gap between experimental study and theoretical study like DFT, machine learning (ML) [51, 52] operando, and in situ studies (outside the scope of this chapter).

Electrode stability: To date, the vast majority of research has been focused on the synthesis of electrocatalyst, activity, and selectivity for CO_2 reduction into various value-added chemicals; however, research on electrode preparation and their long-term stability including the catalyst and supporting composite in the

ECDRR/ECORR is still in an early stage. The engineering of the electrode design would be an urgent need to avoid flooding of GDEs.

Device engineering: H-type cells are studied for gas product detection in aqueous solutions, while the detection of liquid production is still challenging. However, modified H-type cells have been reported to overcome these issues, but this strategy is still at an early stage. In recent years, various fuel-type cells such as suitable PEM and microfluidic structure, GDE, SOEC, and DEMS cells need to be considered carefully for further industrial applications.

References

1 Liang, S.Y., Altaf, N., Huang, L. et al. (2020). *J. CO2 Util.* 35: 90–105.
2 Hori, Y., Kikuchi, K., and Suzuki, S. (1985). *Chem. Lett.* 14: 1695–1698.
3 Liu, G.Y., Pang, Y., Zhang, B. et al. (2016). *Nature* 537 (7620): 382–386. https://doi.org/10.1038/nature19060.
4 Lee, S.Y., Jung, H., Kim, N.K. et al. (2018). *J. Am. Chem. Soc.* 140: 8681–8689.
5 Zhao, C. and Wang, J. (2016). *Chem. Eng. J.* 293: 161–170.
6 Narayanan, S.R., Haines, B., Soler, J., and Valdez, T.I. (2011). *J. Electrochem. Soc.* 158: A167–A173.
7 Bevilacqua, M., Filippi, J., Lavacchi, A. et al. (2014). *Energy Technol. (Weinheim, Ger.)* 2: 522–525.
8 Salvatore, D.A., Weekes, D.M., He, J. et al. (2017). *ACS Energy Lett.* 3: 149–154.
9 Rosen, B.A., Salehi-Khojin, A., Thorson, M.R. et al. (2011). *Science* 334: 643–644.
10 Dinh, C.T., Burdyny, T., Kibria, M.G. et al. (2018). *Science* 360: 783–787.
11 Ma, S.C., Sadakiyo, M., Luo, R. et al. (2016). *J. Power Sources* 301: 219–228.
12 Ebbesen, S.D., Knibbe, R., and Mogensen, M. (2012). *J. Electrochem. Soc.* 159: F482–F489.
13 Uhm, S. and Kim, Y.D. (2014). *Curr. Appl. Phys.* 14: 672–679.
14 Lv, H.F., Zhou, Y.J., Zhang, X.M. et al. (2019). *J. Energy Chem.* 35: 71–78.
15 Clark, E.L., Singh, M.R., Kwon, Y., and Bell, A.T. (2015). *Anal. Chem.* 87: 8013–8020.
16 Clark, E.L. and Bell, A.T. (2018). *J. Am. Chem. Soc.* 140: 7012–7020.
17 Weekes, D.M., Salvatore, D.A., Reyes, A. et al. (2018). *Acc. Chem. Res.* 51: 910–918.
18 Sun, Z.Y., Ma, T., Tao, H.C. et al. (2017). *Chem-Us* 3: 560–587.
19 Kuhl, K.P., Cave, E.R., Abram, D.N., and Jaramillo, T.F. (2012). *Energy Environ. Sci.* 5: 7050–7059.
20 Garg, S., Li, M.R., Weber, A.Z. et al. (2020). *J. Mater. Chem. A* 8: 1511–1544.
21 Wang, Y.F., Han, P., Lv, X.M. et al. (2018). *Joule* 2: 2551–2582.
22 Jayashree, R.S., Yoon, S.K., Brushett, F.R. et al. (2010). *J. Power Sources* 195: 3569–3578.
23 Nitopi, S., Bertheussen, E., Scott, S.B. et al. (2019). *Chem. Rev.* 119: 7610–7672.
24 Li, Y.G.C., Zhou, D.K., Yan, Z.F. et al. (2016). *ACS Energy Lett.* 1: 1149–1153.
25 Song, J.T., Song, H., Kim, B., and Oh, J. (2019). *Catalysts* 9: 224.

26 Gutiérrez-Guerra, N., Moreno-López, L., Serrano-Ruiz, J.C. et al. (2016). *Appl. Catal. B* 188: 272–282.

27 Wang, X., Wang, Z.Y., de Arquer, F.P.G. et al. (2020). *Nat. Energy* 5: 478–486.

28 Del Castillo, A., Alvarez-Guerra, M., Solla-Gullón, J. et al. (2017). *CO2 Util.* 18: 222–228.

29 Pătru, A., Binninger, T., Pribyl, B., and Schmidt, T.J. (2019). *J. Electrochem. Soc.* 166: F34–F43.

30 Luo, W., Zhang, J., Li, M., and Zuttel, A. (2019). *ACS Catal.* 9: 3783–3791.

31 Möller, T., Ju, W., Bagger, A. et al. (2019). *Energy Environ. Sci.* 12: 640–647.

32 Yang, H., Lin, Q., Zhang, C. et al. (2020). *Nat. Commun.* 11: 593.

33 Garcia de Arquer, F.P., Bushuyev, O.S., De Luna, P. et al. (2018). *Adv. Mater.* 30: e1802858.

34 Verma, S., Hamasaki, Y., Kim, C. et al. (2017). *ACS Energy Lett.* 3: 193–198.

35 Zhuang, T.T., Liang, Z.Q., Seifitokaldani, A. et al. (2018). *Nat. Catal.* 1: 421–428.

36 Lv, J.J., Jouny, M., Luc, W. et al. (2018). *Adv. Mater.* 30: e1803111.

37 Whipple, D.T., Finke, E.C., and Kenis, P.J.A. (2010). *Electrochem. Solid-State Lett.* 13: D109–D111.

38 Hoang, T.T.H., Ma, S.C., Gold, J.I. et al. (2017). *ACS Catal.* 7: 3313–3321.

39 Fan, L., Xia, C., Yang, F. et al. (2020). *Sci. Adv.* 6, eaay3111.

40 Verma, S., Lu, X., Ma, S. et al. (2016). *Phys. Chem. Chem. Phys.* 18: 7075–7084.

41 Wu, J., Ma, S., Sun, J. et al. (2016). *Nat. Commun.* 7: 13869.

42 Ma, S.C., Luo, R., Gold, J.I. et al. (2016). *J. Mater. Chem. A* 4: 8573–8578.

43 Wu, J.J., Sharma, P.P., Harris, B.H., and Zhou, X.D. (2014). *J. Power Sources* 258: 189–194.

44 Park, S., Lee, J.W., and Popov, B.N. (2006). *J. Power Sources* 163: 357–363.

45 Wu, J., Sharifi, T., Gao, Y. et al. (2019). *Adv. Mater.* 31: e1804257.

46 Murata, A. and Hori, Y. (1991). *Bull. Chem. Soc. Jpn.* 64: 123–127.

47 Resasco, J., Lum, Y., Clark, E. et al. (2018). *ChemElectroChem* 5: 1064–1072.

48 Kas, R., Kortlever, R., Yılmaz, H. et al. (2015). *ChemElectroChem* 2: 354–358.

49 Liu, Y., Zhang, Y., Cheng, K. et al. (2017). *Angew. Chem. Int. Ed. Engl.* 56: 15607–15611.

50 Li, N., Chen, X., Ong, W.J. et al. (2017). *ACS Nano* 11: 10825–10833.

51 Zhong, M., Tran, K., Min, Y.M. et al. (2020). *Nature* 581: 178.

52 Daiyan, R., Saputera, W.H., Masood, H. et al. (2020). *Adv. Energy Mater.* 10: 1902106.

Part VIII

Computations-Guided Electrocatalysis

Although the arrays of experimental techniques show their ability to measure the activity and selectivity of an electrocatalyst – by quantifying the electrical current density and product distribution – besides getting into kinetic information via Tafel measurements, the demand to (i) interpret the experimentally observed electrocatalytic behavior, (ii) find the active site characteristics–performance relationship, (iii) impact our ability to embrace the fine details of catalytic reactions including the stability for included reactants, intermediates, and products, and (iv) calculate the kinetic and thermodynamic parameters (such as activation barriers of the involved elementary steps and their free energies), (v) predict the performance and activity of new catalysts or optimize already existing ones become required in catalysis science. In this sense, computational methods are a valuable tool to achieve a comprehensive understanding of a catalytic process. This part is arranged around computations in electrocatalytic cycles. In Chapters 19–22, we provide some needed background on the catalytic process addressing in the interfacial electrode–electrolyte structure, electrochemical-related kinetics and thermodynamics, and potential-related energetic changes. Section 24 reviews computational-based catalytic concepts such as catalyst screening, descriptors, and Sabatier principle, as well as density functional theory (DFT) and its modeling approaches such as the "Computational Hydrogen Electrode" model, solvation, and kinetic simulations before directly addressing the rational design toward catalyst development (Section 25) followed by DFT-based computation-driven insights of selected electrochemical systems (Section 26). Finally, we provide DFT perspectives and challenges in future directions.

Electrocatalysis in Balancing the Natural Carbon Cycle, First Edition. Yaobing Wang.
© 2021 WILEY-VCH GmbH. Published 2021 by WILEY-VCH GmbH.

19

Insights into the Catalytic Process

Catalysis is a process of affecting the rate of chemical transformation and controlling the reaction productivity using a catalyst [1, 2]. An appropriate catalyst should affect the rate of the reaction without modifying its equilibrium position in terms of the overall standard Gibbs energy change [3]. Common catalysts are solids or liquids. If the media of the chemical transformation on these catalysts occurs in one phase, then the catalytic process is called homogeneous catalysis. On the other hand, heterogeneous catalysis includes the presence of a solid surface as a different phase, which is easy to be separated from the reaction (as an essential advantage). In this section, we focus on the nature of this solid interface in electrochemical catalysis, kinetics, and thermodynamics of catalytic process, the effects of applying potential on such physical relations over electron transfer-involved reactions, and the challenges to face the understanding of the electrocatalytic process.

19.1 Electric Double Layer

Electrochemical reactions – that we are interested in – are considered as a type of a heterogeneous catalytic system; however, it involves a kinetic parameter in addition to the reactant concentration and temperature ones. This parameter is the potential which relates to the electrode surfaces, where chemical bond dissociation and formation as electron and ion transfer occur. The overall rate of the electrochemical reactions is determined by the electrochemical reaction kinetics on these interfacial structures and it varies according to their chemical properties. Thus, the question arises: What is the nature of this interface in electrocatalytic reactions? The phenomena at the solid–liquid interface are much more complicated. The electrode–electrolyte interface is considered to be an "electrified interface" where electrocatalytic reactions occur between a charged surface and the countercharge that distributed in the electrolyte [4, 5]. This charge separation produces a potential gradient across the interfacial region. For a solid–solution interface, a potential of zero charge appears, at which the countercharge in the electrolyte does not affect the electrode surface. If the electrode potential positively shifted from the potential of zero charges, the electrode surface then evolves electrons into the circuit allowing the electrode to carry a positive charge. The positive electrode will

Electrocatalysis in Balancing the Natural Carbon Cycle, First Edition. Yaobing Wang.
© 2021 WILEY-VCH GmbH. Published 2021 by WILEY-VCH GmbH.

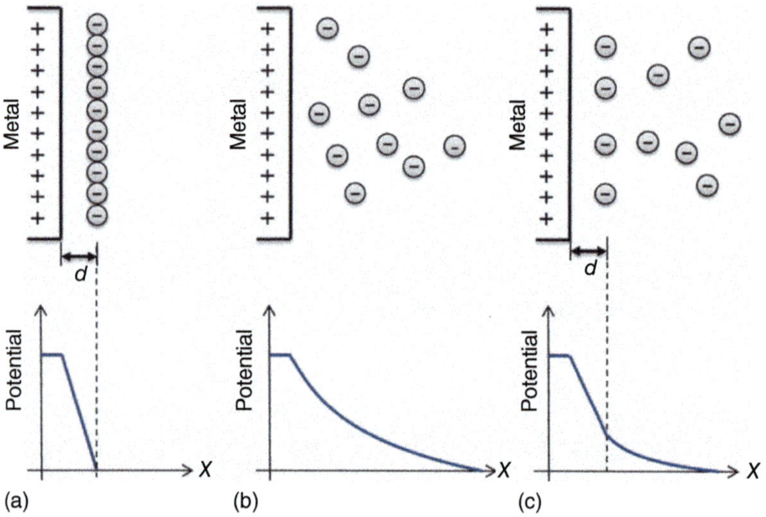

Figure 19.1 Double-layer models, according to (a) Helmholtz, (b) Gouy–Chapman, and (c) Stern models with their potential profiles. Source: Asthagiri and Janik [4]. © 2014, Royal Society of Chemistry.

attract the negatively charged ions in the electrolyte. This charged interfacial region is called the "Electric double layer." Several theoretical models have been proposed to describe this phenomenon; Figure 19.1.

A) *Helmholtz model*: The double layer approximated that the charged electrode will be neutralized by the countercharged ion in a parallel plane at an increment of distance *d*. This model predicts a linear potential decline across this interfacial region resulting in a constant capacitance and constant electric field. This approximation does not interpret the real system.
B) *The Gouy–Chapman model*: This model takes into account the dynamic feature of the counterion by applying Maxwell–Boltzmann statistics. The ion concentration fluctuates within the diffusion layer, with the net charge decreasing exponentially, moving away from the surface. This approach represents the behavior of a diluted system with a small range of potential. However, it does not express the constant capacitance nearby the electrode surface at high potentials.
C) *The Stern model*: The diffuse double layer is modified by the combination of the former models. It integrates the fixed sheet of countercharge to partially neutralize the surface with a diffusion layer next to the sheet.

These approximations help to capture the general nature of the electrochemical interfacial layer at a macroscopic level, not at the atomistic level, getting rid of the solvation structure and dynamic feature of the dissolved ions as well as the chemical interactions between these ions and the electrode surface. Taking into account that the key elementary steps of the catalytic reaction parallelly occur on specific positions (active sites) of this catalytic electrode surface. Recently, the significance of the electric double layer has been presented using computational modeling to understand the local environment of the electrocatalytic reactions [6–8], such as CO_2 and O_2 reduction and the factors which affect it, such as cations, solvents, and electrolyte concentration [9–14]. Accurate representation of the electrode structure of particular interest at the atomic level is very important; thus, the computational electrocatalysis is required to probe such surface considerations, as a primary step to realize the environment and mechanisms of the electrocatalytic reaction.

19.2 Kinetics and Thermodynamics

Catalysis lies in increasing the rate of the chemical reaction and guiding it toward maximizing the selectivity by enhancing the probability of interactions between reactants affecting the rate equation [3, 15]. The kinetics rate and selectivity depend on the catalyst characteristics. In this regard, the fundamental understanding of kinetic and thermodynamic parameters is pivotal for the design of new catalysts. These parameters provide the connection between the microscopic properties of the catalyst, adsorption energies, and activation barriers of elementary steps and experiments [16]. In terms of the Gibbs free energy, a chemical transformation can be represented as a reaction in which reactants (R) derive to products (P) over transition state (TS). As shown in Figure 19.2, the free energy of the activation barrier (G_c^*) is decreased in catalyst-based reaction compared to the non-catalyst-based one (G_{nc}^*). Consequently, the rate of constant of catalyzed reaction increases according to the Arrhenius equation:

$$K = Ae^{-\frac{G^*}{RT}} \tag{19.1}$$

where K is the rate constant, A is the pre-exponential factor, G^* is the activation energy, R is the gas constant, and T is the absolute temperature. Kinetics are related to the position of activation barriers, while thermodynamics are associated with the stability of intermediates [17]. Thus, the catalyst acts as a promoter affecting the activation energy, not the overall reaction energy (G_R) [3]. If the reaction of the Gibbs free energy change is negative (positive) value, then the reaction is called spontaneous (non-spontaneous) process and vice versa.

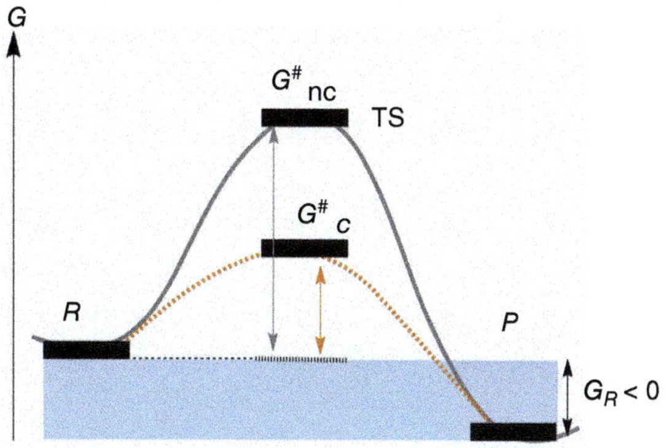

Figure 19.2 Schematic representation of catalyzed and non-catalyzed chemical reactions in terms of the Gibbs free energy. Source: Marken and Fermin [3]. © 2018, Royal Society of Chemistry.

Many catalytic reactions involve multiple steps through different intermediates species over their TSs. Accordingly, different thermodynamics appears depending on the nature of the interactions between these intermediates and the catalyst. The reaction mechanism is determined by the overall minimum energy route in which the mechanism could be governed by kinetic or thermodynamic control [3]. By Gibbs free energy diagrams, researchers representing the mechanisms of various catalyzed reactions. Understanding and visualization of such mechanisms require basic knowledge of physico-chemistry fundamentals of the reaction. In this regard, three energy quantities are widely used: the Gibbs free energy (G), the electronic energy (E), and the enthalpy, which is related by:

$$G^{\circ} = H^{\circ} - TS \tag{19.2}$$

where $G°$ and $H°$ refer to the standard Gibbs free energy and enthalpy, and T and S denote temperature and entropy, respectively.

At constant pressure and temperatures of >0 K, the enthalpy term can be expressed by the heat capacity (C_P); besides, the entropy can be expressed as the sum of the translational, rotational, vibrational, and electronic contributions. Thus, the Gibbs free energy (G) is modified to:

$$G = H^{\circ} + \int C_P dT - T(S_t + S_r + S_V + S_e) \tag{19.3}$$

Taking into account the intrinsic nature, vibrational zero-point energy (ZPE) term should be added and also the extrinsic dispersion (D) correction. Further, the Gibbs free energy (G) is represented by:

$$G = H^{\circ} + \int C_P dT - T(S_t + S_r + S_V + S_e) + \text{ZPE} + D \tag{19.4}$$

For any heterogeneous catalytic reaction, three different Gibbs free energy changes are typically calculated. These energy changes are the binding (adsorption), activation barrier, and reaction Gibbs free energy changes. Designing thermodynamic–kinetic models and gaining related mechanistic insights for different electrochemical reactions have been recently reported [18–22]. These calculations provide fundamental information to further propose the reaction mechanism of multiple steps and define the reactivity and stability parameters toward rational insights for developing the desired catalyst [23].

19.3 Electrode Potential Effects

Consider the electron transfer reaction where an electron (e^-) is passed from donor to adsorbed acceptor (A^*). At the equilibrium potential, $U°$, the chemical potentials of A and ($A^+ + e^-$) are equal. By applying potential U greater than $U°$, the potential change becomes ($\Delta U° = U - U°$). Consequently, the reaction free energy change, as shown in Figure 19.3, is given by:

$$\triangle G^{\circ} = nFe^{-} \triangle U^{\circ} \tag{19.5}$$

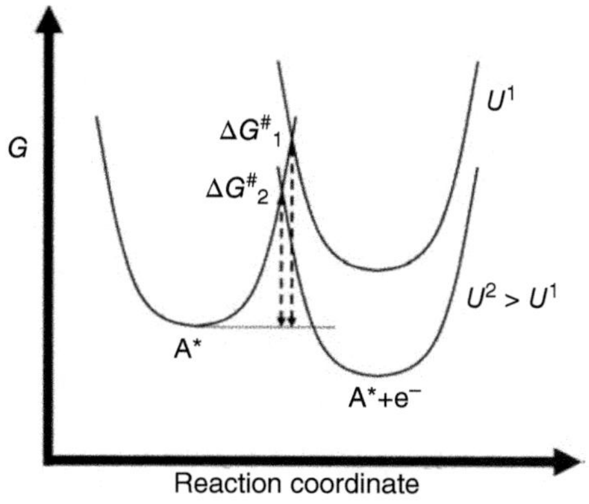

Figure 19.3 A hypothetical free energy relationship illustrating the reaction barrier change with the applied potential for A* species oxidation. Source: Asthagiri and Janik [4]. © 2014, Royal Society of Chemistry.

where F represents Faraday's constant and e^- is the elementary charge of the electron. The integer n in Eq. (19.5) refers to the number of electrons evolved by the reaction. This equation provides a relationship between the reaction free energy and the electrode potential [3, 4].

Now, the TS location for such a reaction is somewhere between R and P involving potentially electron transfer. Thus, the activation energy at potential U^2 is given by:

$$\Delta G_2^* = \Delta G_1^* + \beta F(U^2 - U^1) \tag{19.6}$$

where β is a factor referring to the relationship between the activation barrier change and the reaction energy change. $F(U^2 - U^1)$ refers to the elementary reaction free energy change due to the altered electrode potential. Thus, the activation barrier can be changed by applying potential and also determined at any potential (U^2) once the barrier is known at one potential (U^1). Thus, the potential is the important key factor affecting the electrocatalytic reaction [24–28], and potential dependence modeling has attracted great interest [29].

In light of the previous catalytic insights, physical studies and thermodynamic relations provide essential information about (i) the chemical potentials of intermediates and (ii) the electrode potential-based electron energy changes. Such meso- and the macro-level-based studies could provide mass and heat transport effects of heterogeneous catalysis. On the other hand, deep understanding requires efforts based on the atomistic scale for the physical and electronic properties of the entire electrocatalytic reactions. Notably, the electrochemical catalysis faces significant challenges such as (i) the interaction of the catalyst–intermediates with the ionic species of the electrolyte, (ii) the consequences that related to the electrified catalyst and charge separation between its surface and electrolyte, (iii) making clear composition\structure\function–activity\selectivity\catalyst stability connections, and (iv) unrevealing the mechanisms of electrocatalytic reaction over long time and length scales (Figure 19.4) to evaluate the dynamics of potential electrocatalytic systems. In this sense, the computational calculations, especially principles based on the electronic structure, demonstrate their power as highlighted in Chapters 20–22.

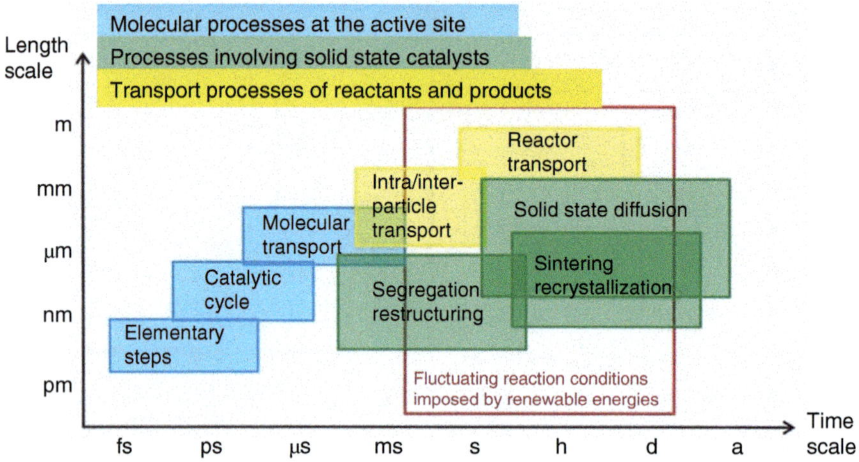

Figure 19.4 Typical time length scales in catalysis. Source: Kalz et al. [30]. © 2016, John Wiley & Sons.

References

1 Deutschmann, O., Knözinger, H., Kochloefl, K., and Turek, T. (2009). *Ullmann's Encyclopedia of Industrial Chemistry*. Wiley-VCH.

2 Pašti, I.A., Skorodumova, N.V., and Mentus, S.V. (2014). *React. Kinet. Mech. Catal.* 115: 5–32.

3 Marken, F. and Fermin, D. (2018). *Electrochemical Reduction of Carbon Dioxide: Overcoming the Limitations of Photosynthesis*, vol. 21. Croydon, UK: Royal Society of Chemistry.

4 Asthagiri, A. and Janik, M. (2014). *Computational Catalysis*, vol. 14. Royal Society of Chemistry.

5 Maheshwari, S., Li, Y., Agrawal, N., and Janik, M.J. (2018). Density functional theory models for electrocatalytic reactions. In: *Advances in Catalysis*, vol. 63 (ed. C. Song), 117–167. Academic Press.

6 Bohra, D., Chaudhry, J.H., Burdyny, T. et al. (2019). *Energy Environ. Sci.* 12: 3380–3389.

7 Lopes, P.P., Strmcnik, D., Jirkovsky, J.S. et al. (2016). *Catal. Today* 262: 41–47.

8 Ramaswamy, N., Ghoshal, S., Bates, M.K. et al. (2017). *Nano Energy* 41: 765–771.

9 Waegele, M.M., Gunathunge, C.M., Li, J., and Li, X. (2019). *J. Chem. Phys.* 151: 160902.

10 Dunwell, M., Yan, Y., and Xu, B. (2018). *Curr. Opin. Chem. Eng.* 20: 151–158.

11 Favaro, M., Jeong, B., Ross, P.N. et al. (2016). *Nat. Commun.* 7: 12695.

12 Brown, M.A., Abbas, Z., Kleibert, A. et al. (2016). *Phys. Rev. X* 6: 011007.

13 Brown, M.A., Goel, A., and Abbas, Z. (2016). *Angew. Chem. Int. Ed.* 55: 3790–3794.

14 Gauthier, J.A., Ringe, S., Dickens, C.F. et al. (2019). *ACS Catal.* 9: 920–931.

15 Toulhoat, H. (2010). Heterogeneous catalysis: use of density functional theory. In: *Encyclopedia of Materials: Science and Technology* (eds. K.H.J. Buschow, R. Cahn, M. Flemings, et al.), 1–7. Elsevier.
16 Norskov, J.K., Abild-Pedersen, F., Studt, F., and Bligaard, T. (2011). *Proc. Natl. Acad. Sci. U.S.A.* 108: 937–943.
17 Exner, K.S. and Over, H. (2017). *Acc. Chem. Res.* 50: 1240–1247.
18 Zhang, J., Tao, H.B., Kuang, M. et al. (2020). *ACS Catal.* 10: 8597–8610.
19 Wang, V.C.C. (2016). *Phys. Chem. Chem. Phys.* 18: 22364–22372.
20 Vincent, W. (2020). Beyond the Active Site: Mechanistic Investigations of the Role of the Secondary Coordination Sphere and Beyond on Multi-electron Electrocatalytic Reactions and Their Relations Between Kinetics and Thermodynamics. ChemRxiv.
21 Exner, K.S. (2019). *ACS Catal.* 9: 5320–5329.
22 Ceballos, B.M. and Yang, J.Y. (2018). *Proc. Natl. Acad. Sci. U.S.A.* 115: 12686–12691.
23 Tao, H.B., Zhang, J., Chen, J. et al. (2019). *J. Am. Chem. Soc.* 141: 13803–13811.
24 Steinmann, S.N., Michel, C., Schwiedernoch, R., and Sautet, P. (2015). *Phys. Chem. Chem. Phys.* 17: 13949–13963.
25 Rountree, E.S. and Dempsey, J.L. (2015). *J. Am. Chem. Soc.* 137: 13371–13380.
26 Zong, Y., Chakthranont, P., and Suntivich, J. (2020). *J. Electrochem. Energy Conv. Stor.* 17: 041007, 7 pages https://doi.org/10.1115/1.4046552.
27 Ruffman, C., Gordon, C.K., Skúlason, E., and Garden, A.L. (2020). *J. Phys. Chem. C* 124: 17015–17026.
28 Wang, L., Nitopi, S.A., Bertheussen, E. et al. (2018). *ACS Catal.* 8: 7445–7454.
29 Chan, K. and Nørskov, J.K. (2016). *J. Phys. Chem. Lett.* 7: 1686–1690.
30 Kalz, K.F., Kraehnert, R., Dvoyashkin, M. et al. (2017). *ChemCatChem* 9: 17–29.

20

Computational Electrocatalysis

Rational design of electrocatalytic reactions and their mechanisms at the atomic level would significantly help to improve its sustainable development in terms of energy efficiency and selectivity. To understand, investigate, and predict aspects of the electrocatalytic systems, it is essential to establish structure–reactivity relationships. This requires to identify the nature of intrinsic active sites and nature electrocatalytic interface through (i) preparing and optimizing a catalyst revealing homogeneous active sites, (ii) counting these sites by which activity can be normalized, and (iii) determining and measuring the reaction energetics and rates in addition to product distributions. Taking into account that synthesis protocols to configure catalytic active sites that could be restricted by electrical connectivity requirements of electrocatalytic reaction. Besides, gas adsorption or adsorption/desorption of ionic species to count the active sites faces additional requirements of ion transport accessibility and electrode potential conditions. Furthermore, to estimate electrocatalyst activity, electrical current density is typically quantified and normalized by the "electrochemically active surface area" (ECSA) or the number of active sites at a macroscopic level associated with a geometric surface area. Verifying that the transport process is not affected in reaction rate measurements is also an additional challenge. What is more, the distribution of products over different phases, such as gas and electrolyte phases, limits selectivity quantifying [1–4]. Thus, establishing the design principles that meet the above challenges and correlate the atomic structure to the catalytic activity is highly demanded. Computational calculations models and methods can help to study and interpret the observed electrocatalytic behavior and correlate structure properties with performance variables. Electronic structure calculations can overcome time and length scales by generating a model on the order of ~100 atoms and, further, tailoring the elementary reaction steps leading to energetics measurements of isolated active sites. Computational algorithm-based calculations can play a crucial role in understanding reaction mechanisms and designing desirable catalysts. In this section, we provide a theoretical overview of computational calculation starting the framework of density functional theory (DFT), recent descriptors and scaling relationships, and modeling, simulations, and methods of DFT.

Electrocatalysis in Balancing the Natural Carbon Cycle, First Edition. Yaobing Wang.
© 2021 WILEY-VCH GmbH. Published 2021 by WILEY-VCH GmbH.

20.1 Computational Screening Toward Calculation Theories

Research toward rational design of catalysts is ranging from trial-and-error investigations and high-throughput target-oriented screening initiatives to insight-driven and bottom-up approaches. Mapping synthesis–structure–property relationships through trial-and-error experimentation is time/labor/resources consuming resulting in an acceptable catalyst, not an efficient\desired one, taking into account many chemical transformations within the catalysis process. On the other hand, bottom-up developments through computational catalyst screening overcome such obstacles with gaining the advances in calculation speed as well as the accuracy of computational methods [1, 5, 6]. Catalyst screening counts on engineering reactivity patterns across various catalysts via several descriptors rather than predicting selectivities and quantitative rates for each catalyst. Once the screening database is built, a very short amount of time is required to identify catalyst candidates within the optimal range for a given reaction.

Computational screening would be possible in the presence of calculation-based theories that enable us to access the mechanistic details of chemical reactions and then establish key structure\activity relationships, subsequently leading to predictions of reactivity. Fortunately, the reactivity trends would be understood in terms of the electronic structure calculations. In this case, the electron (and spin) distribution details are important to explain bond cleavage and formation mechanism in catalytic reactions [7]. The electronic structure of a many-body system would be obtained by solving the time-independent Schrödinger equation:

$$\hat{H}\Psi(r,R) = E\Psi(r,R) \tag{20.1}$$

where $\hat{H}$ is the Hamiltonian operator, $\Psi(r, R)$ is the many-body wavefunction, E is the total energy of the system, and (r, R) refer to the summarized spin and spatial coordinates of electrons and nuclei, respectively. By separating the nuclear and electronic parts through an ansatz for the wave function of the form; the Born–Oppenheimer (BO) approximation [8, 9], to be:

$$\Psi(r,R) = \psi(r;\{R\})\chi(R) \tag{20.2}$$

where $\chi(R)$ denotes the nuclear wave function and $\psi(r; \{R\})$ refers to the electronic wave function; the electronic wave function relies parametrically on the nuclei positions. This works in case of that the first and second variants of the electronic wave function compared to the nuclear coordinates are small. To address electronic structure properties, all coordinates of all the electrons then are needed for electronic wave functions, besides, the number of atoms is very limited [10, 11].

Electronic structure properties would be obtained with rather three spatial coordinates of the electron density approach. Two theorems were formulated for nondegenerate ground states toward the development of DFT. First-principles DFT modeling has become a widespread tool in catalysis research. The two theorems can be summarized as

1) The ground-state properties of a many-electron system are uniquely directed by the electron density of states (DOS).
2) The total energy of a system has a minimum for the ground-state electron density.

The mathematical form of DFT which formulated by Hohenberg and Kohn (HK) can be written as:

$$E[\rho] = T[\rho] + \int \rho(r)V_{\text{ext}}(r)\text{d}r + E_{\text{ee}}[\rho] \tag{20.3}$$

where $E[\rho]$ accounts for the electron density, $T[\rho]$ refers to the kinetic energy functional, the second term donates the potential energy arising from the interaction with the external potential, and the last term refers to the electron–electron interaction [12]. To make this foundation applicable, the exact functionals that electron density depending on are unknown, Kohn and Sham (KS) developed the density function with an approximation; then, the total electronic energy is giving by

$$E[\rho] = T_s[\rho] + \int \rho(r)V_{\text{ext}}(r)dr + J[\rho] + E_{\text{XC}}[\rho] \tag{20.4}$$

where $T_s[\rho]$ and $J[\rho]$ are introduced to describe the kinetic energy of noninteracting electrons and classical Coulomb energy for a system, respectively [13]. The last term, an exchange correlation (XC), addresses the remaining contributions and corrections. Through the definition of one-electron KS orbitals ψ_i^{KS}, where the electron density is similar in the noninteracting and interacting systems. Thus, the ground-state electron density (Eq. (20.4)) could be reduced and calculated as [11]:

$$\rho(r) = \sum_i |\psi_i^{\text{KS}}(r)|^2 \tag{20.5}$$

Therefore, a new density can be solved by assuming an initial electron density, then constructing the Coulombic and exchange-correlation potential in the Hamiltonian. Therefore, valence electrons can be represented using KS calculations. The exact XC contribution functional is unknown; thus, it needs to be approximated. Many approaches were proposed such as local density approximation (LDA) and the generalized gradient approximation (GGA). The former depends on the ground-state density at a single point of a homogeneous electron gas. This works in bulk materials. The latter functional is rather advanced where it depends on the local density and the density gradient and works for metallic systems containing more physical information. Several improvements to the GGA have been suggested (e.g. DFT + U, DFT – D, meta-GGA, hybrid-GGA) [1]. The DFT framework presented above has been implemented in many popular software packages; such as ABINIT [14], CASTEP [15], Quantum espresso [16], and VASP [17].

To this end, tracking the electronic structure using DFT first principles provides the ability to define and calculate activation barrier energetics, interaction energies, reaction mechanisms in addition to their kinetics. The advanced atomistic insight-driven approaches can preferably help to solve some of the electrocatalysis challenges described in the introduction of Chapter 20. Undoubtedly, DFT calculations play an important role in interpreting practically observed catalytic behavior and helping to identify reactivity trends as well as predict the performance of new catalysts rather than analyze quantitative rates and selectivities for every single catalyst.

20.2 Reactivity Descriptors

To rationalize a trend for catalytic reactivity over different catalysts, a design concept\identifier is demanded. This identifier is termed as a catalytic reactivity descriptor. It could be an intrinsic or extrinsic property of catalysts such as the electronic structure, geometric feature, or the binding strength between the reactive intermediates and its surface. The utility of these descriptors lies in its ability to affect the kinetics of catalytic processes and further describe the catalytic performance [1, 3]. A detailed study of kinetic and atomistic thermodynamic functionals became possible upon taking account of the electronic-based DFT power allowing us to describe the microscopic characteristics of a catalyst. Doubtless, descriptor guidance is rational rather than computing the entire catalytic reaction. Further, such a reactivity descriptor could be scaled with another feature often in a linear trend helping to predict the catalytic performance of a given catalyst. Descriptors-based scaling relationships enable the facility toward high-throughput catalysts design and screening in addition to the increase our knowledge of the catalytic process. Herein, an overview is presented on various descriptors that have been used in electrocatalysis reactions.

To start a descriptors-based design, identifying the proper descriptor is required. For a specific reaction, many properties influence the catalytic performance of a presented catalyst ranging from chemical composition and texture or structure to interconnected characteristics such as electronic conductivity and redox potential, etc. Such variety further raises the question: Which is the most suitable descriptor that can control a given macroscopic reaction rate? To address this question, two requirements are outlined [1] as follows:

(I) The descriptor has to be sufficient enough to be estimated and achieve the aim of the study such as product selectivity not only activity.
(II) The descriptor should be uncomplicated and measurable to enable fast and efficient screening leading to scale a trend.

This could be singled out from underlying the dominant reaction mechanism and identifying the kinetically relevant steps and possible reaction intermediates and transition states. The energy correlations using modern DFT, the observed scaling relations, the kinetic measurements in combination with the educated guesses can guide the descriptor refining process. Most often descriptors are the chemisorption energies of key intermediates, the geometry of the catalyst surface, or the intrinsic parameter of the catalyst (Figure 20.1).

Once the appropriate descriptor(s) has been pinpointed, the computational calculation can be used to scan many possible catalysts of several materials classes leading to scaling relations. Consequently, experimentation work would verify the proposed design.

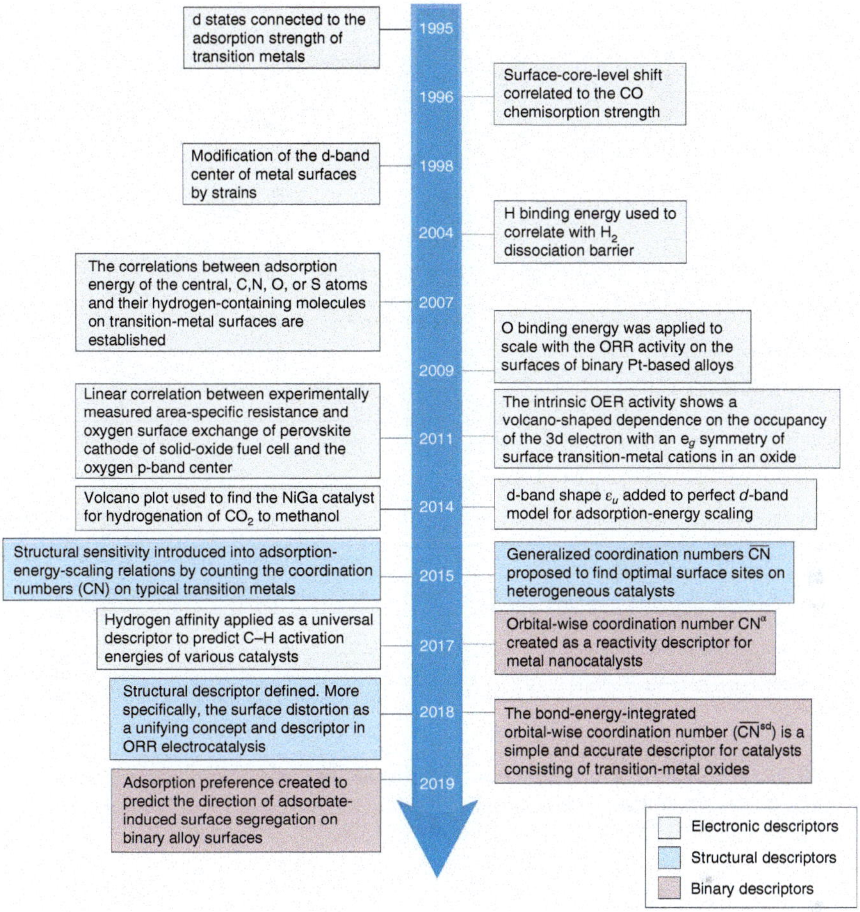

Figure 20.1 A timeframe of the development of reactivity descriptors and scaling relationships in heterogeneous catalysis. Source: Zhao et al. [18]. © 2019, Springer Nature.

20.2.1 d-band Theory Motivates Electronic Descriptor

Investigating the change in adsorbate\active site bond energies offers a valuable strategy for predicting the catalytic trend in electrolysis. Transition metals are dominant catalysts among the various materials for electrocatalytic systems. Hammer and Nørskov [19] developed a metal-based concept that the adsorption strength of key intermediate will vary from element to the next based on their d-band position resulting in the d-band model. Within this picture, the model helps to clarify the catalytic activity descriptor concept. The interaction between the molecular orbitals of the adsorbate and the electronic structure of the metal includes coupling to s then d states, Figure 20.2a. For the former coupling, all transition metals approximately contribute the same way as they have broad and half-filled s-bands getting the molecular adsorbate states to be widened. On the other hand,

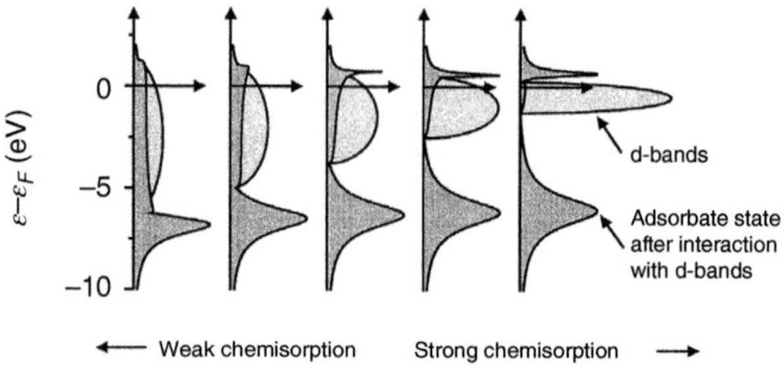

(a)

Idealized d-band filling

V^2_{ad} [Relative to Cu]

0.1 20.8 Ca 4.12	0.2 7.90 Sc 3.43	0.3 4.65 Ti 1.50 3.05	0.4 3.15 V 1.06 2.82	0.5 2.35 Cr 0.16 2.68	0.6 1.94 Mn 0.07 2.70	0.7 1.59 Fe −0.92 2.66	0.8 1.34 Co −1.17 2.62	0.9 1.16 Ni −1.29 2.60	1.0 1.0 Cu −2.67 2.67	1.0 0.46 Zn 2.65
0.1 36.5 Sr 4.49	0.2 17.3 Y 3.76	0.3 10.9 Zr 1.95 3.35	0.4 7.73 Nb 1.41 3.07	0.5 6.62 Mo 0.35 2.99	0.6 4.71 Tc −0.60 2.84	0.7 3.87 Ru −1.41 2.79	0.8 3.32 Rh −1.73 2.81	0.9 2.78 Pd −1.83 2.87	1.0 2.26 Ag −4.30 3.01	1.0 1.58 Cd 3.1
0.1 41.5 Ba 4.65	0.2 17.1 Lu 3.62	0.3 11.9 Hf 2.47 3.30	0.4 9.05 Ta 2.00 3.07	0.5 7.27 W 0.77 2.95	0.6 6.04 Re −0.51 2.87	0.7 5.13 Os 2.83	0.8 4.45 Ir −2.11 2.84	0.9 3.90 Pt −2.25 2.90	1.0 3.35 Au −3.56 3.00	1.0 2.64 Hg 3.1

Bulk Wigner–Seitz radius, s (a.u)

ε_d (eV)

(b)

Figure 20.2 (a) The local density of states (LDOSs) change where an adsorbate state interacts with the d bands at a metal surface [20]. (b) The 3d, 4d, and 5d transition metals with calculated d-band. Source: Hammer and Nørskov [20]. © 2000, Elsevier.

coupling to d-electrons can describe this interaction making a trend for different metals. The broadened adsorbate states combine with the narrow d-states of a given metal causing bonding and anti-bonding states splitting. Since the transition metals are considered as an electron reservoir, a specific amount of those electrons will occupy these new orbitals. The more anti-bonding states (occupied electrons) are located lower to the Fermi level, the weaker the resulting adsorption bond and vice versa [4, 20]. Thus, the state's location at Fermi level is determined by the d-band position (center), the average energy of the band, and varies upon adsorption strength. Transition metals with higher occupying d-bands have stronger adsorption strength reflecting adsorption characteristics of the catalyst. Figure 20.2b shows a theoretically calculated d-band center of different metals. The d band center concept is considered to be a highly valuable insight in the catalysis field.

A few approximations are applied in DFT calculation to measure the difference in adsorption energy of similar adsorbate (a) interact with two different active sites (s1 and s2). The adsorption energy of s1 is defined as

$$E_{ad} = E_{(s1+a)} - E_{s1} - a \tag{20.6}$$

Therefore, the change of adsorption energy between s1 and s2, following the derivation of Hammer and Nørskov, is given by

$$\delta E_{\text{ad}} = \delta E_{1\text{el}} + \delta E_{\text{es,A-s}} \tag{20.7}$$

The first term donates the one-electron energy difference between the a-s1 and the a-s2 systems. The second term refers to the difference in the adsorbate electrostatic interactions between the two systems. Using the Newns-Anderson model, Nørskov and coworkers calculated these terms [21]. The d-band model works along periodic table row of transition metals; however, it does not work for comparing metals along the periodic table column. This is due to different subshells; 3d and 4d, the Pauli repulsion will be shown between the adsorbate and d band of the metal resulting in the wrong indication of the strength of the bond. Therefore, the adsorption energy could be in trend with the same d-band filling of metals. Thus, predicting a trend using calculating the electron and adsorption energy differences is valid with a d-band descriptor proving the effectiveness of computation in catalytic screening.

20.2.2 Coordination Numbers Motives Structure Descriptor

Recently, the structure descriptor has been raised on the basis of the geometry of optimal active sites. The concept of structure descriptor lies in varying the catalytic activity over different surface sites on the same catalyst. It correlates the adsorption energies with the coordination numbers (CNs) of the binding atoms. By counting the CNs of the atoms of the active sites, the binding strength toward reaction intermediates can be predicted [18]. However, this restricts by finite-size effects in nanoscale systems, where CNs are sometimes ineffective for small nanoparticles. Consequently, generalized coordination number CN was introduced to extend the counting of nearest neighbors to weighting each first-nearest neighbor (j) corresponding to its CN [22]. Figure 20.3 represents the differences in CN over various sizes of Pt(111). Using DFT calculation, it was found that the geometry of optimal sites affects the adsorption energy of *OH intermediates addressing in oxygen reduction reaction (ORR). Therefore, such a structure effect is recommended to be used as a descriptor.

Based on the findings of introducing CNs as a descriptor, the orbital wise CN^{α}, where $\alpha = s$ or d, is considered as a better descriptor for nanoparticle catalysts. The CN^{α} defines the degree of coordinative saturation of transition metal atoms and to what extend their α orbitals able to bind the adsorbate [18, 22]. This development offers descriptor set for complex systems, as they connect the geometry and the electronic structure characteristics of given active sites.

20.3 Scaling Relationships: Applications of Descriptors

A series of correlations; namely scaling relationships, between adsorption or activation energetics; as descriptors, is often sufficient to describe the free energy surfaces

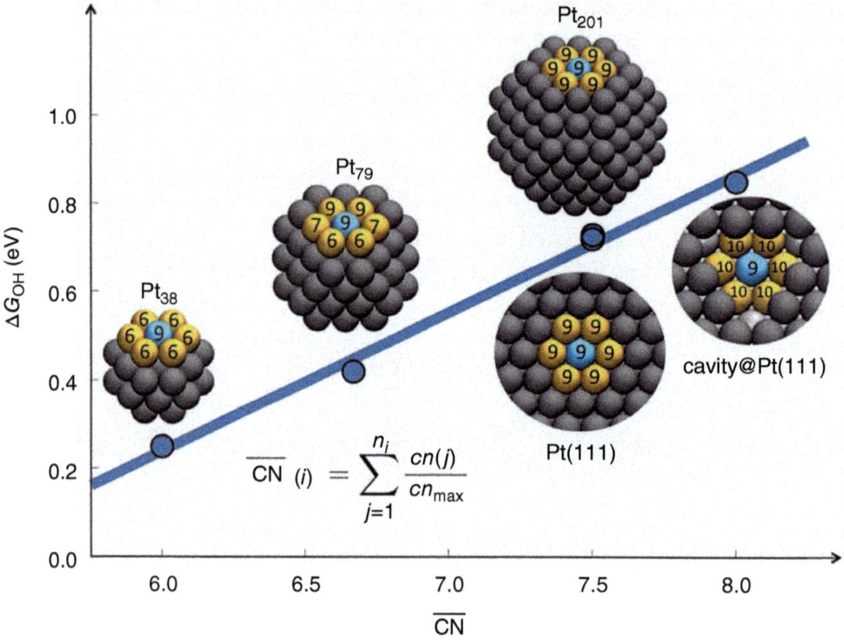

Figure 20.3 Calculated generalized coordination numbers describing the adsorption energy trend of *OH for different sizes of Pt(111). Source: Garlyyev et al. [23]. Licensed under CC BY 3.0.

and allow visualization of trends in catalytic performance. They help to reduce the complexity of calculation by identifying a few adsorption energies that only have a major impact on design studies. Scaling relations are generally linear. They classify into thermodynamic scaling and Brønsted–Evans–Polanyi (BEP), also known as transition state scaling (TSS), relations. They both share the prediction of adsorption energies as linear correlations of adsorption energies but in a different form. The development of TSS is essentially to treat the transition state as an adsorbate. It is useful to predict the kinetics of the key elementary steps from corresponding thermodynamics, where the binding energies of transition states correlate with that of either initial or final states for a given elementary step over different catalytic surfaces.

The power of scaling relations typically relies on adsorption energies correlations commonly on transition metal surfaces, thus physical theories to describe the binding energy of an adsorbate authorized. Generally, the binding energy is governed by the strength and the number of chemical bonds between the adsorbate molecule and active surface. The participated electron density (N_M) from the surface (M) to the atom A is given by [24]:

$$N_M = \frac{x_{max} - x}{x_{max}} N_A \tag{20.8}$$

where N_A donates an optimal electron density of atom A which is bonded with x and x_{max} bonds; the number of hydrogen bonds and the maximum number of bonds that atom A may form based on standard bond valency analyses, respectively. Further, a linear relation is found between N_M and the d electron hybridization energy (ΔE_d) with the A orbitals as following:

$$\Delta E_d(x) \propto \frac{x_{max} - x}{x_{max}} \tag{20.9}$$

$$\Delta E_{AH_x} = \frac{x_{max} - x}{x_{max}} \Delta E_A + \text{constant} \tag{20.10}$$

By arranging Eq. (20.10) to consider transition metals, the adsorption energy of hydrogen-containing-molecule AH_x on metal M_2 is given by [1, 25]:

$$\Delta E_{\left(\frac{AH_x}{M_2}\right)} = \Delta E_{\left(\frac{AH_x}{M_1}\right)} + \frac{x_{max} - x}{x_{max}} \cdot \left(\Delta E_{\left(\frac{A}{M_2}\right)} - \Delta E_{\left(\frac{A}{M_1}\right)}\right) \tag{20.11}$$

Equations (20.10) and (20.11) are common scaling relations used for adsorption energies correlations across a variety of catalytic surfaces, Figure 20.4a, which is known as thermodynamic relations.

On the other hand, an empirical relationship (TSS), Figure 20.4b, was established where the activation energy (E_A) is a function of the reaction heat (ΔE) energy represented by

$$E_A = \alpha \cdot \Delta E + \beta \tag{20.12}$$

where β is an arbitrary energy offset and α donates the slope of the relation. Both are parameters identified by the kinetics of the elementary steps and governed by the geometric and electronic structures of a given catalyst. Consequently, this provides an avenue to manipulate catalytic functionality for catalyst design for electrocatalytic reactions [27].

In summary, scaling relations, either thermodynamic or TSS, are very versatile and broadly applicable to set a small number of reactivity descriptors. A more complete picture, the linear relationships are effective to accelerate the understanding of a wide range of molecular adsorptions across different surfaces.

20.4 The Activity Principles and the Volcano Curve

To address the activity of the catalyst for a catalytic cycle, well-known principles have been established. The catalyst center is regenerated after multiple key elementary steps. The interaction strength between initials and intermediates over the active center has to be optimal, the Sabatier principle. In Sabatier's sense, the adsorption energy of that optimal center is not too weak or too strong, which leads to reaction hindering or surface poisoning, Figure 20.5a [1, 28]. Although this principle sounds clear, it opens the door to find the optimal catalyst for a given catalytic reaction. Calling the BEP principle, the difference in activation energy between two catalytic reactions is proportional to the difference of their enthalpies. In other words, the pre-exponential factor of the Arrhenius equation and the transition state energy should be similar [3]. An advantage emerged from Sabatier and BEP principles, volcano-shaped plots visualize the reactivity as a function of

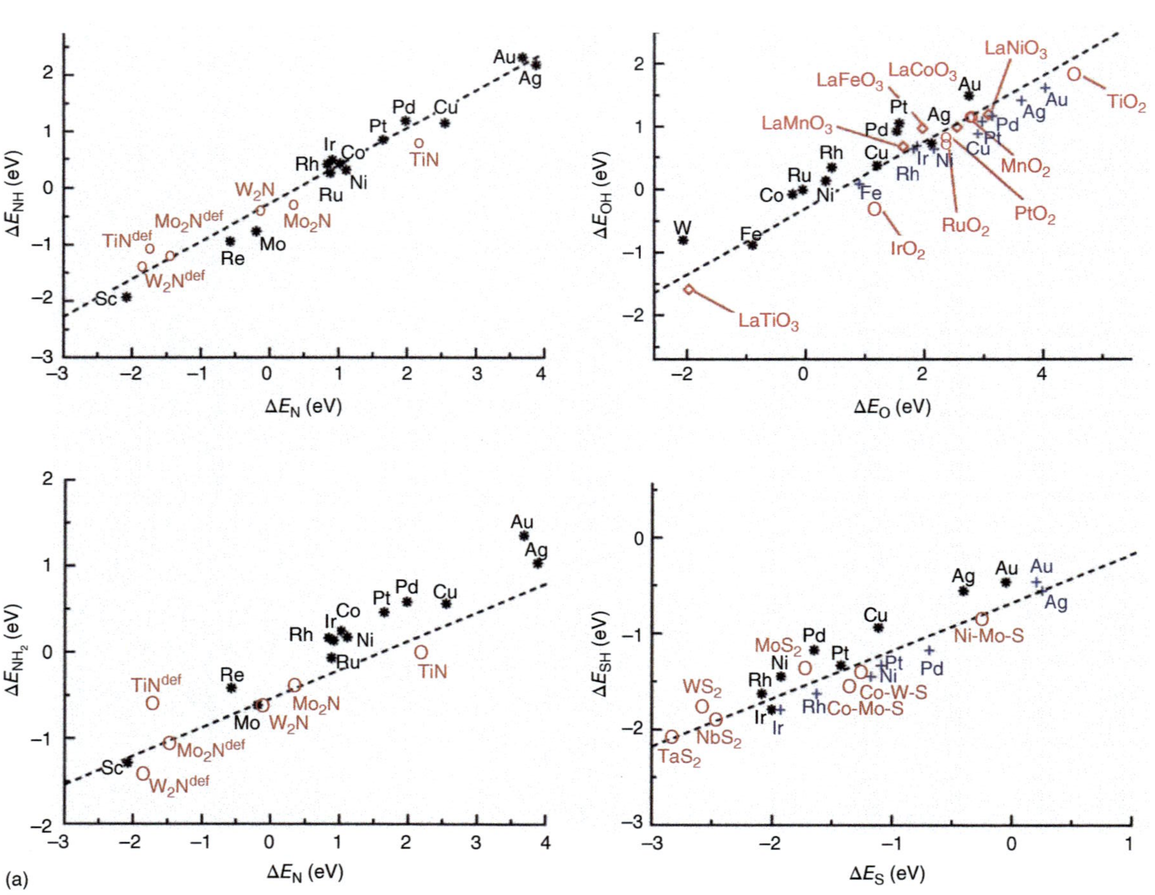

Figure 20.4 (a) Adsorption scaling relationships for adopted adsorbates on nitride, oxide, and sulfide surfaces (red marks), compared with corresponding relationships on metal surfaces (black and blue marks). Source: Fernández et al. [26]. © 2008, John Wiley & Sons. (b) Transition state scaling relations for adsorbed CO* on selected metal surfaces. Source: Liu et al. [27]. Licensed under CC BY 4.0.

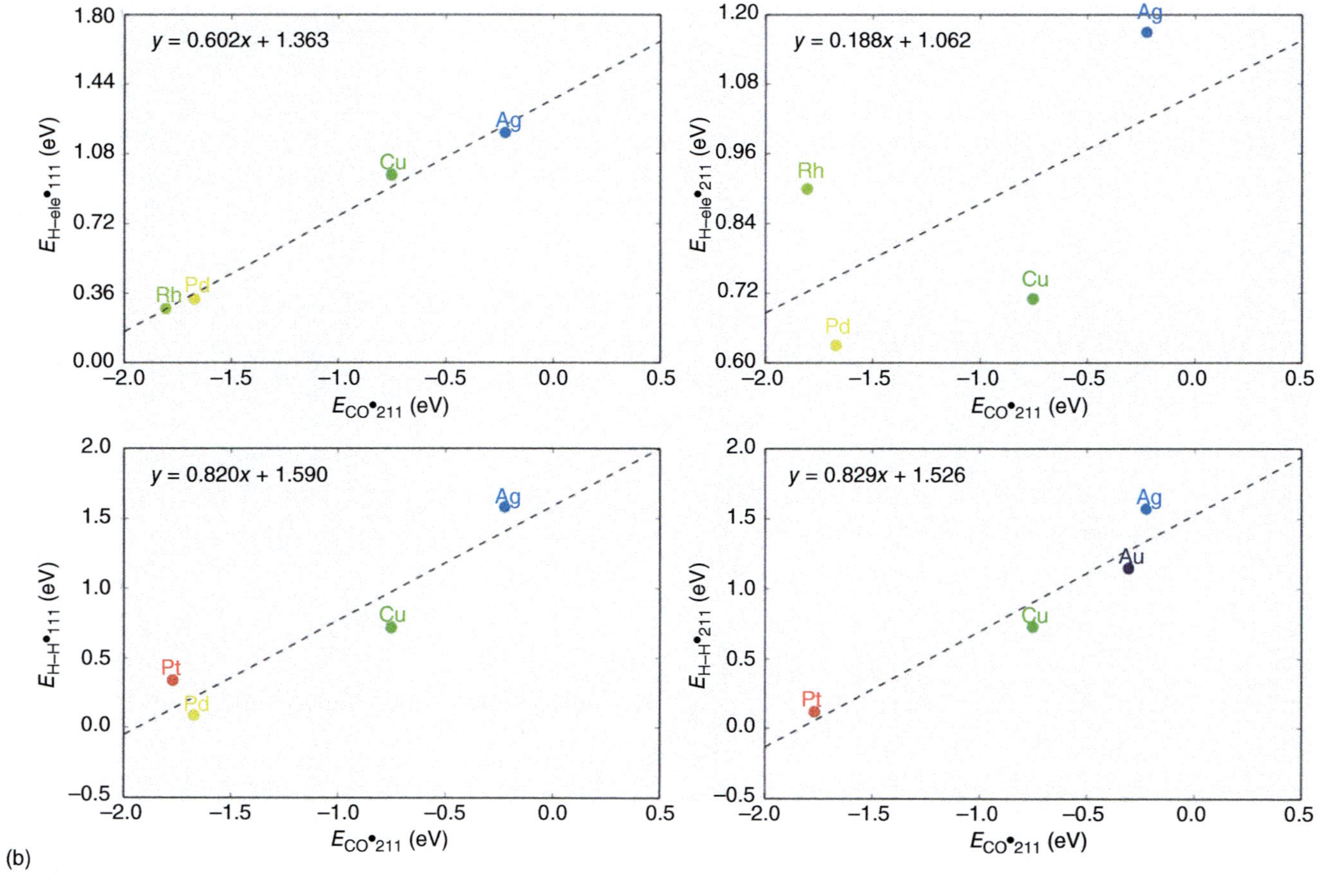

Figure 20.4 *(Continued)*

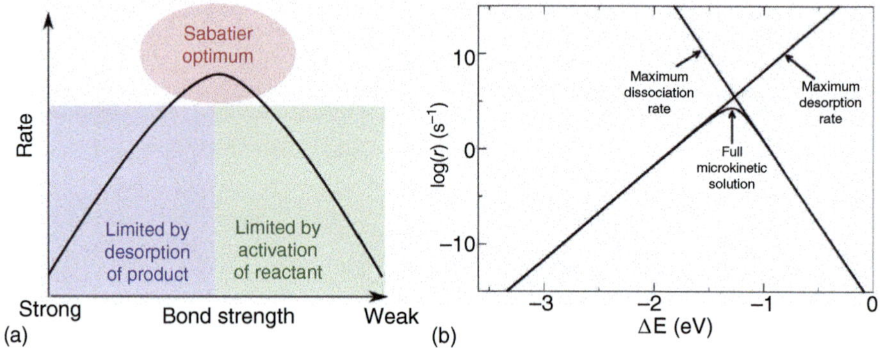

Figure 20.5 (a) Representative illustration of Sabatier principle. Source: Medford et al. [28]. © 2015, Elsevier. (b) The Sabatier volcano plot for the general catalytic reaction. Source: Nørskov et al. [29]. © 2008, Royal Society of Chemistry.

an independent descriptor (predicted by scaling relations) with the optimal center located at its apex. Thus, the volcano curve would correlate the activation energy (a kinetic variable) with the adsorption energy (a thermodynamic one). Figure 20.5b represents a volcano plot where the rate of activation and that of product desorption has a balance [29, 30]. Nørskov and his coworkers have improved volcano curves based on scaling relations for different electrocatalytic reactions on several material classes such as metals and their oxides, which have summarized by Greeley in Ref. [24]. In most cases, the correlation between the adsorption energies allows predicting activity trend in terms of a special descriptor, which leading to volcano plots [31, 32].

20.5 DFT Modeling

Modeling can be used to integrate fundamental insights into the catalytic cycle. It allows elucidating the critical aspects of the catalytic reaction that improve activity or selectivity controlling the catalyst performance. The mechanism of such reaction on catalyst surface involves multiple elementary steps comprising species adsorption, desorption, diffusion, and reaction, Figure 20.6a. Few of these steps affect the kinetics of a catalytic cycle. In this respect, approximate models have been constructed to identify these pivotal steps with the advantages of DFT theories. DFT modeling becomes a powerful tool available for this rationale, measuring thermodynamic parameters of main species, predicting the rate constants over-involved transition states, describing the adsorbate–adsorbate interactions, and identifying the overall pathway, Figure 20.6b, [34]. Catalytic modeling can take various common forms, including the computational hydrogen electrode (CHE) model to calculate the reaction free energies of the elementary reactions, solvation models to estimate solvation energy of liquid–solid interface, and kinetic simulations to access thermodynamic properties of adsorption systems. Their integration is useful to complete a catalytic cycle design.

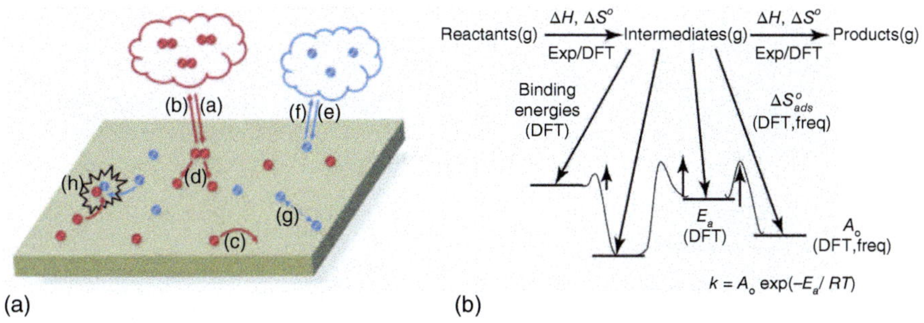

Figure 20.6 (A) A representation of a catalyst surface-involved processes: molecular (a) adsorption, (b) desorption, (c) diffusion, (d) dissociation, atomic (e) adsorption, (f) desorption, (g) adsorbate–adsorbate interactions, and (h) surface reaction. Source: Asthagiri and Janik [1]. © 2014, Royal Society of Chemistry (B) Strategy for predicting the kinetic parameters of a catalytic model. DFT, freq: functional theory calculations of vibrational frequencies. Source: Thomas and Thomas [33]. © 2015, John Wiley & Sons.

20.5.1 CHE Model

In 2004, Nørskov et al. introduced the CHE approach into catalytic cycle studies. They formulated that each of the elementary steps involves a proton-coupled electron transfer process and then calculate the free energy of each step as a function of the operating electrode potential. With an assumption that (i) the free energy is exergonic as protonation process is favorable for adsorbed O_2 and OH and (ii) the rate-limiting step is the most positive value of overall free energy change. The chemical potentials, μ_{H^+} and μ_{e^-}, can be identified with the free energy G_{H_2} of a hydrogen molecule by:

$$\mu_{H^+} + \mu_{e^-} = \frac{1}{2}G_{H_2} \tag{20.13}$$

By applying a potential U, the chemical potential of the electron will be decreased with a certain amount of eU, where e is the elementary positive charge [2]. Thus, Eq. (20.13) will be arranged to calculate the total chemical potential by

$$\mu_{H^+} + \mu_{e^-} = \frac{1}{2}G_{H_2} - eU \tag{20.14}$$

Consequently, by replacing $\mu_{H^+} + \mu_{e^-}$ in the free energy of a proton–electron transfer to a^* forming b^* is given by:

$$\Delta G = \mu(b^*) - \mu(a^*) - \left[\frac{1}{2}\mu(H_2) - eU\right] \tag{20.15}$$

The free energy changes can be easily computed using DFT calculation with its three main contributions: the energy E, the entropy S, and a zero-point energy (ZPE) correction [35]. The more positive step in free energy change calculation ΔG_{max} is considered to be the rate-determining step. While the limiting potential U_{lim} is the smallest applied potential required to make the limiting step is the least favorable step having a negative energy change [2]. Thus, the overpotential is given by

$$\eta = \left|U_{eq} - U_{lim}\right| = \left|U_{eq} + \frac{\Delta G_{max}}{e}\right| \tag{20.16}$$

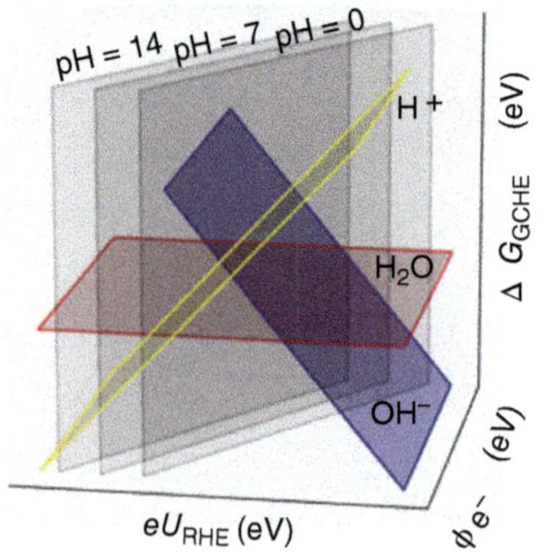

Figure 20.7 The ΔG_{GCHE} energy as a function of work function and eU_{RHE}, with planes representing different pH values (gray) and an added proton (yellow) or a hydroxy (blue) and water (red). Source: Bagger et al. [35]. © 2020, Springer Nature

where U_{eq} donates the equilibrium potential of a given reaction. Generally, the limiting potential is estimated by making use of the scaling relations addressing in the adsorption energies of the reaction intermediates. This model is called CHE because the evaluation of the chemical potential of a proton–electron couple is with respect to the reversible, normal, or standard hydrogen electrode. Recently, an approximation made for the CHE model. Generalization of the computational hydrogen electrode (GCHE) was provided by Rossmeisl and coworker to simulate the pH environment in the catalytic cycle [36, 37]. They considered two reference electrodes referring to the chemical potentials of the proton and electron species. Taking into account the work functions ϕ_{H^+} and ϕ_{e^-} for the proton and electron, respectively, the free energy of the GCHE model is calculated as a function of work function and the chemical potential eU_{RHE} of coupled proton–electron for different pH, Figure 20.7 [35, 36].

The CHE model shows significant advances in catalytic cycle reactions that involve multiple proton–electron couple steps such as ORR [38], oxygen evolution reaction (OER) [39, 40], and electrochemical carbon dioxide reduction (ECR) [41, 42]. However, the CHE model has its limitations [2, 41]: (i) the model counts the charge on the catalyst surface as zero but, in reality, the surface has a charge affecting the reactivity. (ii) The interactions between solvent molecules and the catalyst are neglected. (iii) The model only explores thermodynamic not kinetics factors.

20.5.2 Solvation Models

To tackle the energetics of solid–liquid interfacial reactions, thermodynamic effects of solvents need to be considered. The environmental impact of solvents on the reaction rates is up to 3 orders of magnitude [43]. Solvation is an interaction process of a solute (ions) with the solvent to stabilize the solute species in the solution. More approximate models were formulated for this purpose. Solvent manipulations can be classified into four categories: "implicit," "explicit," "embedded," and "hybrid cluster-continuum." An implicit model is an approach in which the solvent is described as a homogeneous constant dielectric continuum surrounding the solute. An explicit model represents the solvent as an adsorbed species in DFT calculations. An embedded model includes different levels of calculations in which some number of solvent molecules is calculated by DFT and the rest by force-field calculations. A hybrid cluster-continuum model combines solvent explicitly and implicitly in processing. On the other hand, the H-shuttling model is developed as an alternative to model water in electrochemical reactions [2, 35, 44]. Hereafter, we highlight the most common models in electrocatalytic cycles.

The implicit solvation model describes the solvent behavior in which the concept of a "reaction field" is employed [45]. The media surrounding the electrode surface is a homogeneous and isotropic solvent embedded in continuum level with specific dielectric constant ε and the solute is located at a proper cavity, Figure 20.8. By applying an electrostatic perturbation, the total electrostatic potential $\phi(r)$ of a solute is determined by a generalized Poisson equation [46], as follows:

$$\nabla\varepsilon(r)\nabla\phi(r) = \rho_{\text{sol}}(r) \tag{20.17}$$

where $\rho_{\text{sol}}(r)$ is the charge distribution of the solute and $\nabla\varepsilon(r)$ is the dielectric permittivity of the system. The linear response of the solvent happens within a finite distance losing a degree of freedom, thus forming a solvation cavity [35]. The permittivity parameter has a significant role in the solvation model being a hard step function or a soft transition functionals to measure the free energy of solvation in DFT calculations [35]. These functionals can be defined in several ways such as early models; iso-surfaces of the density [47], soft spheres [48], or modern one; overlapping spheres with van der Waals radii [45]. This equation is the basis of the polarizable continuum model (PCM). In the standard continuum model, the free energy of system response can be mathematically expressed by [49]:

$$G = G_{\text{cavity}} + G_{\text{electrostatic}} + G_{\text{dispersion}} + G_{\text{repulsion}} + G_{\text{thermal motion}} \tag{20.18}$$

where G_{cavity} donates the contributed energy for cavity formation, $G_{\text{electrostatic}}$ treats with the polarization of the solute and solvent, $G_{\text{dispersion}}$ refers to the interaction

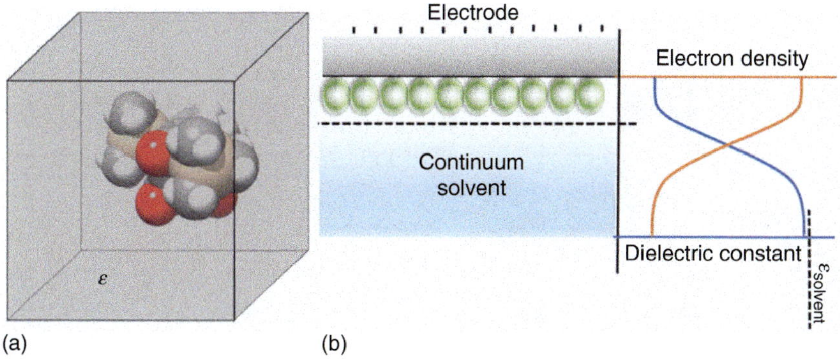

Figure 20.8 Schematic representation of reaction field concept (a), Source: Francke et al. [45]. © 2018, American Chemical Society., and a continuum implicit model (b), showing the cutoff electron density (dotted line) at which a smoothly shifting dielectric continuum is added to interact with the DFT portion of the model. Source: Maheshwari et al. [2]. © 2018, Elsevier.

energy in small molecular clusters model, and $G_{repulsion}$ is related to the electron exchange functional of interactions between molecules. Based on this model, the solvation effects were revealed in some electrocatalytic reactions such as ORR [50], ECR [51].

However, the model meets some limitations. The interactions of the solvent and electrode surface within solvation shells are difficult to represent taking into account the electron transfer and the covalent bonds of a given molecule. Further, the dielectric constant of some solvents varies depending on their distance from the electrode surface, for example, the dielectric constant of water molecule decreases with getting closer to the metal electrode surface. Moreover, the side reactions between the solvent molecule and the solute could not be approximated.

On the other hand, the explicit solvation model considers molecular insights of each solvent molecule. This approach used to consider local geometry of the water molecule in different forms such as a single molecule or an ice-like highly ordered multilayer model. It describes a distribution of water molecules with possibly hydrogen bonding either pointing toward or away from the surface. Water molecules are simulated with a wide (more tetrahedral) H—O—H angle and longer H—O bond length even in the case of ion existence, Figure 20.9. The physical and electronic properties of some of the water models are calculated. Consequently, the solvation free energy is given for solute–solvent interactions with a static approximation in case the binding energy between the electrode surface and water is significantly stronger than water–water interactions. DFT-based explicit calculations are successfully applied to investigate the kinetic barriers in electrochemical cycles [27, 53].

Recently, the H-shuttling model is developed in which a water molecule is used to shuttle an adsorbed *H from the surface to the adsorbate, Figure 20.10. The shuttling process contributes to calculating the included proton–electron transfer barrier. Further, the potential is determined by the CHE model where *H in the initial state is equilibrated with a proton–electron couple [44]. This model is commonly applied in ECR reactions [30, 44, 54, 55].

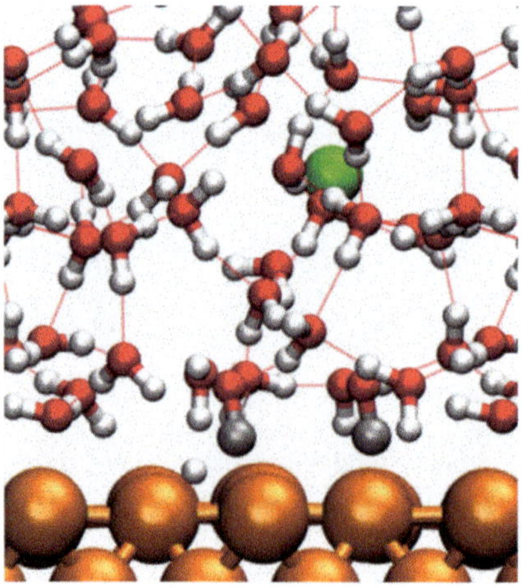

Figure 20.9 Water/Cu simulation with hydrogen bonds and Na^+ ion's inclusion [53]. Source: Cheng et al. [52]. © 2017, National Academy of Sciences.

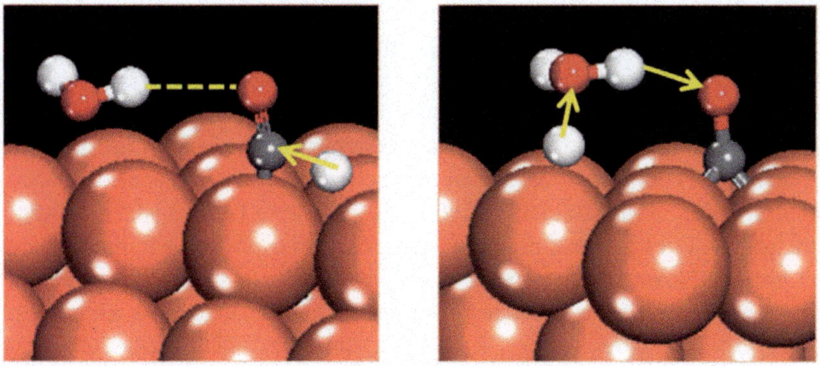

Figure 20.10 Water shuttling simulation for proton–electron transfer to *CO forming (a) *CHO and (b) *COH species. Source: Nie et al. [54]. © 2013, John Wiley & Sons.

Solvation models provide a way to simulate the liquid–solid interface. Rather than neglect solvent effects in DFT calculations in Nørskov sense, recent efforts are effective by which the DFT calculations got improved to propose better mechanisms considering the effects of solvent response on charged intermediates through potential applying.

20.5.3 Kinetic Modeling

Although DFT-based calculations allow to map of a reaction mechanism and identify the rate-determining step, studying the kinetic behavior of an electrocatalytic cycle under operation conditions is also demanded. Two approaches for kinetic modeling of the catalytic surface are microkinetic modeling and dynamic Monte Carlo simulations. The main difference between them is the simulation of the catalytic surface [56]. The former treats the adsorbates via the mean-field approximation, describing reaction rates in terms of species coverages while the later construct the catalytic surface as a lattice of reaction sites, explicitly simulating each reaction on the lattice [57]. Due to the more detailed information provided by Monte Carlo simulations, we are going to focus on its fundamental domains.

Monte Carlo approach stimulates the dynamics of an electrocatalytic cycle [58–61]. As mentioned in Section 20.5 (Figure 20.6a), the catalytic reaction includes multiple elementary events, such as diffusion, adsorption, chemical reactions, and desorption, taking place on special sites on the catalyst surface. The rate of the elementary transformations can be estimated from transition state theory (TST) in combination with DFT computations. However, during catalysis, the surface structure is updated with time after each event due to its freedom degree. The time scale of these events is much longer than atomic vibrations [60]. Thus, molecular dynamics that help to describe the reaction kinetics based on the atomic time scale is restricted to tackle the longer elementary scale. To overcome the time scale limitation and estimate the dependence of elementary reactions on the surface structure as well as the reactivity of catalyst (e.g. activity, selectivity, and performance), Kinetic Monte Carlo (KMC) simulations is a powerful approach for this purpose.

The Monte Carlo approach is a promising tool toward capturing detailed energetics and complexities relevant to reaction mechanism addressing in the effect of multiple adsorbate configurations, lateral interactions, and the presence of spectator species. The idea of the simulation is to statistically treat the involved reactions, where the probabilities per unit time that such reaction transition will happen are obtained from TST-based rate constants k_{TST}. These constants cannot provide the time at which the reaction will occur but rather suggest that within a time interval; $t_{initial}$:t_{final}, on average $k_{TST} \times t_{final} - t_{initial}$ reaction event will take place (Poisson statistics) [60, 62].

In KMC modeling, a reaction mechanism is considered to include the elementary transactions represented as Poisson processes under constant reaction conditions such as temperature, pressure, and lattice compositions [60]. For the state space of the system W, the probability of finding the system at state w at time t $P(w, t)$ is given by namely "Master equation":

$$\frac{dP(w,t)}{dt} = \sum_{w' \neq w} [k_{w' \to w}(w') \cdot P(w',t) - k_{w \to w'}(w) \cdot P(w,t)] \tag{20.19}$$

This equation describes the dynamics of a system in which $k_{w' \to w}$ and $k_{w \to w'}$ donate the rate constant for the forward and reverse transition states w' and w, respectively, of an elementary event. The sum is taken over all the possible states of the given system. KMC continues the simulation by generating exponentially distributed random times for the incidents of the events and then executing them in the order they arise.

To determine the rate constants of an elementary event, TST assumes quasi-equilibrated reactants, and transition states. The kinetic constant is then calculated by considering the probability of the system visiting the transition state and averaging over the "speeds" with which it is traveling toward the products state [58, 60]. Thus, the result of the TST rate constant is given by

$$k_{TST} = k \cdot \frac{k_B T}{h} \cdot \frac{q_w}{q_{w'}} \cdot \exp\left(-\frac{\Delta E^{w' \to w}}{k_B T}\right) \tag{20.20}$$

where q_w and $q_{w'}$ are quasi-partition functions for the products and reactants, k_B and h denote Boltzmann's and Planck's constants, k is a fudge factor termed the transmission coefficient which accounts for re-crossings of the barrier. The activation barrier $\Delta E^{w' \to w}$ is obtained from DFT calculations. Imposing the balance condition and introducing the microscopic energy of the system, the change in the electronic energy of the system is given by

$$\Delta E^{w' \to w} = E(w) + E^{gas}(w) - E(w') + E^{gas}(w') \tag{20.21}$$

where $E(w)$ and $E(w')$ refer to the electronic energies of the given lattice configuration, while $E^{gas}(w)$ and $E^{gas}(w')$ represent that of the gas species at transition states, w

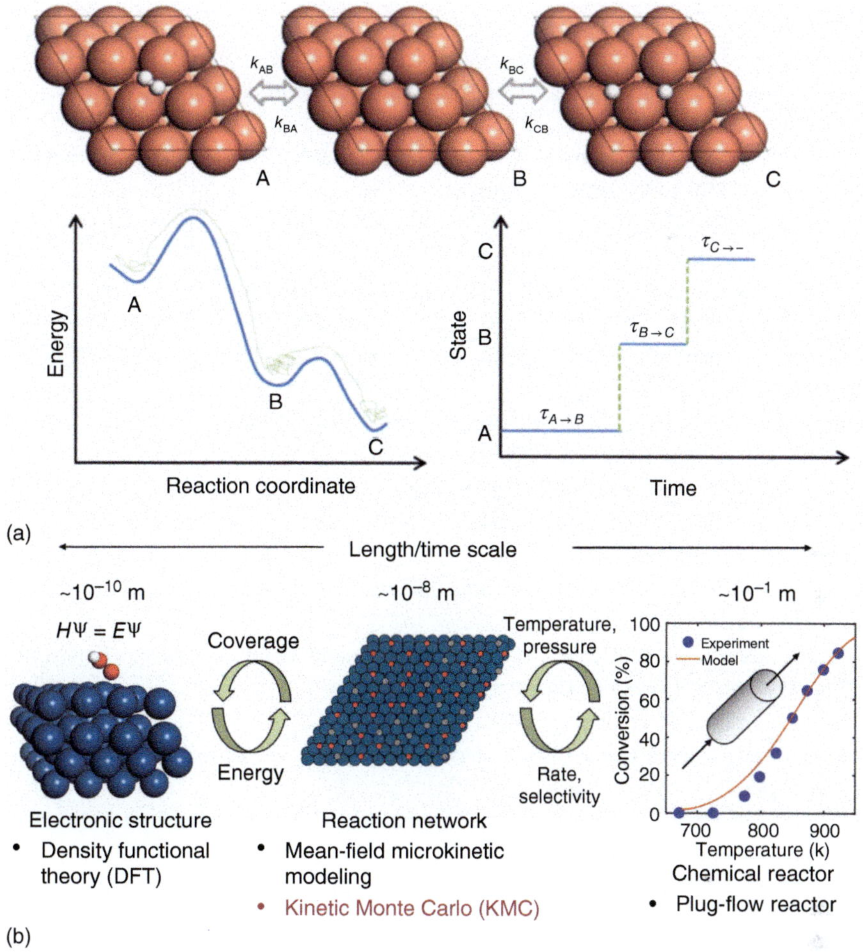

Figure 20.11 (a) Schematic simulation of hydrogen adsorption on Cu(111) (A → B), further its diffusion (B ↔ C) with a representative profile for PES and coarse-grained KMC modeling. Source: Darby et al. [58]. © 2016, IOP Publishing. (b) Overview of the multiscale modeling paradigm. Source: Nunez [63]. © 2018, University of Delaware.

and w'. KMC algorithms are focusing on the barrier crossings and treating minimum potential energy surface (PES) in a coarse-grained sense providing accurate values for the rate constants, Figure 20.11a.

At this end, the multiscale framework, KMC mesoscale modeling incorporation with DFT calculation, Figure 20.11b, can enhance the fundamental understanding toward rationally designing the catalytic cycle. However, the behavior of catalytic systems on the basis of KMC is rarely reported in the literature such as CO oxidation on O-pre-covered Pd(111) surfaces [64], ECR using cobalt porphyrin [65] and peanut-shaped carbon nanotubes [66]. Therefore, it is considered to be an attractive subject for the electrocatalytic theorists.

References

1 Asthagiri, A. and Janik, M. (2014). *Computational Catalysis*, vol. 14. Royal Society of Chemistry.
2 Maheshwari, S., Li, Y., Agrawal, N., and Janik, M.J. (2018). *Advances in Catalysis*, vol. 63, 117–167. Academic Press.
3 Pašti, I.A., Skorodumova, N.V., and Mentus, S.V. (2014). *React. Kinet. Mech. Catal.* 115: 5–32.
4 Chen, Z.W., Chen, L.X., Wen, Z., and Jiang, Q. (2019). *Phys. Chem. Chem. Phys.* 21: 23782–23802.
5 Cheng, D., Negreiros, F.R., Apra, E., and Fortunelli, A. (2013). *ChemSusChem* 6: 944–965.
6 Ananikov, V.P. (2014). *Understanding Organometallic Reaction Mechanisms and Catalysis: Computational and Experimental Tools.* Wiley-VCH.
7 Schlögl, R. (2013). *Chemical Energy Storage.* Walter de Gruyter.
8 Atkins, P.W., De Paula, J., and Keeler, J. (2018). *Atkins' Physical Chemistry.* Oxford University Press.
9 Stalke, D. (2012). *Electron Density and Chemical Bonding II: Theoretical Charge Density Studies*, vol. 147. Springer.
10 Deutschmann, O., Knözinger, H., Kochloefl, K., and Turek, T. (2009). *Ullmann's Encyclopedia of Industrial Chemistry*. Wiley-VCH.
11 Shan, N., Zhou, M., Hanchett, M.K. et al. (2017). *Mol. Simul.* 43: 861–885.
12 Elke, S. and Carlos, C.J. (2017). *Molecular Electronics: An Introduction to Theory and Experiment*, vol. 1. World Scientific.
13 Zinola, C.F. (2010). *Electrocatalysis: Computational, Experimental, and Industrial Aspects.* Taylor & Francis Group.
14 Gonze, X., Beuken, J.-M., Caracas, R. et al. (2002). *Comput. Mater. Sci.* 25: 478–492.
15 Clark, S.J., Segall, M.D., Pickard, C.J. et al. (2005). *Z. Kristallogr. Cryst. Mater.* 220: 567–570.
16 Giannozzi, P., Baroni, S., Bonini, N. et al. (2009). *J. Phys. Condens. Matter* 21: 395502.
17 Kresse, G. and Furthmüller, J. (1996). *Comput. Mater. Sci.* 6: 15–50.
18 Zhao, Z.-J., Liu, S., Zha, S. et al. (2019). *Nat. Rev. Mater.* 4: 792–804.
19 Hammer, B. and Norskov, J.K. (1995). *Nature* 376: 238–240.
20 Hammer, B. and Norskov, J. (2000). *Adv. Catal.* 45: 71–129.
21 Nilsson, A., Pettersson, L.G.M., and Norskov, J. (2008). *Chemical Bonding at Surfaces and Interfaces.* Elsevier.
22 Calle-Vallejo, F., Tymoczko, J., Colic, V. et al. (2015). *Science* 350: 185–189.
23 Garlyyev, B., Fichtner, J., Piqué, O. et al. (2019). *Chem. Sci.* 10: 8060–8075.
24 Greeley, J. (2016). *Annu. Rev. Chem. Biomol. Eng.* 7: 605–635.
25 Montemore, M.M. and Medlin, J.W. (2014). *Catal. Sci. Technol.* 4: 3748–3761.
26 Fernández, E.M., Moses, P.G., Toftelund, A. et al. (2008). *Angew. Chem. Int. Ed.* 47: 4683–4686.
27 Liu, X., Xiao, J., Peng, H. et al. (2017). *Nat. Commun.* 8: 15438.

28 Medford, A.J., Vojvodic, A., Hummelshøj, J.S. et al. (2015). *J. Catal.* 328: 36–42.
29 Norskov, J.K., Bligaard, T., Hvolbaek, B. et al. (2008). *Chem. Soc. Rev.* 37: 2163–2171.
30 Nie, X., Luo, W., Janik, M.J., and Asthagiri, A. (2014). *J. Catal.* 312: 108–122.
31 Seh, Z.W., Kibsgaard, J., Dickens, C.F. et al. (2017). *Science* 355: eaad4998.
32 Nørskov, J.K., Rossmeisl, J., Logadottir, A. et al. (2004). *J. Phys. Chem. B* 108: 17886–17892.
33 Thomas, J.M. and Thomas, W.J. (2015). *Principles and Practice of Heterogeneous Catalysis*. Wiley-VCH.
34 Ertl, G., Knözinger, H., Schüth, F., and Weitkamp, J. (2008). *Handbook of Heterogeneous Catalysis*. Wiley-VCH.
35 Andreoni, W. and Yip, S. (2020). *Handbook of Materials Modeling: Applications: Current and Emerging Materials*. Springer International Publishing.
36 Hansen, M.H. and Rossmeisl, J. (2016). *J. Phys. Chem. C* 120: 29135–29143.
37 Rossmeisl, J., Chan, K., Ahmed, R. et al. (2013). *Phys. Chem. Chem. Phys.* 15: 10321–10325.
38 Zhong, L. and Li, S. (2020). *ACS Catal.* 10: 4313–4318.
39 Yao, Y., Hu, S., Chen, W. et al. (2019). *Nat. Catal.* 2: 304–313.
40 Gong, L., Zhang, D., Lin, C.Y. et al. (2019). *Adv. Energy Mater.* 9: 1902625.
41 Zhao, X. and Liu, Y. (2020). *J. Am. Chem. Soc.* 142: 5773–5777.
42 Tian, Z., Priest, C., and Chen, L. (2018). *Adv. Theory Simul.* 1: 1800004.
43 Saleheen, M. and Heyden, A. (2018). *ACS Catal.* 8: 2188–2194.
44 Nitopi, S., Bertheussen, E., Scott, S.B. et al. (2019). *Chem. Rev.* 119: 7610–7672.
45 Francke, R., Schille, B., and Roemelt, M. (2018). *Chem. Rev.* 118: 4631–4701.
46 Tomasi, J., Mennucci, B., and Cammi, R. (2005). *Chem. Rev.* 105: 2999–3093.
47 Sinstein, M., Scheurer, C., Matera, S. et al. (2017). *J. Chem. Theory Comput.* 13: 5582–5603.
48 Fisicaro, G., Genovese, L., Andreussi, O. et al. (2017). *J. Chem. Theory Comput.* 13: 3829–3845.
49 Mennucci, B. and Cammi, R. (2007). *Continuum Solvation Models in Chemical Physics: From Theory to Applications*. Wiley.
50 Reda, M., Hansen, H.A., and Vegge, T. (2018). *Catal. Today* 312: 118–125.
51 Isegawa, M. and Sharma, A.K. (2019). *Sustain. Energy Fuels* 3: 1730–1738.
52 Cheng, T., Xiao, H., and Goddard, W.A. (2017). *Proc. Natl. Acad. Sci. U.S.A.* 114: 1795–1800.
53 Liu, L., Liu, Y., and Liu, C. (2020). *J. Am. Chem. Soc.* 142: 4985–4989.
54 Nie, X., Esopi, M.R., Janik, M.J., and Asthagiri, A. (2013). *Angew. Chem. Int. Ed.* 52: 2459–2462.
55 Luo, W., Nie, X., Janik, M.J., and Asthagiri, A. (2015). *ACS Catal.* 6: 219–229.
56 Dooling, D.J. and Broadbelt, L.J. (2001). *AlChE J.* 47: 1193–1202.
57 Li, J., Croiset, E., and Ricardez-Sandoval, L. (2015). *J. Catal.* 326: 15–25.
58 Darby, M.T., Piccinin, S., and Stamatakis, M. (2016). *Physics of Surface, Interface and Cluster Catalysis*, 38. IOP.
59 Andersen, M., Panosetti, C., and Reuter, K. (2019). *Front. Chem.* 7: 202.
60 Stamatakis, M. (2015). *J. Phys. Condens. Matter* 27: 013001.

61 Reuter, K. (2011). *Modeling Heterogeneous Catalytic Reactions: From the Molecular Process to the Technical System*, vol. 1, 23. Wiley-VCH.

62 Kopač, D., Huš, M., Ogrizek, M., and Likozar, B. (2017). *J. Phys. Chem. C* 121: 17941–17949.

63 Nunez, M. (2018). Uncertainty quantification in multiscale stochastic models of catalytic reactions. PhD thesis. University of Delaware.

64 Piccinin, S. and Stamatakis, M. (2014). *ACS Catal.* 4: 2143–2152.

65 Leung, K., Nielsen, I.M., Śai, N. et al. (2010). *J. Phys. Chem. A* 114: 10174–10184.

66 Zhao, T., Wang, Q., Kawazoe, Y., and Jena, P. (2018). *Carbon* 132: 249–256.

21

Theory-Guided Rational Design

The rational approach to catalyst design would be supported by computational modeling addressing in descriptors, scaling relations, and models. Density functional theory (DFT) theories have furthered our knowledge of the several electrocatalytic processes. In this section, we provide the recent reactivity descriptors in different catalytic systems, taking into account how some descriptors could successfully be applied to formulate a trend helping to rationally predict the catalytic performance, and in addition, the feasibility of density of state (DOS) analysis to design models and identifying the active sites of a given catalyst.

21.1 Descriptors-Guided Screening

Many studies have been calculated the free energy of all possible intermediates on numerous materials. However, this would offer more complexities to DFT-based design. Thereby, the prediction of descriptors facilitates screening a wide range of catalysts rather than applying DFT computation for the full catalytic reaction as well as the discovery of structure–activity relationships. Several examples of different descriptors for numerous electrocatalytic reactions have been identified and modeled on the basis of DFT calculations, Table 21.1. Indeed, the d-band model has received widespread attention of design researchers; however, many descriptors are recently identified for other material classes such as transition metal oxides, alloy, perovskites, doped graphene, and graphene-based single-atom catalysts. Descriptors-based screening has been effectively identified to replace trial-and-error investigations. Thus far, such descriptors are limited to a single material class. The d-band model and generalized $\overline{\mathrm{CN}}$ have been successfully employed only for transition metals. The e_g occupancy descriptor was applied in metal and perovskite oxides. Surface distortion accounts for alloy class. With this picture, establishing a "universal reactivity descriptor" that can be appropriate for several types of material and reaction systems is crucial. A remarkable example of this principle is the local environment of the metal center descriptor addressing in its coordination number (CN), electronegativity, and electronegativity of the nearest neighbor atoms of graphene-based single-atom catalyst class for oxygen reduction reaction (ORR), oxygen evolution reaction (OER), and hydrogen evolution reaction

Electrocatalysis in Balancing the Natural Carbon Cycle, First Edition. Yaobing Wang.
© 2021 WILEY-VCH GmbH. Published 2021 by WILEY-VCH GmbH.

Table 21.1 An overview of descriptors sets found in different electrocatalytic reactions.

Material class	Catalytic descriptor	Calculation	Catalytic reaction	References
Metal surface	d-band center (O and OH adsorption energy)	$\frac{\int_{-\infty}^{0} \rho_d E\, dE}{\int_{-\infty}^{0} \rho_d\, dE}$	ORR	[1]
	H* binding energy		HER	[2–4]
	Average 2p-state energy	$\frac{\int_{\varepsilon_{min}}^{\varepsilon_{max}} \rho_{2p} E\, dE}{\int_{\varepsilon_{min}}^{\varepsilon_{max}} \rho_{2p}\, dE}$	OER	[5]
	Generalized coordination number	$\sum_{j=1}^{n_s} \frac{CN(j)}{CN_{max}}$	ORR	[6]
	*COOH binding energy		ECR	[7]
	*CHO adsorption energy		ECR	[8]
	Two-parameter descriptor; H* and *CO binding energies		ECR	[9]
Alloy	*CO adsorption energy	Surrogate-based optimization and active machine learning	ECR	[10, 11]
	H binding energy		HER	[10, 12, 13]
	Surface distortion	$\frac{1}{Nh} \sum_{k=1}^{N} K\left(\frac{\delta - \delta_k}{h}\right)$	ORR	[14]
Metal oxide	Average 2p-state energy	$\frac{\int_{\varepsilon_{min}}^{\varepsilon_{max}} \rho_{2p} E\, dE}{\int_{\varepsilon_{min}}^{\varepsilon_{max}} \rho_{2p}\, dE}$	OER	[5]
	e_g occupancy	$\int_{-\infty}^{0} \rho_{e_g}\, dE$	OER\ORR	[15]
	t_{2g} occupancy	$\int_{-\infty}^{0} \rho_{t_{2g}}\, dE$	OER	[16]

Perovskites	e_g occupancy	$\int_{-\infty}^{0} \rho_{e_g} dE$	OER	[17, 18]
	e_g occupancy		ORR	[19]
	O p-band center	$\frac{\int_{-\infty}^{0} \rho_p E\, dE}{\int_{-\infty}^{0} \rho_p\, dE}$	ORR	[20]
	O p-band center		OER	[21]
Doped graphene	*OH adsorption energy	$\Delta G_{OH*} = (G_{OH@G}) - (G_G + G_{OH^-})$	ORR\OER	[22]
	Product of relative electron affinity (A) and electronegativity (E)	$\Phi = \left(\frac{E_x}{E_c}\right) \times \left(\frac{A_x}{A_c}\right)$	ORR\OER	[23]
	E_p	Position of the highest peak of the active site's DOS	ORR	[24]
	p_z occupancy	$\left[\int_{-\infty}^{0} \rho_{p_z} dE\right]_{\text{active site}} - \left[\int_{-\infty}^{0} \rho_{p_z} dE\right]_{\text{C atom}}$	ORR	[25]
Graphene-based single-atom catalysts	d-band center	$\frac{\int_{-\infty}^{0} \rho_d E\, dE}{\int_{-\infty}^{0} \rho_d dE}$	ORR	[26]
	Coordination number and electronegativity and the electronegativity of the nearest neighbor atoms	$\theta_d \times \frac{E_M + \alpha \times (n_N \times E_N + n_C \times E_C)}{E_{O/H}}$	ORR\OER\HER	[27]
M–N–C single atom catalysts (SACs)	Intrinsic descriptor	$V_M \times E_M / r_M$	ECR\HER	[28]
	*COOH binding energy		ECR	[7]

(HER) electrocatalytic reactions [27]. In a word, attention should be paid to discover a universal reactivity descriptor offering a strategy to design and screen highly active catalysts for different electrocatalytic systems.

21.2 Scaling Relationship-Guided Trends

Scaling relationships visualize reactivity trends. As the reactivity is identified by the rate of catalyzed reactions or closely related variables, DFT calculations are required. Exclusively, the presence of adsorption energies' scaling relationships better facilitates the simultaneous analysis of different materials through Sabatier-type volcano plots resulting in a reduction of the complexity of DFT-based models. Herein, we highlight some specific achievements of DFT computation in the term of scaling trends for key intermediates of selected electrochemical reactions.

21.2.1 Reactivity Trends of ECR

Based on the d-band descriptor, the linear relationship over metals explains the nature of the overpotential toward C1 products and helps to classify electrochemical carbon dioxide reduction (ECR) selectivity to groups: (i) carbon monoxide (e.g. Au, Ag), (ii) formate (e.g. Pb, Sn), (iii) hydrocarbons (e.g. Cu), and (iv) hydrogen (e.g. Pt, Ni), as shown in Figure 21.1a [32]. A volcano plot describes the metal classification on the basis of combining the theoretical DFT and experimental frameworks as shown in Figure 21.1b [32]. The challenge of ECR-related scales is that the binding intermediates have similar C-bond such as *COOH, *CO, and *COH affecting each other [31]. In the CH_4 pathway, for example, the weak *CHO binding compared to *CO binding energy is considered to be rate-limiting step (RLS), as shown in Figure 21.1c. To overcome this activation barrier by reducing RLS energy, the catalytic surface would have to bind *CO weaker than *CHO while it is difficult according to scaling relationships, as shown in Figure 21.1d. The volcano plot of the two largest free energy barriers; *CO_2 to *COOH and *CO to *CHO, for the calculated limiting potentials shows that Cu is the optimal catalyst for CH_4 generation, as shown in Figure 21.1e.

On the other hand, the linear scaling is restricted to optimize metals toward C2 products but it partially explains the reason. The kinetic linear scaling relations between *CO activation and binding energies over various metals show that the stronger the binding energy is, the higher the barrier shows, as shown in Figure 21.1f [33]. Therefore, C–C coupling is somehow difficult on metal catalysts. Theoretical guidance suggests that the reduction of the activation energy of the coupled proton–electron transfer to *CO can potentially be achieved by breaking strategies for the traditional linear scaling relationship. These strategies take advantage of the electronic characteristics underlying the scaling relations such as reducing coordination numbers (CNs), introducing p-states, inducing adsorption sites with oxophilicity, modifying functionalized surfaces (tethering active ligands), or using different material classes [34].

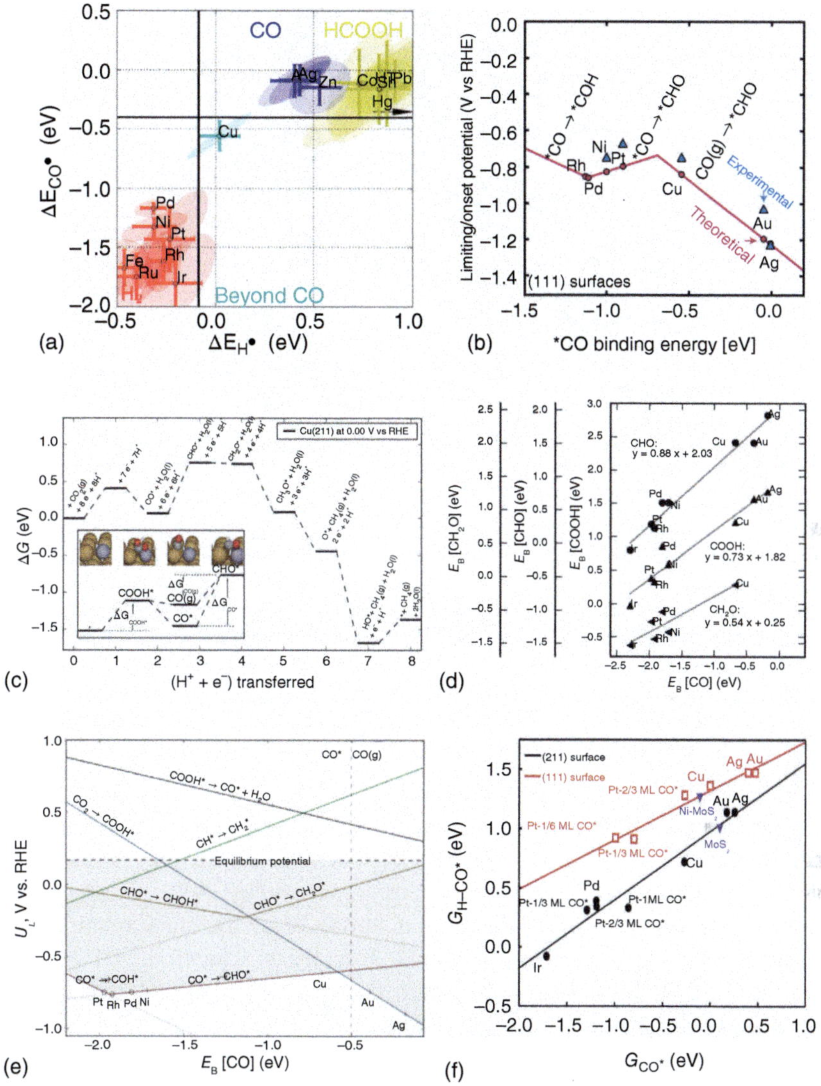

Figure 21.1 (a) The adsorption scaling of metal classification according to CO* and H* binding energies. Source: Bagger et al. [29]. (b) Volcano plot for ECR on metals. Source: Shi et al. [30]. Reproduced with permission of Royal Society of Chemistry. (c) Proposed CH_4 pathways on Cu(211) through the *CHO intermediate. Source: Vasileff et al. [31]. Reproduced with permission of Elsevier. (d) Scaling relationships for C-bond adsorbates relevant to the ECR pathway on several metal surfaces. Source: Vasileff et al. [31]. Reproduced with permission of Elsevier. (e) The limiting potentials of elementary steps for the CH_4 mechanism. Source: Vasileff et al. [31]. Reproduced with permission of Elsevier. (f) Kinetic relations of reduced CO transition state energies vs. adsorbed *CO energies on metal surfaces (111) and (211). Source: Liu et al. [8]. Licensed under CC-BY-4.0.

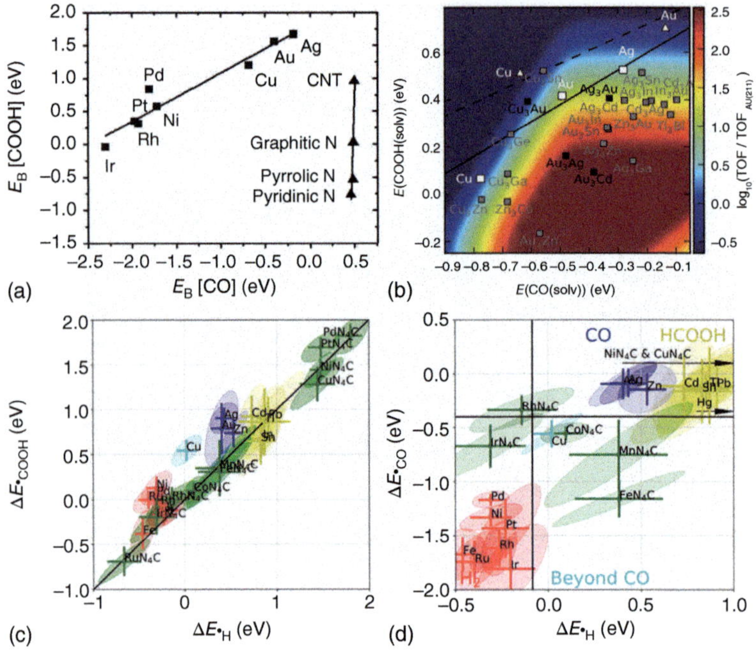

Figure 21.2 Illustration for breaking the scaling relationships by (a) Nitrogen-doped carbon nanotubes (NCNTs). Source: Wu et al. [33]. Reproduced with permission of John Wiley & Sons, Inc. (b) Various alloy surfaces. Source: Vasileff et al. [31]. Reproduced with permission of Elsevier. (c) $M-N_4-C$ structures. Source: Varela et al. [7]. Reproduced with permission of American Chemical Society.

A number of material classes such as bimetallic, metal-free, and single-atom materials have been designed to tune the ECR activity [7, 31, 33, 35, 36]. For example, as shown in Figure 21.2, N-doped carbon nanotube (CNT), various transition metals alloys, and $M-N_4-C$ single-atom materials break the linear relationship opening an avenue to high-performance ECR catalysts. Therefore, DFT-based calculation plays an important role to screen plenty of catalysts and reveal their reactivity even before the experimental validation.

21.2.2 Reactivity Trends of O-included Reactions

Many descriptors are identified for O-included reactions: ORR and OER, as shown in Table 21.1. The volcano activity as a function of such descriptor provides a reactivity trend for O-related reactions. A significant finding from thermodynamic features, OER and ORR mechanistic insights, is that the O^*, OH^*, and OOH^* species are considered as the key intermediates over many catalysts. A significant correlation is found between the last two with a constant difference of 3.2 ± 0.2 eV over a wide range of catalysts using DFT computations. This free energy offset favorably matches with 3.4 eV of the experimental energy difference between OH^- and OOH^- [37]. Thus, the free energy value of OH^* can be predicted from that of OOH^*, and vice versa. Additionally, as the optimal binding energy difference for O-related reactions is 2.46 eV, the 0.37 eV of minimum theoretical overpotential can be obtained. On the basis of this scaling, the activity trend of ORR and OER can be predicted from volcano overpotentials as a function of the O^* and OH^* adsorption energies, or their difference as descriptors [38, 39]. As shown in Figure 21.3A, it represents volcano-related activity of ORR and OER over metals and metal oxides, respectively [32]. Recently, a different scaling relationship: $\Delta G(2OH^*) = \Delta G(O^*) + 1.5$ eV is correlated over carbon-based single-atom class, as shown in Figure 21.3B [40]. This study revealed the structure dependence scaling relations toward ORR activity. Further, literature studies enable rational O-involved reactions activity trends as a function of thermodynamic descriptors over the metal-free material class [41, 42].

Alternatively, electronic, geometric, and binary descriptors are also used to describe the activity of O-confined reactions. In particular, the activity of OER-catalyzed materials is predicted on the basis of volcano plots as a function of such descriptors, as shown in Figure 21.4 [43]. A number of excellent studies and discussions verify the volcano-activity advances through intrinsic identifiers

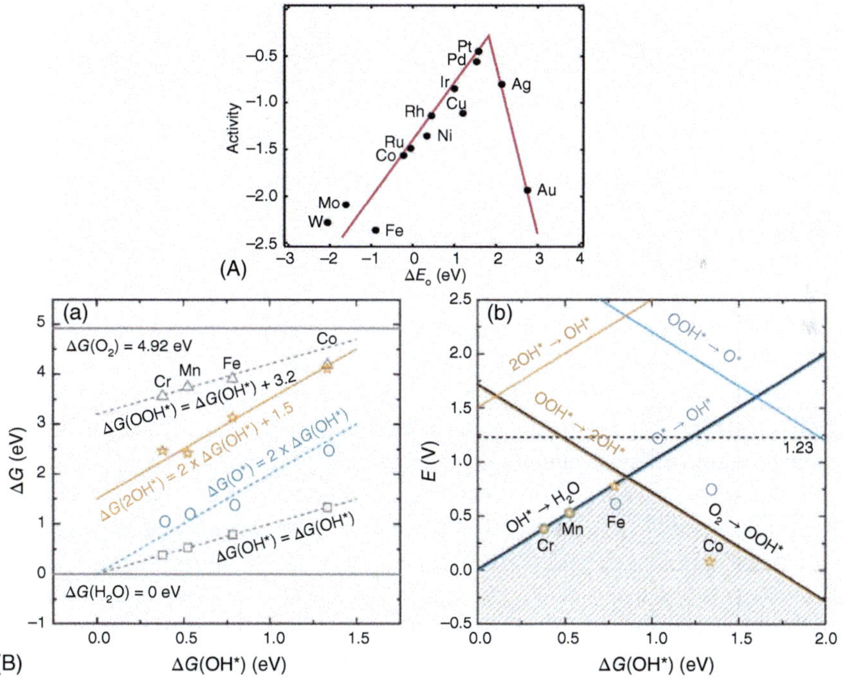

Figure 21.3 (A) ORR volcano plot for metals. Source: Nørskov et al. [1]. Reproduced with permission of American Chemical Society. (B) Scaling relations between the free energies of selected adsorbates OH^*, O^*, $2OH^*$, and OOH^* on $M-N_4-C$ (M = Cr, Mn, Fe, Co) single porphyrin catalysts (a) and volcano plot of the equilibrium ORR potentials as a function of the free energy of OH^* (b). Source: Zhong and Li [40]. Reproduced with permission of American Chemical Society.

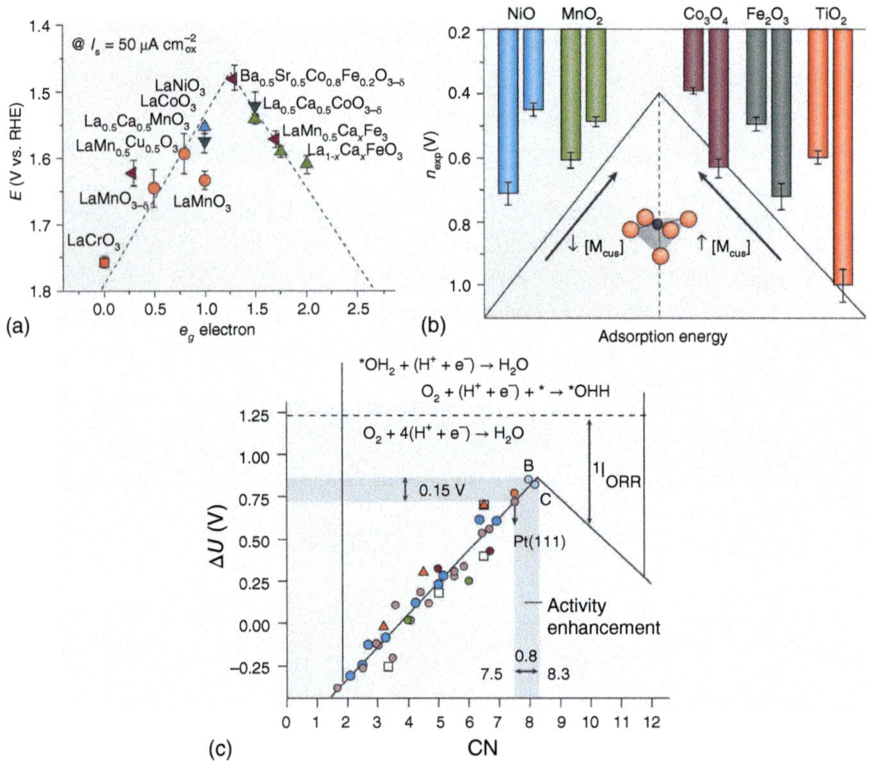

Figure 21.4 OER volcano plots according to various descriptors, such as (a) e_g electrons, (b) adsorption energy, and (c) generalized coordination number. Source: Zhao et al. [43]. Reproduced with permission of Nature Publishing Group.

[5, 27, 43–45]. Based on molecular orbital principles, it was found that the lower the e_g occupancy of metal in selected perovskites is, the stronger the binding of oxygen shows, and vice versa. The optimal OER catalyst is obtained when the e_g electron occupancy is close to unity, as shown in Figure 21.4a. To regulate this occupancy, metal oxidation and spin states should be modified [44]. Moreover, the adjusted coordination number was suggested to predict the performance of metal oxides toward OER, as shown in Figure 21.4c. It is important to say that the intrinsic descriptors are used according to the anion exchange membrane (AEM) mechanism of OER, while the dual-site lattice oxygen-mediated mechanism (LOM) is considered to break the scaling relationships [46]. On the other hand, surface distortion and intrinsic strain are identified as structure descriptors for dual-site metal ORR catalysts, respectively [14, 47]. Additionally, surface strain and p_z occupancy are reported for 2D-doped graphene and transition metals sheets to optimize their activity toward ORR [25, 48]. Hence, in this zone, the DFT-based scaling relationships prove their advantages toward the rational search of more-effective catalysts and further help to predict the optimal one according to volcano-related activity trend.

21.2.3 Reactivity Trends of H-included Reactions

As previously mentioned, the adsorption–activity correlation is related to the key intermediates of given reactions. According to mechanistic insights of HER and hydrogen oxidation reaction (HOR), the free energy, ΔG_{H*}, and the reaction enthalpy, ΔE_H, of the adsorbed hydrogen are considered to describe the reactivity trend over their catalysts, respectively. Ideally, the higher activity of HER theoretically identified by the exchange current density is obtained with zero hydrogen adsorption energy, as shown in Figure 21.5a. Thus, the volcano-related activity trend is suggested as a function of DFT-calculated energies, as shown in Figure 21.5b. The Gibbs free energy of hydrogen adsorption is described by $\Delta G_{H*} = \Delta E_H + 0.24\,\text{eV}$ [49]. Thus, the optimal catalyst for HOR is identified by 0.24 eV of reaction enthalpy. Interestingly, the reaction enthalpy descriptor plays a major role in the HOR activity. In this respect, Cong et al. and Davydova et al. discussions are highly recommended reviews for understanding the reaction enthalpy aspect, and further the rational

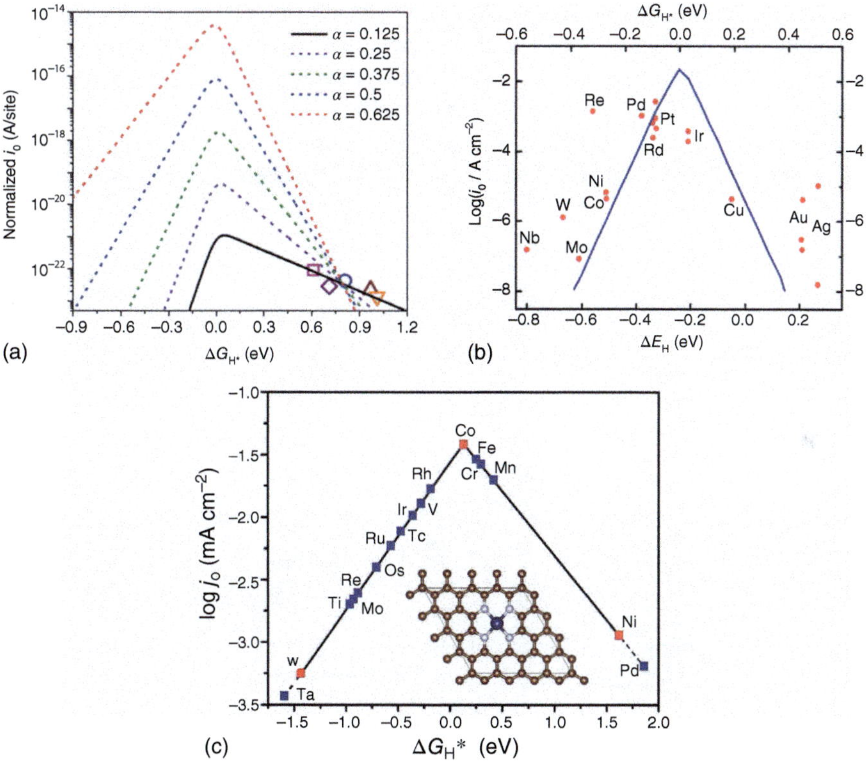

Figure 21.5 (a) Theoretical volcano–activity curve of HER where α is the transfer coefficient in one-electron transfer reaction. Source: Zhang et al. [45]. Reproduced with permission of John Wiley & Sons, Inc. (b) Theoretical kinetic model (blue line) vs. experimental HER/HOR exchange current densities (pointed as dots) over different metal surfaces. Source: Davydova et al. [49]. Reproduced with permission of American Chemical Society. (c) Volcano–activity plot for $M{-}N_4{-}C$ toward HER. Source: Hossain et al. [50]. Reproduced with permission of John Wiley & Sons, Inc.

design insights of HOR [49, 51]. Recently, the electronic–activity relationships are described for HER along with the single-atom class. The binding energy ΔG_{H*} is found to be sensitive to the d-band center of metal with respect to the Fermi level over M–N_x-graphene edges, Figure 21.5c [52]. Separately, based on DOS analysis, the catalytic activity of HER is identified by the influence of the energy states of active valence (d_z^2) orbitals and their resulting antibonding orbitals (σ^*) onΔG_{H*} [50]. Therefore, DFT calculations are feasible means to formulate the volcano–activity curve toward identifying the best performance of given catalysts for H-reactions.

21.3 DOS-Guided Models and Active Sites

Unveiling the origin of the active structure is the main advantage of applying computational DFT in electrocatalytic systems due to its responsibility for the catalyst activity. DFT calculations help to describe and evaluate which sites should be the potential ones affecting the adsorption properties of intermediates. Fundamental understanding of the geometric and electronic structure of the key surface states leads to more accurate thermodynamic and kinetic calculations and further to develop a better model of included interactions toward a well-defined catalytic mechanism. We have discussed the electronic, geometric, thermodynamic descriptors, such as the d-band model, coordination number, and binding energies,

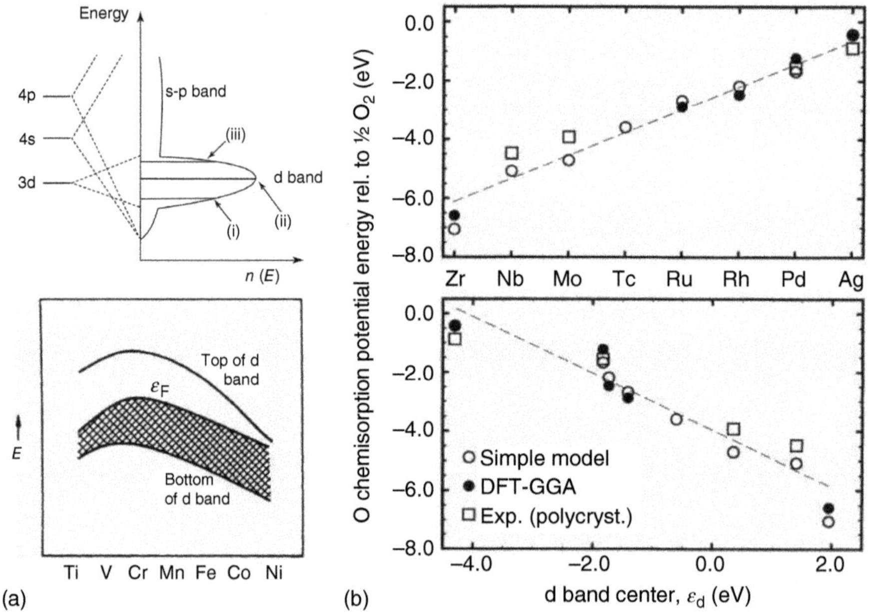

Figure 21.6 (a) Representation of DOS plot (top) and a qualitative indication of Fermi levels and d-band energies changes for transition metal (bottom) [54]. (b) Estimation of the bond strengths via comparison of DFT-based oxygen adsorptions energies experimental values and model for the various metal surfaces. Source: Thomas and Thomas [54]. Reproduced with permission of Wiley-VCH Verlag GmbH & Co. KGaA.

respectively. The electronic distributions in molecules and solids are mostly utilized to identify the catalytic sites using the DOS analysis which is the basis of DFT theory. DOS analysis investigates the charge distribution near the Fermi level by which the valence maximum and the conduction minimum bands' positions are related to electron-donating and accepting mechanisms; thereby, it can be utilized for the selection of potential active sites [53]. In handling metals as catalysts, electronically, they have an incompletely filled energy band. As shown in Figure 21.6a, a trend appears across d-band of the first series of transition metals. Additionally, Figure 21.6b represents how the d-band position affects the adsorption energy relative to the Fermi level. Thus, DOS analysis as a viable tool was used to formulate the d-band model as previously mentioned.

Simultaneously, theoreticians have recognized that the local density of states (LDOSs) is used as a measure of (or to monitor) a catalyst's activity by which it reflects the weight of electronic states, at given energy and position, available for interacting with reactants [54, 55], while the projected/partial density of states (PDOSs) is the relative contribution of a particular atom/orbital to the total DOS [56]. In this respect, the catalytic interaction occurs at sites that possess a high electron DOS just at or below the Fermi level, as shown in Figure 21.7. Recently,

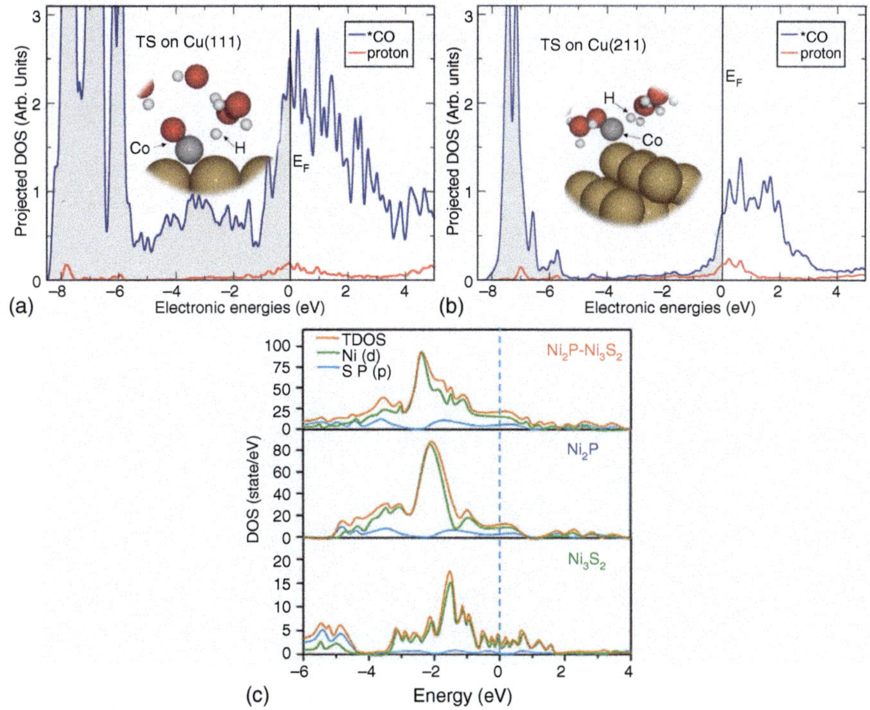

Figure 21.7 Calculated DOS over different surfaces, such as (a and b) CO protonation to CHO on Cu(111) and Cu(211) surfaces toward ECR. Source: Liu et al. [8]. Reproduced with permission of Nature Publishing Group. (c) Ni-based HER catalysts. Source: Zeng et al. [57]. Reproduced with permission of Elsevier.

most experimental findings of electrocatalytic reactions are supported by the DOS diagram to consider the intrinsic activity of each active site and provide a piece of evidence for the catalytic activity of a given catalyst done by DFT computations. Interestingly, single-atom material class is an attractive catalyst due to its atomic-level precision which provides an active site model closer to real conditions. On the other hand, DOS analysis in the metal-free class plays an important role to evaluate different configurations of similar active sites. DOS is sensitive to the type, configuration, and location of sites. Literature has good resources to figure out the electronic structure–catalytic activity relationship endorsed by the DOS tool [28, 58–66].

References

1 Nørskov, J.K., Rossmeisl, J., Logadottir, A. et al. (2004). *J. Phys. Chem. B* 108: 17886–17892.
2 Nørskov, J.K., Bligaard, T., Logadottir, A. et al. (2005). *J. Am. Chem. Soc.* 152: J23–J26.
3 Zhang, Y.-J., Sethuraman, V., Michalsky, R., and Peterson, A.A. (2014). *ACS Catal.* 4: 3742–3748.
4 Cave, E.R., Shi, C., Kuhl, K.P. et al. (2018). *ACS Catal.* 8: 3035–3040.
5 Dickens, C.F., Montoya, J.H., Kulkarni, A.R. et al. (2019). *Surf. Sci.* 681: 122–129.
6 Calle-Vallejo, F., Tymoczko, J., Colic, V. et al. (2015). *Science* 350: 185–189.
7 Varela, A.S., Ju, W., Bagger, A. et al. (2019). *ACS Catal.* 9: 7270–7284.
8 Liu, X., Xiao, J., Peng, H. et al. (2017). *Nat. Commun.* 8: 15438.
9 Hussain, J., Jónsson, H., and Skúlason, E. (2018). *ACS Catal.* 8: 5240–5249.
10 Tran, K. and Ulissi, Z.W. (2018). *Nat. Catal.* 1: 696–703.
11 Zhao, Z. and Lu, G. (2018). *ACS Catal.* 8: 3885–3894.
12 Greeley, J., Jaramillo, T.F., Bonde, J. et al. (2006). *Nat. Mater.* 5: 909–913.
13 Greeley, J., Norskov, J.K., Kibler, L.A. et al. (2006). *ChemPhysChem* 7: 1032–1035.
14 Chattot, R., Le Bacq, O., Beermann, V. et al. (2018). *Nat. Mater.* 17: 827–833.
15 Wei, C., Feng, Z., Scherer, G.G. et al. (2017). *Adv. Mater.* 29: 1606800.
16 Toyoda, K., Hinogami, R., Miyata, N., and Aizawa, M. (2015). *J. Phys. Chem. C* 119: 6495–6501.
17 Hong, W.T., Risch, M., Stoerzinger, K.A. et al. (2015). *Energy Environ. Sci.* 8: 1404–1427.
18 Suntivich, J., May, K.J., Gasteiger, H.A. et al. (2011). *Science* 334: 1383–1385.
19 Suntivich, J., Gasteiger, H.A., Yabuuchi, N. et al. (2011). *Nat. Chem.* 3: 546–550.
20 Lee, Y.-L., Kleis, J., Rossmeisl, J. et al. (2011). *Energy Environ. Sci.* 4: 3966–3970.
21 Grimaud, A., May, K.J., Carlton, C.E. et al. (2013). *Nat. Commun.* 4: 2439.
22 Zhang, J., Zhao, Z., Xia, Z., and Dai, L. (2015). *Nat. Nanotechnol.* 10: 444–452.
23 Zhao, Z., Li, M., Zhang, L. et al. (2015). *Adv. Mater.* 27: 6834–6840.
24 Jiao, Y., Zheng, Y., Davey, K., and Qiao, S.-Z. (2016). *Nat. Energy* 1: 16130.
25 Sinthika, S., Waghmare, U.V., and Thapa, R. (2018). *Small* 14: 1703609.

26 Yang, W., Xu, S., Ma, K. et al. (2019). *Nano Mat. Sci.* 1 (5): 339–348.
27 Xu, H., Cheng, D., Cao, D., and Zeng, X.C. (2018). *Nat. Catal.* 1: 339–348.
28 Gong, L., Zhang, D., Lin, C.Y. et al. (2019). *Adv. Energy Mater.* 9: 1902625.
29 Bagger, A., Ju, W., Varela, A.S. et al. (2017). *ChemPhysChem* 18: 3266–3273.
30 Shi, C., Hansen, H.A., Lausche, A.C., and Norskov, J.K. (2014). *Phys. Chem. Chem. Phys.* 16: 4720–4727.
31 Vasileff, A., Xu, C., Jiao, Y. et al. (2018). *Chem* 4: 1809–1831.
32 Seh, Z.W., Kibsgaard, J., Dickens, C.F. et al. (2017). *Science* 355: eaad4998.
33 Wu, J., Sharifi, T., Gao, Y. et al. (2019). *Adv. Mater.* 31: e1804257.
34 Li, Y. and Sun, Q. (2016). *Adv. Energy Mater.* 6: 1600463.
35 Nitopi, S., Bertheussen, E., Scott, S.B. et al. (2019). *Chem. Rev.* 119: 7610–7672.
36 Tian, Z., Priest, C., and Chen, L. (2018). *Adv. Theory Simul.* 1: 1800004.
37 Christensen, R., Hansen, H.A., Dickens, C.F. et al. (2016). *J. Phys. Chem. C* 120: 24910–24916.
38 Kulkarni, A., Siahrostami, S., Patel, A., and Norskov, J.K. (2018). *Chem. Rev.* 118: 2302–2312.
39 Greeley, J. (2016). *Annu. Rev. Chem. Biomol. Eng.* 7: 605–635.
40 Zhong, L. and Li, S. (2020). *ACS Catal.* 10: 4313–4318.
41 Zhang, J. and Dai, L. (2015). *ACS Catal.* 5: 7244–7253.
42 Li, M., Zhang, L., Xu, Q. et al. (2014). *J. Catal.* 314: 66–72.
43 Zhao, Z.-J., Liu, S., Zha, S. et al. (2019). *Nat. Rev. Mater.* 4: 792–804.
44 Song, J., Wei, C., Huang, Z.F. et al. (2020). *Chem. Soc. Rev.* 49: 2196–2214.
45 Zhang, L., Lin, C.Y., Zhang, D. et al. (2019). *Adv. Mater.* 31: e1805252.
46 Chen, Z.W., Chen, L.X., Wen, Z., and Jiang, Q. (2019). *Phys. Chem. Chem. Phys.* 21: 23782–23802.
47 Escudero-Escribano, M., Malacrida, P., Hansen, M.H. et al. (2016). *Science* 352: 73–76.
48 Wang, L., Zeng, Z., Gao, W. et al. (2019). *Science* 363: 870–874.
49 Davydova, E.S., Mukerjee, S., Jaouen, F., and Dekel, D.R. (2018). *ACS Catal.* 8: 6665–6690.
50 Hossain, M.D., Liu, Z., Zhuang, M. et al. (2019). *Adv. Energy Mater.* 9: 1803689.
51 Cong, Y., Yi, B., and Song, Y. (2018). *Nano Energy* 44: 288–303.
52 Gao, G., Bottle, S., and Du, A. (2018). *Catal. Sci. Technol.* 8: 996–1001.
53 Jiao, S., Fu, X., Zhang, L. et al. (2020). *Nano Today* 31: 100833.
54 Thomas, J.M. and Thomas, W.J. (2015). *Principles and Practice of Heterogeneous Catalysis*. Wiley-VCH.
55 Maclaren, J.M., Pendry, J.B., Joyner, R.W., and Meehan, P. (1986). *Surf. Sci.* 175: 263–275.
56 Gao, Y., Lockart, M., Kispert, L.D., and Bowman, M.K. (2019). *J. Phys. Chem. C* 123: 24634–24642.
57 Zeng, L., Sun, K., Wang, X. et al. (2018). *Nano Energy* 51: 26–36.
58 Chen, J., Liu, J., Xie, J.-Q. et al. (2019). *Nano Energy* 56: 225–233.
59 Zhou, P., Lv, X., Xing, D. et al. (2020). *Appl. Catal., B* 263: 118330.
60 Diao, J., Qiu, Y., Liu, S. et al. (2020). *Adv. Mater.* 32: e1905679.
61 Zhi, X., Jiao, Y., Zheng, Y. et al. (2020). *Nano Energy* 71: 104601.

62 Wu, Q., Gao, J., Feng, J. et al. (2020). *J. Mater. Chem. A* 8: 1205–1211.
63 Guan, C., Xiao, W., Wu, H. et al. (2018). *Nano Energy* 48: 73–80.
64 He, W., Han, L., Hao, Q. et al. (2019). *ACS Energy Lett.* 4: 2905–2912.
65 Zou, X., Liu, M., Wu, J. et al. (2017). *ACS Catal.* 7: 6245–6250.
66 Quaino, P., Santos, E., Soldano, G., and Schmickler, W. (2011). *Adv. Phys. Chem.* 2011: 1–14.

22

DFT Applications in Selected Electrocatalytic Systems

Density functional theory (DFT) modeling is considered to be the heart of computational electrocatalysis. In this regard, electrocatalytic modeling can be initiated by representing the electrode surface, Figure 22.1a, after selecting a proper approximation of DFT approaches, such as generalized gradient approximation (GGA). Once the active site model is optimized, physicochemical contact is built between the site and reactive species. From this step, the electronic structure of the catalyst can be calculated resulting in the band structure, density of states (DOS), and binding free energy diagrams, Figure 22.1b. Due to the electrocatalytic cycle involves proton–electron pair transfer, the next step is to introduce the proton–electron pair. At this point, the reaction free energy can also be measured. Consequently, the location of the activation barrier, transition state (TS), is required. By employing activation barrier approaches, such as Brønsted–Evans–Polanyi (BEP) relations, the free energy of the transition step is hypothesized. This run should be done for all elementary steps toward getting the minimum energy path, Figure 22.1c. Finally, the free energy of the electrochemical barriers is proposed to elucidate an accurate mechanism describing the catalytic process, Figure 22.1d. As we discussed, the nature of the interfacial layer and the chemical potential of participated electrons and ionic species should be considered. Thus, computational hydrogen electrode (CHE) and solvation models are incorporated to determine the reaction-related thermodynamics, Figure 22.1e. Moreover, the DFT-microkinetic model is applied to estimate reaction-related kinetics, Figure 22.1f. DFT calculations allow mechanistic understudying of a given catalytic cycle resulting in a well-defined descriptor, such descriptor is used to screen class of materials to find an optimal catalyst located at the top of the volcano. Real experimentation tests its validation toward the rational design of understudying reaction. Hereafter, we aim to describe the feasibility of DFT calculations in electrocatalytic reactions; such as electrochemical carbon dioxide reduction (ECR), oxygen evolution reaction (OER), ORR, hydrogen evolution reaction (HER), and hydrogen oxidation reaction (HOR).

22.1 Unveiling the Electrocatalytic Mechanism

DFT modeling aims to understand the atomistic insights, reveal the reaction pathway, and evaluate a given catalyst, and finally drive the rational design

Electrocatalysis in Balancing the Natural Carbon Cycle, First Edition. Yaobing Wang.
© 2021 WILEY-VCH GmbH. Published 2021 by WILEY-VCH GmbH.

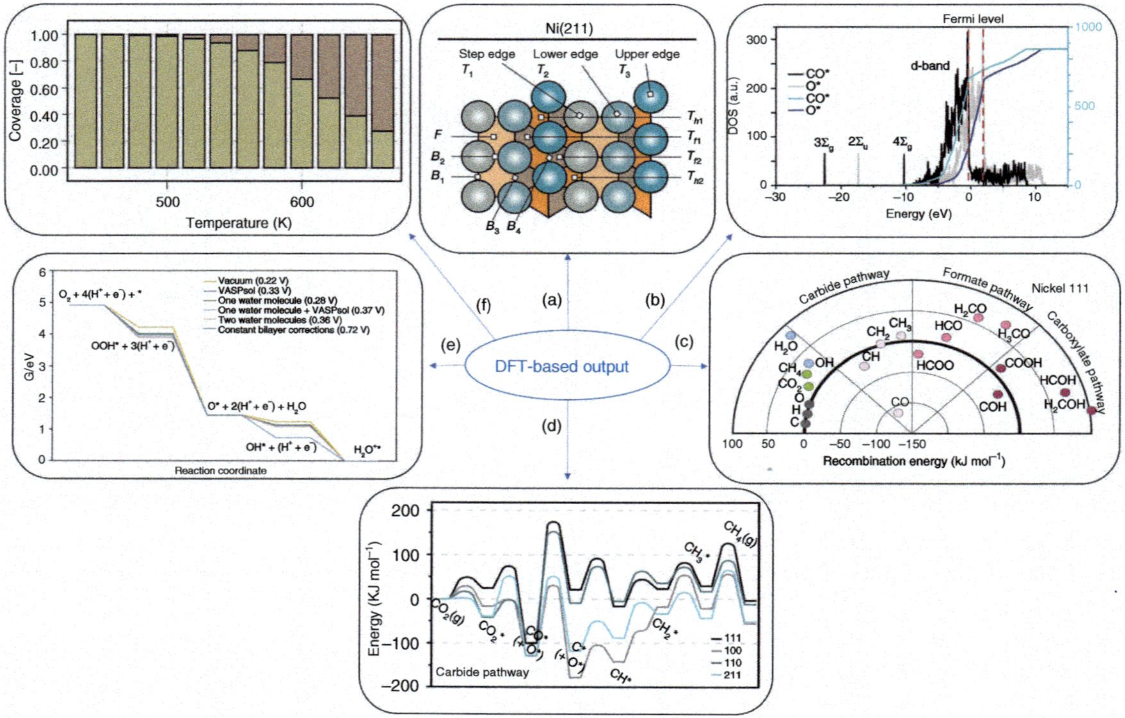

Figure 22.1 Schematic presentation for DFT output; such as catalyst representation (a), the electronic structure variation along with different intermediates (b), different pathways evaluation (c), free energy profile of elementary steps (d), free energy profile under different solvation conditions (e), and microkinetic modeling (f) [1, 2]. Source: Ref. Vogt et al. [1]. Licensed under CC-BY-4.0. Ref. Zhang and Asthagiri [2]. Reproduced with permission of Elsevier.

approach to meet the commercial demand. To achieve the rationally designed reaction target, possible pathways should be simulated including reactive sites and intermediates, taking into account: (i) the large kinetic barrier of rate-limiting step (RLS), (ii) the existence of competing reactions, and (iii) the interfacial region of the electrode and electrolyte at which electrocatalytic reaction occurs. Using DFT in combination with the CHE model, the optimal pathway, which represented by free energy diagram, is proposed by comparing the binding/adsorption free energies of different intermediates at different sites. As mentioned in the CHE subsection 20.5.1, the free energy of proton–electron pairs can be calculated. Besides, thermodynamic analysis using overpotential is measured. The overpotential correlates to the difference of free energy between two adsorbed intermediates involved in the RLS. The lower overpotential value is the better performance of the catalyst shows. To uncover to what extend the overpotential will be, the reaction activation barrier is required to be simulated accounting the potential; ΔG_{field}, solvent; ΔG_{solv}, and ions effects. Using some approximation, mentioned above, such as transition state, solvation, and kinetic models, the desired mechanism becomes close to be accurate.

22.1.1 ECR Reaction

The general form of cathodic reaction of ECR is represented by:

$$xCO_2 + nH^+ + ne^- \rightarrow \text{carbon product} + yH_2O \quad (22.1)$$

ECR standard potentials (thermodynamics) is simply calculated using gas-phase thermochemistry data according to the reaction Gibbs free energy, (pH = 7). However, the kinetics of the ECR is not as easy as thermodynamics, where it proceeds at more negative potentials than the equilibrium ones listed in Table 22.1. ECR involves multi-step paths depending on the number of catalyst-boosted electrons toward different products.

Table 22.1 Equilibrium potentials of different ECR possibilities.

Reaction	E^0 (V vs RHE)
$CO_2 + 2H^+ + 2e^- \rightarrow HCOOH(aq)$	−0.12
$CO_2 + 2H^+ + 2e^- \rightarrow CO(g) + 2H_2O$	−0.10
$2CO_2 + 2H^+ + 2e^- \rightarrow (COOH)_2(s)$	−0.47
$CO_2 + 4H^+ + 4e^- \rightarrow C(s) + 2H_2O$	0.21
$CO_2 + 6H^+ + 6e^- \rightarrow CH_3OH(aq) + H_2O$	0.03
$CO_2 + 8H^+ + 8e^- \rightarrow CH_4(g) + 2H_2O$	0.17
$2CO_2 + 8H^+ + 8e^- \rightarrow CH_3COOH(aq) + 2H_2O$	0.11
$2CO_2 + 10H^+ + 10e^- \rightarrow CH_3CHO(aq) + 3H_2O$	0.06
$2CO_2 + 12H^+ + 12e^- \rightarrow C_2H_5OH(aq) + 3H_2O$	0.09
$2CO_2 + 12H^+ + 12e^- \rightarrow C_2H_4(g) + 4H_2O$	0.08
$2CO_2 + 14H^+ + 14e^- \rightarrow C_2H_6(g) + 4H_2O$	0.14

Among available ECR catalysts, only pure copper shows a superior ability to catalyze CO_2 to C1 and C2+ products, so that theoretical scientists gave it more attention to be designed for the ECR system. The current state of pathway possibilities on polycrystalline Cu for ECR is summarized in Figure 22.2a. Generally, most pathways agree that the RLS of ECR process is adsorption step, i.e. $CO_2 \rightarrow CO_2^-$ [15]. In addition, *CO and *COOH are considered to be crucial intermediates in ECR. Their binding energies identify the product selectivity of various electrode surfaces. Analyzing such a proposed pathway using DFT results in a free energy diagram representing a complete series of elementary steps including their activation energies. The free energy profile varies according to a given catalyst with different facets, dimensions, defects, vacancies, size, and interface modifications ranging from metal-based to metal-free materials, Figure 22.2b–d, [1, 16–21]. In a word, the electrochemical ECR pathway can be tuned by geometric, electronic, and local environmental modifications using selective catalyst toward a certain product.

22.1.2 OER Reaction

OER is the complement half-reaction of ECR, the reverse reaction of the O_2 reduction reaction (ORR), the half-reaction of electrochemical water splitting, and a crucial part in rechargeable metal-air batteries whereas it is limited by sluggish kinetics. The anodic OER half-cell is a pH-dependent reaction as represented by following equations [22, 23]:

$$\text{Low pH}: \quad 2H_2O(l) \rightarrow O_2(g) + 4H^+ + 4e^- \quad E^0 = -1.23\,V \tag{22.2}$$

$$\text{High pH}: \quad 4OH^- \rightarrow O_2 + 2H_2O(l) + 4e^- \quad E^0 = -0.40\,V \tag{22.3}$$

The stepwise OER includes four concerted proton-coupled electron-transfer (CPET) steps which slow the kinetics. According to the number of reactive sites; i.e. in general one or two metal, OER mechanism varies to widely accepted adsorbate evolution mechanism (AEM) and lattice oxygen-mediated mechanism (LOM), Figure 22.3a,b, respectively [24]. The proposed AEM pathway is represented in Scheme 22.1. At each step, a proton is released to the electrolyte, eventually combining with a transferred electron to generate hydrogen (HER) at the cathode.

$H_2O + * \rightarrow OH^* + H^+ + e^-$ or $OH^- + * \rightarrow OH^* + e^-$; according to pH
$OH^* \rightarrow O^* + H^+ + e^-$
$H_2O + O^* \rightarrow OOH^* + H^+ + e^-$
$OOH^* \rightarrow O_2 + * + H^+ + e^-$

Scheme 22.1 Elementary steps of OER half-reaction.

On the other hand, LOM occurs on two neighboring metal sites in which OER initiates by deprotonation of two HO* on two metal sites, resulting in metal–oxo species

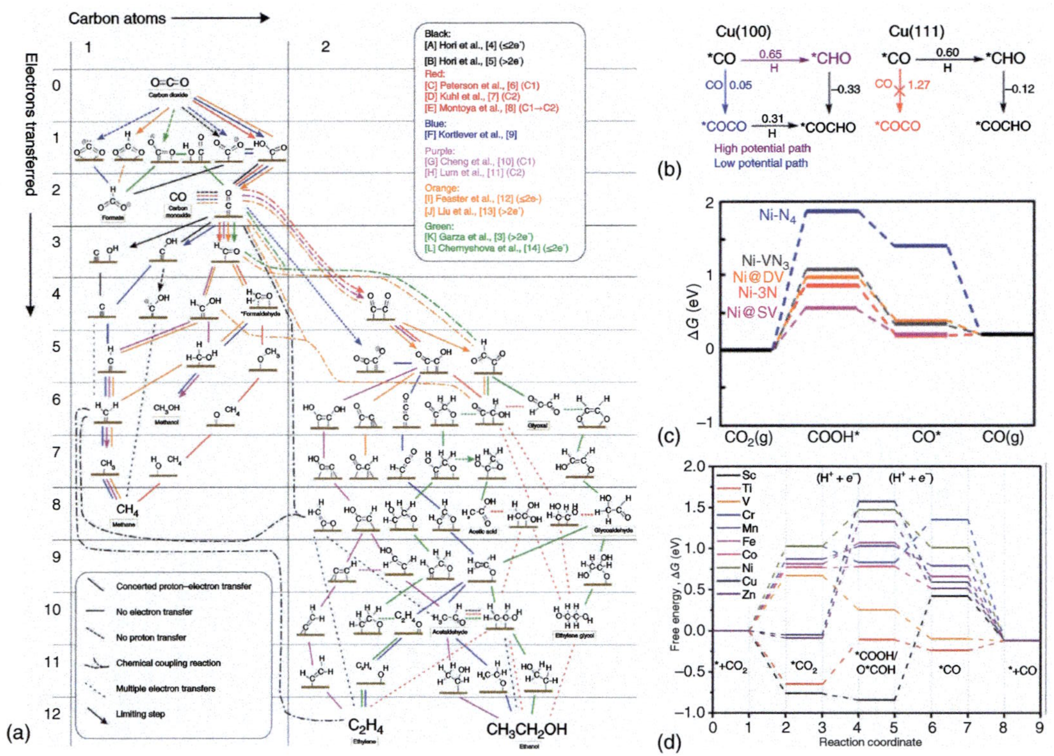

Figure 22.2 (a) Overview of possible pathways for ECR toward various products on polycrystalline copper [15]. (b) Calculated energies along with various catalysts; describing the potential paths to C−C bond formation and *COCHO on Cu(100) and Cu(111) (b) [3], and ECR diagram over different Ni-based active sites (c) [16], and TM-SACs surfaces (d) [17]. Source: Refs Nitopi et al. [3, 15, 16]. Reproduced with permission of American Chemical Society. Source: Ref. Gong et al. [17]. Reproduced with permission of John Wiley & Sons, Inc.

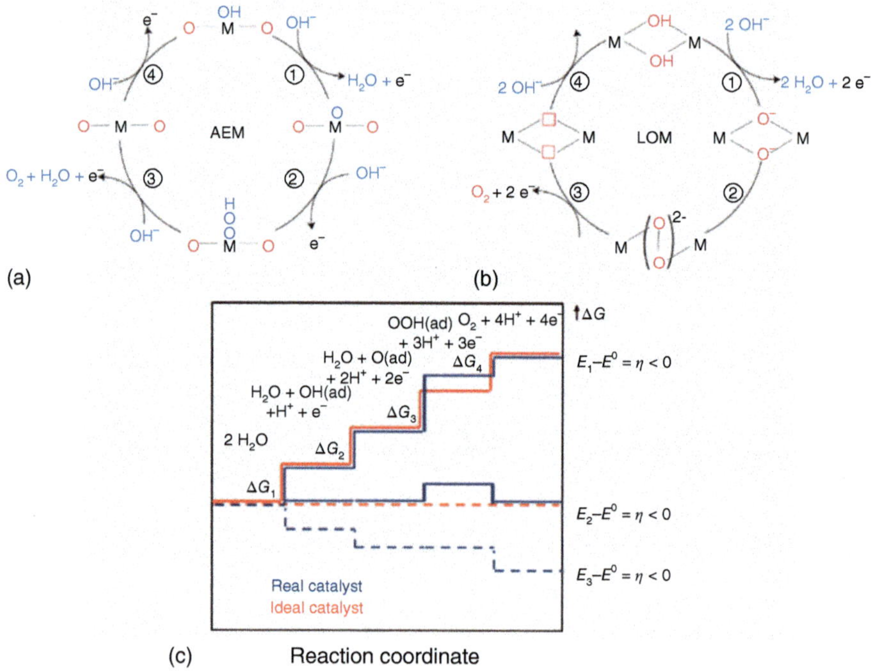

Figure 22.3 Schematic representation of two possible OER mechanisms; AEM (a) and LOM (b) [24]. (c) Gibbs free energies profile at $U = 0$ for ideal and real catalysts. Source: Song et al. [24]. Reproduced with permission of Royal Society of Chemistry.

[24]. Then, these oxo species combine as an O—O bond leading to HOO* formation, Figure 22.3b. The free energy diagram based on DFT studies is used to assess the catalyst activity requiring large overpotentials that are needed to overcome the reaction barriers [24, 25]. An ideal OER catalyst needs to overcome the Gibbs free energy barrier of each elemental step, Figure 22.3c. Thus, the catalyst might be limited by any of the key steps [26]. Moreover, both HOO* and HO* bind through an oxygen atom via a single bond, resulting in a strong linear correlation of their adsorption free energies [24]. IrO_2/RuO_2 has been regarded as the best OER catalyst in practical devices [27]. A reasonable DFT means dramatically improve our understanding of the OER mechanism. Literature has good studies using DFT to uncover mechanistic insights of OER under different pH conditions along with numerous catalysts [28–39].

22.1.3 ORR Reaction

ORR involves multiple electrochemical reactions including both the cleavage and formation of the O—O and O—H bonds, respectively. As OER, ORR is a pH-dependent reaction; however, it produces different products; H_2O_2 and H_2O, via direct and indirect four-electron processes depending on acidic or alkaline conditions [40, 41]. Overall ORR processes are summarized in Scheme 22.2 with their thermodynamic potentials. Regarding the mechanism of ORRs, it has different approaches; i.e. dissociation, association, and peroxo paths (Scheme 22.2) [42]. The former belongs to direct $4e^-$ process at low pH while the last two is included in indirect $2e^-$ steps processes. Pathway selection depends on catalyst ability to dissociate the O_2 molecule before reduction.

			E^0 (V)
Direct processes	Low pH	$O_2 + 4H^+ + 4e^- \rightarrow H_2O$	1.229
	High pH	$O_2 + H_2O + 4e^- \rightarrow 4OH^-$	0.401
Indirect processes	Low pH	$O_2 + 2H^+ + 2e^- \rightarrow H_2O_2$	0.70
		$H_2O_2 + 2H^+ + 2e^- \rightarrow 2H_2O$	1.76
	High pH	$O_2 + H_2O + 2e^- \rightarrow HO_2^- + OH^-$	−0.065
		$HO_2^- + H_2O + 2e^- \rightarrow 3OH^-$	0.867
Dissociation path	$O_2^* \rightarrow O^* + O^*$		
	$O^* + H^+ + e^- \rightarrow OH^*$		
	$OH^* + H^+ + e^- \rightarrow H_2O^*$		
Association path	$O_2^* + H^+ + e^- \rightarrow OOH^*$		
	$OOH^* \rightarrow O^* + OH^*$		
	$O^* + H^+ + e^- \rightarrow OH^*$		
	$OH^* + H^+ + e^- \rightarrow H_2O^*$		
Peroxo path	$O_2^* \rightarrow HOOH^*$		
	$HOOH^* \rightarrow OH^* + OH^*$		
	$OH^* + H^+ + e^- \rightarrow H_2O^*$		

Scheme 22.2 ORR processes with thermodynamic potentials and elementary steps of included pathways.

DFT calculations describe the adsorption energies of ORR intermediates on different surfaces. Among them, platinum is the best and well-known ORR catalyst. Figure 22.4a represents the free energy profile for the associative $4e^-$ ORR path over Pt(111). The first and last key steps are considered to be an activity measure for various catalysts as they show uphill activation barriers (green path of energy diagram). The maximum potential of downhill free energy paths is c. 0.8 V, which agrees with the experimentally observed voltage for large ORR current densities [41]. Figure 22.4b shows slightly uphill formation energy of OOH^* toward hydrogen peroxide generation. Understanding the mechanism insights – as advantage of theoretical computations – and product selectivity would help to enhance the catalyst performance for potential applications which are related to ORR processes.

22.1.4 HER Reaction

Mechanistically, HER holds pH-dependent intermediates. The potential intermediates are protons and water molecules in low and high pH solutions, respectively.

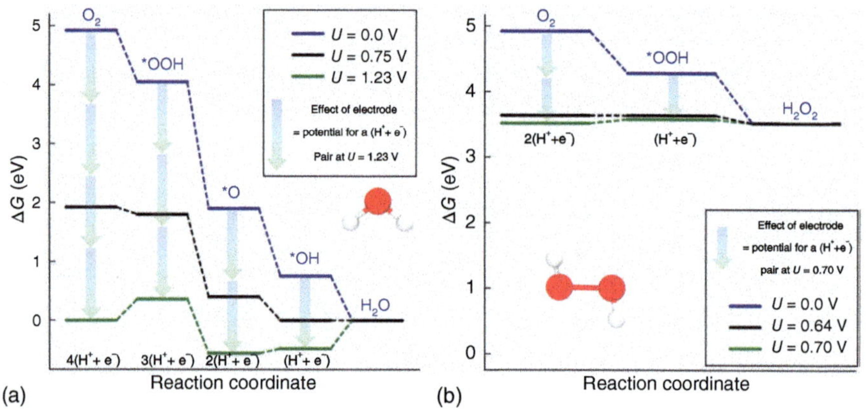

Figure 22.4 Free energies diagram for four-electrons (a) and two-electrons (b) ORR on Pt(111) and $PtHg_4$, respectively. Source: Kulkarni et al. [41]. Reproduced with permission of American Chemical Society.

HER occurs through three possible reactions named by Volmer, Heyrovsky, and Tafel steps, Figure 22.5a, [43, 45]. The Volmer reaction represents the interaction between a proton (hydronium cation) or water molecule and an electron on the electrode surface forming adsorbed proton. The Heyrovsky and Tafel reactions stand for H_2 production via another proton diffusion either directly or jointly to the first one on the surface. The HER kinetics is identified by the Tafel slope. When the desorption or combination is RLS, fast, or slow step, the Tafel slope should be 29, 39, 116 mV dec^{-1} [43]. Based on DFT modeling, the free energy change is provided considering that $\Delta G_{H^*}$is the key energy for HER process as H* is the only involved intermediate, Figure 22.5b, [44, 46]. The activity of a given catalyst is evaluated according to this parameter. Thus, the excellent catalytic performance toward HER is authorized to platinum group transition metals (including Pt, Pd, Ru, Ir, and Rh) [44, 47, 48].

22.1.5 HOR Reaction

The dihydrogen molecule is oxidized in alkaline media to form water via the following overall HOR equation:

$$H_2 + 2OH^- \rightarrow 2H_2O + 2e^- \qquad (22.4)$$

HOR mechanism is generally assumed to proceed via three typical mechanisms Tafel–Volmer, Heyrovsky–Volmer, and Horiuti reactions, Figure 22.6a. The elementary steps of each reaction are depicted in Scheme 22.3. Tafel reaction represents the dissociative adsorption of dihydrogen while Horiuti suggested nondissociative dihydrogen adsorption, see Scheme 22.3, [49]. Further, an electron release that occurs from dihydrogen to the catalyst surface is called Heyrovsky reaction. Fundamental understanding of the Heyrovsky step is under considerable debate in which two categories are proposed; direct attack from a solution of OH^- or joint combination of

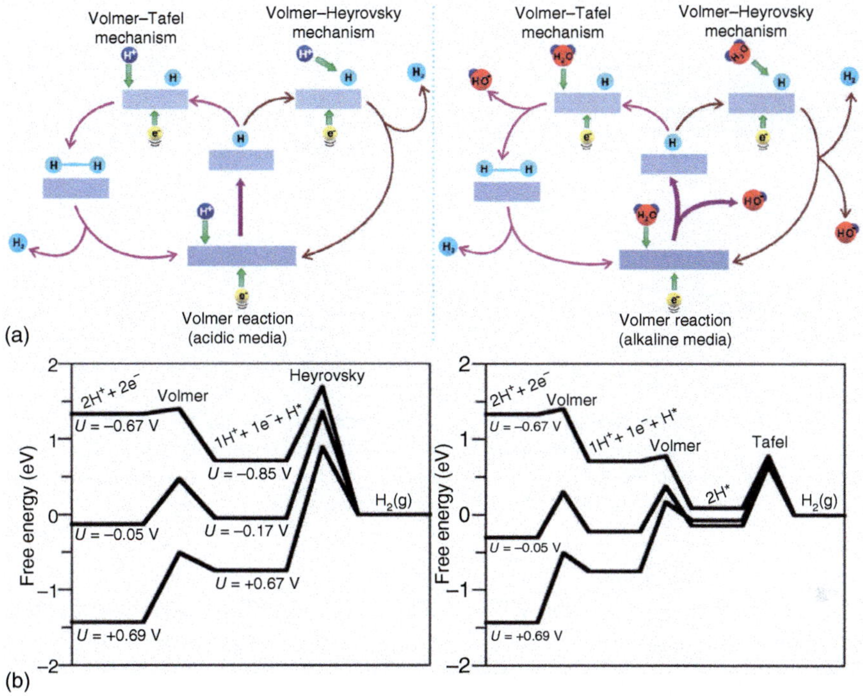

Figure 22.5 (a) Schematic illustration of HER mechanism possibilities in acidic (left) and basic (right) media. Source: Zhu et al. [43]. Reproduced with permission of American Chemical Society. (b) Free energies profiles of Volmer–Heyrovsky (left) and Volmer–Tafel (right) routes on Pt(111) under different potentials. Source: Zheng et al. [44]. Reproduced with permission of John Wiley & Sons, Inc.

OH^* [50]. Finally, HOR ends with a Volmer reaction which stands for the discharge of the adsorbed hydrogen atom.

Tafel reaction		$H_2 + 2^* \rightarrow 2H^*$
Horiuti reaction	Step (I)	$H_2 + * \leftrightarrows H_2M \rightarrow H_2^+M + e^-$
	Step (II)	$H_2^+M \leftrightarrows MH + H^+$
Heyrovsky reaction	Direct OH^- attack	$H_2 + OH^- + * \rightarrow H^* + H_2O + e^-$
	Indirect OH^- attack (I)	$OH^- + * \leftrightarrows OH^* + e^-$
	Indirect OH^- attack (II)	$H_2 + OH^* \leftrightarrows H^* + H_2O$
Volmer reaction		$H^* + OH^- \rightarrow * + H_2O + e^-$

Scheme 22.3 The elementary steps of HOR possible mechanisms.

On the basis of DFT, the provided binding energy of possible intermediates and via the comparison of their kinetic parameters, the mechanistic insights have been

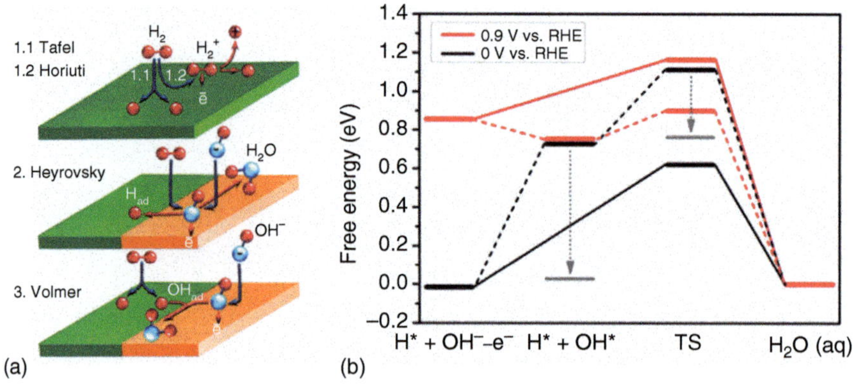

Figure 22.6 (a) Schematic representation of HER mechanism possibilities. Source: Davydova et al. [49]. Reproduced with permission of American Chemical Society. (b) Free energy profile of the competing OH^- and OH^* pathways on the Pt(111). Source: Liu et al. [47]. Reproduced with permission of American Chemical Society.

revealed on HOR catalysts. Recently, Heyrovsky's reaction has been studied on Pt(111) comparing the two mentioned mechanisms via a free energy diagram, Figure 22.6b. OH^* species are found to influence the HOR kinetics; thus, the OH^*-involved mechanism is an acceptable one. Several works explain the activity of different types of materials toward HOR using DFT [51–57].

22.1.6 CO Oxidation Reaction

In general, oxidation of CO molecule by dioxygen forms gaseous CO_2. Mechanistically, it follows three possible mechanisms, namely, Eley–Rideal (ER), bimolecular Langmuir–Hinshelwood (LH), and trimolecular (termolecular) Eley–Rideal (TER), according to molecule attack mode [58]. Scheme 22.4 represents the elementary steps of each pathway. ER includes adsorbed O directly attacked by gaseous CO to form a carbonate-like intermediate state leading to desorption of gaseous CO_2 [58]. On the other hand, LH initiates by simultaneous co-adsorption of CO and oxygen molecules to come by peroxo-type intermediate completing by CO_2 desorption. A TER mechanism firstly reported by Mao et al. become considered in CO oxidation's theoretical studies [59]. It starts with co-adsorption of two CO molecules that undergo direct dioxygen attack to generate OCOOCO intermediate desorbing two

ER	$O_2 + * \rightleftharpoons O_2^* CO(g) + O_2^* \rightleftharpoons CO_3{}^* \rightleftharpoons CO_2 + O^*$
LH	$CO(g) + * \rightleftharpoons CO^* \; O_2 + * \rightleftharpoons O_2^* \; CO^* + 2O^* \rightleftharpoons OOCO^* \rightleftharpoons CO_2 + O^*$
TER	$2CO(g) + * \rightleftharpoons 2CO^* \; O_2 + 2CO^* \rightleftharpoons OCOOCO^* \rightleftharpoons 2CO_2$

Scheme 22.4 Elementary steps of several mechanisms for CO oxidation.

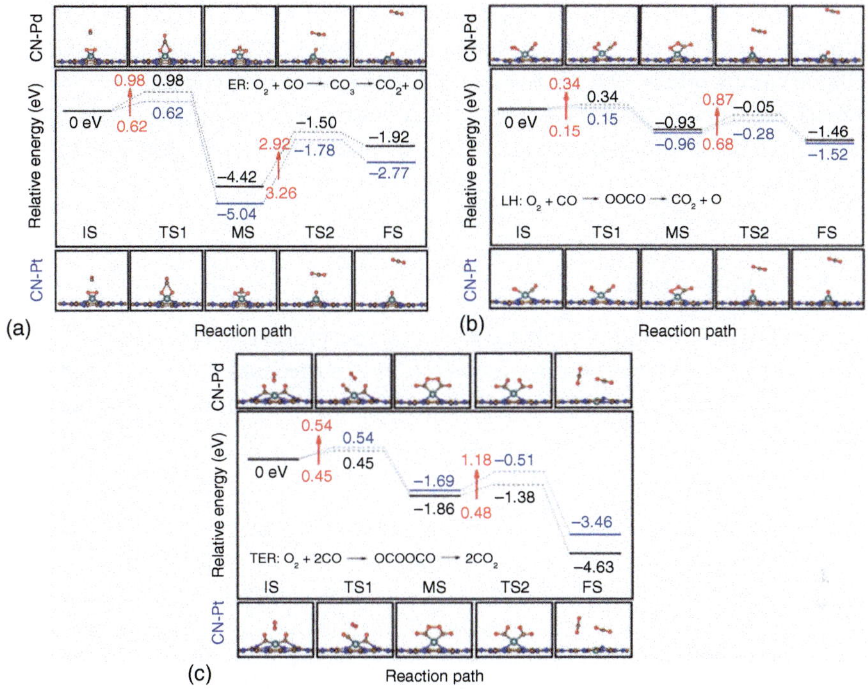

Figure 22.7 Free energies diagrams of CO oxidation via different possible (a) ER, (b) LH, (c) TER mechanisms on Pd@CN, and Pt@CN. Source: Rao et al. [65]. Reproduced with permission of Royal Society of Chemistry.

CO_2 molecules. Most studies examine the mechanism of CO oxidation for a given catalyst by DFT [60–64].

Depending on thermodynamic and kinetic-related calculations, the catalyst prefers one of the possible mechanisms which is suggested by the author. For example, CO oxidation was examined on Pd@CN and Pt@CN employing DFT calculations. They found that they prefer TER and LH mechanisms, respectively, as shown in Figure 22.7, [65]. Single atom-based studies support the newly TER mechanism [59, 61, 66]. However, LH and ER are considered for most research work [62, 63, 67–70]. Another advantage of DFT is to reveal the different effects of water in CO oxidation in the presence or absence of oxygen [71, 72]. The oxidation mechanism under water inclusion is given by the following equations:

$$CO(g) + * \rightleftharpoons CO^* \tag{22.5}$$

$$H_2O + * \Longleftrightarrow OH^* + H^+ + e^- \tag{22.6}$$

$$CO^* + OH^* \rightarrow COOH^* \tag{22.7}$$

$$COOH^* + OH^* \rightarrow CO_2 + H_2O \tag{22.8}$$

In which, the *COOH is included. However, the reaction in Eq. (22.8) was a matter of discussion [72]. Recently, Liu et al. reported that this intermediate prohibits the oxidation completion as $COOH^*$ is very stable and cannot dissociate to release CO_2. Whereas introducing water molecule in CO oxidation by dioxygen promotes it and decreases the reaction barriers due to the steric effect of H_2O molecules [71]. By this end, DFT calculations show its ability to screen the mechanistic potential among various pathways for CO oxidation reaction.

22.1.7 FAOR Reaction

Formic acid undergoes 2-electron electro-oxidation process to form gaseous CO_2. A dual-path approach is generally accepted from the mechanistic point; direct and indirect paths [73, 74]. The indirect path, dehydration process, involves CO intermediates as following:

$$\mathrm{HCOOH} \rightarrow \mathrm{CO_{ad}} + \mathrm{H_2O} \rightarrow \mathrm{CO_2} + 2\mathrm{H^+} + 2\mathrm{e^-} \tag{22.9}$$

The further reaction between adsorbed CO and water requires very positive potentials; thus, it is considered as an electrocatalytic poison resulting invalid reaction [75, 76]. On the other hand, the intermediate in the direct path is controversial, either CHO, COH, COOH, $HCOO^-$, or HCOOH. The common possibilities of the direct path are C—H (direct pathway) and O—H (formate pathway) activation, as shown in Figure 22.8a. Figure 22.8b represent different intermediate's free energy under different conditions. In the direct pathway, the C—H bond is cleaved to obtain COOH which is further dissociated at O—H bond to generate CO_2 while the formate pathway starts with O–H cleavage to form bi-site formate ($HCOO_B$) adsorption via its two O adsorption on the catalyst surface. This bi-site formate dissociates twice to finally desorb CO_2 through mono-site formate ($HCOO_M$) intermediate. The proposed free energy profiles for formic acid oxidation reaction (FAOR) on Pt with different models and PdCu@Cu nanowires (NW) are shown in Figure 22.8c based on DFT computations [74, 78].

Additionally, the DFT studies uncover the role of water on the FAOR via LH and ER mechanisms. The LH-type mechanism in FAOR contains adsorbed H and hydronium intermediates, while the ER-type mechanism involves H atom dissociation as a solvated proton in electrolyte and electron donation to electrode affecting on its potential [74]. The solvation free energy profiles under possible mechanisms are shown in Figure 22.9. Hydrogen bond that is provided by water existence stabilizes dissociated H of C—H bond indirect pathway of LH-type mechanism resulting in barrier energy reduction. Besides, a little but significant activation reduction appears in the formate pathway. At last, by means of DFT calculations, FAOR is studied providing significant insights from the mechanistic point of view [75, 77–81].

22.1.8 MOR Reaction

Electrooxidation of methanol generates CO_2 via six-electron process under acidic or alkaline conditions [82]. The overall oxidation reaction is given by the following equations:

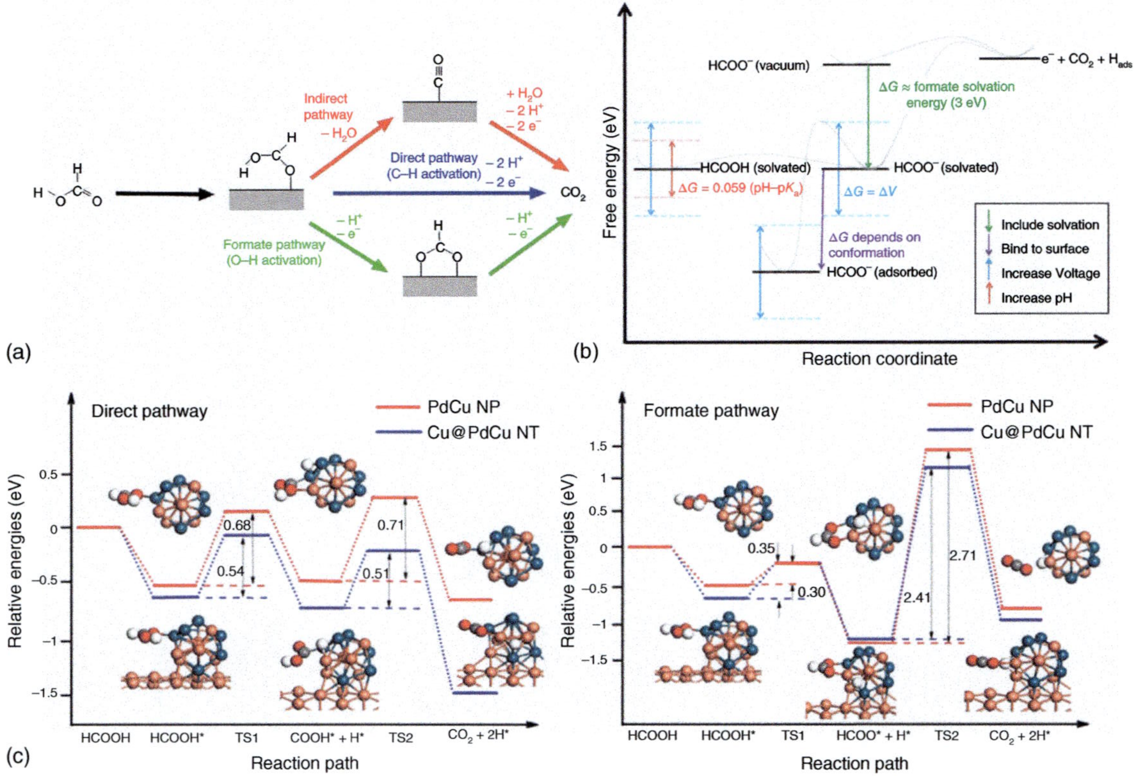

Figure 22.8 (a) Schematic diagram of possible FAOR mechanisms. Source: Alkire et al. [74]. Reproduced with permission of Wiley-VCH Verlag GmbH & Co. KGaA. (b) Illustration of the relative free energies of possible products and reactants in FAOR on a clean Pt surface. Source: Schwarz et al. [77]. Reproduced with permission of Royal Society of Chemistry. (c) Free energy profiles indirect (left) and formate (right) pathways of FAOR on PdCuNPs and PdCu@CuNTs. Source: Wu et al. [78]. Reproduced with permission of IOP Publishing.

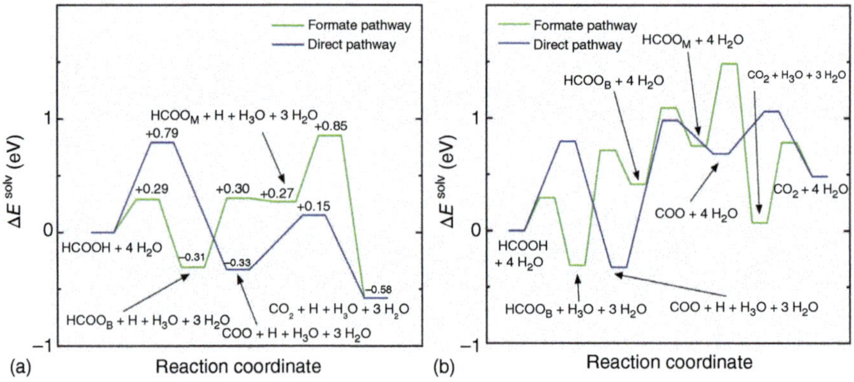

Figure 22.9 Solvation profiles of Langmuir–Hinshelwood (a) and Eley–Rideal (b) FAOR mechanisms. Source: Alkire et al. [74]. Reproduced with permission of Wiley-VCH Verlag GmbH & Co. KGaA.

$$\text{Low pH}: CH_3OH + H_2O \rightarrow CO_2 + 6H^+ + 6e^- \quad (22.10)$$

$$\text{High pH}: CH_3OH + 6OH^- \rightarrow CO_2 + 5H_2O + 6e^- \quad (22.11)$$

Lots of controversies exist toward explaining methanol oxidation reaction (MOR) mechanism either via OH- or CH-bond scission, Figure 22.10a. To complete its oxidation forward to CO_2, several intermediates are involved; such as CHO, CO, and COOH species [82, 85]. The proposed mechanism through any of such intermediates depends on used electrolyte and free energy calculations [86, 87]. The methanol oxidation on Pt at low pH involve stable CHO and COOH intermediates [85]. DFT calculations explain the MOR mechanism over several catalysts [88–91]. For example, initiation methanol oxidation on Ni-based clusters via OH- or CH-bond cleavage was compared and uncovered [92]. In addition, the free energy profile of different compositions of the Ni–Co catalyst was proposed as shown in Figure 22.10b, [83]. Separately, the thermodynamic calculations show that the electro-oxidation occurring in the presence of water is exothermic reaction through HCOH intermediate, Figure 22.10c, [84]. Lastly, mechanistic insights have been conducted proving the potential of DFT modeling in electrocatalytic MOR.

22.1.9 EOR Reaction

Ethanol oxidation reaction (EOR) is a mechanistically complex reaction being a multiple-electron process involving various intermediates and products [93]. Understanding the EOR mechanism is still a challenge; however, a dual path; C1 and C2, is proposed to describe it by

$$\text{C1 pathway}: CH_3CH_2OH + 3H_2O \rightarrow 2CO_2 + 12H^+ + 12e^- \quad (22.12)$$

$$\text{C2 pathway (I)}: CH_3CH_2OH \rightarrow CH_3CHO + 2H^+ + 2e^- \quad (22.13)$$

$$\text{C2 pathway (II)}: CH_3CH_2OH + H_2O \rightarrow CH_3COOH + 4H^+ + 4e^- \quad (22.14)$$

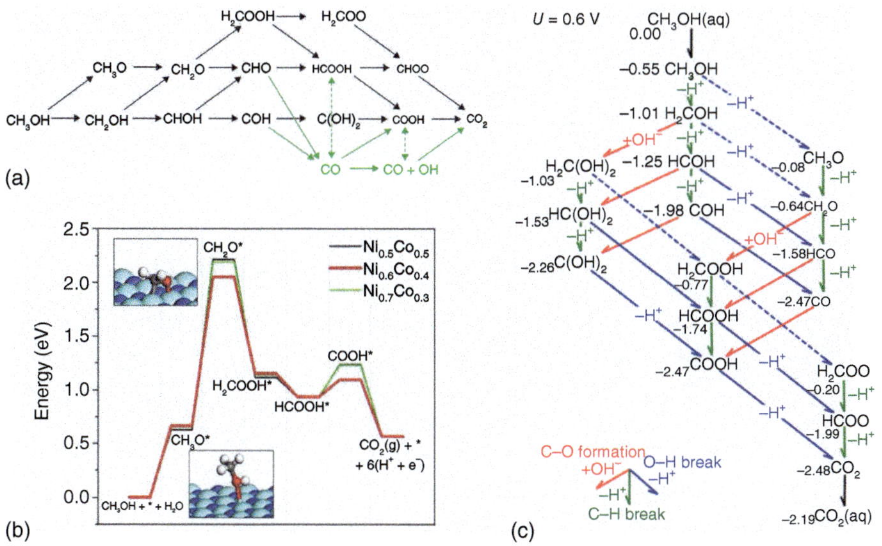

Figure 22.10 (a) Schematic diagram for MOR under CH and OH cleavage pathways. Source: Das et al. [82]. Reproduced with permission of Elsevier. (b) Free energies profile of MOR on selected catalysts. Source: Que et al. [83]. Reproduced with permission of John Wiley & Sons, Inc. (c) thermodynamic calculations of implicit water model for MOR. Source: Sakong and Groß [84]. Reproduced with permission of Springer.

Ethanol completely oxidizes via undergoing C—C bond cleavage to form CO_2 or carbonate in alkaline conditions [93]. On the other hand, partial EOR proceeds without cleavage to acetaldehyde or acetic acid; acetate in alkaline media, via two- or four-electrons, respectively [94]. The RLS is questionable because it can involve more than 40 possible intermediates, for example, EOR on Pt(111), Scheme 22.5.

Advances in theoretical studies continue to add some mechanistic insights toward EOR over several catalysts. On one hand, Wang and Liu suggested the sensitivity of EOR to catalyst structure where they studied it on Pt(111), (211), and (100) surfaces using DFT calculations [96]. they recommended that two key elementary steps are the first ethanol dehydrogenation step and the acetyl; CH_3CO, oxidation step. On the other hand, Asiri and Anderson described an identifier for EOR by disclosing its mechanism based on bond strength calculations using DFT [95]. Separately, the adsorption and dehydrogenation of ethanol were revealed on alloy catalysts via the energy profile on the bases of DFT calculations [97].

Ultimately, the DFT computations are useful tools that contribute to mechanistic insights of electrochemical cycles. Thermodynamically, the adsorption energies of active H-, C-, and O-containing intermediates on atomic level surfaces are successfully measured. What is more, the activation energies-related kinetics of the elementary steps and RLS of the relevant reactions are determined in the catalytic cycle. Further, with the help of CHE modeling, the computed Gibbs free energy profile represents all of these features at several electrode potentials. In a word, interpreting most of the catalytic mechanisms has been accomplished thanks to DFT-based calculations.

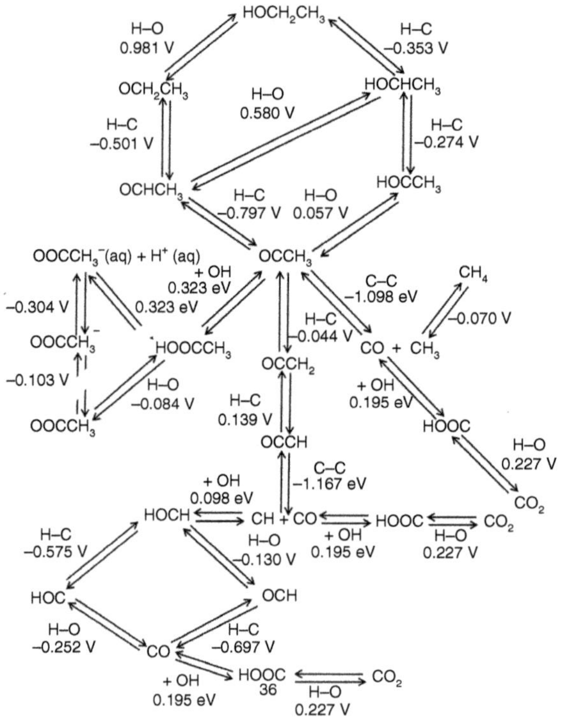

Scheme 22.5 Calculated reaction network and reaction barriers (units, eV) for two possible ethanol oxidation mechanisms on Pt(111). Source: Asiri et al. [95]. Licensed under CC-BY-4.0.

22.2 Understanding the Electrocatalytic Environment

Understanding the electrolyte effects would facilitate the accuracy of DFT calculations at a liquid–electrode interfacial site and liquid–adsorbate interactions and further enhance the rational design of electrolyte as complemental part of the electrocatalytic cell. However, this interface is much more complicated as previously discussed in the double-layer interfacial subsection 19.1. Environmental electrolyte including specific ion concentrations [98], buffer interaction [99], solvent type, [100], transport properties, [101, 102], and pH [103, 104] needs to be accounted as these properties factors experimentally affect the selectivity and activity of catalytic reactions. Here we proceed with solvation and pH insights according to DFT calculation and their effects in electrocatalysis.

22.2.1 Solvation Effects

While considering solvation in DFT calculations is usually disregarded, it recently gets more attraction due to its effects on free energy surface, binding energies of key intermediates, and consequently the mechanism through the free energy diagram.

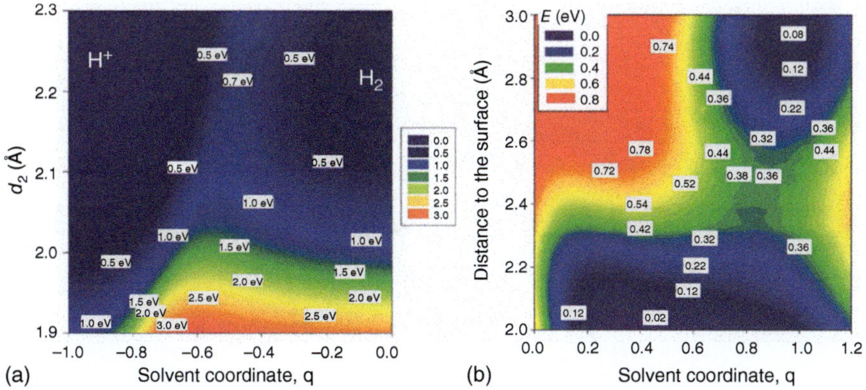

Figure 22.11 Free energy surfaces as (a) a function of the position d_2 of the approaching hydrogen on Ag(111) and (b) as a function of the distance and the solvent coordinate during OH adsorption on Pt(111). Source: Santos et al. [105]. Reproduced with permission of Royal Society of Chemistry.

The free energy surface of the hydrogen recombination step of HER depends on their location and the solvent coordinates [105]. DFT simulation is conducted on the fcc hollow site of Ag(111) uncovering its mechanism, Figure 22.11a. The energy of hydrogen adsorption on the surface is endergonic by 0.4 eV at equilibrium. Toward the hydrogen combination, the slight uphill of the reaction energy is represented where the adsorbed atom is at 1.4 Å. The binding distance between approaching atom and the adsorbed one is about 0.8 Å resulting that the approaching occurs before electron transfer. Finally, the newly dihydrogen molecule gains energy to leave the electrode surface. The occurred increase in the energy reflects the decreasing solvation as the approached atom closes to the surface. Thus, the free energy surface is derived by solvent distribution, i.e. solvation. Another impact of solvation is an increase of charge for adsorbed species. Simultaneously, the free energy surface modeling is conducted on Pt(111) for OH adsorption, Figure 22.11b. The surface shows two minimum coordinates for solvated OH^- ion and adsorbed OH^* on the metal, respectively. The latter minimum with an excess charge of about −0.3 electrons. Apparently, this charge is induced by solvent–ion interactions. In a word, solvation has a significant effect on the free energy surfaces which explains some insights toward kinetic-involved reactions.

Furthermore, the influence of solvent on the intermediates binding energy and its mechanism is lately studied by an explicit water model for ECR reaction on Cu(111) [106]. About 5 and 10 water molecules were simulated containing nine surface atoms per simulation supercell using DFT in combination with global minimization. The difference between the binding free energies in solvent and vacuum models was calculated namely as the solvation free energy ΔG_{solv}. First, the local solvent structure shows some changes induced by adsorbates existence, Figure 22.12a. Further, a small but effective difference in solvation energy is to observe for coupling adsorbates, OCCO and OCCHO, according to the number of the water molecule, Figure 22.12b leading to the invalidity of universal modeling correction for solvation

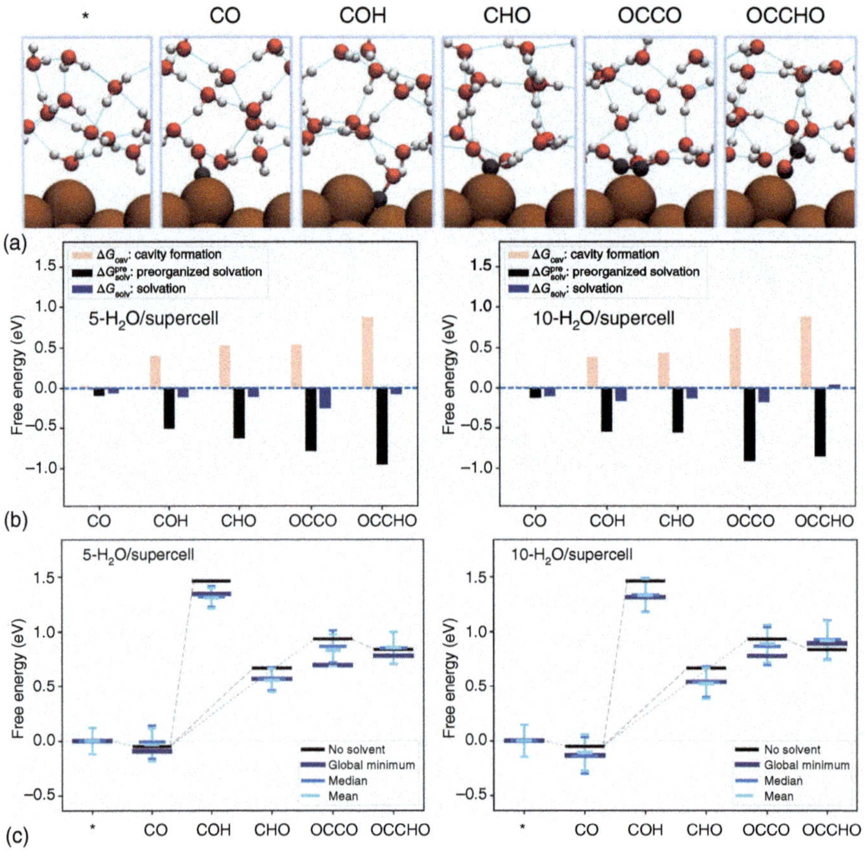

Figure 22.12 (a) Representation of global minima structures of ECR possible adsorbates for 10 water molecules. (b) Contributions to the solvation energy of ECR adsorbates of 5 (left) and 10 (right) water models. (c) Free energies of ECR adsorbates of 5 (left) and 10 (right) water models. Source: Ludwig et al. [106]. Reproduced with permission of American Chemical Society.

modeling. These considerations suggest relative changes in the kinetics associated with relevant elementary steps, especially proton transfer. By comparing the free energies diagram, the effect of solvent can be evaluated resulting in mechanistic insights, Figure 22.12c. In the presence of a solvent, the barriers energies show downhill along with diagram, besides that adsorbate intermediates with 10 water molecules are more stabilized compared to 5 water molecules. Although this study is limited to local water structure changes without kinetic-related understanding, it helps to reveal that solvation is required to be included in mechanistic studies of different electrochemical reactions. Solvation-based insights have been achieved in which explicit, implicit, and H-shuttling models are used to describe the critical role of solvents in several electrocatalytic ECR studies [3, 107–110].

It has been used not only to address the changes of surface and barriers energies but also to predict the onset potentials as reported for ORR study [2]. Generally,

DFT-based solvation provides a better understanding of the liquid-electrode effect making it more likely to be attracted by the electrocatalytic community.

22.2.2 pH Effects

Although the pH can alter the electrochemical interface environment resulting in the different activity of catalytic materials, the studies addressing the nature of pH-dependent kinetics on electrocatalytic reactions is limited. The current state of pH kinetics is mostly reported for H-involved reactions where their kinetic in alkaline solutions is 2 orders of magnitude slower than that in acidic one on Pt [104]. Hypothetically, the origin of pH-related kinetics lies in shifts in hydrogen binding, proton donor, and water reorganization energy aspects [111]. As previously discussed, H-binding energy has been used as a descriptor for activity trends of HER\HOR reactions. As a result, the origin of pH kinetics lies in this thermodynamic descriptor which is proved by shifts of the voltammetric peaks in the so-called "underpotential deposition" (UPD) hydrogen region (H_{UPD}) of Ir/C and polycrystalline Pt [103, 112, 113]. However, an argument is conducted around the nature of the relationship between the "hydrogen" peaks and hydrogen adsorption [103, 114]. It is found that "hydrogen" peaks also affected by oxygenated species on Pt(110) and (100) sites.

Separately, it is suggested that proton donor is considered as the origin of pH reactivity where it can convert from hydronium ions in acidic media into the water in alkaline media, in addition to the configurational entropy changes of the proton [115–117]. On the other hand, interfacial water reorganization energy is reported as a pH-dependent aspect where it would be greater as a result of the increase of interfacial fields under alkaline solution [103]. Recently, by means of DFT in combination with microkinetic modeling, the proton donor changes compared with other aspects is supported as the intrinsic larger barriers changes with respect to hydronium for the splitting of water explaining the reason of slowing HER kinetic in alkaline media, Figure 22.13 [111].

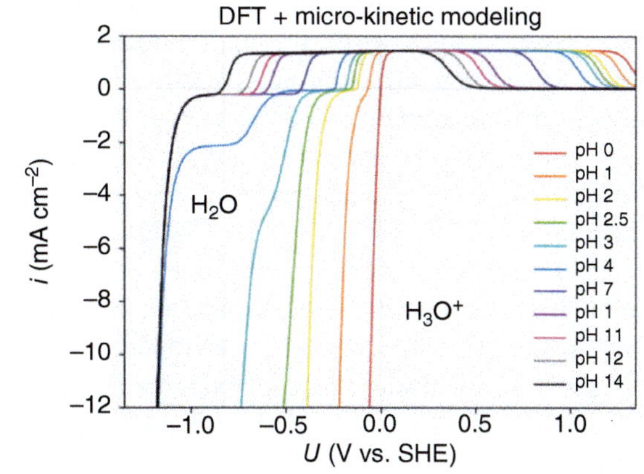

Figure 22.13 Computed HER/HOR polarization plots using the DFT incorporation with mean-field microkinetic modeling. Source: Lamoureux et al. [111]. Reproduced with permission of American Chemical Society.

Although the hydrogen bonding between the adsorbed hydrogen and water can affect the pH kinetics, it is not theoretically studied yet [118]. In this respect, the DFT-based solvation modeling could be invited to study the hydrogen bonding effect revealing more insights of H-involved reactions. In this end, the pH-dependent kinetics on other electrocatalytic reaction has not received much attention opening a door to study the kinetic gap along with catalytic processes.

22.3 Analyzing the Electrochemical Kinetics

For catalysis control and optimization, it is necessary to estimate the reaction rates and kinetic parameters of an electrocatalytic reaction. DFT computations including kinetic models, dynamic Monte Carlo simulations and microkinetic modeling, that allow us to build approximate models to focus on the critical characteristics of the surface chemistry, examine the reversibility of elementary steps to know which is kinetically significant, and to quantify extents of surface coverage to determine degrees of rate control [119]. In a word, the kinetic insights deal with activity in terms of reaction rate and selectivity in terms of the ratio of the rate of generation of the desired product and that of consumption of the initial reactants. Herein, we are going to highlight some kinetic insights for selected reactions over specific catalysts.

As well known, the overpotential is applied to overcome the highest transition barrier of a given reaction. Via DFT-microkinetic analysis, transition barrier insights have been supported the origin of low overpotential for ORR and ECR reactions on transition metals and Mo_2C, respectively. An activity volcano is theoretically constructed by Nørskov et al. [120]. They connected the ORR activity with the binding energy of key intermediates for a series of transition metals, as shown in Figure 22.14a. It shows that the ORR activity is not only correlated with O* binding energy but also *OH binding energy resulting that Pt and Pd are the best catalysts in terms of the lowest overpotential applied. Separately, as $^*CO_2$ protonation is a crucial step for ECR, the kinetics of this step should be considered. Once CO_2 adsorption on Mo_2C, C–O cleavage would occur leaving adsorbed O* attacked by a proton to form strongly adsorbed *OH [121]. The protonation of *OH is considered as the potential limiting step because this step allows regenerating the adsorption site for the next reaction cycle. Then, the effect of *OH coverage was studied resulting in weakening its binding energy. By applying potential, *CO protonation slightly increases with significant less *OH coverage, Figure 22.14b as desired. These studies reveal coverage-related insights through DFT-kinetic calculations.

Additionally, the favorable selectivity of transition metal phosphide and sulfide catalysts toward HER has been understood compared to ECR using DFT-kinetic analysis [122]. Regardless of lower kinetic barriers for the two-electron pathway of HER, the higher kinetic barrier for the proton–electron transfer to *CO is determined as a crucial reason for poor ECR selectivity [122]. The theoretical limiting potential (U_L) values were calculated for the HER and ECR, Figure 22.15a.

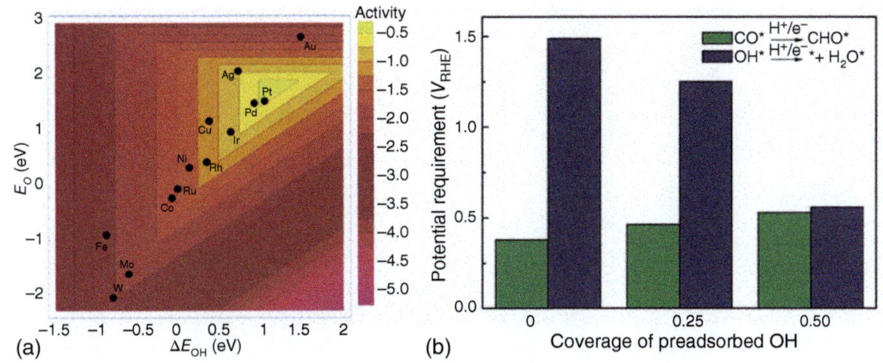

Figure 22.14 (a) Trends in ORR activity plotted as a function of the O and the OH adsorption energies. Source: Nørskov et al. [120]. Reproduced with permission of American Chemical Society. (b) Coverage effect of pre-adsorbed OH on electrochemical transitions. Source: Kim et al. [121]. Reproduced with permission of American Chemical Society.

For all considered sulfides and phosphides, the difference in U_L; $|U_{L_{ECR}} - U_{L_{HER}}|$, lies above the 0.5 V, while it is less than 0.5 V for transition metals. This would suggest high selectivity for HER. Comparing MoS_2 to $NiMoS_2$, it needs higher applied potential to select ECR. On the other hand, with kinetic consideration via potential-dependent barriers calculation of protonation step, as shown in Figure 22.15b, the activation barrier of Ni-incorporated sulfide is higher than Ni-free sulfide. This results that ECR is facile on Ni-free sulfide; however, the binding free energy of adsorbed *CO is week limiting further reduction and lowering its coverage. While the binding free energy on Ni-incorporated sulfide is more negative for higher coverage, Figure 22.15c. Thus, it is concluded that Ni-introduced sulfide is thermodynamically ECR promising catalyst; however, it kinetically restricted with higher protonation (transition state) barrier. Further, DFT-kinetic computation provides a better understanding of the reaction rate and further selectivity along with transition metals for ECR [123]. Although, Cu is the best common metal catalyst; however, the lack of *CO coverage limits the reaction rate and selectivity due to its weak binding energy according to microkinetic-based activity volcanoes, Figure 22.15d. To increase the reaction rate, the transition state should be stabilized for a given *CO adsorption energy, Figure 22.15e. These insights demonstrate significant characteristics of the optimal catalyst toward the rational design approach. Understanding the kinetics would help to get more insights about the selectivity aspect.

In summary, DFT-kinetic analysis refine the reaction selectivity factors exceeding the thermodynamic binding energies to considerable transition state ones. In addition, it explained the effect of applied potential on specific adsorbate coverage. Interestingly, microkinetic analysis can simulate Tafel relations for several reactions such as HER, HOR, and OER showing that its slopes are also coverage dependent [124]. Although a significant complexity is added using DFT-kinetic modeling, insights from it seem to be a worthy endeavor.

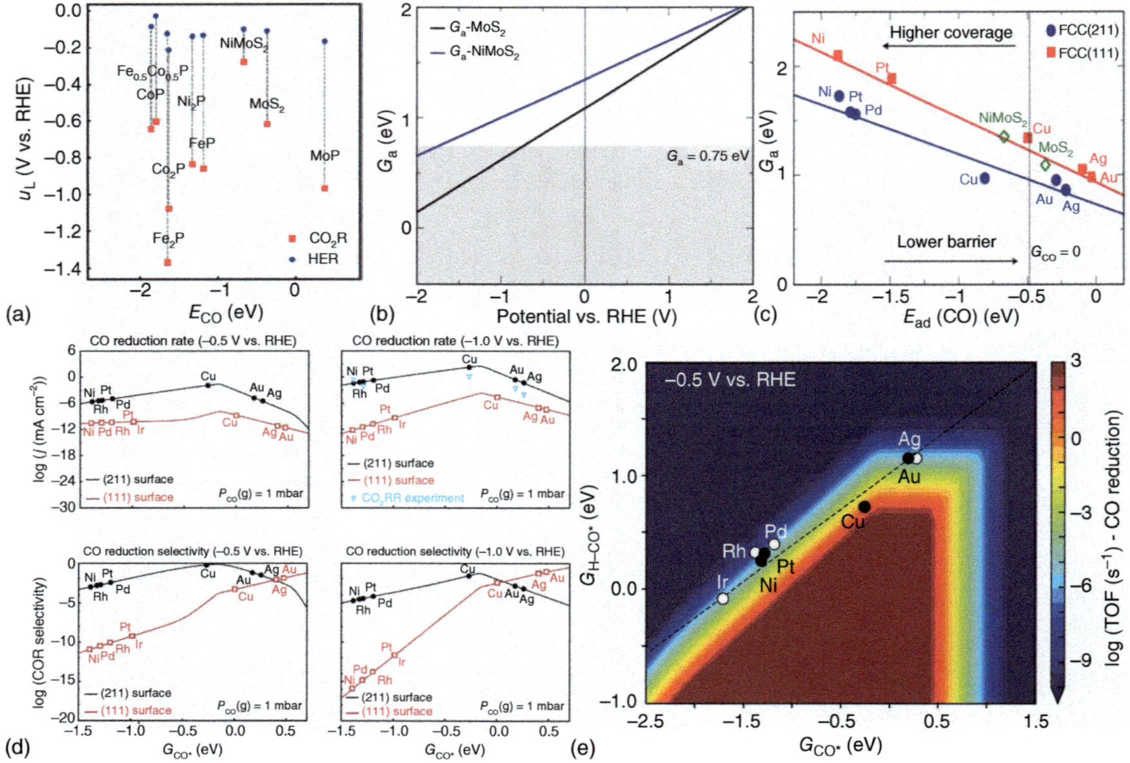

Figure 22.15 (a) Corresponding theoretical limiting potentials for HER and ECR [122]. (b, c) Potential- and adsorption-dependent barriers for CO protonation on MoS_2 catalyst. Source: Landers et al. [122]. Reproduced with permission of American Chemical Society. (d) Activity volcanoes for rate and selectivity of CO reduction for flat (111) and stepped (211) transition metal surfaces [123]. (e) Transition state map of the rate of CO reduction. Source: Liu et al. [123]. Licensed under CCBY 4.0.

22.4 Perspectives, Challenges, and Future Direction of DFT Computation in Electrocatalysis

Having shown considerable prospect toward providing invaluable mechanistic insights in electrocatalytic systems as well as the potential of the rational design over trial and error approach via predicting promising catalysts: this is of considerable value to DFT computation. The perspective of DFT computations can be divided into three aspects: (i) explaining the catalytic mechanism, (ii) predicting catalytic reactivity features, and (iii) guiding experimental advances. Collaboratively, with the aid of DFT calculations, the active sites can be identified using DOS analysis. In addition, the catalytic mechanism of many reactions can be understood from atomistic and electronic structures directly to the reaction mechanism proposal. Moreover, the in-depth insights through the formulated analytical theories based on DFT, such as Sabatier principle and scaling relations, are conducive via overall catalytic activity – the interaction energy of reactants, activation–reaction energies, and thermodynamic correlations which are useful to reasonably predict the catalytic reactivity of designed catalysts. Besides, the embedded DFT results in kinetic models demonstrate the feasibility to predict the activity and selectivity patterns of the electrocatalytic systems. Further, an excellent balance provided by stable DFT codes in the accuracy-cost trade-off would encourage its application to electrocatalysis searching for advanced electrocatalysts and further guiding its experimentation. However, these perspectives represent an initial platform to rational design expanding research opportunities to derive the fast development of high-performance catalysts for a wide range of electrocatalytic reactions.

Although DFT computation has been proved its impact on catalytic understanding progress, it faces some challenges. Challenges in the application of DFT calculations in electrocatalysis can be classified into computing-related and simulating-related issues. The demand for DFT software is to be more accurate and efficient. On one hand, modeling approaches include many inherent approximations to probe complex catalytic interactions. These approximations have intrinsic errors not only get emerged in the multiscale strategies producing unrealistic rates for a given catalytic process but also remain challenging in recognizing them. On the other hand, developing functionals, such as local density approximation (LDA), GGA, meta-GGAs, and B3LYP, is considered as the basis of DFT calculations to describe the physics of catalysts. However, by using LDA functional, the chemical bonds are overestimated and the barrier heights are underestimated. GGA is the common functional used in electrocatalytic computation where it more accurate than LDA and less computationally demanded. The other high lever functionals are more accurate than GGA; however, they are costly and quite time consuming to be applied. Moreover, the university of density functionals is limited. More efficient codes are demanded to stop using oversimplified entropic contributions such as harmonic transition state measurements for various free energy states of the reactants [125]. Therefore, the improvement of computing DFT systems is required in terms of accuracy and efficiency for exploring a wide range of catalysts.

Concurrently, the simulation process meets unresolved issues as follows: (i) Although the effects of hydrogen bonding interactions and the charge capacity of the active site can resolve the catalytic origin, DFT approaches so far are not able to simulate them but it used to be neglected or oversimplified [126]. (ii) Many elementary steps that parallelly take place at different sites of the catalyst are required to be considered in the simulation. Furthermore, the time of catalytic events and length scale gap between macroscopic catalytic performance and molecular simulations are also needed to be overcome. (iii) Other than the important activity and selectivity requirements for discovering new catalysts, long-term stability (i.e. durability, undiminished catalytic performance), resistance to poisons, lack of side-products, susceptibilities to promoters and cost of production are also equivalent demand. To some extent, these elements should also be simulated. (iv) Morphological and structural effects have been demonstrated to be significant in electrocatalysis, yet it needs to be taken into account. It is critical to not only understand the catalytic reactivity-catalyst composition and structure relationship but also the effect of gas-phase reactants on the active site structure. Besides, developing universal descriptors to determine such effects could be desired with much input of external insights to DFT modeling. (v) As many electrocatalysts are supported by overlayer\underlayer structure, their thermodynamic changes, potential reconstruction, acidity, basicity, coverage dependence, and surface reactivity are meaningful to be modeled and predicted. (vi) The catalyst synthesis affects the structure of the desired optimal catalyst; notwithstanding, DFT ability is limited to screen, and then predict, for example, the size and morphology to form the catalytically relevant phase. It is a standpoint not to limit DFT computation to identify radically new catalysts but to improve the performance of known catalysts. Overcoming such challenges and requirements will let DFT computation dominate the design of much higher performance catalysts.

As we look to the future, it seems that creating an accessible structured database of DFT-simulated materials properties is a demand inspired by Brookhaven database of biological structures. Such a computational-characteristics database would potentially improve catalyst design strategy to a new era. Cooperatively, a new degree of realism in material design is provided by machine learning. High accuracy and efficiency could be achieved from the combination of machine learning and high-throughput DFT computations [127]. They are currently emerging as a powerful toolbox for catalytic materials in which a database of typical structural models is assembled, energies and forces are calculated using DFT, the atomic structure in "machine-readable" is expressed using descriptor, and finally, the potential-energy surface could be then understood [128–132]. This would create important opportunities for DFT theory to have a major impact in the future of catalysis.

References

1 Vogt, C., Monai, M., Sterk, E.B. et al. (2019). *Nat. Commun.* 10: 5330.
2 Zhang, Q. and Asthagiri, A. (2019). *Catal. Today* 323: 35–43.
3 Garza, A.J., Bell, A.T., and Head-Gordon, M. (2018). *ACS Catal.* 8: 1490–1499.

4 Hori, Y., Wakebe, H., Tsukamoto, T., and Koga, O. (1994). Electrocatalytic process of CO selectivity in electrochemical reduction of CO_2 at metal electrodes in aqueous media. *Electrochim. Acta* 39: 1833–1839. https://doi.org/10.1016/0013-4686(94)85172-7.

5 Hori, Y., Takahashi, R., Yoshinami, Y., and Murata, A. (1997). Electrochemical reduction of CO at a copper electrode. *J. Phys. Chem. B.* 101: 7075–7081. https://doi.org/10.1021/jp970284i.

6 Peterson, A.A., Abild-Pedersen, F., Studt, F. et al. (2010). How copper catalyzes the electroreduction of carbon dioxide into hydrocarbon fuels. *Energy Environ. Sci.* 3: 1311–1315. https://doi.org/10.1039/C0EE00071J.

7 Kuhl, K.P., Cave, E.R., Abram, D.N., and Jaramillo, T.F. (2012). New insights into the electrochemical reduction of carbon dioxide on metallic copper surfaces. *Energy Environ. Sci.* 5: 7050–7059. https://doi.org/10.1039/C2EE21234J.

8 Thorson, M.R., Siil, K.I., and Kenis, P.J.A. (2013). Effect of cations on the electrochemical conversion of CO_2 to CO. *J. Electrochem. Soc.* 160: F69–F74. https://doi.org/10.1149/2.052301jes.

9 Kortlever, R., Shen, J., Schouten, K.J.P. et al. (2015). Catalysts and reaction pathways for the electrochemical reduction of carbon dioxide. *J. Phys. Chem. Lett.* 6: 4073–4082. https://doi.org/https://doi.org/10.1021/acs.jpclett.5b01559.

10 Cheng, T., Xiao, H., and Goddard, W.A. (2015). Free-energy barriers and reaction mechanisms for the electrochemical reduction of CO on the Cu(100) surface, including multiple layers of explicit solvent at PH 0. *J. Phys. Chem. Lett.* 6: 4767–4773. https://doi.org/https://doi.org/10.1021/acs.jpclett.5b02247.

11 Lum, Y., Cheng, T., Goddard, W.A. et al. (2018). Reduction builds solvent water into oxygenate products. *J. Am. Chem. Soc.* 140: 9337–9340. https://doi.org/10.1021/jacs.8b03986.

12 Feaster, J.T., Shi, C., Cave, E.R. et al. (2017). Understanding selectivity for the electrochemical reduction of carbon dioxide to formic acid and carbon monoxide on metal electrodes. *ACS Catal.* 7: 4822–4827. https://doi.org/10.1021/acscatal.7b00687.

13 Liu, X., Schlexer, P., Xiao, J. et al. (2019). PH effects on the electrochemical reduction of CO(2) towards C2 products on stepped copper. *Nat. Commun.* 10: 32. https://doi.org/10.1038/s41467-018-07970-9.

14 Chernyshova, I.V., Somasundaran, P., and Ponnurangam, S. (2018). On the origin of the elusive first intermediate of CO_2 electroreduction. *Proc. Natl. Acad. Sci. U. S. A.* 115: E9261–E9270. https://doi.org/10.1073/pnas.1802256115.

15 Nitopi, S., Bertheussen, E., Scott, S.B. et al. (2019). *Chem. Rev.* 119: 7610–7672.

16 Varela, A.S., Ju, W., Bagger, A. et al. (2019). *ACS Catal.* 9: 7270–7284.

17 Gong, L., Zhang, D., Lin, C.Y. et al. (2019). *Adv. Energy Mater.* 9: 1902625.

18 Yang, F., Ma, X., Cai, W.B. et al. (2019). *J. Am. Chem. Soc.* 141: 20451–20459.

19 Liu, J.-H., Yang, L.-M., and Ganz, E. (2019). *J. Mater. Chem. A* 7: 3805–3814.

20 Huang, Q., Liu, H., An, W. et al. (2019). *ACS Sustainable Chem. Eng.* 7: 19113–19121.

21 Zhao, Z. and Lu, G. (2018). *ACS Catal.* 8: 3885–3894.

22 Hunter, B.M., Gray, H.B., and Muller, A.M. (2016). *Chem. Rev.* 116: 14120–14136.
23 Alarawi, A., Ramalingam, V., and He, J.-H. (2019). *Mater. Today Energy* 11: 1–23.
24 Song, J., Wei, C., Huang, Z.F. et al. (2020). *Chem. Soc. Rev.* 49: 2196–2214.
25 Montoya, J.H., Seitz, L.C., Chakthranont, P. et al. (2016). *Nat. Mater.* 16: 70–81.
26 Dau, H., Limberg, C., Reier, T. et al. (2010). *ChemCatChem* 5: 7081–7089.
27 Yang, W., Wang, Z., Zhang, W., and Guo, S. (2019). *Trends Chem.* 1: 259–271.
28 Zhou, P., Lv, X., Xing, D. et al. (2020). *Appl. Catal., B* 263: 118330.
29 Qu, M., Jiang, Y., Yang, M. et al. (2020). *Appl. Catal., B* 263: 118324.
30 Yao, Y., Hu, S., Chen, W. et al. (2019). *Nat. Catal.* 2: 304–313.
31 Xiong, Y., Xu, L., Jin, C., and Sun, Q. (2019). *Appl. Catal., B* 254: 329–338.
32 Wu, T., Vegge, T., and Hansen, H.A. (2019). *ACS Catal.* 9: 4853–4861.
33 Liu, F.Q., Liu, J.W., Gao, Z. et al. (2019). *Appl. Catal., B* 258: 117973.
34 Ji, L., Wang, J., Teng, X. et al. (2019). *ACS Catal.* 10: 412–419.
35 Hou, Y., Qiu, M., Kim, M.G. et al. (2019). *Nat. Commun.* 10: 1392.
36 Zhang, L., Jia, Y., Gao, G. et al. (2018). *Chem* 4: 285–297.
37 Zeng, L., Sun, K., Wang, X. et al. (2018). *Nano Energy* 51: 26–36.
38 Hajiyani, H. and Pentcheva, R. (2018). *ACS Catal.* 8: 11773–11782.
39 Guan, C., Xiao, W., Wu, H. et al. (2018). *Nano Energy* 48: 73–80.
40 Zhang, J. and Dai, L. (2015). *ACS Catal.* 5: 7244–7253.
41 Kulkarni, A., Siahrostami, S., Patel, A., and Norskov, J.K. (2018). *Chem. Rev.* 118: 2302–2312.
42 Zhao, D., Zhuang, Z., Cao, X. et al. (2020). *Chem. Soc. Rev.* 49: 2215–2264.
43 Zhu, J., Hu, L., Zhao, P. et al. (2020). *Chem. Rev.* 120: 851–918.
44 Zheng, Y., Jiao, Y., Jaroniec, M., and Qiao, S.Z. (2015). *Angew. Chem. Int. Ed.* 54: 52–65.
45 Morales-Guio, C.G., Stern, L.A., and Hu, X. (2014). *Chem. Soc. Rev.* 43: 6555–6569.
46 Zhu, Y., Sokolowski, J., Song, X. et al. (2019). *Adv. Energy Mater.* 10: 1902844.
47 Liu, L., Liu, Y., and Liu, C. (2020). *J. Am. Chem. Soc.* 142: 4985–4989.
48 Wang, Y., Mao, J., Meng, X. et al. (2019). *Chem. Rev.* 119: 1806–1854.
49 Davydova, E.S., Mukerjee, S., Jaouen, F., and Dekel, D.R. (2018). *ACS Catal.* 8: 6665–6690.
50 Cong, Y., Yi, B., and Song, Y. (2018). *Nano Energy* 44: 288–303.
51 Tang, H., Brothers, E.N., Grapperhaus, C.A., and Hall, M.B. (2020). *ACS Catal.* 10: 3778–3789.
52 Yang, Y., Sun, X., Han, G. et al. (2019). *Angew. Chem. Int. Ed.* 58: 10644–10649.
53 Yang, F., Bao, X., Li, P. et al. (2019). *Angew. Chem. Int. Ed.* 131: 14317–14321.
54 Cong, Y., McCrum, I.T., Gao, X. et al. (2019). *J. Mater. Chem. A* 7: 3161–3169.
55 Xiao, W., Lei, W., Wang, J. et al. (2018). *J. Mater. Chem. A* 6: 11346–11352.
56 Qiu, Y., Xin, L., Li, Y. et al. (2018). *J. Am. Chem. Soc.* 140: 16580–16588.
57 Giles, S.A., Yan, Y., and Vlachos, D.G. (2018). *ACS Catal.* 9: 1129–1139.
58 Esrafili, M.D. and Asadollahi, S. (2019). *Appl. Surf. Sci.* 463: 526–534.
59 Mao, K., Li, L., Zhang, W. et al. (2014). *Sci. Rep.* 4: 5441.

60 Therrien, A.J., Hensley, A.J.R., Marcinkowski, M.D. et al. (2018). *Nat. Catal.* 1: 192–198.

61 Xu, G., Wang, R., Yang, F. et al. (2017). *Carbon* 118: 35–42.

62 Sinthika, S., Vala, S.T., Kawazoe, Y., and Thapa, R. (2016). *ACS Appl. Mater. Interfaces* 8: 5290–5299.

63 Wang, Y., Wu, G., Yang, M., and Wang, J. (2013). *J. Phys. Chem. C* 117: 8767–8773.

64 Gong, X.Q., Raval, R., and Hu, P. (2004). *Phys. Rev. Lett.* 93: 106104.

65 Rao, Y.-C. and Duan, X.-M. (2019). *Phys. Chem. Chem. Phys.* 21: 25743–25748.

66 Cheng, C., Zhang, X., Wang, M. et al. (2018). *Phys. Chem. Chem. Phys.* 20: 3504–3513.

67 Jiang, Q., Zhang, J., Ao, Z. et al. (2018). *Front. Chem.* 6: 187.

68 Esrafili, M.D. and Rad, F.A. (2018). *ChemistrySelect* 3: 7402–7409.

69 Yang, P.Y., Ju, S.P., Lai, Z.M. et al. (2016). *Nanoscale* 8: 2041–2045.

70 Lyu, P., He, J., and Nachtigall, P. (2017). *RSC Adv.* 7: 19630–19638.

71 Liu, S.P., Zhao, M., Sun, G.E. et al. (2018). *Phys. Chem. Chem. Phys.* 20: 8341–8348.

72 Iwasita, T. and Ciapina, E.G. (2010). *Handbook of Fuel Cells*. Wiley.

73 Jeon, H., Jeong, B., Joo, J., and Lee, J. (2014). *Electrocatalysis* 6: 20–32.

74 Alkire, R.C., Kolb, D.M., and Lipkowski, J. (2013). *Electrocatalysis: Theoretical Foundations and Model Experiments*, vol. 14. Wiley-VCH.

75 Cuesta, A. (2018). *Encyclopedia of Interfacial Chemistry*, 620–632. Oxford: Elsevier.

76 Alkire, R.C., Kolb, D.M., and Lipkowski, J. (2013). *Electrocatalysis: Theoretical Foundations and Model Experiments*. Wiley.

77 Schwarz, K.A., Sundararaman, R., Moffat, T.P., and Allison, T.C. (2015). *Phys. Chem. Chem. Phys.* 17: 20805–20813.

78 Wu, D., Xu, H., Cao, D. et al. (2016). *Nanotechnology* 27: 495403.

79 Busó-Rogero, C., Ferre-Vilaplana, A., Herrero, E., and Feliu, J.M. (2019). *Electrochem. Commun.* 98: 10–14.

80 Wang, Y.-Y. (2019). *Prog. React. Kinet. Mech.* 44: 67–73.

81 Zhang, N., Chen, F., and Guo, L. (2019). *Phys. Chem. Chem. Phys.* 21: 22598–22610.

82 Das, S., Dutta, K., Nessim, G.D., and Kader, M.A. (2020). *Direct Methanol Fuel Cell Technology*, 1–12. Elsevier.

83 Que, R., Li, M., Yao, H. et al. (2020). *ChemSusChem* 13: 964–973.

84 Sakong, S. and Groß, A. (2017). *Electrocatalysis* 8: 577–586.

85 Petrii, O.A. (2019). *Russ. J. Electrochem.* 55: 1–33.

86 Bhunia, P., Dutta, K., and Kader, M.A. (2020). *Direct Methanol Fuel Cell Technology*, 443–494. Elsevier.

87 Tomaschun, G. and Kluner, T. (2019). *Phys. Chem. Chem. Phys.* 21: 18227–18239.

88 You, G., Jiang, J., Li, M. et al. (2017). *ACS Catal.* 8: 132–143.

89 Xu, J., Guo, S., Hou, F. et al. (2018). *Int. J. Quantum Chem.* 118: e25491.

90 Chen, Z., Mao, Y., Chen, J. et al. (2017). *ACS Catal.* 7: 4281–4290.

91 Choksi, T. and Greeley, J. (2016). *ACS Catal.* 6: 7260–7277.

92 Lian, X., Guo, W., He, B. et al. (2019). *Mol. Phys.* 118: 1–11.

93 Wang, Y., Zou, S., and Cai, W.-B. (2015). *Catalysts* 5: 1507–1534.

94 Monyoncho, E.A., Ntais, S., Brazeau, N. et al. (2016). *ChemElectroChem* 3: 218–227.

95 Asiri, H.A. and Anderson, A.B. (2015). *J. Electrochem. Soc.* 162: F115–F122.

96 Wang, H.F. and Liu, Z.P. (2008). *J. Am. Chem. Soc.* 130: 10996–11004.

97 Wang, Y. (2016). *Nanomaterials for Direct Alcohol Fuel Cell*. Jenny Stanford Publishing.

98 Resasco, J., Chen, L.D., Clark, E. et al. (2017). *J. Am. Chem. Soc.* 139: 11277–11287.

99 Holst-Olesen, K., Silvioli, L., Rossmeisl, J., and Arenz, M. (2019). *ACS Catal.* 9: 3082–3089.

100 Konig, M., Vaes, J., Klemm, E., and Pant, D. (2019). *iScience* 19: 135–160.

101 Dumitrescu, I. and Crooks, R.M. (2012). *Proc. Natl. Acad. Sci. U.S.A.* 109: 11493–11497.

102 Raciti, D., Mao, M., and Wang, C. (2018). *Nanotechnology* 29: 044001.

103 Ledezma-Yanez, I., Wallace, W.D.Z., Sebastián-Pascual, P. et al. (2017). *Nat. Energy* 2: 17031.

104 Zhu, S., Qin, X., Yao, Y., and Shao, M. (2020). *J. Am. Chem. Soc.* 142: 8748–8754.

105 Santos, E., Quaino, P., and Schmickler, W. (2012). *Phys. Chem. Chem. Phys.* 14: 11224–11233.

106 Ludwig, T., Gauthier, J.A., Brown, K.S. et al. (2019). *J. Phys. Chem. C* 123: 5999–6009.

107 Xiao, H., Cheng, T., and Goddard, W.A. (2017). *J. Am. Chem. Soc.* 139: 130–136.

108 Cheng, T., Xiao, H., and Goddard, W.A. (2017). *Proc. Natl. Acad. Sci. U.S.A.* 114: 1795–1800.

109 Nie, X., Esopi, M.R., Janik, M.J., and Asthagiri, A. (2013). *Angew. Chem. Int. Ed.* 52: 2459–2462.

110 Nie, X., Luo, W., Janik, M.J., and Asthagiri, A. (2014). *J. Catal.* 312: 108–122.

111 Lamoureux, P.S., Singh, A.R., and Chan, K. (2019). *ACS Catal.* 9: 6194–6201.

112 Zheng, J., Zhuang, Z., Xu, B., and Yan, Y. (2015). *ACS Catal.* 5: 4449–4455.

113 Sheng, W., Zhuang, Z., Gao, M. et al. (2015). *Nat. Commun.* 6: 5848.

114 van der Niet, M.J.T.C., Garcia-Araez, N., Hernández, J. et al. (2013). *Catal. Today* 202: 105–113.

115 Strmcnik, D., Uchimura, M., Wang, C. et al. (2013). *Nat. Chem.* 5: 300–306.

116 Intikhab, S., Snyder, J.D., and Tang, M.H. (2017). *ACS Catal.* 7: 8314–8319.

117 Rossmeisl, J., Chan, K., Ahmed, R. et al. (2013). *Phys. Chem. Chem. Phys.* 15: 10321–10325.

118 Hamada, I. and Morikawa, Y. (2008). *J. Phys. Chem. C* 112: 10889–10898.

119 Thomas, J.M. and Thomas, W.J. (2015). *Principles and Practice of Heterogeneous Catalysis*. Wiley-VCH.

120 Nørskov, J.K., Rossmeisl, J., Logadottir, A. et al. (2004). *J. Phys. Chem. B* 108: 17886–17892.

121 Kim, S.K., Zhang, Y.-J., Bergstrom, H. et al. (2016). *ACS Catal.* 6: 2003–2013.

122 Landers, A.T., Fields, M., Torelli, D.A. et al. (2018). *ACS Energy Lett.* 3: 1450–1457.

123 Liu, X., Xiao, J., Peng, H. et al. (2017). *Nat. Commun.* 8: 15438.

124 Shinagawa, T., Garcia-Esparza, A.T., and Takanabe, K. (2015). *Sci. Rep.* 5: 13801.

125 Norskov, J.K., Abild-Pedersen, F., Studt, F., and Bligaard, T. (2011). *Proc. Natl. Acad. Sci. U.S.A.* 108: 937–943.

126 Zhao, X. and Liu, Y. (2020). *J. Am. Chem. Soc.* 142: 5773–5777.

127 Zhu, X., Yan, J., Gu, M. et al. (2019). *J. Phys. Chem. Lett.* 10: 7760–7766.

128 Goldsmith, B.R., Esterhuizen, J., Liu, J.X. et al. (2018). *AlChE J.* 64: 2311–2323.

129 Toyao, T., Maeno, Z., Takakusagi, S. et al. (2019). *ACS Catal.* 10: 2260–2297.

130 Schlexer Lamoureux, P., Winther, K.T., Garrido Torres, J.A. et al. (2019). *ChemCatChem* 11: 3581–3601.

131 Kitchin, J.R. (2018). *Nat. Catal.* 1: 230–232.

132 Zhang, J., Hu, P., and Wang, H. (2020). *J. Phys. Chem. C* 124: 10483–10494.

Part IX

Potential of In Situ Characterizations for Electrocatalysis

Theorizing is essential, but the reality is inevitable. In social life, people admit that they should not adjudicate others before understanding and analyzing their circumstances; such a principle seems to be valid for the catalyst. During electrocatalysis, scientists deal with the macroscopic inputs and outputs. However, to be fully aware of the whole process and its circumstances, they should seek to analyze the catalyst while in real-time reaction conditions. In this event, one can understand the catalyst evaluation/stability and its effects for whether interpreting the catalyst performance for further development or recognizing the defects that should be avoided.

Developing the heterogeneous catalysts has always been receiving intensive efforts in research and industry; however, the key aspects of designing a new catalyst, namely activity, selectivity, and durability, still face challenges in terms of its fundamental understanding. So as to obtain superior energy management and excellent control of suppressing the side reactions along with a sustained regeneration manner, providing an in-depth insight under realistic reaction conditions into the nature of the active sites and its interaction with reactants, intermediates, and products is inevitable. Accordingly, introducing in situ characterization techniques allows researchers to bring the catalysis research into its real reaction conditions as well as provides molecular-level information about the states of the catalyst and surface reaction intermediates during these reactions. Hence, the real-time detection of the active sites through in situ techniques is extremely promising in disclosing the reaction mechanism and the electronic environment throughout the catalytic processes [1–3]. However, there are as always rules and restrictions needed to be realized and to act upon it. In other words, both electrocatalytic systems and characterization techniques have several different parameters to deal with, not to mention the overlaps that will happen when combined, which could limit the detection process.

This chapter provides the reader with an overview of the common and the most crucial in situ characterization techniques, which have been summoned to help the

Electrocatalysis in Balancing the Natural Carbon Cycle, First Edition. Yaobing Wang.
© 2021 WILEY-VCH GmbH. Published 2021 by WILEY-VCH GmbH.

development of the catalysis research. Then, a discussion according to their feasibility to study different electrocatalytic reactions will be covered.

References

1 Bañares, M.A. (2011). Operando spectroscopy: the knowledge bridge to assessing structure–performance relationships in catalyst nanoparticles. *Adv. Mater.* 23 (44): 5293–5301. https://doi.org/10.1002/adma.201101803.

2 Li, X., Yang, X., Zhang, J. et al. (2019). In situ/operando techniques for characterization of single-atom catalysts. *ACS Catal.* 9 (3): 2521–2531. https://doi.org/10.1021/acscatal.8b04937.

3 Rodriguez, J.A., Hanson, J.C. and Chupas, P.J. (2013). Front Matter in In-situ characterization of heterogeneous catalysts. *Wiley*, 1–22. https://doi.org/10.1002/9781118355923.fmatter

23
In Situ Characterization Techniques

Since merging the characterization techniques with the electrocatalysis research, the need to adjust their requirements to acquire the most value became a necessity. Therefore, different designs for the in situ cells are demanded in order to conform to the analysis environment. Hence, in this chapter, the reader will find an informative but brief background on the available in situ characterization tools used with different electrocatalytic reactions.

23.1 Optical Characterization Techniques

Optical characterization instruments use light in the range of electromagnetic radiation ranging from the deep ultraviolet to the far-infrared (IR) (Figure 23.1) [1, 2]. Due to the dual nature of the light and its broad range of frequencies, wavelengths, and energies, it interacts with materials resulting in molecular vibrations or electronic transitions. Subsequently, by selecting the proper photon energy, useful information can be obtained, particularly when merging these techniques with the real-time experiment conditions [1, 3]. On the one hand, in situ applied IR and Raman spectroscopies have been widely used on monitoring the catalyst under the realistic reaction conditions, which both considered as vibrational spectroscopy. Throughout, they offer substantial structural information that provides a chemical or a molecular "fingerprint" of the material. Thus, this information will be of value to the analysis of the double-layer, soluble oxidation or reduction products, and the surface adsorbents. On the other hand, Ultraviolet–Visible (UV–vis) spectroscopy addresses the electronic transitions that happen within the electronic states of the materials; accordingly, it was mentioned in the literature as the electronic spectroscopy [3, 4]. Herein, a preview of in situ IR, Raman, and UV–vis spectroscopies will be presented.

23.1.1 Infrared Spectroscopy

An electromagnetic beam, in the range of 400–4000 cm^{-1}, is enough to give the molecules sufficient energy to excite its molecular vibrations and rotations depending on the materials' atomic structure. Indeed, the material is being active for the IR absorption, if a change in its dipole moment occurs. Thus, the IR radiation

Electrocatalysis in Balancing the Natural Carbon Cycle, First Edition. Yaobing Wang.
© 2021 WILEY-VCH GmbH. Published 2021 by WILEY-VCH GmbH.

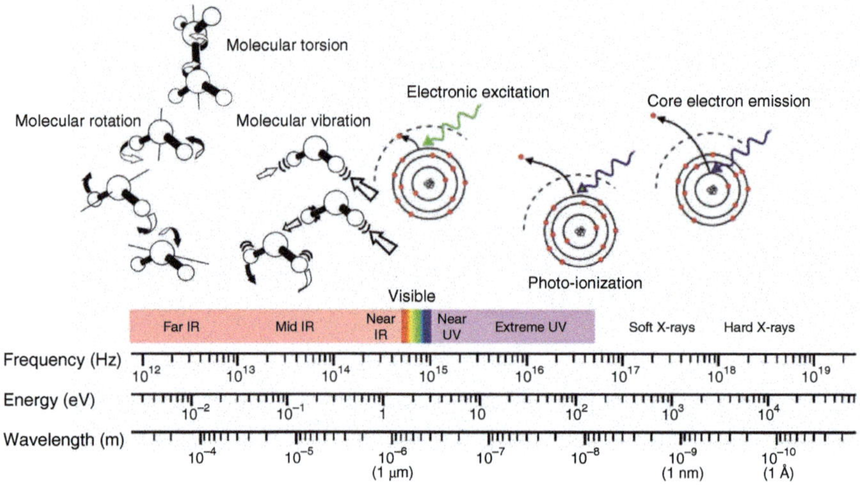

Figure 23.1 Illustration of electromagnetic radiation range expressing its interactions with the material in molecular and atomic levels based on the estimated frequencies of energies. Source: Soares [1]. Reproduced with permission of Springer.

will be absorbed [1, 5]. In the electrochemical cell, a rearrangement of the ionic species in the electrolyte occurs as the potential is applied; then, a double layer will be formed near the electrode. Using the IR radiation will provide valuable information on the existing chemical species in that double layer, including the reaction products, intermediates, adsorbed, and desorbed species by revealing the structural composition, identifying the function groups, determining the bond strength, etc. [5–7]. Therefore, more understanding of different electrocatalytic reactions will be attained.

Despite constructive advances, IR spectro-electrochemical approach faces two major issues, the strong solvent absorption and the sensitivity for the received absorption beam of interest [8]. Therefore, a variety of different mode designs of the IR in situ cells have been introduced and discussed in the literature [1, 2, 5, 6, 8–11], which are different from the conventional cells used by the electrochemists [5]. In Figure 23.2, the reader can find some typical schematic designs of different IR configurations to meet the electrochemistry requirements and reduce the effect of the mentioned problems.

23.1.2 Raman Spectroscopy

Different from IR, the material is being Raman active when the incident beam vibrates the molecule, and further, a change in its polarizability (the degree of deformation of the electron cloud) occurs. Thus, the inelastically scattered light will be detected. Since water is a weak scatterer, unlike the IR, Raman has become a great option to probe the structural information about the reacting species. In this sense, Raman spectroscopy has considered as a complementary tool for the IR

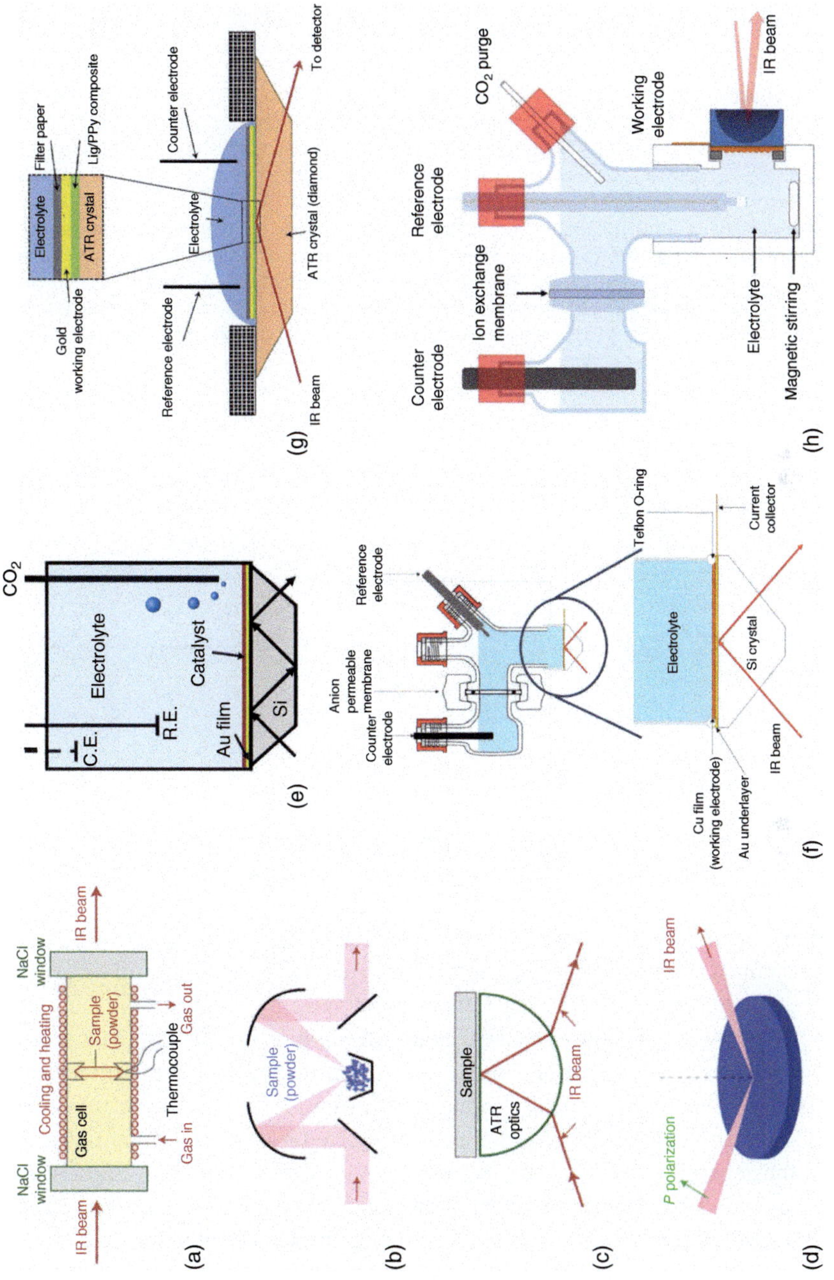
(a)
IR beam
NaCl window
Gas cell
Cooling and heating
Sample (powder)
Thermocouple
Gas in
Gas out
(b)
Sample (powder)
(c)
Sample
ATR optics
IR beam
(d)
P polarization
IR beam
(e)
C.E.
R.E.
Electrolyte
Catalyst
Au film
Si
CO_2
(f)
Counter electrode
Anion permeable membrane
Reference electrode
Teflon O-ring
Electrolyte
Si crystal
Cu film (working electrode)
Au underlayer
IR beam
Current collector
(g)
Gold working electrode
Electrolyte
Filter paper
ATR crystal
Lig/PPy composite
Reference electrode
Counter electrode
ATR crystal (diamond)
To detector
(h)
Counter electrode
Ion exchange membrane
Reference electrode
CO_2 purge
Working electrode
Electrolyte
Magnetic stirring

Figure 23.2 Schematic designs of different in situ IR configurations and cells used for the characterization of different electrocatalytic reactions. (a) The transmission mode is the traditional setup where the IR beam is directed through the catalyst inside the reactor and collected immediately. (b) By using high-area parabolic mirrors, diffuse reflectance infrared Fourier transform spectroscopy (DRIFTS) mode is used to analyze the opaque materials. (c) Attenuated total reflection (ATR) or (d) reflection-absorption infrared spectroscopy (RAIRS) modes can be used with studying flat surface and catalysts immersed in electrolytes [2, 12]. Source of ref. Handoko et al. [2]: Reproduced with permission of John Wiley & Sons, Inc. Source of ref. [12]: Reproduced with permission of Nature Publishing Group. (e) A cross-sectional view of the ATR spectro-electrochemical chamber shows the deposited catalyst over Si crystal as a working electrode [13]. Source: Mandal et al. [14]. Reproduced with permission of AIP Publishing. (f) Schematic design for H-cell setup used in surface-enhanced infrared absorption spectroscopy (SEIRAS) measurements with an ATR configuration. Source: Heyes et al. [15]. Reproduced with permission of American Chemical Society. (g) Schematic view of the ATR-Fourier-transform infrared (ATR-FTIR) in situ cell setup. Source: Ajjan et al. [16]. Reproduced with permission of Royal Society of Chemistry. (h) A modified design of ATR-SEIRAS spectro-electrochemical in situ cell with magnetic stirring [17]. Source: Reproduced with permission of Elsevier.

(Figure 23.3a,c) [18, 22, 23]. Surface-Enhanced Raman Spectroscopy (SERS) and Tip-Enhanced Raman Spectroscopy (TERS) (Figure 23.3b,d) both have considered as developed modes of Raman to enhance sensitivity and intensity and spatial resolution, respectively [2, 24]. Evidently, combining obtained data from Raman and that of the electrochemical systems could provide real-time spectroscopic information that could help realize more insights about various dynamic electrochemical reactions [18].

23.1.3 UV–vis Spectroscopy

One of the oldest and most popular inquiries to scientists is to investigate how much light the material reflects, transmits, and absorbs. UV–vis spectroscopy uses a wide range of the electromagnetic radiation wavelength (Figure 23.1), from ~200 nm (near-ultraviolet) to ~3 μm (near-IR) [1, 25]. Recently, UV–vis spectroelectroscopy has become a suitable technique for studying the redox reactions and investigating the reactants, products, and reaction intermediates by determining the electron transfer pathways and the molecular states at the interface [3, 25]. In a typical in situ UV–vis cell, as shown in Figure 23.4, a light beam, in the mentioned range, is directed toward the optically transparent electrode (OTE) in the transmission mode, and the changes in absorbance are detected as a result of light absorption from the generated chemical species. While opaque electrodes, transmission mode, internal, or external reflection modes are applied, taking into account the electrode design used in the transmission mode, either mini-grids or fine mesh [31]. Under those circumstances, the negative consequences of the light source and its broad frequencies may limit its use against electrochemical processes [3, 7].

As can be seen, in situ optical characterization techniques are successfully developed to be more appropriate for studying the electrocatalytic processes and further

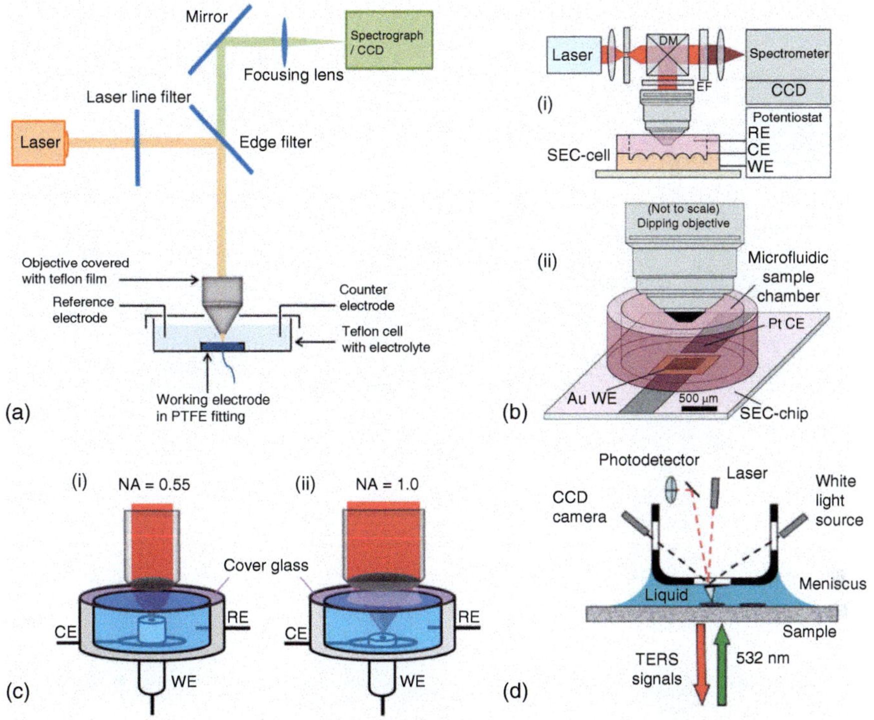

Figure 23.3 Different adjustments of Raman spectroscopy setup for in situ electrochemical experiments. (a) Schematic diagram of a combination between Raman spectroscopy and electrochemistry setup used for in situ measurements [18]. (b) An integrated surface-enhanced Raman scattering (SERS) spectro-electrochemical analysis system: (i) schematic diagram of in situ SERS spectro-electrochemical cell, (ii) SERS spectro-electrochemical cell with a microfluidic sample chamber [19]. (c) Conventional electrochemical Raman setup, (i) Long-working distance air objective, (ii) Long working distance water immersion objective [20]. (d) Schematic diagram of the TERS setup for real-time measurement in aqueous electrolytes [21]. Source of ref. [18–20]: Reproduced with permission of American Chemical Society. Source of ref. [21]: Reproduced with permission of John Wiley & Sons, Inc.

provide useful information that could help us monitor the catalyst surface and identify possible intermediates. However, they have some limitations regarding their surface sensitivities that other techniques can overcome them.

23.2 X-Ray Characterization Techniques

In light of our brief discussion of light characteristics in Section 23.1, X-ray interacts with the matter, like all forms of electromagnetic radiation, but in different principles, and it can be either scattered or absorbed [32]. In situ X-ray tools are considered being complemental to those in situ optical techniques since they successfully overcome their limitations [2]. The X-ray can answer the most ambiguous questions

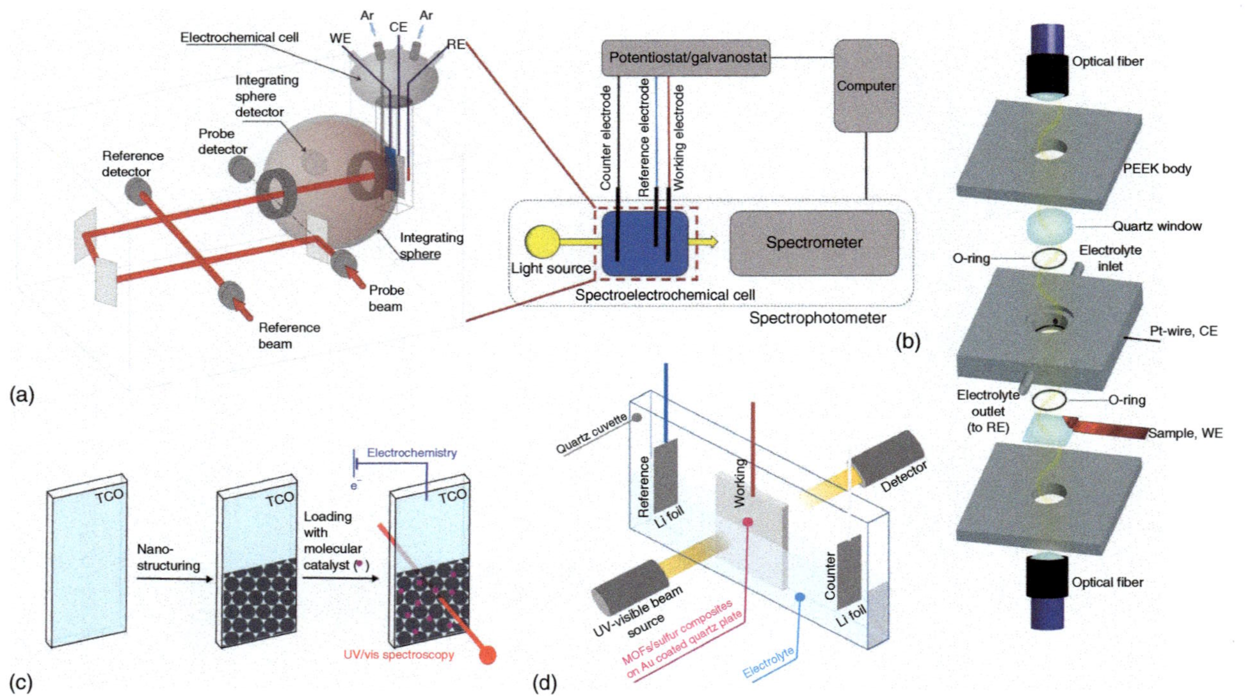

Figure 23.4 UV–vis spectroscopy for in situ electrochemical measurements. (a) Schematic representation of UV–vis spectrophotometer equipped with a diffuse reflectance accessory for the in situ spectro-electrochemical experiments [26, 27]. (b) A flexible combined electrochemistry and fiber optics-based in situ UV–vis spectroscopy setup [28]. (c) A schematic representation of transparent conducting oxide (TCO) substrate used for in situ analysis for a molecular catalyst [29]. (d) Simple illustration of in situ cell used for UV–visible spectroscopy measurements [30]. Source of ref. [26, 28]: Reproduced with permission of Royal Society of Chemistry. Source of ref. [27, 29]: Reproduced with permission of American Chemical Society. Source of ref. [30]: Reproduced with permission of Nature Publishing Group.

about catalyst states. It could provide information about the electronic and geometric evolution of the probed materials and give us the ability to possess the structural insights of the chemical species [32, 33]. Further, different X-ray-based in situ techniques such as X-ray diffraction (XRD), X-ray absorption spectroscopy (XAS), and X-ray photoelectron spectroscopy (XPS) will be introduced.

23.2.1 X-Ray Diffraction (XRD)

Depending upon Bragg diffraction, in situ XRD performs as real-time detection of the crystal phase of the working electrode (Figure 23.5a). As a result, lattice parameters and crystal orientation could be detected. Accordingly, the catalyst evolution during the electrochemical process, the doping effect, and the grain size could be studied by analyzing its diffraction pattern. Unfortunately, due to the nature of XRD, which can only detect the crystalline phase, the amorphous phases cannot be recognized and analyzed, resulting in its limitation [7, 33].

23.2.2 X-Ray Absorption Spectroscopy (XAS)

One of the most commonly used X-ray methods is XAS because of the ease of spectrum collection. It can provide valuable information about the catalyst during the catalytic reaction [2, 41]. Unlike XRD, XAS can investigate the amorphous materials because the material does not require to be in an ordered structure [33]. Furthermore, both hard and soft X-rays could be applied during in situ analysis, which is useful in probing the catalyst atoms and the intermediates species [41].

Hard XAS contains X-ray absorption near edge structure (XANES) spectrum and extended X-ray absorption fine structure (EXAFS) spectrum [33]. XANES helps to identify the active sites by detecting the changes in the elements' oxidation and chemical states since its nuclear electron energy is influenced by the valence electron distribution [7, 33]. Moreover, XANES allows us to get information about electron transfer within the reacting species [41]. On the other hand, EXAFS can probe the local atomic environment into the catalyst and provide information about the interatomic distance and the coordination number [33, 41]. As has been noted, a detailed understanding of the mechanism of the catalytic process could be achieved by the information probed via XAS (Figure 23.5b–d) [33]. Indeed, XAS is not only for the structural investigation but also, with the assistance of computer post-processing, used to get 3D chemical tomographic analysis of the working electrodes using the XAS signals [2]. Nevertheless, hard XAS techniques are not only surface-sensitive, but they can provide chemical information about the bulk also [11].

Due to the short attenuation lengths of soft X-rays compared with hard X-rays, the acquisition of the catalyst near-surface information can be obtained [33]. Consequently, in situ soft X-rays cells are designed to be used in an ultrahigh vacuum (UHV) environment, with a thin Si_3N_4 window due to the low energy of the incident soft X-ray (Figure 23.5e) [33, 41]. Under those circumstances, the soft X-ray is used to probe the L and M edges of certain metals and K edges of lighter chemical elements [41]. Therefore, the changes caused by the adsorbed intermediates, such as valence

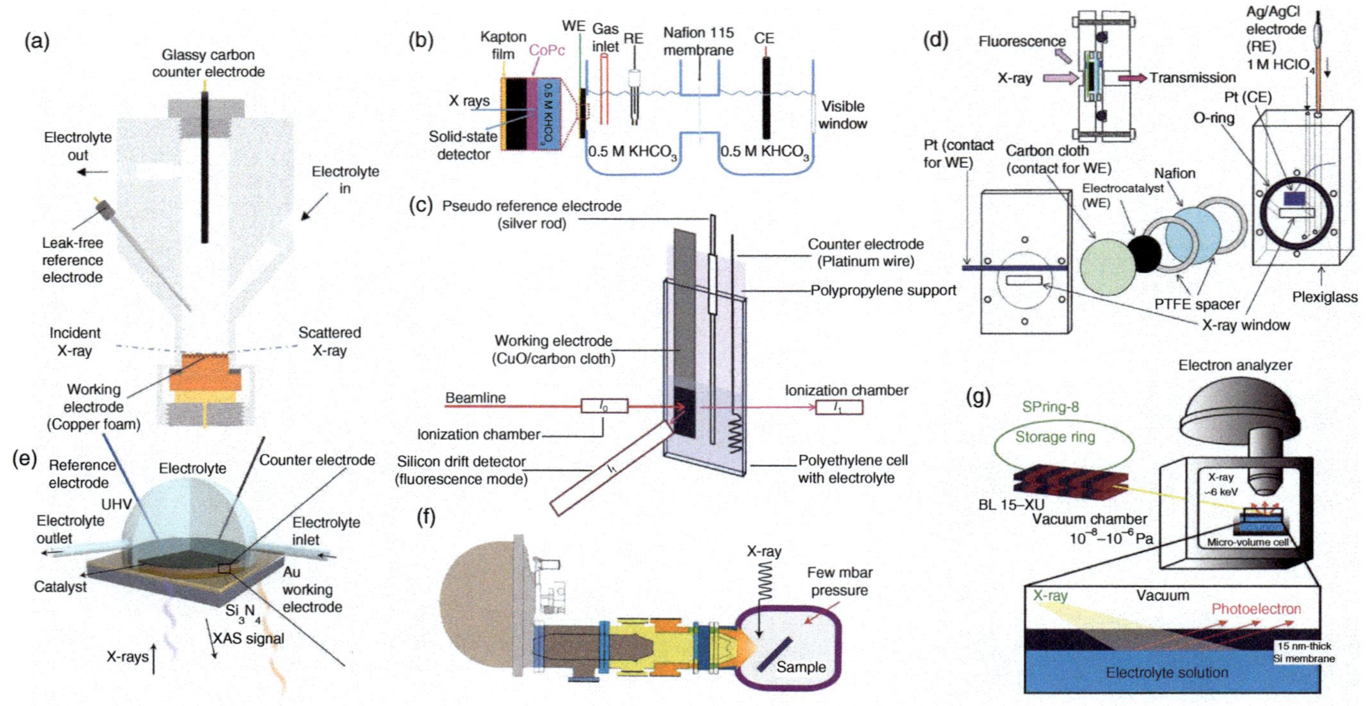

Figure 23.5 In situ X-ray techniques. (a) Schematic illustration of the custom-made in situ electrochemical flow-cell for XRD measurements. Source: Ahn ety al. [34]. Reproduced with permission of American Chemical Society. Different schematic designs for in situ XAS analysis shown as (b) a conventional H-cell attached to the in situ XAS setup for electrocatalytic reactions, Source: Zhang et al. [35]. Reproduced with permission of John Wiley & Sons, Inc., (c) A custom-made polyethylene cell for in situ XANES measurements, Source: Deng et al. [36]. Reproduced with permission of American Chemical Society, and (d) in situ electrochemical EXAFS cell. Source: Sasaki et al. [37]. Reproduced with permission of Elsevier. (e) Schematic diagram of in situ liquid cell experimental setup for soft-XAS measurements. Source: Zheng et al. [38]. Reproduced with permission of Nature Publishing Group. Schematic illustrations for in situ XPS setup in (f) pumping system for AP-XPS [2, 39] Source of ref. [2]: Reproduced with permission of John Wiley & Sons, Inc., Source of ref. Roy et al. [39]. Reproduced with permission of Nature Publishing Group. (g) Conceptual design of in situ electrochemical XPS cell. Source: Masuda et al. [40]. Reproduced with permission of AIP Publishing.

changes of the catalyst and charge transfer, could provide insights into the catalytic process mechanisms [33]. Additionally, the study of the catalytic mechanisms with respect to d orbitals interaction/hybridization is addressed through high-energy resolution fluorescence detected X-ray absorption spectroscopy (HERFD-XAS) [33]. The UHV is not required when analyzing the catalyst with in situ HERFD-XAS [42]. However, this technique is seldom used to investigate catalyst characteristics [33].

23.2.3 X-Ray Photoelectron Spectroscopy (XPS)

Generally speaking, when a photon with high enough energy hits an electron, the electron is ejected, resulting in a photoelectron creation, and then it can be detected using XPS. Since XPS has been described as a surface-sensitive tool, the kinetic energy and the ejected electrons could be detected in order to provide information on the composition of a specific element and give insights about chemical and electronic state of the catalyst (Figure 23.5g) [2]. Due to the inelastic mean free path of electrons emitted from the sample in the ambient conditions, XPS must operate in the UHV environment, making a challenge to perform the in situ analysis of solid–liquid interface [2, 41]. Synchrotron-based ambient pressure X-ray photoelectron spectroscopy (AP-XPS) is an improved tool to overcome the mentioned challenge (Figure 23.5f) [41]. Typically, the electrode is dipped into the electrolyte solution to allow the formation of an ultrathin layer of the electrolyte on the catalyst surface [33, 41]. However, the elements near the surface can only be detected through the in situ XPS due to the limited detection depth; thus, the changes within the catalyst cannot be probed [7].

The history of X-ray always shows great success in supplying the scientific community with the most valuable insights about the materials. In situ X-ray tools can detect active sites and track the evolution of the catalytic reaction obtaining strong and deep chemical and structural information. In addition, it complements the role of in situ optical techniques in drawing the complete picture of the mechanisms to develop the catalytic processes.

23.3 Mass Spectrometric Characterization Techniques

The mass-to-charge ratio (m/z) of ions is the passcode of this type of characterization tools. That is to say, the charged ion with a charge z has a specific ion mass m, can be separated from another ion with different m/z values. Since ions or fragments are charged, they can be controlled by using the electric and magnetic fields. Accordingly, because of their unique features, they can be grouped and analyzed. However, a vacuum environment is needed to maintain the existence of these ions, in which ions must be in gas phase. Therefore, mass spectrometry (MS) is a suitable tool to accomplish this mission because it has the capability to analyze any molecule that can be ionized in different models.

Electron ionization (EI) and chemical ionization (CI) approaches are used to get the gas-phase ions from the gas-phase molecules [43]. EI is considered a hard ionization technique due to the high-energy electrons (70 eV) that ionize the molecular ions into fragments for qualitative analysis [44, 45]. In contrast, soft technique or CI forms ions by reacting the analyte with reagent gas ions to quantitatively analyze the compounds of interest [46, 47]. Without any doubt, analyzing the catalytic products has no less important than monitoring the catalyst evolution since they also provide valuable information in order to understand the dynamics of the catalytic process. So, real-time detection of the catalytic products can be achieved via the in situ electrochemical MS [7].

Differential electrochemical mass spectrometry (DEMS) and online-electrochemical mass spectrometry (OLEMS) are the potent candidates within MS techniques that use the EI technique [23]. The DEMS technique basically is an electrochemical setup that is connected to the MS vacuum system (Figure 23.6a,b), which helps in the detection of the volatile and gaseous products from the electrocatalytic reaction [3]. In that case, the reaction rate of a specific product formation could be obtained from the relevant data by correlating the relevant mass ion currents with the faradaic currents [23, 55]. On the other hand, a small porous Teflon tip in the OLEMS technique is placed near the catalyst to detect the reaction intermediates and the formed products, which provide valuable mechanistic insights (Figure 23.6c). However, the obtained data from the detected gaseous species is limited to the tip sensitivity since it is related to the distance between the tip and the working electrode [23].

Another technique under this category based on the CI approach is named by selected-ion flow tube mass spectrometry (SIFT-MS) (Figure 23.6d,e). SIFT-MS can quantitatively detect the concentrations of the produced gases and liquids formed within the catalytic reaction [51, 56]. As a result, the simultaneous detection through SIFT-MS during the reaction provides real-time analysis of complex multicomponent mixtures; thus, the catalytic reaction dynamics can be captured [14]. But when it comes to studying the catalyst stability during the real-time reaction, electrochemical online inductively coupled plasma mass spectrometry (ICP-MS) outshines in turn of detecting the dissolved traces into the electrolyte in minimal amounts that reach parts per trillion (ppt). Generally, electrochemical online ICP-MS is an electrochemical cell attached to the ICP-MS setup (Figure 23.6f). Hence, the simultaneous quantifying of the dissolved elements provides a multielement analysis that could help study the attitude of the working catalyst in order to get information on its stability during the changes within the electrochemical cell. However, the oxidation states of the detected elements cannot be provided by ICP-MS [53].

As has been noted, MS techniques have a set of great tools that provide the chemists valuable information about their running experiments. In fact, MS techniques prove how important to detect the catalytic reaction products in real-time analysis in order to get clear and complete mechanistic insights.

23.4 Electron-Based Characterization Techniques

Since catalysis is a surface process, it is vital to observe the changes that happen at the catalyst surface. Electron-based techniques, including electron microscopes,

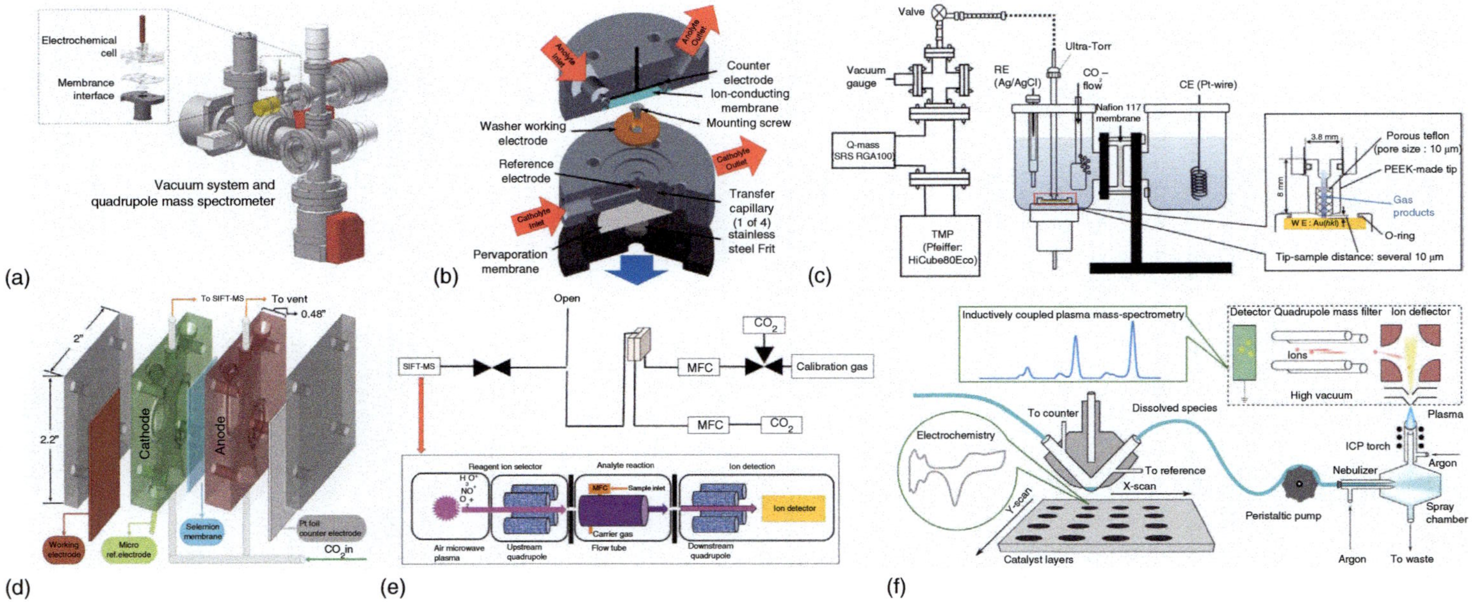

Figure 23.6 Different designs of electrochemical MS cells for in situ analysis. A technical computer-aided design (CAD) of (a) the DEMS instrument showing the electrochemical cell attached to the MS vacuum system [48] and (b) the in situ electrochemical cell used with the MS [49]. (c) A custom-made electrochemical H-cell with OLEMS setup for in situ measurements [50]. A schematic representation of the experimental setup of in situ SIFT-MS is shown as (d) an electrochemical cell design used for the in situ experiments [51, 52] with (e) an overview of the analysis operation for real-time detection [14]. (f) A schematic diagram showing the basic operational principles of a V-shaped electrochemical scanning flow cell (SFC) whose outlet is directly connected to the sample introduction system of ICP-MS [53, 54]. Source of ref. [48]: Reproduced with permission of Springer. Source of ref. [14, 49, 50]: Reproduced with permission of American Chemical Society. Source of ref. [51, 54]: Reproduced with permission of John Wiley & Sons, Inc. Source of ref. [52]: Reproduced with permission of Royal Society of Chemistry. Source of ref. [53]: Reproduced with permission of Elsevier.

provide imaging records of dynamics for the catalytic process using electronic signals either by static or scanning probes [2, 57]. Herein, we will highlight the popular surface-based tools used by the electrochemists to visualize the catalyst's surface-related reactions.

23.4.1 Transmission Electron Microscopy (TEM)

In situ transmission electron microscopy (TEM) is a powerful tool to display the catalyst surface, track the morphological changes during the electrocatalytic process, provide detailed information on kinetics, mechanisms, and structure of produced materials, and look for the catalytically active site (Figure 23.7a) [2, 7, 57]. Environmental transmission electron microscopy (E-TEM) is a more recent development of TEM, which offers the study of the catalyst with a high surface area such as nanoparticles and enables us to visualize the dynamic state of atoms while the proceeding the reaction [2, 67]. Nevertheless, obtaining information about the atomic arrangement of the catalyst surfaces is a little difficult [67]. On the other hand, liquid-phase transmission electron microscopy (LP-TEM), under the electrochemical reaction conditions, can track physical migration, morphological evolution, and structural changes of the catalytic process at high resolution, which is valuable to understand the transformation and degradation behavior of the catalyst surface (Figure 23.7b) [2]. For both E-TEM and LP-TEM, SiN_x material, an electron-penetrable window, is the superior option because of its electron transparency to provide real-time observation of the catalyst [2, 7, 41].

23.4.2 Scanning Probe Microscopy (SPM)

A powerful package for surface imaging is entitled scanning probe microscopes. In general, an interaction force between a fine-tipped probe and surface is atoms used to create an atomic-scale resolution image. Scanning tunneling microscopy (STM) uses an atomically sharp conductive probe to image the surface at the atomic level by facilitating the flow of quantum tunneling current from the tip to the sample (Figure 23.7c). This allows the detection of the changes on the surface under real-time reaction [2, 68]. Similar to STM, the atomic force microscope (AFM) enables the study of the catalyst morphology and its electronic properties [41]. However, there are some differences between STM and AFM based on their working principles. First, the probing tip. STM tip consists of non-contact-rigid metal filament, while AFM uses a flexible cantilever binding when a force is detected between a substrate and the mechanical tip attached to the cantilever (Figure 23.7d). Second, the nature of the substrate. A conductive substrate required while imaging with STM, but AFM applied with both conductive and nonconductive substrates [23, 68].

Electrochemical scanning tunneling microscopy (EC-STM), a modification to STM, can measure tunneling electrons and detect the atomic surface changes (Figure 23.7e). Furthermore, EC-STM can track, in controlled potential, the dissolution, degradation, and sintering phenomena of catalytic material [2, 41]. It has

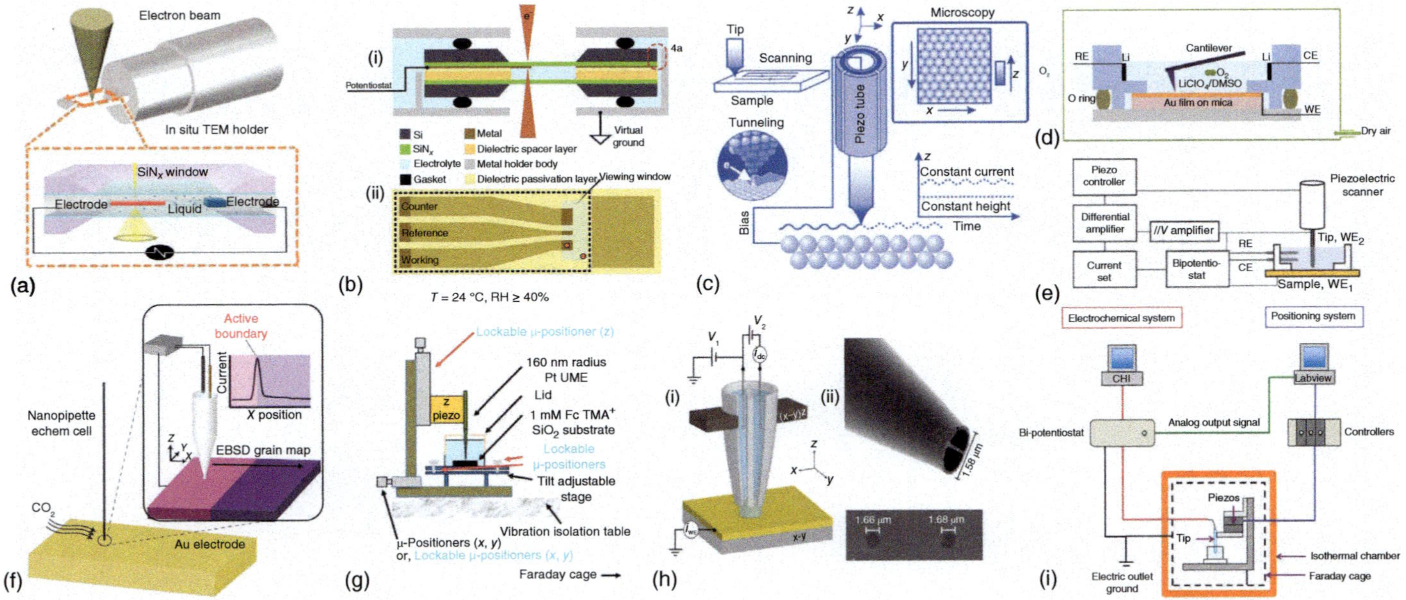

Figure 23.7 In situ electron-based characterization techniques. (a) A schematic drawing showing the in situ transmission electron microscopy [7, 58]. (b) A schematic illustration of LP-TEM cell configurations used for in situ electrochemical setup with (i) cross-sectional and (ii) plan view schematics of the cell [2, 59]. (c) An Illustration of the in situ STM setup showing the scanning tip on a sample with different operation modes [60]. (d) Home-made three electrodes cell with the configuration of a commercially available AFM setup [61]. (e) The in situ visualization cell setup for the EC-STM system [62]. (f) Activity measurement using SECCM with nanopipette electrochemical cell [2, 63]. (g) The necessary components of the SECM stage placed in the isothermal chamber [64]. (h) The in situ SECCM with (i) an illustration of the basic operational principles and (ii) field emission scanning-electron microscopy (FE-SEM) images of the pipette used as the probe. Source: Zhang etr al. [23, 65]. Reproduced with permission of Elsevier. (i) A schematic drawing of the in situ nanometer-scale SECM [64, 66]. Source of ref. [7, 58, 61, 64, 66]: Reproduced with permission of American Chemical Society. Source of ref. [2]: Reproduced with permission of IOP Publishing. Source of ref. [59, 63]: Reproduced with permission of Nature Publishing Group. Source of ref. [60, 62]: Reproduced with permission of Springer. Source of ref. [23]: Reproduced with permission of Elsevier. Source or ref. [65]: Reproduced with permission of Annual Reviews.

also been used to get useful information about electrode surface reconstruction and the adsorption/desorption of the intermediates; thus, mechanistic insights about the electrocatalytic systems could be obtained [23]. Another variation of STM is the scanning electrochemical microscopy (SECM), in which the electrochemical reaction occurs on the probing tip electrode (Figure 23.7g,i). In the meantime, the SECM technique can provide useful information about the chemical and topographic of the surfaces in the electrochemical reactions, where it creates a chemical reactivity map of the catalyst in the operating reaction. Subsequently, SECM is an efficient technique to study the electrocatalytic process rates and mechanisms [64, 69]. Moreover, the innovative development of SECM called the scanning electrochemical cell microscopy (SECCM) had been introduced since 2010 (Figure 23.7f,h) [23, 65]. Unlike SECM, SECCM provides much more stable responses, tracks the surface topography, and maps activity simultaneously and directly [23, 65, 70].

Given these points, one can realize how much these techniques help watch and track the changes that happened on the catalyst surface, besides providing morphological and topographical insights about the working catalyst. Generally, the variety of in situ techniques are considered to be of value to be integrated with predicted theoretical insights of various catalytic processes helping in advancing these important technologies.

References

1 Soares, J.A.N.T. (2014). Introduction to optical characterization of materials. In: *Practical Materials Characterization* (ed. M. Sardela), 43–92. New York: Springer.

2 Handoko, A.D., Wei, F., Jenndy et al. (2018). *Nat. Catal.* 1: 922–934.

3 Pérez-Rodríguez, S., García, G., Lázaro, M.J., and Pastor, E. (2018). Probing CO_2 reduction intermediates employing in situ spectroscopy and spectrometry. In: *Electrochemical Reduction of Carbon Dioxide* (eds. F. Marken and D. Fermin), 212–243. The Royal Society of Chemistry.

4 Sealy, C. (2020). *Nano Today* 31: 100852.

5 Bron, M. (2014). Infrared spectroelectrochemistry. In: *Encyclopedia of Applied Electrochemistry* (eds. G. Kreysa, K.-i. Ota and R.F. Savinell), 1071–1075. New York: Springer.

6 Li, M. and Marinkovic, N.S. (2013). In situ infrared spectroelectrochemistry: principles and applications. In: *Infrared Spectroscopy: Theory, Developments and Applications* (ed. D. Cozzolino), 307–332. Hauppauge, NY: Nova Science Publishers.

7 Li, X., Wang, S., Li, L. et al. (2020). *J. Am. Chem. Soc.* 142: 9567–9581.

8 Christensen, P.A. (2003). *Encyclopedia of Electrochemistry*. Wiley-VCH Verlag GmbH & Co. KGaA.

9 Mudiyanselage, K. and Stacchiola, D.J. (2013). In situ infrared spectroscopy on model catalysts: principles and applications. In: *In-Situ Characterization*

of Heterogeneous Catalysts (eds. J.A. Rodriguez, J.C. Hanson and P.J. Chupas), 209–239. Wiley.
10 Stavitski, E. (2013). Infrared spectroscopy on powder catalysts: principles and applications. In: *In-Situ Characterization of Heterogeneous Catalysts* (eds. J.A. Rodriguez, J.C. Hanson and P.J. Chupas), 241–265. Wiley.
11 Foster, A.J. and Lobo, R.F. (2010). *Chem. Soc. Rev.* 39: 4783–4793.
12 Zaera, F. (2012). *ChemCatChem* 4: 1525–1533.
13 Zhong, D., Zhang, L., Zhao, Q. et al. (2020). *J. Chem. Phys.* 152: 204703.
14 Mandal, L., Yang, K.R., Motapothula, M.R. et al. (2018). *ACS Appl. Mater. Interfaces* 10: 8574–8584.
15 Heyes, J., Dunwell, M., and Xu, B. (2016). *J. Phys. Chem. C* 120: 17334–17341.
16 Ajjan, F.N., Jafari, M.J., Rębiś, T. et al. (2015). *J. Mater. Chem. A* 3: 12927–12937.
17 Dong, B., Li, W., Huang, X. et al. (2019). *Nano Energy* 55: 37–41.
18 Deng, Y. and Yeo, B.S. (2017). *ACS Catal.* 7: 7873–7889.
19 Yuan, T., Le Thi Ngoc, L., van Nieuwkasteele, J. et al. (2015). *Anal. Chem.* 87: 2588–2592.
20 Zeng, Z.-C., Hu, S., Huang, S.-C. et al. (2016). *Anal. Chem.* 88: 9381–9385.
21 Schmid, T., Yeo, B.-S., Leong, G. et al. (2009). *J. Raman Spectrosc.* 40: 1392–1399.
22 Bilal, S. (2014). Raman spectroelectrochemistry. In: *Encyclopedia of Applied Electrochemistry* (eds. G. Kreysa, K.-i. Ota and R.F. Savinell), 1761–1765. New York: Springer.
23 Zhang, Y., Guo, S.-X., Zhang, X. et al. (2020). *Nano Today* 31: 100835.
24 Le Ru, E.C. and Etchegoin, P.G. (2009). A quick overview of surface-enhanced Raman spectroscopy. In: *Principles of Surface-Enhanced Raman Spectroscopy* (eds. E.C. Le Ru and P.G. Etchegoin), 1–27. Amsterdam: Elsevier.
25 Crayston, J.A. (2003). *Encyclopedia of Electrochemistry*. Wiley-VCH Verlag GmbH & Co. KGaA.
26 Siuzdak, K., Szkoda, M., Sawczak, M. et al. (2015). *RSC Adv.* 5: 50379–50391.
27 Samu, G.F., Scheidt, R.A., Kamat, P.V., and Janáky, C. (2018). *Chem. Mater.* 30: 561–569.
28 Wickman, B., Fredriksson, M., Feng, L. et al. (2015). *Phys. Chem. Chem. Phys.* 17: 18953–18960.
29 Rosser, T.E. and Reisner, E. (2017). *ACS Catal.* 7: 3131–3141.
30 Park, J.H., Choi, K.M., Lee, D.K. et al. (2016). *Sci. Rep.* 6: 25555.
31 Shah, A.-u.-H.A. (2014). UV–Vis spectroelectrochemistry. In: *Encyclopedia of Applied Electrochemistry* (eds. G. Kreysa, K.-i. Ota and R.F. Savinell), 2099–2102. New York: Springer.
32 Coppens, P. and Penner-Hahn, J. (2001). *Chem. Rev.* 101: 1567–1568.
33 Hung, S.-F. (2020). In-situ X-ray techniques for non-noble electrocatalysts. *Pure Appl. Chem.* 92 (5): 733–749, https://doi.org/https://doi.org/10.1515/pac-2019-1006.
34 Ahn, S., Klyukin, K., Wakeham, R.J. et al. (2018). *ACS Catal.* 8: 4132–4142.
35 Zhang, Z., Xiao, J., Chen, X.J. et al. (2018). *Angew. Chem. Int. Ed.* 57: 16339–16342.
36 Deng, Y., Handoko, A.D., Du, Y. et al. (2016). *ACS Catal.* 6: 2473–2481.

37 Sasaki, K., Wang, J.X., Naohara, H. et al. (2010). *Electrochim. Acta* 55: 2645–2652.

38 Zheng, X., Zhang, B., De Luna, P. et al. (2018). *Nat. Chem.* 10: 149–154.

39 Roy, K., Artiglia, L., and van Bokhoven, J.A. (2018). *ChemCatChem* 10: 666–682.

40 Masuda, T., Yoshikawa, H., Noguchi, H. et al. (2013). *Appl. Phys. Lett.* 103: 111605.

41 Choi, Y.W., Mistry, H., and Roldan Cuenya, B. (2017). *Curr. Opin. Electrochem.* 1: 95–103.

42 Singh, J., Tromp, M., Safonova, O.V. et al. (2009). *Catal. Today* 145: 300–306.

43 Watson, J.T. and Sparkman, O.D. (2007). Introduction. In: *Introduction to Mass Spectrometry: Instrumentation, Applications, and Strategies for Data Interpretation* (eds. J.T. Watson and O.D. Sparkman), 1–52. Chichester, England: John Wiley & Sons Ltd.

44 Hori, Y. and Suzuki, S. (1982). *Bull. Chem. Soc. Jpn.* 55: 660–665.

45 Watson, J.T. and Sparkman, O.D. (2007). Electron ionization. In: *Introduction to Mass Spectrometry: Instrumentation, Applications, and Strategies for Data Interpretation* (eds. J.T. Watson and O.D. Sparkman), 315–448. Chichester, England: John Wiley & Sons Ltd.

46 Mizuno, T., Naitoh, A., and Ohta, K. (1995). *J. Electroanal. Chem.* 391: 199–201.

47 Watson, J.T. and Sparkman, O.D. (2007). Chemical ionization. In: *Introduction to Mass Spectrometry: Instrumentation, Applications, and Strategies for Data Interpretation* (eds. J.T. Watson and O.D. Sparkman), 449–484. Chichester, England: John Wiley & Sons Ltd.

48 Ashton, S.J. (2012). Design and construction of the DEMS instrument. In: *Design, Construction and Research Application of a Differential Electrochemical Mass Spectrometer (DEMS)* (ed. S.J. Ashton), 29–80. Berlin, Heidelberg: Springer Berlin Heidelberg.

49 Clark, E.L., Singh, M.R., Kwon, Y., and Bell, A.T. (2015). *Anal. Chem.* 87: 8013–8020.

50 Todoroki, N., Tei, H., Tsurumaki, H. et al. (2019). *ACS Catal.* 9: 1383–1388.

51 Lobaccaro, P., Mandal, L., Motapothula, M.R. et al. (2018). *Energy Technol.* 6: 110–121.

52 Lobaccaro, P., Singh, M.R., Clark, E.L. et al. (2016). *Phys. Chem. Chem. Phys.* 18: 26777–26785.

53 Kasian, O., Geiger, S., Mayrhofer, K.J.J., and Cherevko, S. (2019). *Chem. Rec.* 19: 2130–2142.

54 Cherevko, S. and Mayrhofer, K.J.J. (2018). On-line inductively coupled plasma spectrometry in electrochemistry: basic principles and applications. In: *Encyclopedia of Interfacial Chemistry* (ed. K. Wandelt), 326–335. Oxford: Elsevier.

55 Ashton, S.J. (2012). Differential electrochemical mass spectrometry. In: *Design, Construction and Research Application of a Differential Electrochemical Mass Spectrometer (DEMS)* (ed. S.J. Ashton), 9–27. Berlin, Heidelberg: Springer.

56 Španěl, P. and Smith, D. (2017). Selected Ion Flow Tube (SIFT) applications in mass spectrometry. In: *Encyclopedia of Spectroscopy and Spectrometry*, 3e (eds. J.C. Lindon, G.E. Tranter and D.W. Koppenaal), 56–68. Oxford: Academic Press.

57 Ross, F.M. (2007). In situ transmission electron microscopy: basic principles and applications. In: *Science of Microscopy* (eds. P.W. Hawkes and J.C.H. Spence), 445–534. New York: Springer.

58 Gu, M., Parent, L.R., Mehdi, B.L. et al. (2013). *Nano Lett.* 13: 6106–6112.

59 Fahrenkrug, E., Alsem, D.H., Salmon, N., and Maldonado, S. (2017). *J. Am. Chem. Soc.* 164: H358–H364.

60 Pobelov, I.V., Li, C., and Wandlowski, T. (2014). Electrochemical scanning tunneling microscopy. In: *Encyclopedia of Nanotechnology* (ed. B. Bhushan), 1–16. Netherlands, Dordrecht: Springer.

61 Liu, C. and Ye, S. (2016). *J. Phys. Chem. C* 120: 25246–25255.

62 Nishino, T. (2018). Electrochemical scanning tunneling microscopy. In: *Compendium of Surface and Interface Analysis* (ed. The Surface Science Society of Japan), 87–90. Singapore: Springer Singapore.

63 Mariano, R.G., McKelvey, K., White, H.S., and Kanan, M.W. (2017). *Science* 358: 1187–1192.

64 Kim, J., Renault, C., Nioradze, N. et al. (2016). *Anal. Chem.* 88: 10284–10289.

65 Ebejer, N., Güell, A.G., Lai, S.C.S. et al. (2013). *Annu. Rev. Anal. Chem.* 6: 329–351.

66 Kim, J., Renault, C., Nioradze, N. et al. (2016). *J. Am. Chem. Soc.* 138: 8560–8568.

67 Nguyen, L., Zhang, S., Ye, Y. et al. (2013). Visualization of surface structures of heterogeneous catalysts under reaction conditions or during catalysis with high-pressure scanning tunneling microscopy. In: *In-Situ Characterization of Heterogeneous Catalysts* (eds. J.A. Rodriguez, J.C. Hanson and P.J. Chupas), 191–207. Wiley.

68 Wallace, A.F. (2019). Scanning probe microscopy: basic principles and applications. In: *Analytical Geomicrobiology: A Handbook of Instrumental Techniques* (eds. J.P.L. Kenney, H. Veeramani and D.S. Alessi), 121–147. Cambridge: Cambridge University Press.

69 Jantz, D.T. and Leonard, K.C. (2018). *Ind. Eng. Chem. Res.* 57: 7431–7440.

70 Bentley, C.L., Kang, M., and Unwin, P.R. (2017). *Curr. Opin. Electrochem.* 6: 23–30.

24

In Situ Characterizations in Electrocatalytic Cycle

A series of in situ characterizations have been designed to identify the dynamic changes, capture the intermediate states, and detect the catalytic products of the catalytic process. In addition, by considering an efficient electrocatalyst – which stands for an active, selective, stable, abundant, and affordable catalyst – these techniques are also used to verify the structural stability and reveal reasonable insights into the environmental factors at the interfacial electrode. In this section, we will discuss how these techniques help to (i) observe the real active centers via understanding the catalyst reconstruction during the catalytic reaction including the electronic structure, atomic structure, and phase transformation, (ii) investigate the reaction mechanism addressing in understanding the reactants adsorption/activation, probing the reaction intermediates, and detecting the catalytic products, (iii) evaluate the catalyst stability/decay, and finally (iv) describe the environment of electrode/electrolyte interface.

24.1 Investigating the Real Active Centers

Studies of active sites of a given catalyst have attracted much attention where the nature of the active structure dependence of electrocatalytic performance is questionable [1]. To describe the surface reactivity, theoreticians have developed semi-quantitative density functional theory (DFT)-based computations [2]. They provided several advances: descriptors, scaling relationships, catalytic principles, and density of states (DOS) analysis to predict the optimal catalyst with high reactive sites toward different classes of reactions, as highlighted in the Chapters 20:22. On the other hand, the correlation between Faradic efficiencies and structural ex situ characterization outputs, such as standard X-ray photoelectron spectroscopy (XPS), has been suggested as an integral role with theoretical calculations to define the active center, which is responsible for detected performance. Although theoretical and standard characterizations can help to identify the active centers of a given catalyst, observing the real sites would be more accurate. As known, the catalyst structure could be modified during the catalytic process [3]. We can categorize the structural evolutions into three groups: (i) electronic structure where a change in

Electrocatalysis in Balancing the Natural Carbon Cycle, First Edition. Yaobing Wang.
© 2021 WILEY-VCH GmbH. Published 2021 by WILEY-VCH GmbH.

valence band or oxidation states could happen under different applied voltages, (ii) atomic structure in which coordination environment could be varied affected by reactant adsorption, and (iii) phase transformation of the catalyst caused by material reconstruction. Precisely identifying the actively real locations and their deviation allows us to feed theoretical software with accurate data for suggesting a rational mechanism as a forward step to new catalyst engineering. Advanced in situ analysis contributes to characterize such evolutions and further to determine the nature of real active sites. Herein, we will show how in situ tools can help in monitoring the structural modifications.

24.1.1 Monitoring the Electronic Structure

As the catalysis proceeds under applied potential, the electronic structure, such as oxidation and valence states, could be altered. By monitoring such modifications, scientists can identify the real site state of a given catalyst. Gu et al. reported that as negative potential is higher than −0.5 V vs. reversible hydrogen electrode (RHE), the catalyst performance is correlated to electronic change [4]. For example, in electrochemical carbon dioxide reduction (ECR), by X-ray absorption spectroscopy (XAS) measurements, they demonstrated that the energy of Fe K-edge is shifted to be lower as same as FeO edge at −0.5 V and beyond indicating that the Fe^{3+}–N–C was reduced to Fe^{2+}–N–C, as shown in Figure 24.1a. When applying potential as high as 80 mV, the catalyst deactivates defining the potential limitation of Fe^{3+}–N–C catalyst for ECR due to the oxidation state changes. In addition, by examining Cu catalysts under in situ XAS, Weng et al. clearly revealed that Cu oxidation state alters among Cu(II), Cu(I), and Cu(0) with different applied potential, as shown in Figure 24.1b [5]. The maximum CH_4 FE was attributed to forming dominant Cu(0) of CuPc. These studies provide that potential-dependent change in the electronic states of catalyst could be correlated with the catalytic performance resulting in value-added insights.

Separately, in oxygen evolution reaction (OER), Zheng et al. tracked the electronic changes of Ni sites in nanoporous NiCoFeP [7]. Using soft XAS, they demonstrated the formation of high oxidation state Ni^{4+} during catalysis, which is favorable for OER. Besides, the Ni^{4+}/Ni^{2+} ratio has been controlled on the basis of applied potential, as shown in Figure 24.2a,b. Similarly, in oxygen reduction reaction (ORR) catalytic reaction, the activity was controlled by redox transitions of $Fe^{2+/3+}$-based catalyst [9]. On the other hand, employing potential- and time-dependent Raman spectroscopy, Dutta et al. reported that tin oxide nanoparticles undergo electronic switches during the ECR process affecting the formate production, as shown in Figure 24.2c. In the same direction, the SnO_x shell switched to Sn–SnO_x interface in the SnCu alloy/Sn core catalyst, under ECR conditions, at −0.93 V vs. RHE observed by in situ X-ray absorption fine structure (EXAFS) examinations cooperated with ex situ techniques [10]. Disappearing of Sn–O peak upon applied potential suggests its reduction to be Sn under given conditions. It was concluded that the reconstructed interface due to partial reduction is the real active phase for formate production.

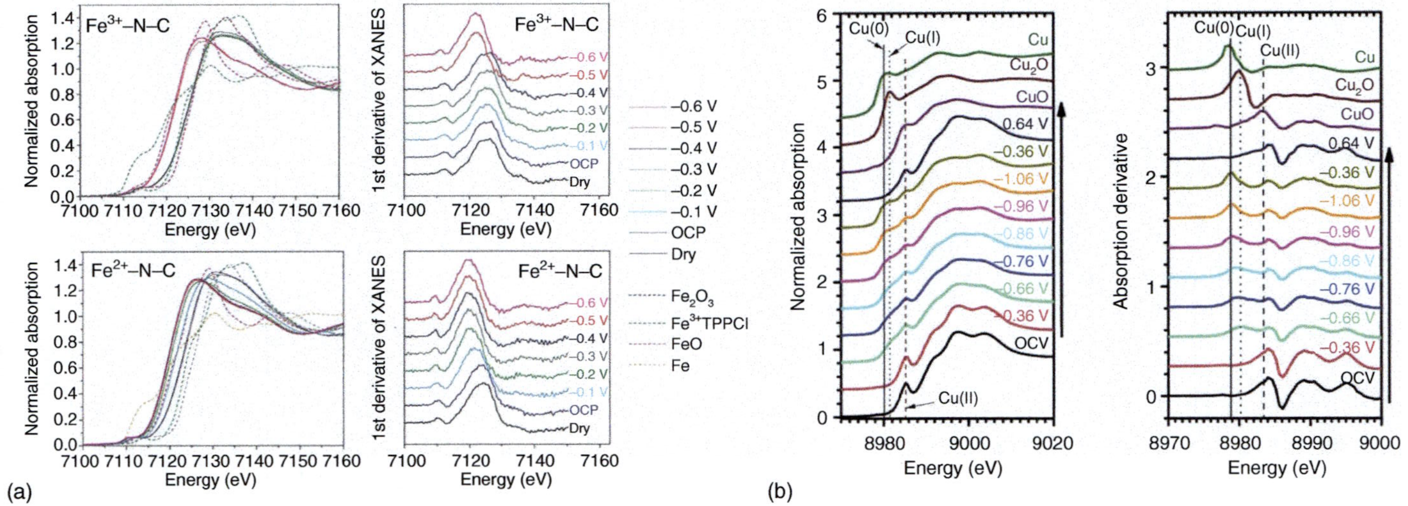

Figure 24.1 Exploring the electronic changes by XAS characterization: (a, b; left) Fe and Cu K-edge XANES spectra and their first-order derivatives (a, b; right) for different oxidation states of Fe- and Cu-based catalysts [5, 6]. Source of Ref. Weng et al. [5]. Licensed under CC-BY-4.0. Source of Ref. Huang et al. [6]. Reproduced with permission of Royal Society of Chemistry.

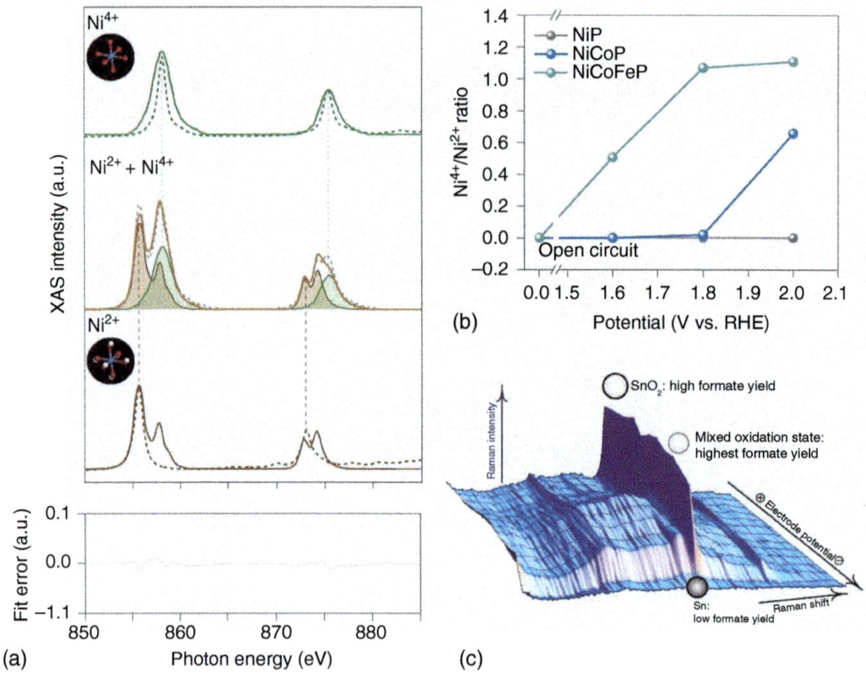

Figure 24.2 Monitoring the electronic change using sXAS and Raman in situ techniques: (a) in situ experimental and theoretical sXAS; solid and dashed lines, respectively, at the Ni L-edge of various Ni^{2+} and Ni^{4+} states; brown and green, respectively. Source: Zheng et al. [7]. Reproduced with permission of Nature Publishing Group. (b) Ni^{4+}/Ni^{2+} ratio vs. each potential of NiCoFeP catalyst [7]. (c) Raman spectra on SnO_2 with its modifications under different potentials affecting formate production. Source: Dutta et al. [8]. Reproduced with permission of American Chemical Society.

Several efforts have been conducted proving the correlation between the electronic changes and the catalyst performance under a given reaction [3, 11–14].

Recently, using diffuse reflectance UV–vis (DRUV), the reversibility of electronic change can be tracked under OER working conditions [15]. The results at anodic and cathodic potentials indicate the oxidation of Co to +4 species in the $CoAl_2O_4$ catalyst surface; however, it seems to be a reversible process, as shown in Figure 24.3. Such effort offers the ability of different in situ techniques to monitor the changes of electronic catalytic structure. In electrocatalysis literature, a number of studies have been described the electronic change during reaction [16–18]. By virtue of different in situ techniques, the electronic structural evolution can be monitored in real time, which further helps to improve our understanding of the nature of real active centers.

24.1.2 Monitoring the Atomic Structure

The local atomic/coordination structure could undergo rearrangement during catalysis. In this sense, to gain a deeper understanding of the atomic structure–activity

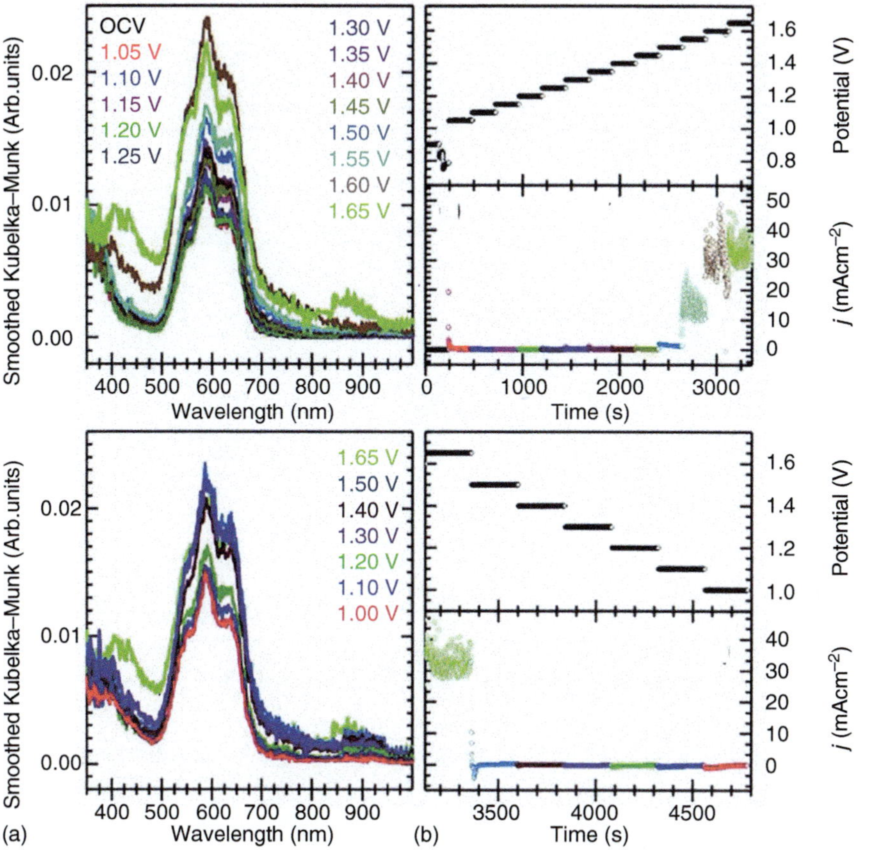

Figure 24.3 The electronic reversibility of $CoAl_2O_4$ by DRUV spectroelectrochemistry. (a) DRUV spectra for oxidation (top) and reduction (bottom) in 1 M KOH. (b) Corresponding applied potentials and chronoamperometry (CA) measurements. Source: Wahl et al. [15]. Reproduced with permission of Royal Society of Chemistry.

relationship, investigating the local coordination environment would be required. Fortunately, by means of in situ characterizations, this would be achieved. Recently, Cao et al. [19] followed the deviation in bond length in single Co atoms under hydrogen evolution reaction (HER) conditions using EXAFS cooperated with Fourier transform infrared (FTIR) and X-ray absorption near-edge structure (XANES), as shown in Figure 24.4. They suggested that such deviation is induced by the “oxo” group adsorption forming HO–Co–N_2 in an alkaline solution. When the applied potential gets higher, further, water molecules chemically adsorbed to be promoted for H_2 evolution. Thus, the real active site was attributed to HO–Co–N_2, which is formed during reaction with a higher oxidation state confirmed by XANES. The same aspect was reported by Lu et al. studying CO oxidation catalyzed by Ir/$MgAl_2O_4$. They found that while exposure of $Ir(CO)_2$ intermediates to O_2, it gets oxidized to form Ir(CO)(O) as stable intermediate with local atomic variation [20].

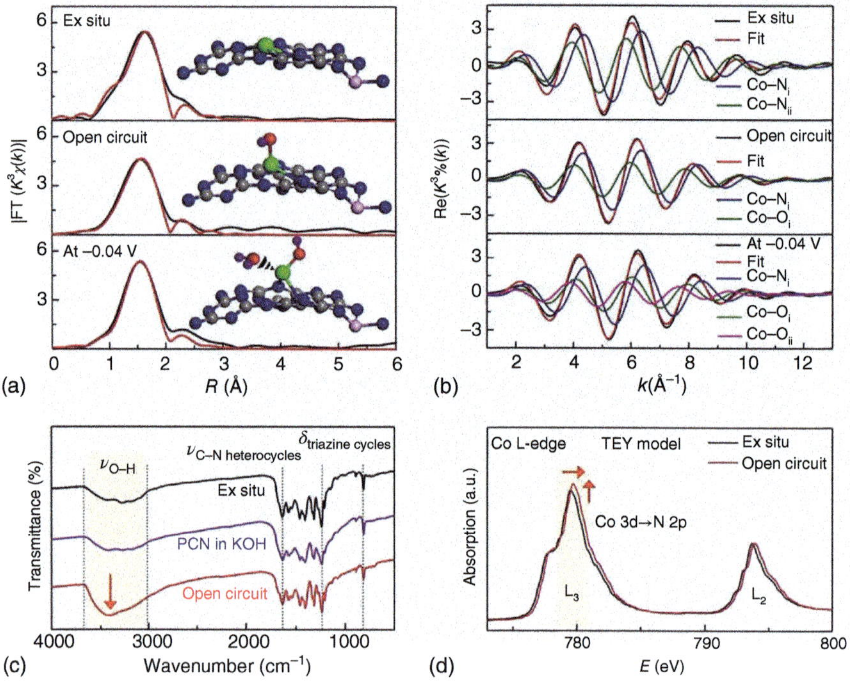

Figure 24.4 Capturing the atomic structure evaluations: fitted first shell of EXAFS spectra (a) and corresponding Re($k^3\chi(k)$) oscillations (b) of different paths, (c) and (d) refer to FTIR spectra and Co L-edge XANES spectra of Co-based catalyst, respectively. Source: Cao et al. [19]. Reproduced with permission of Nature Publishing Group.

In the same direction, the structure–activity relationship was tracked under ECR conditions using in situ XAS [21]. The low coordination number of metal–organic framework (MOF)-derived Cu clusters was governed by the amount of Cu dimer distortion caused by heat treatment of MOF, as shown in Figure 24.5. However, this is a specific feature for this structure, i.e. Cu dimer with benzene tricarboxylates. This variation in the local atomic environment shifts ECR products to C2+. Such results reflect the importance of monitoring the coordination variations for understanding the impact of deviated structure on product selectivity. Separately, with similar coordination numbers, the ligand type as a local atomic structure also could affect the catalyst activity. It was found that coordination of the Fe^{3+} ions to pyrrolic-N configuration was attributed to being the real active sites instead of common pyridinic-N type, gained by XAS measurements under ECR catalysis [4]. Pinpointing like these factors is successfully achieved by the help of advanced in situ characterizations [22].

24.1.3 Monitoring the Catalyst Phase Transformation

The catalytic conditions, including gaseous reactants under applied potential, could influence the phase of the catalyst. New insights into the restructuring of the catalyst

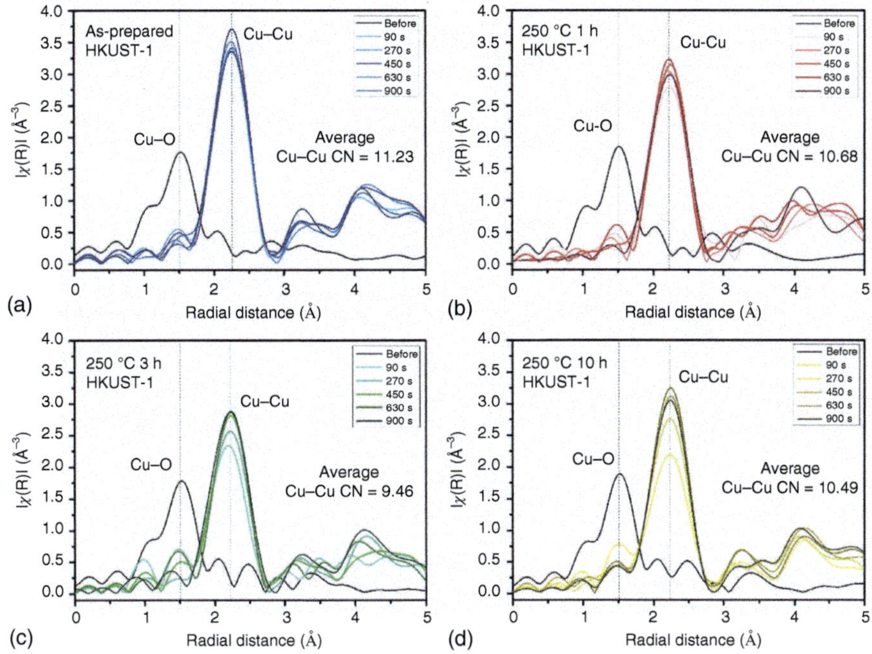

Figure 24.5 Tracking the atomic structure of HKUST-1-derived Cu clusters in different temperature by in situ EXAFS during ECR at −1.38 V vs. RHE in 1 M KOH: (a) as-prepared MOF, (b) 250 °C, one hour calcined MOF, (c) 250 °C, three hours calcined MOF, and (d) 250 °C, 10 hours calcined MOF. Source: Nam et al. [21]. Reproduced with permission of American Chemical Society.

phase become possible using real-time techniques. Product selectivity of the catalyst could be shifted by phase restructuring governed by applied potential. For instance, by in situ Pd K-edge and L3-edge XAS measurements on Pd NPs under ECR, as shown in Figure 24.6, it was concluded that a mixture of the α- and β-phases of PdH_x is formed at −0.3 V facilitating formate production whereas with increasing the potential to reach −0.7 V, the β-phase of palladium hydride becomes predominant promoting CO selectivity [23]. A separately prepared Pd-based catalyst shows similar behavior where it undergoes phase shift but with syngas production [24]. Hence, in situ investigations during catalysis could provide significant insights toward product selectivity for further enhancing our catalytic understanding of the rational design toward new catalysts.

Additionally, the catalytic phase change can be monitored by in situ X-ray diffraction (XRD) in which the diffraction intensity is calculated as a function of voltage. Tung et al. investigate the OER system on Co_3O_4@CoO single catalyst [25]. As potential increases, β-CoOOH was formed, indicating catalyst conversion, as shown in Figure 24.7a. With higher voltage, a new phase was observed, which was identified as α-CoOOH. Surprisingly, the structure between metal oxyhydroxide and amorphous phases is reversible process confirmed by voltage cycling in the application of in situ XRD, as shown in Figure 24.7b. Phase transformation and

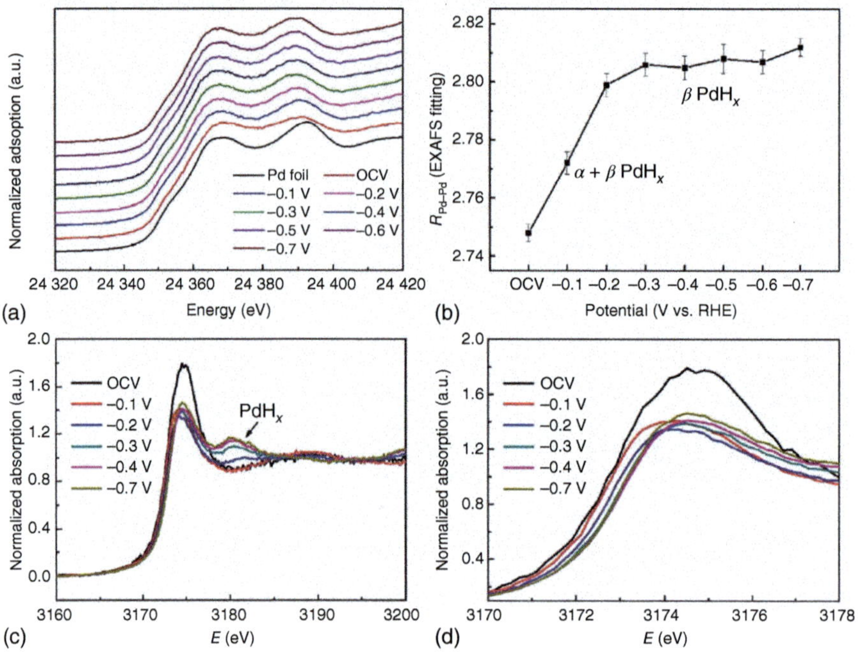

Figure 24.6 Detecting phase changes by in situ XAS measurements on Pd catalyst in 1 M $KHCO_3$ solution under various ECR applied potentials. (a and c) In situ Pd K-edge and L3-edge XANES spectra of Pd. (b) Average interatomic distance of Pd–Pd atoms according to EXAFS fitting. (d) Enlarged in situ L3-edge XANES plot. Source: Gao et al. [23]. Reproduced with permission of Springer.

structure reversibility are significant insights that cannot be achieved without the help of such real-time means.

Phase transition of copper has been detected by imaging at fixed ECR potential. Kim et al. clarified that the polycrystalline Cu surface has reconstructed to Cu(111) after passing 30 minutes of applying −0.9 V vs. standard hydrogen electrode (SHE) in KOH electrolyte [26]. Then, Cu(111) showed another phase change forward to be Cu(100) after 30 minutes. These continuous transformations were detected by electrochemical scanning tunneling imaging, as shown in Figure 24.8a. It is interesting to say that the perspectives of imaging are not only limited to track phase transformation but to monitor surface-related morphological changes also, which affects the catalyst phase. In this regard, Grosse et al. used electrochemical atomic force microscopy (EC-AFM) to testify the phase stability of Cu(100) facets under ECR operating conditions [27]. The disappearance of the (100) phase was observed with increasing roughness surface under the potential of −1.1 V vs. RHE in 0.1 M $KHCO_3$ catholyte, as shown in Figure 24.8b. Phase evolutions have been adequately captured by in situ imaging applications.

In summary, it can be remarked that the in situ tools are widely used to monitor the catalyst evolutions involving different reconstructions at electronic, atomic, and configurational levels [16, 28–34]. Insights regarding these structural dynamics and

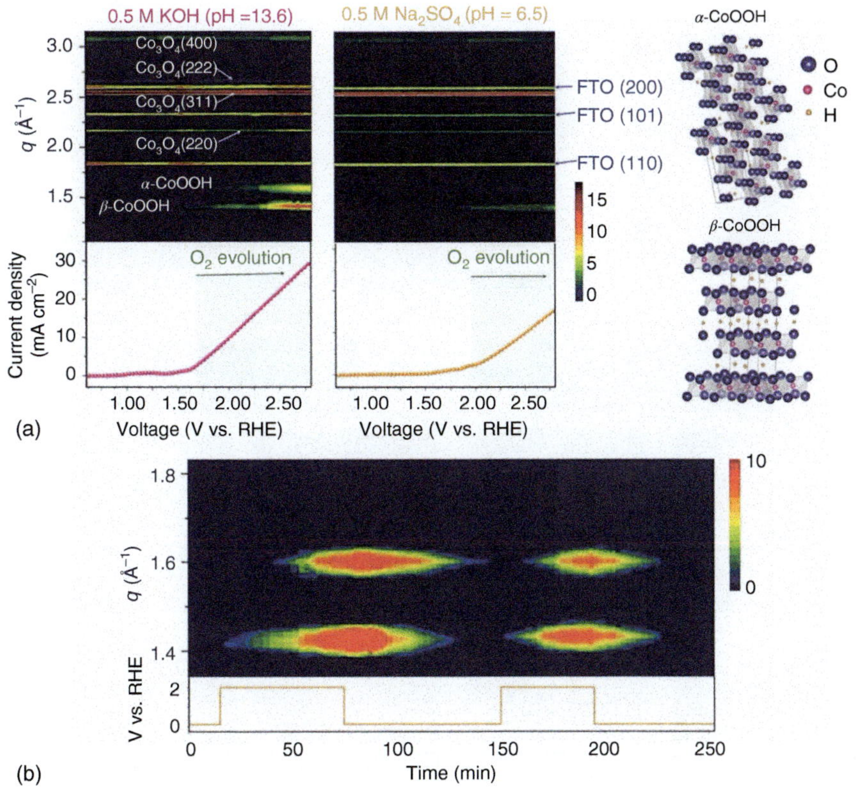

Figure 24.7 Phase evaluation by in situ grazing-angle X-ray diffraction: (a) contour plots of XRD signals of Co_3O_4@CoO in 0.5 M KOH and 0.5 M Na_2SO_4 and corresponding current density. (b) Contour plots of XRD signals of Co_3O_4@CoO in 0.5 M KOH under-voltage cycling (between 2.0 and 0.1 V vs. RHE). Source: Tung et al. [25]. Reproduced with permission of Nature Publishing Group.

their correlation with activity/selectivity motivate the catalysis community toward a deeper understanding of the real nature of active centers under the electrocatalytic cycle.

24.2 Investigating the Reaction Mechanism

Although simulating the reaction pathway using DFT provides kinetics- and thermodynamics-related valuable insights, deep understanding of dynamic adsorption is real, recognizing the proposed intermediates are limited in the first case and dismissed by default in the second one. Fortunately, in situ analysis shows their potential and feasibility to help to an in-depth description of catalytic activation and intermediates observation at reality level and support the theoretically proposed mechanism. In general, the reaction pathway of any catalytic process involves reactants adsorption, intermediates reaction, and product desorption. According to these three aspects, we will highlight how being the in situ analysis is

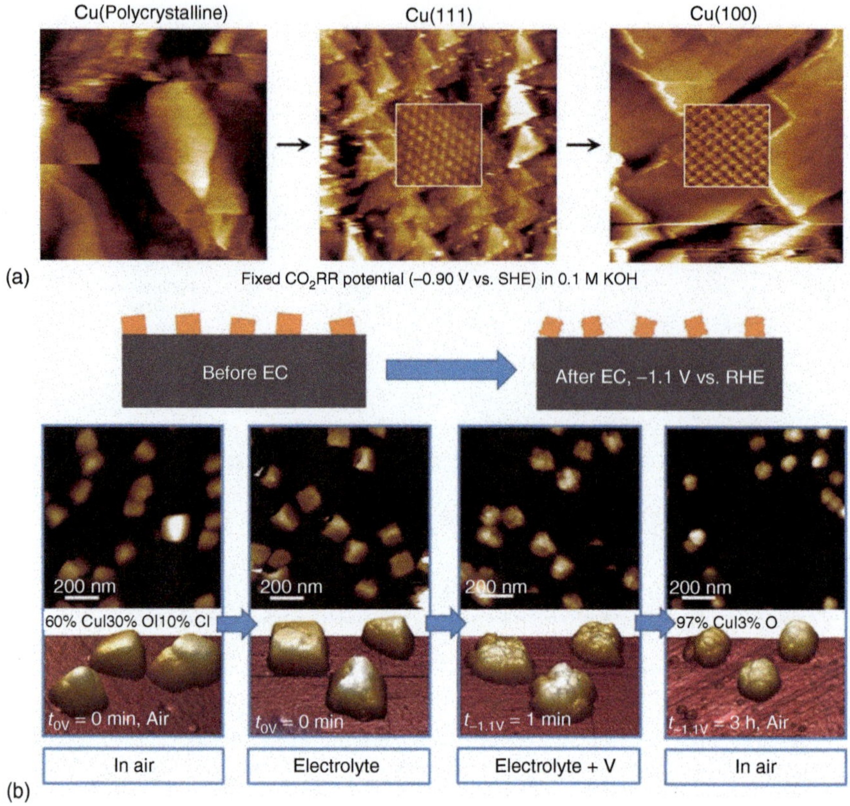

Figure 24.8 Electrochemical scanning tunneling microscopy (ECSTM) and EC-AFM imaging for phase and morphological deviations: (a) ECSTM images of Cu phases varying with time at −0.9 V in 0.1 M KOH. Source: Kim et al. [26]. Reproduced with permission of American Chemical Society (b) ESAFM images of Cu cubes evolution in air, at open-circuit potential, at −1.1 V vs. RHE for one minute, and after three hours in order from left to right. Source: Grosse et al. [27]. Reproduced with permission of John Wiley & Sons, Inc.

a promising strategy to interpret the adsorption-structural evolution process, detect the generated species, and finally measure the desorbed products through most recent advances of different electrocatalytic systems.

24.2.1 Through Adsorption/Activation Understanding

Among various catalyst requirements, the ability of adsorption, which allows the catalytic activation, is considered to be a vital characteristic of a given surface, including reactants and intermediate adsorption. In this regard, it is a high demand to understand how reactive species such as CO_2, O_2, H_2, and methanol get adsorbed on different structures. By tracking the evolution of the electrode surface during electrocatalysis, considerable adsorption-related insights can be achieved. For instance, Yang et al. reported a comprehensive study using in situ XAS, electron paramagnetic resonance (EPR), and XPS to interpret the CO_2

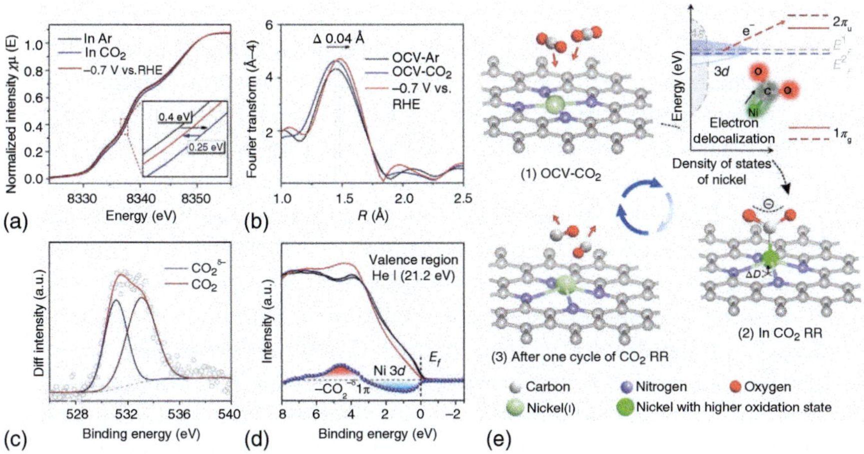

Figure 24.9 Molecular adsorption and activation by in situ XAS and XPS investigations of atomically Ni–NG: (a and b) Ni K-edge XANES and EXAFS spectra at various voltages vs. RHE in 0.5 M $KHCO_3$ catholyte. (c) The XPS O^{1s} intensity is demonstrating CO_2 adsorption. (d) Valence band spectra before (black line) and after (red line) CO_2 gas flow, and after CO_2 desorption (dark blue line), in addition, valence band variation spectra induced by CO_2 adsorption (blue line). (e) Schematic representation shows the structural evolution of the active site in ECR. Source: Yang et al. [12]. Reproduced with permission of Nature Publishing Group.

adsorption and activation mechanism on the single-Ni-atom catalyst during the working conditions [12]. A shift was observed in XANES spectra by 0.4 eV as soon as CO_2 is purged, as represented in Figure 24.9a. This shift refers to the delocalization of unpaired e in Ni, which is boosted the activation process by donating two electrons to the C atom of CO_2 forming Ni—C bond with partially negative chemically adsorbed $^*CO_2$ species. Moreover, a backshift was captured as catalysis continuously indicating the active site recovery after one reduction cycle. Such spectra-derived information gives a hint for further learning about the adsorption mechanism. Transferring to the EXAFS spectrum, as shown in Figure 24.9b, a higher intensity was pointed at the increased potential in which Ni—N bond got expanded due to Ni—C bond contribution. Further, diffraction XPS results verified the electron transfer from Ni active site to CO_2 to form $CO_2^{-\delta}$, as shown in Figure 24.9c. Valence band spectra also confirm electron-donating from the $3d_{x^2-y^2}$ orbital of Ni(i) to the C $2\pi_u$ orbital, as shown in Figure 24.9d. They described the adsorption–activation reaction as a scheme shown in Figure 24.9e. Obviously, this interpretation stems from in situ-driven data. Therefore, to obtain further mechanistic insights about adsorption/activation of other reactants, the make use of in situ analysis is recommended.

24.2.2 Through Intermediates In Situ Probing

A reliable reaction mechanism can be described by several elementary steps, including short time-existing intermediate species. These intermediates cannot be captured by ex situ tools due to insufficient lifetime. To probe such species, highly sensitive and quickly responding techniques are required. Evidently, some in situ advances show their ability under working conditions toward intermediate structures. A representative study reported by Firet et al. [35] helps to find out the surface-bound intermediates under the ECR process on Ag catalyst using attenuating total reflection Fourier transform infrared (ATR-FTIR) measurements in comparison with predicted intermediate's peaks, as shown in Figure 24.10a,b. The nature of intermediates was correlated to applied potential in which the *COOH is observed at potential range of −1.40 to −1.55 V vs. Ag/AgCl. In contrast, the $^*COO^-$ is noticed at −1.6 V. New mechanism-related insight is pinpointed that the ECR mechanism on Ag follows proton-coupled electron transfer (PCET) at low applied potential while it follows sequential electron–proton transfer pathway at higher potential, as shown in Figure 24.10b,c. In a word, by examining the interplay between the applied potential and intermediate type, valuable mechanistic insight can be disclosed using such advanced techniques. Another critical example using the in situ infrared technique is acetate production from ECR on nitrogen-doped nano-diamond. The presence of the OOC–COO peak at 1307 cm^{-1}, as shown in Figure 24.10d, boosts mechanism suggestion to be proposed, including OOC–COO intermediate.

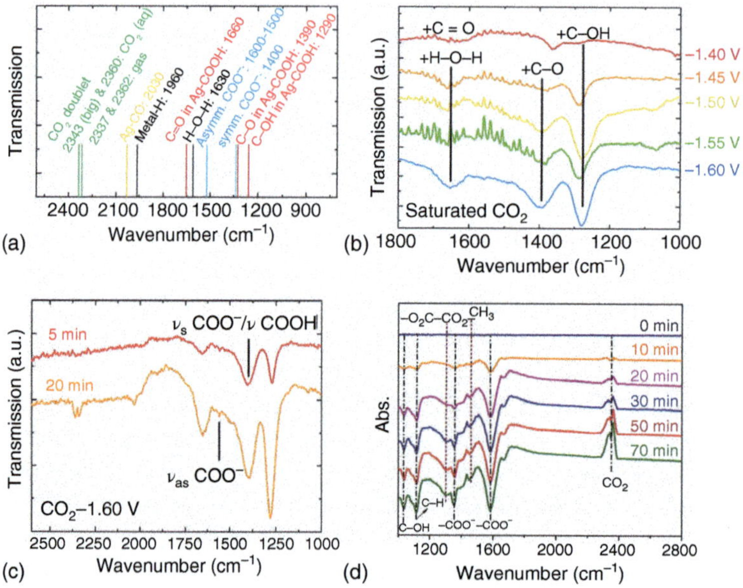

Figure 24.10 Probing ECR intermediate species using in situ FTIR: (a) Predicted peak positions for the various reactants, products, and intermediates of the ECR system. (b) ATR-FTIR spectra at different potentials displaying the peaks that belong to ECR intermediates. (c) ATR transmission spectra at −1.60 V for 5 and 20 minutes after the CO_2 exposure. Source: Firet and Smith [35]. Reproduced with permission of American Chemical Society. (d) FTIR of ECR on the N-doped nanodiamond (NDDL)/Si rod array (Si RA) electrode at −1.0 V with various electrolysis time. Source: Liu et al. [36]. Reproduced with permission of American Chemical Society.

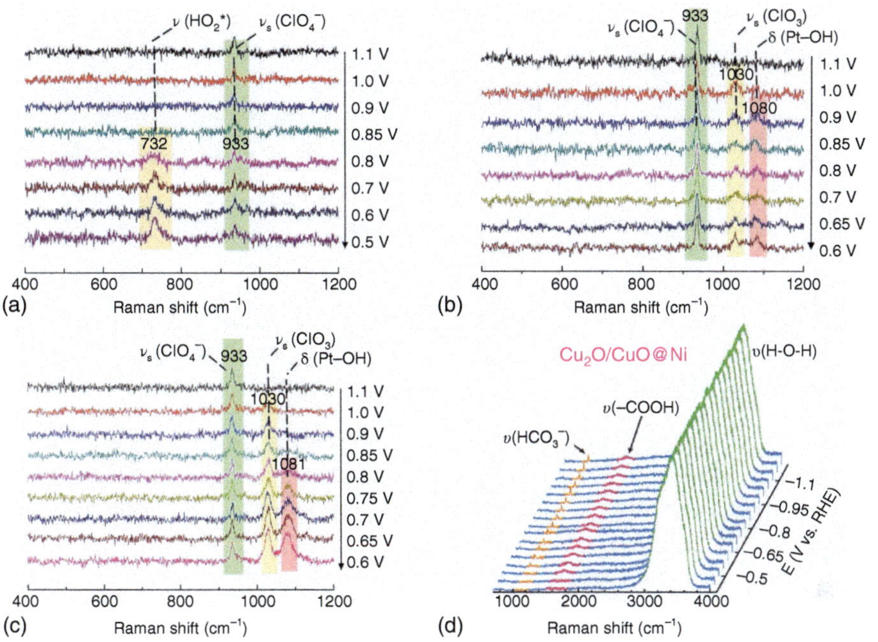

Figure 24.11 Intermediate detection by in situ Raman: Raman shift on Pt(111) (a), Pt(100) (b), and Pt(110) (c) catalytic surfaces in a 0.1 M $HClO_4$ solution during ORR electrolysis [37]. Source: Dong et al. In situ Raman spectroscopic evidence for oxygen reduction reaction intermediates at platinum single-crystal surfaces. Nat Energy 4, 60–67 (2019). https://doi.org/10.1038/s41560-018-0292-z (d) Raman spectra of $Cu_2O/CuO@Ni$ ECR catalyst in a 0.5 m $KHCO_3$ solution at potential range from −0.4 to −1.2 V vs. RHE. Source: Yang et al. [38]. Reproduced with permission of John Wiley & Sons, Inc.

Observing reaction intermediates can be conducted by another potential technique; surface-enhanced Raman scattering (SERS). The advantage of developed SERS, electrochemical (EC)-Shell-Isolated Nanoparticle-Enhanced Raman Spectroscopy (SHINERS) provided by Li group [37], is its ability to consider the adsorption configuration at single-crystal face during electrocatalysis. The EC-SHINERS spectra of ORR catalysis on different Pt surfaces, under acidic conditions, are shown in Figure 24.11a–c. At Pt(111) and by excluding the vibrational peak of chlorate ion, a new peak appears at 732 cm^{-1} when the applied potential reaches 0.8 V, where it becomes predominant with potential increase. By carrying out the deuterium isotopic experiment, they confirmed that the newly formed peak belongs to the O–O stretching vibration mode of HO_2* intermediate. On the other hand, at Pt(100) and Pt(110), different features are observed due to different crystallographic orientations and lower coordination number. The adsorbed *OH intermediate was suggested around 1080 cm^{-1}. Meanwhile, the intermediate at Pt surfaces was only O_2^- under alkaline conditions. This study provides a reasonable description of different intermediates on different single-crystal catalyst under catalytic process using the advanced in situ characterization. In the same direction, using SERS, the possible intermediate can be identified under ECR conditions on

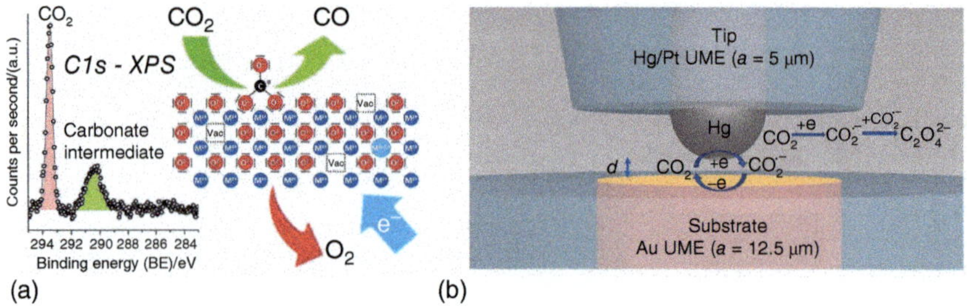

Figure 24.12 Schematic representations of the detection of carbonate (a) and $CO_2^{\bullet-}$ (b) intermediates by XPS on perovskite pattern electrode [40] and SECM on Au ultramicroelectrode, respectively. Source: Kai et al. [41]. Reproduced with permission of American Chemical Society.

Cu_2O/CuO@Ni surface [38]. The only *COOH possible intermediate was observed at potential range from −0.5 to −1.1 V vs. RHE, as shown in Figure 24.11d, proving the PCET mechanism of ECR reaction toward CO selectivity. In situ Raman shows its feasibility for intermediates detection [39].

One further technique named near-ambient pressure X-ray photoelectron spectroscopy (NAP–XPS) is reported to facilitate the identification of the intermediates of CO_2 conversion to CO on perovskite-based catalyst under operation conditions [40]. Interestingly, NAP–XPS measurements confirm the adsorption of a carbonate intermediate formed on oxides surfaces, as shown in Figure 24.12a. They suggested that bidentate radical $(CO_3)^{\bullet 3-}$ is the form of carbonate species at oxygen vacancies sites. This new insight is different from commonly known intermediates of the ECR process, such as *COOH and *COO⁻. Interestingly, another ECR intermediate $^*CO_2^{\bullet-}$ was detected on the Au substrate of the Hg/Pt ultramicroelectrode probe using scanning electrochemical microscope (SECM), as shown in Figure 24.12b. Further detail for SECM design to such detection can be found elsewhere [41]. Numerous studies prove the feasibility of in situ tools for ECR catalytic process [14, 42].

To end up, it is evident that the dynamic evolution characterizations such as in situ FTIR, Raman, XPS, and SECM are valuable tools as it helps to get more insights toward identifying and confirming the nature of intermediates getting closer to deep understanding of the catalytic process.

24.2.3 Through Catalytic Product In Situ Detections

Considerable benefits can be obtained by in situ analysis in detecting the catalytic products under working conditions. Among in situ tools, mass spectrometry (MS)-based techniques show a remarkable superiority in this aspect. Online electrochemical mass spectrometry (OLEMS) was recently applied by Todoroki et al. to study ECR on single-Au-crystal surfaces [43]. Depending on applied potential, gaseous products such as CH_4, CO, and H_2 were detected, as shown in Figure 24.13a Regarding CO product, Au(110) shows the highest Q-mass ion intensity revealing that its activity toward ECR. In contrast, other surfaces promote H_2 generation. The corresponding relative Faradaic selectivity was calculated on the basis of

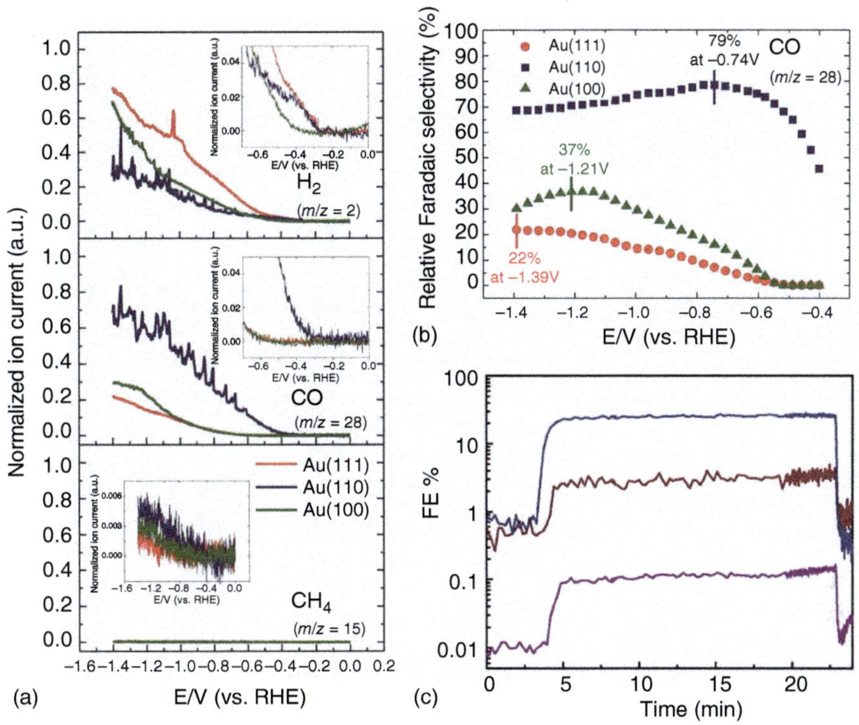

Figure 24.13 Product detection using in situ MS measurements: (a) mormalized Q-mass ion currents for H_2, CO, and CH_4 detected by online electrochemical mass spectrometry (EMS) as a function of the applied potentials. (b) Relative Faradaic selectivity for carbon monoxide on various Au surfaces. Source: Todoroki et al. [43]. Reproduced with permission of American Chemical Society. (c) Semi-log plot of SIFT-MS data for C_2H_4, CH_4, and propene, in order from up Ref. [39]. Source: Mandal et al. [39]. Reproduced with permission of American Chemical Society.

OLEMS results, as presented in Figure 24.13b. On the other hand, higher-order hydrocarbons could be detected using selected-ion flow tube mass spectrometry (SIFT-MS). Mandal et al. provided the real-time evaluation of FEs for CH_4, C_2H_4, and C_3H_6 on copper oxide nanoneedles, as shown in Figure 24.13c [39]. These studies clarified the benefits of such techniques to qualify and quantify the catalytic performance for a given catalyst.

One further study of in situ MS is to compare the ECR products on Ag from gas chromatography (GC) and differential electrochemical mass spectrometry (DEMS) measurements [44]. It shows good agreement with GC measurements toward activity and performance in which DEMS was conducting during linear sweep voltammetry, as shown in Figure 24.14a,b. Besides, the consumption percentage of CO_2 along potential rage was provided, as shown in Figure 24.14c. Moreover, they detected nine products on Cu catalyst during cyclic voltammetry, as shown in Figure 24.14d. It was concluded that acetaldehyde is considered to be critical species for further reduction toward ethanol and propionaldehyde productions. Besides, propionaldehyde is a precursor toward *n*-propanol generation. These efforts show significant benefits of

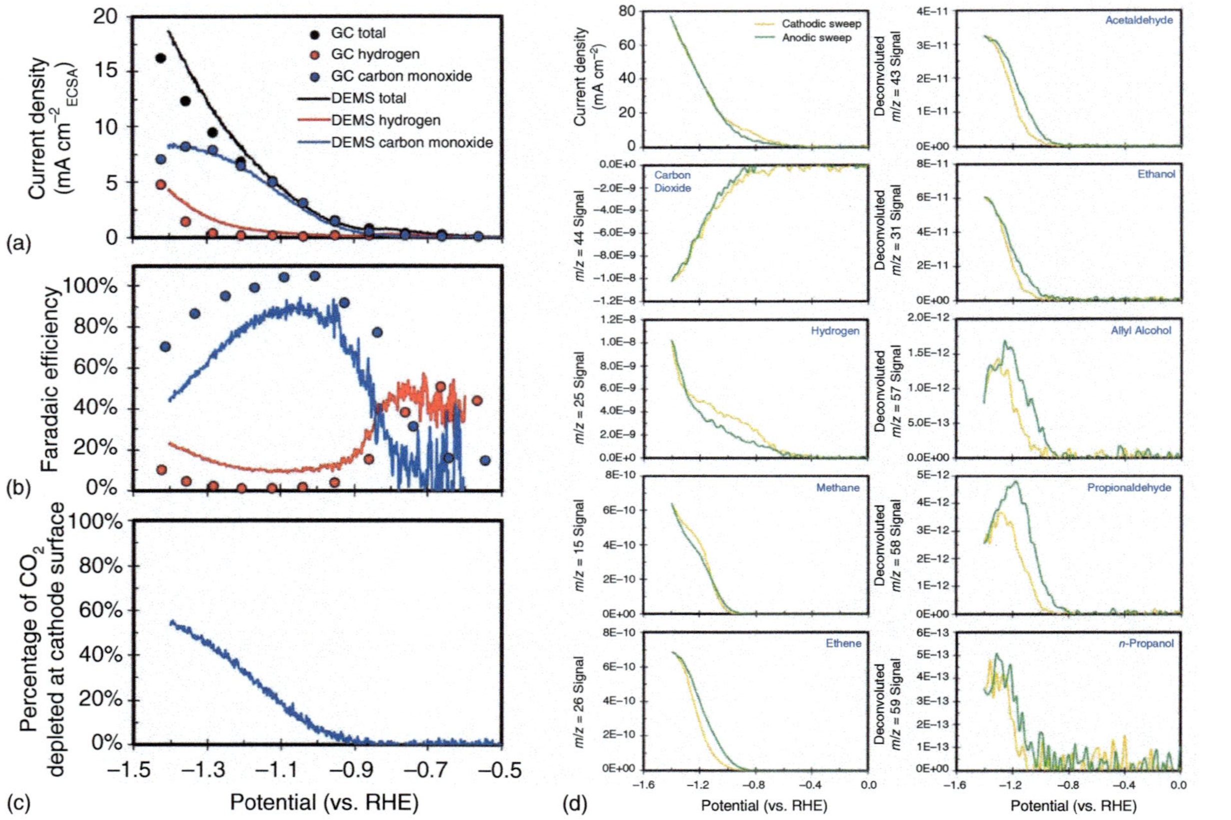

Figure 24.14 In situ ECR product quantification: (a and b) Comparison of the ECR activity; electrochemical specific surface area (ECSA)-normalized partial current densities and FE, respectively, on Ag evaluated by GC and DEMS in 0.1 M $CsHCO_3$. (c) The proportion of CO_2 depleted measured by DEMS during linear sweep voltammetry (LSV). (d) Total mass-ion current panels of the nine primary contributing products. Source: Clark and Bell [44]. Reproduced with permission of American Chemical Society.

using DEMS for the ECR process. Generally, despite the attractive value of in situ MS in product detection, to date, there is relatively little attention for its utilization under other reduction reactions [45, 46].

Clearly, in situ FTIR, Raman, and MS-based techniques could investigate intermediate configurations and catalytic products resulting in an accurate suggestion of the electrocatalytic mechanism under experimental circumstances. Moreover, real-time characterizations can verify the reaction mechanism, which is proposed by the help of theoretical calculations.

24.3 Evaluating the Catalyst Stability/Decay

Beyond the stable performance factor of the catalyst over long-time electrolysis, structural stability of the catalyst during reaction dynamics is also considered as a critical factor for further commercial catalysis. Therefore, it is imperative to investigate to what extend the catalyst surface is stable. The most important sites of catalyst surface are real active centers. As mentioned above, these sites can undergo structural reversibility; thus, using conventional characterization cannot able to figure out their stability. Hence, in situ techniques comes to monitor the stability of the real center concerning time-on-stream. XAS analysis is widely used to testify the stability of the active site. For example, the phase stability of $1T'$ MoS_2 in the FeCoNi-HNT (hybrid nanotube) catalyst was analyzed using the EXAFS method during OER and HER [47]. As presented in Figure 24.15, the Mo—Mo bond peak positions are similar across the potential range during catalysis in comparison with the as-prepared sample. This would indicate that MoS_2 phase; $1T'$, is unchanged to the typical $2H$ phase as a forward step toward the real use of this hybrid. On the other hand, the long-time structural stability suffers from the partial oxidation of MoS_2. They suggest that a small amount of added hydrazine hydrate can form a coordinate complex with hybrid resulting $1T'$ phase protection. This was confirmed by XRD patterns integrated with high-resolution transmission electron microscopy (HRTEM) imaging after 1000 OER cycles. This study successfully evaluated the stability of catalyst structure during working catalysis by making use of in situ XAS means.

One further catalyst structure was investigated in terms of stability in which the M–N–C catalyst acts as an ECR catalyst during catalytic conditions. Zhang et al. probed the electronic and atomic structure of cobalt phthalocyanine (CoPh) [48]. From the Co K-edge spectra of the catalyst in cooperating with the XANS plot, the valence state of Co^{2+} kept unchanged as well as Co configurations, as shown in Figure 24.16a,b. Separately, the stability of the Ni-single catalyst was also reported using in situ XANES under ECR electrolysis, as shown in Figure 24.16c [49]. As we can see, no apparent changes at different applied potentials indicating its structural stability. It is well known that the stability of single-atom catalysts is a crucial factor for its development as a high-performance catalyst. Therefore, such studies show up the critical advantage of in situ techniques for the stability aspect. Recently, some research works are reported offering the perspective of in situ characterizations toward structural stability of several surfaces during electrocatalysis [50, 51].

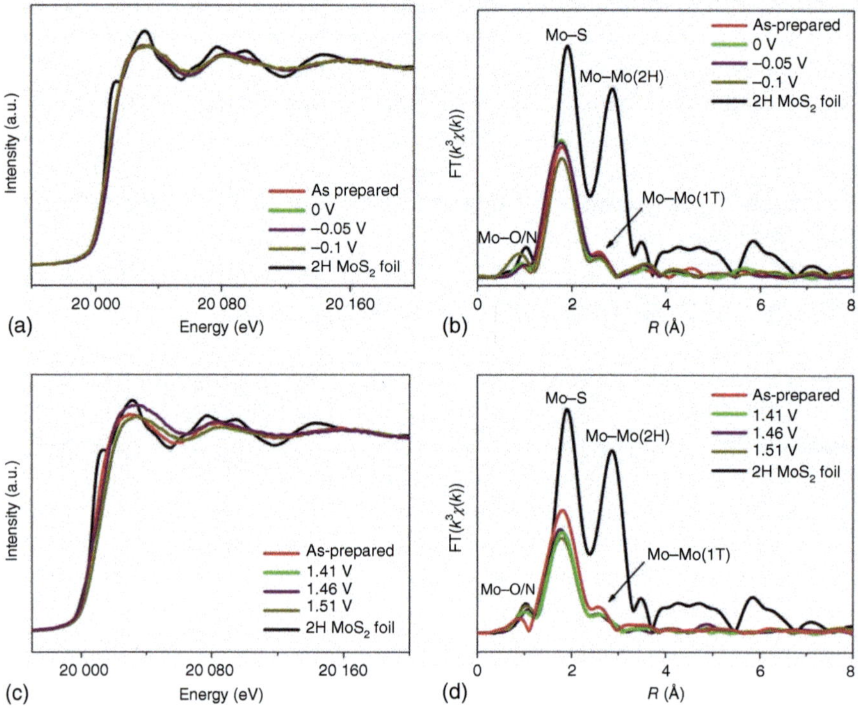

Figure 24.15 Phase stability evaluated by in situ XAX of 1T' MoS_2. The normalized XANES (a and c) and the Mo K-edge EXAFS spectra (b and d) of as-synthesized FeCoNi-HNT hybrid at different voltages: 0, −0.05 and −0.1 V vs. RHE for HER and 1.41, 1.46, and 1.51 V vs. RHE for OER. Source: Li et al. [47]. Licensed under CCBY 4.0.

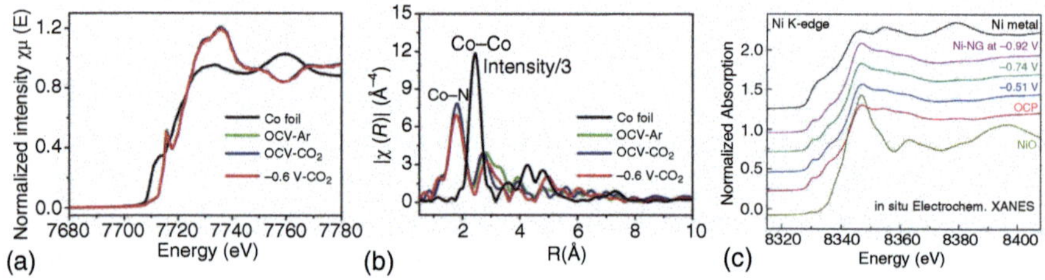

Figure 24.16 M–N–C-like materials stability interpreted by in situ XAS characterizations. (a and b) Co K-edge XANES and corresponding EXAFS profiles for real-time ECR catalysis on the CoPc sample. Source: Zhang et al. [48]. Reproduced with permission of John Wiley & Sons, Inc. (c) XANES spectra of Ni–NG under selected potentials during ECR electrolysis. Source: Jiang et al. [49]. Reproduced with permission of Royal Society of Chemistry.

On the other hand, the catalyst in the acidic or alkaline environment during applying potentials often meets difficulties in the degradation process, which is considered as a stability-related issue. Moreover, lifetimes of metals, oxides, and alloys are difficult to be predicted by typical characterizations. Real-time analysis has assisted in specifying catalyst corrosion [52]. Recently, a potential-dependent dissolution study was conducted for tungsten carbide under different potential windows using in situ inductively coupled plasma mass spectrometry (ICP-MS) [53]. It was concluded that Tungsten carbide (WC) under HER potentials is stable, whereas it suffers from dissolution during ORR/OER potentials, notably higher than 0.80 V vs. RHE, as shown in Figure 24.17a,b. Another work, provided by the same group, is to compare the corrosion of metals and its corresponding carbide, sulfide, and phosphide in HER catalysis, as shown in Figure 24.17c,d. Without current flow, metals have been dissolved at a higher rate compared to other corresponding catalysts while they show similar degradation degree under negative voltages. Overall, presented data suggest the ability of in situ tools, such as XAS and ICP-MS, to detect the dissolution factor to further circumvent this problem leading to pilot-level catalysis.

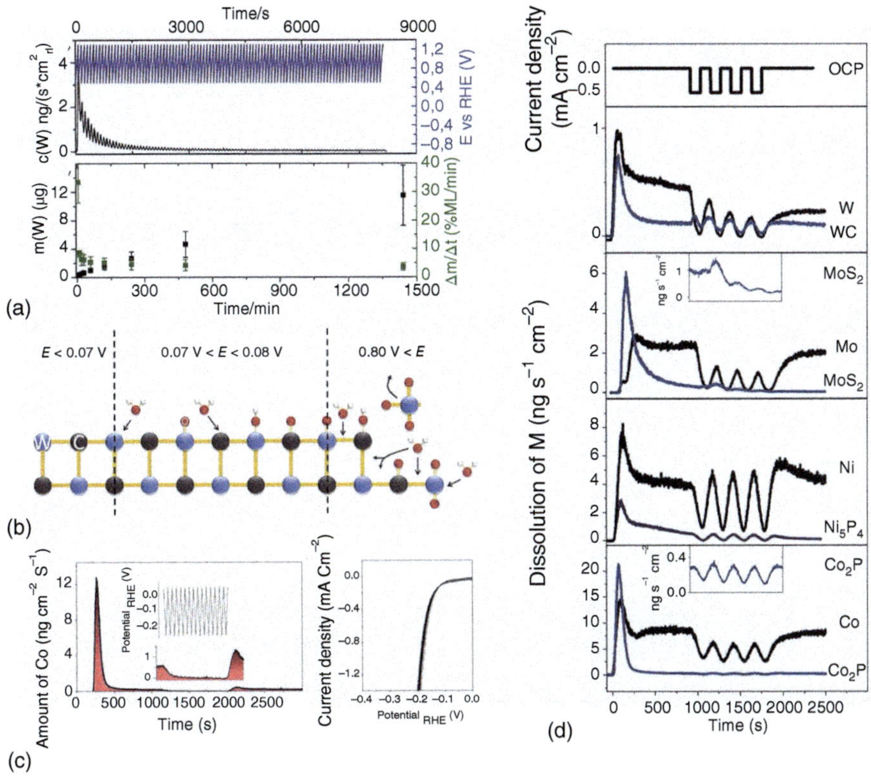

Figure 24.17 In situ probing catalyst decay using ICP-MS measurement: (a) rate profile of dissolved W (black) in WC nanoparticles and differential rate (green) under applying a potential of 1.00 V vs. RHE for 24 hours. (b) Schematic representation of different WC surface states at different potentials. Source: Göhl et al. [53]. Reproduced with permission of Elsevier. (c) The dissolution profile of Co_2P at open-circuit potential and subsequent sweeping potential from 0.05 to −0.25 V vs. RHE and corresponding current density response to the applied potential. (d) The dissolution profiles of WC/W, Mo/MoS_2, Ni/Ni_5P_4, and Co/Co_2P. Source: Ledendecker et al. [54]. Reproduced with permission of John Wiley & Sons, Inc.

24.4 Revealing the Interfacial-Related Insights

The environment around the electrode surface consists of large concentration gradients from reactants and ions. The interfacial pH may be maintained by adding buffer solutions; however, there is no clear evidence about its potential-dependent limitations. In addition, to search for a suitable electrolyte, the concentration difference between local and bulk media should be minimized as much as it does. A most recent study, by in situ developed IR mode, surface enhanced infrared absorption (SEIRA), has described the pH impact near the electrode surface at potentials relevant to hydrocarbon formation in the presence of phosphate buffer

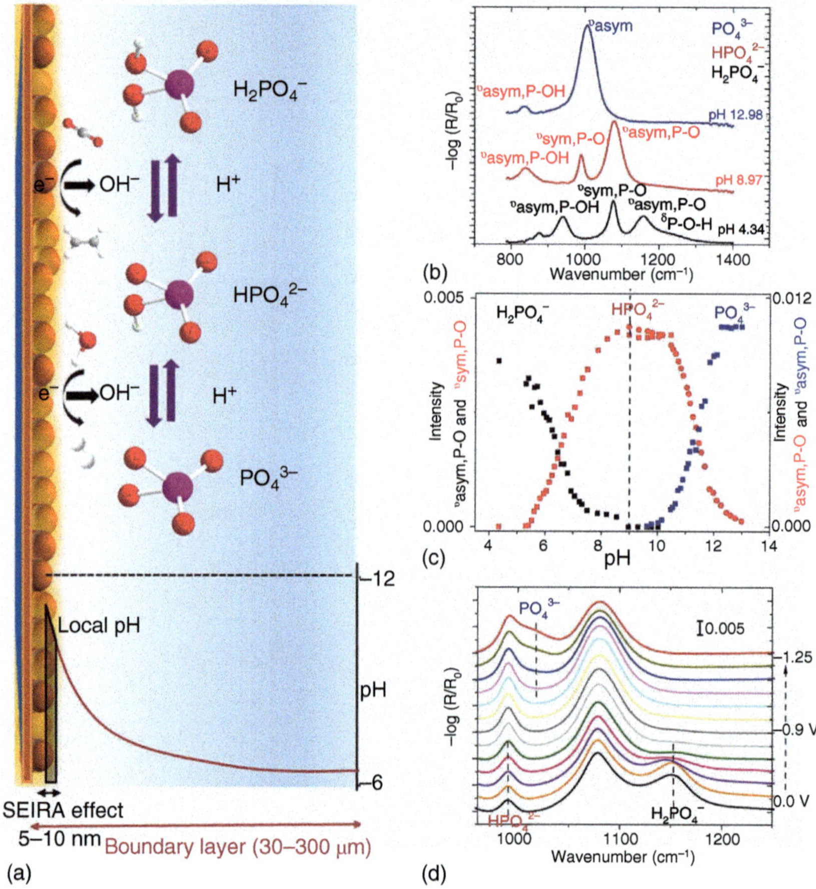

Figure 24.18 (a) Representative scheme of buffer reactions, pH gradient, and probed area by SEIRA. (b) SEIRA spectra at different pH, indicating the dominating species of phosphate solutions. (c) Equilibrium of phosphate species formulated by absorption intensity in SEIRA spectra. (d) Potential-dependent changes in phosphate species concentrations near the electrode surface. Source: Yang et al. [55]. Reproduced with permission of American Chemical Society.

or ECR system [55]. The active thickness for SEIRA measurements was 5–10 nm from the surface, as represented in Figure 24.18a. Different phosphate-related peaks were identified at varied pH, as shown in Figure 24.18b. The ratios of three phosphate species peaks were used for pH record, as shown in Figure 24.18c. The change in spectra then detected at potentials of up to −1.25 V. Local pH was calculated as a function of current density, phosphate concentrations, and potential. It has been concluded that a given buffer concentration shows its limitations to prevent pH change at a current density of less than 10 mA cm^{-2}. In addition, at the higher potential, it gives rise to enhance HER, which is an undesired product under ECR catalysis. Similar work was reported by Dunwell et al. to monitor potential-dependent pH on the ECR model using the SEIRA tool for bicarbonate electrolyte [56]. These investigations support the advantages of in situ analysis to gain more insights about the interfacial layer and its catalytic surroundings.

One another useful advance of in situ characterization is to measure near-surface H^+ ion concentration changes during electrocatalysis. However, proton ions should act as reactants or products as in methanol oxidation methanol oxidation reaction (MOR) or ORR catalysis to be measured using in situ Raman spectroscopy. Chen et al. quantified the produced or consumed proton concentration around Pt nanoparticles in 0.1 M of $HClO_4$ under MOR and ORR conditions, as shown in Figure 24.19 [57]. Protons have no band in Raman shift; however, the counter ion ClO_4^{-1} has a vibrational peak at 930 cm^{-1}. Thus, the idea was formulated to use the concentration change of counter ion as a measured value for proton concentration variations. The study clarifies by representative cases that in situ Raman can be used for $HClO_4$-based electrolyte catalysis to evaluate proton consumption or generation as reaction proceeds. Interfacial environment-driven insights have rarely been described by in situ analysis; therefore, such mentioned findings can open new avenues for in situ application toward this aspect.

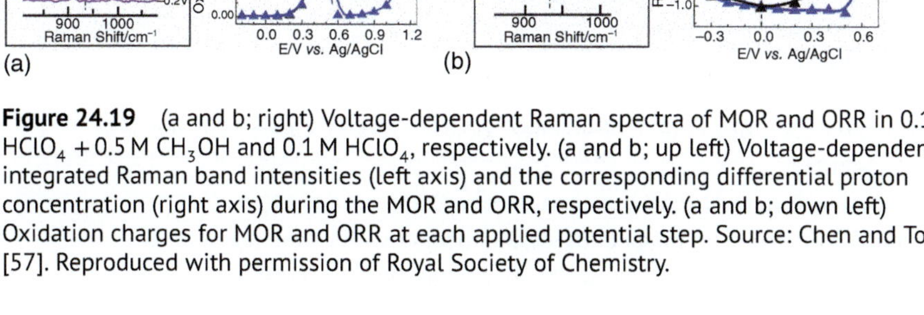

Figure 24.19 (a and b; right) Voltage-dependent Raman spectra of MOR and ORR in 0.1 M $HClO_4$ + 0.5 M CH_3OH and 0.1 M $HClO_4$, respectively. (a and b; up left) Voltage-dependent integrated Raman band intensities (left axis) and the corresponding differential proton concentration (right axis) during the MOR and ORR, respectively. (a and b; down left) Oxidation charges for MOR and ORR at each applied potential step. Source: Chen and Tong [57]. Reproduced with permission of Royal Society of Chemistry.

24.5 Conclusion

Studying the electrocatalytic cycle in the view of in situ techniques is a growing interest since the understanding of potential-induced reactions occurring at, in most cases, gas–solid–liquid interface is still insufficient in many systems. In situ tools that we can use in catalysis include instrumental considerations including experimental design. The instrument-related aspects are governed by the electrochemical setup, operation mode, software, and data recording and handling. As many operation modes have been developed to meet the requirements of catalytic studies, therefore, we have provided an introductory for common in situ techniques simply showing the working definition, their advantages, cell schemes, and available modes related to each one.

Given the complexity of in situ setups, what is inspirational and insightful is the variety of electrocatalytic systems that have been applied in situ. Thus, we have demonstrated (i) how in situ surface-sensitive techniques can be used to monitor and visualize the structural evolution of catalysts under reaction conditions, as they help to provide critical insights toward performance–structure correlations and identify the real activity origin, (ii) their feasibility to qualitatively determine the short-timing intermediate phases and effectively facilitate the interpretation of the reactant's adsorption and activation, (iii) their ability to quantitatively measure some catalytic products with a comparative degree of success, (iv) their validity to indicate to what extend the catalyst is stable or suffering from decay which is useful for commercial usage, and (v) their chance to study the local environment at electrode surface. All such unique information is tightly related to catalytic reactivity and performance; thus, applying these in situ measurements of potential-dependent changes during catalysis would present insightful aspects that help further to explore new efficient catalysts ready for industry usage.

As some groundwork for applying in situ characterization in catalytic field has been laid, it opens a door to be extensively utilized by a wider community. Simply put, the future is in situ techniques. However, the key matter in whether or not you can do an experiment in situ is the cost. To date, there is no commercially available in situ instrument, in addition the design of experimental cell and operating mode should be tailored toward specific reaction process. Overall, the engineering of electrochemical experimentation with highly sensitive in situ design become a demand as it has clearly demonstrated both qualitative and quantitative potential of in situ characterizations leading to our future prosperity.

References

1 Norskov, J.K., Bligaard, T., Hvolbaek, B. et al. (2008). *Chem. Soc. Rev.* 37: 2163–2171.
2 Hammer, B. and Norskov, J. (2000). *Adv. Catal.* 45: 71–129.
3 Pan, Y., Zhang, C., Liu, Z. et al. (2020). *Matter* 2: 78–110.
4 Gu, J., Hsu, C.S., Bai, L. et al. (2019). *Science* 364: 1091–1094.

5 Weng, Z., Wu, Y., Wang, M. et al. (2018). *Nat. Commun.* 9: 415.
6 Huang, D., Luo, Y., Li, S. et al. (2020). *Mater. Horiz.* 7: 970–986.
7 Zheng, X., Zhang, B., De Luna, P. et al. (2018). *Nat. Chem.* 10: 149–154.
8 Dutta, A., Kuzume, A., Rahaman, M. et al. (2015). *ACS Catal.* 5: 7498–7502.
9 Jia, Q., Ramaswamy, N., Hafiz, H. et al. (2015). *ACS Nano* 9: 12496–12505.
10 Ye, K., Zhou, Z., Shao, J. et al. (2020). *Angew. Chem. Int. Ed.* 59: 4814–4821.
11 Pan, Y., Lin, R., Chen, Y. et al. (2018). *J. Am. Chem. Soc.* 140: 4218–4221.
12 Yang, H.B., Hung, S.-F., Liu, S. et al. (2018). *Nat. Energy* 3: 140–147.
13 Luo, M., Wang, Z., Li, Y.C. et al. (2019). *Nat. Commun.* 10: 5814.
14 Li, X., Wang, S., Li, L. et al. (2020). *J. Am. Chem. Soc.* 142: 9567–9581.
15 Wahl, S., El-Refaei, S.M., Amsalem, P. et al. (2020). *Cat. Sci. Technol.* 10: 517–528.
16 Genovese, C., Schuster, M.E., Gibson, E.K. et al. (2018). *Nat. Commun.* 9: 935.
17 Deng, Y., Handoko, A.D., Du, Y. et al. (2016). *ACS Catal.* 6: 2473–2481.
18 Casalongue, H.G.S., Benck, J.D., Tsai, C. et al. (2014). *J. Phys. Chem. C* 118: 29252–29259.
19 Cao, L., Luo, Q., Liu, W. et al. (2019). *Nat. Catal.* 2: 134–141.
20 Lu, Y., Wang, J., Yu, L. et al. (2019). *Nat. Catal.* 2: 149–156.
21 Nam, D.H., Bushuyev, O.S., Li, J. et al. (2018). *J. Am. Chem. Soc.* 140: 11378–11386.
22 Favaro, M., Drisdell, W.S., Marcus, M.A. et al. (2017). *ACS Catal.* 7: 1248–1258.
23 Gao, D., Zhou, H., Cai, F. et al. (2017). *Nano Res.* 10: 2181–2191.
24 Sheng, W., Kattel, S., Yao, S. et al. (2017). *Energy Environ. Sci.* 10: 1180–1185.
25 Tung, C.-W., Hsu, Y.-Y., Shen, Y.-P. et al. (2015). *Nat. Commun.* 6: 8106.
26 Kim, Y.-G., Baricuatro, J.H., Javier, A. et al. (2014). *Langmuir* 30: 15053–15056.
27 Grosse, P., Gao, D., Scholten, F. et al. (2018). *Angew. Chem. Int. Ed.* 57: 6192–6197.
28 Joya, K.S. and Sala, X. (2015). *Phys. Chem. Chem. Phys.* 17: 21094–21103.
29 Deng, Y. and Yeo, B.S. (2017). *ACS Catal.* 7: 7873–7889.
30 Pelliccione, C.J., Timofeeva, E.V., Katsoudas, J.P., and Segre, C.U. (2013). *J. Phys. Chem. C* 117: 18904–18912.
31 Hidalgo, R., Ash, P.A., Healy, A.J., and Vincent, K.A. (2015). *Angew. Chem. Int. Ed.* 54: 7110–7113.
32 Mistry, H., Varela, A.S., Bonifacio, C.S. et al. (2016). *Nat. Commun.* 7: 12123.
33 Qin, H., Lin, L., Jia, J. et al. (2017). *Phys. Chem. Chem. Phys.* 19: 30749–30755.
34 Chang, C.J., Chu, Y.C., Yan, H.Y. et al. (2019). *Dalton Trans.* 48: 7122–7129.
35 Firet, N.J. and Smith, W.A. (2016). *ACS Catal.* 7: 606–612.
36 Liu, Y., Chen, S., Quan, X., and Yu, H. (2015). *J. Am. Chem. Soc.* 137: 11631–11636.
37 Dong, J.C., Zhang, X.G., Briega-Martos, V. et al. (2019). *Nat Energy* 4: 60–67.
38 Yang, H., Hu, Y.-w., Chen, J.-j. et al. (2019). *Adv. Energy Mater.* 9: 1901396.
39 Mandal, L., Yang, K.R., Motapothula, M.R. et al. (2018). *ACS Appl. Mater. Interfaces* 10: 8574–8584.
40 Opitz, A.K., Nenning, A., Rameshan, C. et al. (2017). *ACS Appl. Mater. Interfaces* 9: 35847–35860.

41 Kai, T., Zhou, M., Duan, Z. et al. (2017). *J. Am. Chem. Soc.* 139: 18552–18557.

42 Handoko, A.D., Wei, F., Jenndy et al. (2018). *Nat. Catal.* 1: 922–934.

43 Todoroki, N., Tei, H., Tsurumaki, H. et al. (2019). *ACS Catal.* 9: 1383–1388.

44 Clark, E.L. and Bell, A.T. (2018). *J. Am. Chem. Soc.* 140: 7012–7020.

45 Görlin, M., Chernev, P., Ferreira de Araújo, J. et al. (2016). *J. Am. Chem. Soc.* 138: 5603–5614.

46 Kwon, H.C., Kim, M., Grote, J.-P. et al. (2018). *J. Am. Chem. Soc.* 140: 16198–16205.

47 Li, H., Chen, S., Zhang, Y. et al. (2018). *Nat. Commun.* 9: 2452.

48 Zhang, Z., Xiao, J., Chen, X.J. et al. (2018). *Angew. Chem. Int. Ed.* 57: 16339–16342.

49 Jiang, K., Siahrostami, S., Zheng, T. et al. (2018). *Energy Environ. Sci.* 11: 893–903.

50 Jia, Q., Liang, W., Bates, M.K. et al. (2015). *ACS Nano* 9: 387–400.

51 Gong, M., Zhu, J., Liu, M. et al. (2019). *Nanoscale* 11: 20301–20306.

52 Kasian, O., Geiger, S., Mayrhofer, K.J.J., and Cherevko, S. (2019). *Chem. Rec.* 19: 2130–2142.

53 Göhl, D., Mingers, A.M., Geiger, S. et al. (2018). *Electrochim. Acta* 270: 70–76.

54 Ledendecker, M., Mondschein, J.S., Kasian, O. et al. (2017). *Angew. Chem. Int. Ed.* 56: 9767–9771.

55 Yang, K., Kas, R., and Smith, W.A. (2019). *J. Am. Chem. Soc.* 141: 15891–15900.

56 Dunwell, M., Yang, X., Setzler, B.P. et al. (2018). *ACS Catal.* 8: 3999–4008.

57 Chen, D.J. and Tong, Y.Y. (2015). *Chem. Commun.* 51: 5683–5686.

Part X

Electrochemical Catalytic Carbon Cycle

Modern manual production depends heavily on nonrenewable fossil fuels to provide raw materials and electric power. At the same time, excessive CO_2 emissions derived from increasing consumption of fossil fuels have induced several environmental and ecological risks [1]. A plenty of technologies have been developed to solve these problems, such as renewable energies (accompanied with electricity storage devices) and fuel cells for providing electricity power with reduced carbon footprint, and CO_2 capture and fixation/conversion for lowering CO_2 concentration in atmosphere and water [2–4]. Though excellent achievements have been made, point-by-point solutions with independent techniques going their own ways cannot balance natural carbon cycle (NCC) in a controllable and overall effective way [5].

In an overall and essential view of carbon destiny on the earth, increasing the consumption of fossil fuels and resulting in excessive CO_2 emissions have caused two severe bottlenecks that unbalance the global NCC. In a typical NCC, the capture and fixation of nomadic CO_2 are implemented by plants driven by solar energy. Fixed carbon in plants turns into fossil fuels after a timescale of millions of years. CO_2 is liberated into the atmosphere, mainly through the consumption of plants and fossil fuels by life entities and production activities [6]. However, since the first industrial revolution, the formation rate of fossil fuels had been increasingly lower than its consumption rate, leading to one bottleneck to NCC. Meanwhile, the liberation rate of CO_2 had been increasingly higher than its fixation rate, leading to the other bottleneck that unbalances NCC. The essential reason for the unbalance of NCC is the slow formation but rough consumption of fossil fuels, the latter of which is determined by the increasing requirement of and strong dependence on current human activities. This anthropogenic unbalance of NCC destiny affects not only the normal living and production activities of humans but also the entire economic system on earth, threatening the sustainable survival of human society [7].

A global anthropogenic interference is thus a promising solution to recover NCC more effectively because it takes an overall view to tune the rate of every section, which influences each other in the cycle and forms an organic whole together.

Electrocatalysis in Balancing the Natural Carbon Cycle, First Edition. Yaobing Wang.
© 2021 WILEY-VCH GmbH. Published 2021 by WILEY-VCH GmbH.

Specifically, an anthropogenic supplement by electrochemistry, electrochemical carbon cycle (ECC), shows high potential, which takes advantages of high rate, mild condition, and environmentally friendly of electrochemistry to omit fossil fuels in the entire cycle. A typical ECC includes electrochemical CO_2 conversion to chemicals/fuels (CO_2 fixation) and electrochemical oxidation of chemicals and fuels (CO_2 liberation) along with CO_2 capture and the transport of chemicals/fuels. The inherent characters of using ECC to supplement NCC can thus be described: (i) Captured CO_2 supports high-rate CO_2 fixation in CO_2 electrolytic cell, releasing the CO_2 accumulation bottleneck of NCC; (ii) chemicals and fuels can directly and quickly be produced from CO_2 electrolytic cell rather than be reformed from fossil fuels that need million years to be formed; (iii) the universal application of fuel cells provides electricity that is greener and more controllable compared to current power plants based on fossil fuels and coal; (iv) when carbon is cycled by ECC, renewable energies are stored in chemical bonds and finally utilized by end consumers. Therefore, ECC is of high promise to supplement NCC because of its high controllability, high rate, mild condition, and environmentally friendly.

In this part, we aim to propose an anthropogenic ECC solution to supplement unbalanced NCC from a global viewpoint. The latest advances in four sections of ECC are briefly presented to demonstrate the fundamental feasibility of ECC. The focuses are electrochemical CO_2 reduction (ECDRR) and fuel oxidation that implement key chemical conversions and form CO_2/CO, $CO_2/HCOOH$, CO_2/CH_3OH, and CO_2/CH_3CH_2OH cycles. After a brief mechanism introduction, the advances in critical electrocatalysts and performance for ECDRR and electrochemical fuel oxidation are presented. In addition, the characters in advanced CO_2 capture and fuel transport are briefly discussed. Then, several key indexes are proposed to evaluate the overall efficiency of ECC. In addition, some external managements are introduced to support a high-efficiency ECC. Finally, several general principles are proposed to guide the simple and high-efficiency ECC.

References

1 Kibria, M.G., Edwards, J.P., Gabardo, C.M. et al. (2019). *Adv. Mater.* 31: 1807166.
2 Zhang, B. and Sun, L. (2019). *Chem. Soc. Rev.* 48: 2216–2264.
3 Asset, T. and Atanassov, P. (2020). *Joule* 4: 33–44.
4 Xie, J., Zhou, Z., and Wang, Y. (2020). *Adv. Funct. Mater.* 30: 1908285.
5 Keijer, T., Bakker, V., and Slootweg, J.C. (2019). *Nat. Chem.* 11: 190–195.
6 Ross, M.B. (2019). *Joule* 3: 1814–1816.
7 Rothman, D.H. (2019). *Proc. Natl. Acad. Sci.* 116: 14813–14822.

25

Electrochemical CO_2 Reduction to Fuels

Almost all high-energy carbon-based chemicals and fuels can be generated from electrochemical carbon dioxide reduction reaction (ECDRR) as the carbon in CO_2 holds the lowest energy in nature. Such low energy accompanied by the inertia of CO_2 results in high activation barrier for CO_2 obtaining an electron to form the anion radical, though the theoretical potentials of ECDRR to common chemical and fuels are near 0 V vs. standard hydrogen electrode (SHE) (Table 25.1). One necessity of applying electrocatalysts is that they can tune ECDRR to go a lower-energy pathway based on proton-coupled electron transfer mechanism, promoting energy efficiency and production rate [3]. Another necessity of applying electrocatalysts is derived from utilizing water as a proton source, which introduces a competitive by-reaction, hydrogen evolution reaction (HER), with a standard reduction potential of 0 V vs. SHE [4, 5].

According to these two reasons, an ideal ECDRR catalyst should enable CO_2 electrolysis to achieve high selectivity and current density at low overpotential. There is a classical classification of ECDRR catalysts according to products: (i) for highly selective HCOOH production such as Pb, In, Sn, and Bi; (ii) for dominant CO production such as Au, Ag, and Zn; (iii) for considerable hydrocarbons: only Cu; and (iv) for dominant H_2, including Ni, Fe, Pt, and Ti [6]. With decades of development on mechanism and synthesis studies, we now go deeper into the inherent properties (e.g. electron structure and surface structure) of four catalysts and broaden the ideal ECDRR catalyst candidates to an extent by breaking these inherent properties, which have been summarized by some excellent reviews [7–9]. Here, we present cases to show how advanced design improves carbon-based product generation, specifically including CO, HCOOH, CH_3OH, and CH_3CH_2OH, which have been applied as fuel.

For CO generation from ECDRR, noble metals Au and Ag that are active in bulk phase can significantly enhance their activity and selectivity with lower overpotential when nanostructures are introduced because of abundant nano effects. For example, Au needles promoted CO generation with 95% Faradaic efficiency (FE) and over 14 mA cm^{-2} at only 240 mV overpotential [10]. Sharp tip structure on needles concentrated K^+ around the catalyst surface. In this case, sharp tip created highly active sites, and CO_2 could be quicker stabilized on these sites via a strong interaction with the local electric field, breaking the limit of

Electrocatalysis in Balancing the Natural Carbon Cycle, First Edition. Yaobing Wang.
© 2021 WILEY-VCH GmbH. Published 2021 by WILEY-VCH GmbH.

Table 25.1 The theoretical reduction potential of aqueous ECDRR to several chemicals and fuels at pH = 0, 1 atm, and 25 °C.

Reaction	Potential (V vs. SHE)
$CO_2 + e^- \rightarrow CO_2^{\bullet-}$	−1.9
$CO_2 + 2e^- + 2H^+ \rightarrow CO + H_2O$	−0.1
$CO_2 + 2e^- + 2H^+ \rightarrow HCOOH$	−0.12
$CO_2 + 6e^- + 6H^+ \rightarrow CH_3OH + H_2O$	0.03
$2CO_2 + 8e^- + 8H^+ \rightarrow CH_3COOH + 2H_2O$	0.11
$2CO_2 + 12e^- + 12H^+ \rightarrow CH_3CH_2OH + 3H_2O$	0.09

Sources: Nitopi et al. [1]; Ma et al. [2].

low CO_2 solubility in water [11]. Another example Pd showed a strong activity and selectivity to CO production at a small enough size, typically 3.7 nm enables 91.2% FE and ~9 mA cm^{-2} of current density at 780 mV of overpotential [12]. This size-dependent CO production on Pd was ascribed to the more stable adsorption of *COOH and *CO on the corner site compared to the terrace and step sites, the ratio of which can be tuned by the particle size. An extreme example Ir, which usually favored HER, was recently found to promote CO production with 97.6% of FE at 550 mV of overpotential and 18.6 mA cm^{-2} of current density, when atomic Ir was supported by $Co(OH)_2$ [13]. Transition metals, including Zn and Cu, also enhanced CO generation with high selectivity and activity at lower overpotential by nanostructure design, enabling stable adsorption of intermediates [14–18]. Even transition metal Ni, Fe, and Co can also be excellent candidates for CO generation when introducing atomic dispersion anchored by nitrogen-doped carbon materials [19–24]. Especially Ni-N_x-based carbon materials, which have shown >95% of FE for CO generation at current densities of 100 mA cm^{-2} though relatively high overpotentials (typically 500 mV) and designed flow cells were required [22, 25]. Metal-free materials have also been discovered with excellent catalytic property for CO generation, especially with heteroatom dopants. Nitrogen and sulfur co-doped hierarchically porous carbon nanofiber (NSHCF) showed 94% of CO generation with a high current density of 103 mA cm^{-2} at −0.7 V of overpotential [26]. Authors ascribed this high performance, especially high current densities, to (i) the co-doping of N and S enhancing *COOH formation and (ii) hierarchically porous structure enabling abundant accessible active sites (Table 25.2).

For formic acid/formate generation from ECDRR, Pb, Hg, In, Sn, and Bi have shown catalytic properties in bulk phase with high selectivity of over 75% of FE while high overpotentials were required [12]. Oxide-derived nanocrystalline metals, including Pb, In, Sn, and Bi, have shown higher selectivity to formate generation with inhibited HER compared to polycrystalline metals [14, 29, 30]. For example, oxide-derived nanocrystalline Pb showed 700-fold lower HER activity while similar formate generation activity compared to polycrystalline Pb foil, leading to a nearly quantitative FE for formate generation [31]. Authors ascribed this high selectivity

Table 25.2 Typical catalysts for ECDRR to CO.

Catalyst	Structure–activity relationship	Reactor and catholyte	Applied potential (V_{RHE}[a])	FE (%)	Activity	Stability
Au needles [10]	Sharp tip structure builds strong electric field that concentrates CO_2	H-cell, 0.5 M $KHCO_3$	−0.35	95	15 mA cm^{-2}	8 h
Triangular Ag nanoplates [27]	Dominant Ag(100) enables the low-energy pathway and sufficient active edge sites	H-cell, 0.1 M $KHCO_3$	−0.206	8.0	NM[b]	NM
			−0.856	96.8	~1.3 mA cm^{-2}	14 h
3.7-nm Pd [27]	β-PdH_x stabilizes COOH intermediate	H-cell, 1 M $KHCO_3$	−0.7	93.4	~25 mA cm^{-2}	10 h
Atomic Ir [13]	Atomic Ir on synergic $Co(OH)_2$ support facilitates electrokinetics	H-cell, 0.5 M $KHCO_3$	−0.66	97.6	18.6 mA cm^{-2}	24 h
Copper atom-pair catalyst [14]	Cu_1^{x+} adsorbs H_2O and the neighbor Cu_1^{0} adsorbs CO_2	H-cell, 0.2 M $NaHCO_3$	−0.78	92	9.3 mA cm^{-2}	3 h
Highly porous Zn [15]	Highly porous structure creates abundant active sites and high local pH	H-cell, 0.1 M $KHCO_3$	−0.95	94.4	27 mA cm^{-2}	10 h
		Flow cell, 1 M KOH	−0.64	84	200 mA cm^{-2}	5.5 h
N-anchored Zn single-atom catalyst [16]	Zn-N_4 active site structure enables low-energy *COOH formation	H-cell, 0.5 M $KHCO_3$	−0.43	95	4.8 mA cm^{-2}	75 h

(continued)

Table 25.2 (Continued)

Catalyst	Structure–activity relationship	Reactor and catholyte	Applied potential (V_{RHE} a))	FE (%)	Activity	Stability
C–Zn_1Ni_4 ZIF-8 (Zeolitic imidazolate framework) [19]	Coordinatively unsaturated Ni–N site structure favors ECDRR over HER	H-cell, 1 M $KHCO_3$	−0.43	~85	~4 mA cm^{-2}	NM
			−0.63	97.8	80 mA cm^{-2}	12 h
Fe^{3+}–N–C [14]	Fe^{3+} sites enable faster CO_2 adsorption and weaker CO absorption	H-cell, 0.5 M $KHCO_3$	−0.37	90	8 mA cm^{-2}	12 h
		Flow cell, 0.5 M $KHCO_3$ catholyte, 1 M KOH anolyte	−0.41	90	50 mA cm^{-2}	28 h
Self-assembly of tetra(4-pyridyl) porphyrin (STPyP)-Co [20]	High percentage of Co dz^2 in highest-occupied molecular orbital (HOMOs) lowers CO_2 activation energy	H-cell, 0.5 M $KHCO_3$	−0.62	96	6.5 mA cm^{-2}	48 h
Hydrophobic exfoliated MoS_2 [28]	Fluorosilane decoration lowers CO desorption and inhibits H desorption	H-cell, $EmimBF_4/H_2O$	−0.9	81.2	31 mA cm^{-2}	10 h
NSHCF [26]	N and S dopants enable lower energy demand of *COOH formation; hierarchically porous structure enables abundant accessible active sites	H-cell, 0.1 M $KHCO_3$	−0.7	94	103 mA cm^{-2}	36 h

a) V_{RHE}: V vs. reversible hydrogen electrode (RHE).
b) NM: not mentioned.

to the higher coverage of possibly a thin and kinetically stable Pb oxide or hydroxide layer, which may arise from the abundant defects derived from the reduction of Pd oxide. Recent density functional theory (DFT) calculations with Bi as model catalyst found a much lower ΔG for *OCHO intermediate on defective Bi compared to ideal Bi surface, contributing to a high formate FE of near 100% [32]. Besides the reconstruction of the metal surface, a more thoroughly decreasing size of materials can create highly active and selective sites for ECDRR. Such as bulk Co, on which formate showed the highest FE of ~53% at −0.89 V. When the size of Co decreased to four-atom-thick layer, formate presented a higher maximum FE of up to ~72% at −0.85 V, which further increased to 90% when the partial oxide is introduced [33]. Tafel analysis suggested a stronger stabilization of $^*CO_2$ intermediate on Co site in atomic layers and facilitated proton transfer as the rate-determining step derived from Co oxide. When heteroatoms are introduced, a further higher activity and selectivity of formate generation could be obtained, exemplifying with S-doped oxide-derived catalyst, on which formate generation exhibited high FE over 80% with a wide range of current densities from 0 to 100 mA cm^{-2} [34]. DFT calculations suggested that S sites enabled an easier *H formation, which turned the HCOO* formation to be exergonic compared to still endergonic formation of *H from water alone. Heteroatom doping strategy is specifically effective for metal-free carbon-based materials, such as N doping for PC61BM, a C60 derivative (Table 25.3) [39].

In general, when *CO intermediate is moderately stabilized on active sites, it would further accept proton and electron rather than desorb or block the site, leading to key intermediate *CHO/*CHOH formation and the last generation of products that received over two electrons. Cu has been one of the most promising catalysts that implements these over two-electron reduction processes, though bulk Cu shows low selectivity to single products [36]. Aimed at highly selective CH_4 formation, nanotechnology showed great promise. For example, dispersed Cu nanoparticles with a size of 7 nm showed CO_2-to-CH_4 selectivity up to 80% of FE at −1.25 V [40]. Authors found that a relatively thin layer of initially evaporated Cu film on substrate would result in isolated nanoscale aggregates during catalysis and showed high CH_4 selectivity, while a thicker layer would produce connected networks and show low CH_4 yield. This work implies a way to achieve high CH_4 selectivity by building and keeping isolated Cu nanoparticles during catalysis. On the other hand, facet control also works for CH_4 production on nanoscale Cu. Such as nanotwinned Cu utilized abundant twin boundaries to enhance the conversion rate of absorbed *CO to CH_4 [41]. Moreover, catalysts without Cu have also been investigated. Ultrathin $MoTe_2$ layers took advantage of abundant exposed active sites and total density of state derived from ultrathin morphology to achieve CH_4 selectivity of 83% FE, though the catalysis was conducted in ionic liquid electrolyte [42]. Even metal-free carbon-based materials can catalyze CO_2-to-CH_4 conversion. Perfluorinated covalent triazine framework (fluorinated covalent triazine framework [FN-CTF-400]) catalyzed a record of 99.3% of FE at potentials from −0.7 and −0.9 V in the aqueous electrolyte, though current density was limited compared

Table 25.3 Typical catalysts for ECDRR to $HCOOH/HCOO^-$.

Catalyst	Structure–activity relationship	Reactor and catholyte	Applied potential	FE (%)	Activity ($mA\,cm^{-2}$)	Stability (h)
Oxide-derived (OD)–Pb [31]	High-density defects may induce higher coverage of thin, metastable Pd^{2+} layer that inhibits HER while facilitates formate generation	H-cell, 0.5 M $NaHCO_3$	$-0.8\ V_{RHE}$	95	~0.15	75
Nanotubular-derived Bi [32]	Abundant defective Bi sites stabilize *OCHO intermediates	H-cell, 0.5 M $KHCO_3$	$-0.82\ V_{RHE}$	>98	~36	48
		Flow cell, 1 M KOH	$-0.58\ V_{RHE}$	98	~210	11
Partially oxidized atomic Co layers [33]	Both atomic layer and partial oxidation enhances intrinsic activity	H-cell, 0.1 M Na_2SO_4	$-0.85\ V_{RHE}$	90.1	10.59	60
3.7-nm Pd nanoparticles [35]	Edge sites turned into $\alpha + \beta$ $PdH_x@PdH_x$ at above −0.2 V, facilitating HCOO* intermediates	H-cell, 1 M $KHCO_3$	$-0.1\ V_{RHE}$	98	~17	1[a)]
$CuSn_3$ alloy [36]	Charge transfer from Sn to Cu intermediates	H-cell, 0.1 M $KHCO_3$	$-0.5\ V_{RHE}$	95	33	50
Single-atom $Sn^{\delta+}$ on N-doped graphene [37]	Positively charged Sn atoms stabilize $CO_2^{\bullet -*}$ and $HCOO^-*$, and N dopants weaken $HCOO^-*$ bonding strength with Sn	H-cell, 0.25 M $KHCO_3$	$-1.6\ V_{SCE}$[b)]	74.3	~12	200
S_2-In [34]	Adsorbed S^{2-} help *H formation via interacting with hydrated metal cations in the double layer	H-cell, 0.5 M $KHCO_3$	$-0.98\ V_{RHE}$	93	60	10[c)]
B-doped nanodiamond [38]	Rb^+ may affects double-layer structure	H-cell, 0.075 M Rb^+	NM[d)]	~80	2	48
N-C61-800 [39]	N dopant on fullerene derivatives enhances the kinetics	H-cell, 0.5 M $KHCO_3$	$-0.9\ V_{RHE}$	91	11.6	12

a) Longer stability was shown with purging O_2 in a compact electrolytic cell.
b) V_{SCE}: potential vs. saturated calomel electrode.
c) With 0.02 mM of Na_2S addition.
d) NM: Not mentioned.

to metal catalysts (Table 25.4) [46]. Authors contributed this high selectivity to abundant F-covalent function, increasing the activity and selectivity of N sites.

When aiming at breaking limited activity and selectivity for CH_3OH generation on bulk Cu, modifying Cu with hetero metal or metalloid atoms have shown their great potential, such as Pd–Cu bimetal aerogels, which catalyzed CH_3OH production in ionic liquid/water-electrolyte with 80% of FE at 31.8 mA cm^{-2} of current density and only 240 mV of overpotential [47]. Amorphous Cu with abundant low-coordinated sites accompanied with partial oxidation was regarded as active sites enhanced by the synergistic effect between crystalline Pd and amorphous Cu. Metal-free selenium addition also demonstrated the high catalytic properties of copper selenides to CH_3OH production with 78% of FE at only 285 mV of overpotential and 42 mA cm^{-2} of current density [48]. Authors found that unsaturated Se atoms assisted Cu with lower coordination number in $Cu_{1.63}Se_{(1/3)}$ stabilizing key *CO and *CHO intermediates, leading to enhanced CH_3OH generation. Besides Cu, Co-based carbon materials have shown increasing potential for CH_3OH generation [49]. Recently, cobalt phthalocyanine (CoPc), a common catalyst for CO generation, could promote CH_3OH generation with 41% of FE at −0.94 V and 10 mA cm^{-2} immobilized by carbon nanotubes [50]. Highly conductive carbon nanotubes (CNTs) and their close interaction with CoPc were crucial for the domino pathway with *CO intermediate, which resulted in high-rate CH_3OH generation. In addition, bimetal and even metal-free catalysts also shown the ability for CH_3OH generation, including pulse-depositing (PD)–Zn/Ag foam, Pd/SnO_2, and Boron phosphide nanoparticle (BP), though there is limited current (Table 25.5) [53, 54, 57].

When stabilized *CO is close to each other, C—C bond may form between two *CO (or *CHO), leading to C^{2+} production, ethanol included. Cu is still a mostly investigated metal catalyst for ethanol production from ECDRR because of its moderate adsorption of *CO and *H. However, ethanol production suffers from low selectivity, mainly ascribed to several competing pathways to other products with the same carbon-contained intermediates [58]. Moreover, HER gains more superior than accelerated overpotentials derived from a long sequence of elemental steps to ethanol production. With the adjustment of component and structure, Cu-based catalysts can achieve selective ethanol production with FE over 50%. For example, 3D dendritic copper–cuprous oxide composite exhibited abundant Cu–Cu_2O structure on the surface, and the synergetic effect between CuI and Cu(0) improves CO_2 adsorption and CO dimerization [59]. In addition, a thick film of the 3D structure may facilitate CO dimerization by providing a relatively confined space for lengthened reaction steps. The maximum FE for ethanol production reached 56% at only −0.6 V, along with a relatively high current density of 21 mA cm^{-2}. Cu-free metal catalysts have also been developed with a successful design coupled with another metal or functional carbon materials, as shown in Table 25.6. Exemplifying Ru polypyridyl carbene catalyst on N-doped porous carbon (RuPC/NPC) and Ag-G-NCF (graphene-wrapped nitrogen-doped carbon foam) immobilized active metal sites on N-doped porous carbon [63, 64]. ECDRR to ethanol on both catalysts was assumed to go through a tandem mechanism: *CO was generated on active metal sites, followed by accepting further proton and electron transfer

Table 25.4 Typical catalysts for ECDRR to CH_4.

Catalyst	Structure–activity relationship	Reactor and catholyte	Applied potential	FE (%)	Activity	Stability
7-nm Cu nanoparticles [40]	Cu nanoparticles change the morphology during catalysis, resulting in a reaction between adsorbed CO_2 radical and a second CO_2 molecule	Flow cell, 0.1 M $NaHCO_3$	$-1.25\ V_{RHE}$	80	~9 mA cm^{-2}	1 h
Cu complex [43]	Cu complex changes to ~2-nm Cu clusters, which are active sites	H-cell, 0.5 M $KHCO_3$	$-1.06\ V_{RHE}$	66	13 mA cm^{-2}	NM[a)]
Cu-modified Ag [44]	CO forms on Ag transferred to Cu nanoparticles and is further reduced	H-cell, 0.1 M $NaHCO_3$	$-1.1\ V_{RHE}$	67	~42 mA cm^{-2}	2 h
Nanotwined Cu [41]	Twin boundary promotes absorbed CO conversion to CH_4	H-cell, 0.2 M $KHCO_3$	$-1.6\ V_{RHE}$	59	~12 mA cm^{-2}	12 h
Ultrathin $MoTe_2$ layers [42]	Abundant exposed active sites provide good electrical conductivity and high total density of state (TDOS)	H-cell, $BmimBF_4$-H_2O (mass 96 : 4)	$-1.0\ V_{RHE}$	83	25.6 mA cm^{-2}	45 h
Interconnected SnO_2 nanoparticles [42]	Grain boundary and exposed corner/step sites in interconnected SnO_2 nanoparticles contributes to the unique selectivity	Flow-cell, 1 M KOH	$-1.03\ V_{RHE}$	14	~175 mA cm^{-2}	NM
$Fe_{4.5}Ni_{4.5}S_8$ [45]	$Fe_{4.5}Ni_{4.5}S_8$ acts like CO dehydrogenase ($CODH_{Ni}$) and is sensitive to proton concentration	H-cell, porous carbon (PC)	$-1.2\ V_{NHE}$[b)]	13	~3 mA cm^{-2}	15 h[c)]
FN-CTF derived hybrid [46]	F dopant enhances the conductivity of N sites for CH_4 production	NM, 0.1 M $KHCO_3$	$-0.8\ V_{RHE}$	91.7	~0.2 mA cm^{-2}	5 h

a) NM: not mentioned.
b) V_{NHE}: V vs. normal hydrogen electrode.
c) Stability data were collected at −1.8 V vs. NHE in acetonitrile with water.

Table 25.5 Typical catalysts for ECDRR to CH_3OH.

Catalyst	Structure–activity relationship	Reactor and catholyte	Applied potential	FE (%)	Activity (mA cm^{-2})	Stability (h)
$Cu_{2-x}Se_{(y)}$ nanocatalysts [48]	Structure distortion of $Cu_{1.63}Se_{(1/3)}$ enables moderate binding strength of *CO and *CHO	H-cell, 0.5 M H_2SO_4	−2.1 V_{Ag/Ag^+}	78	42	25
$Pd_{83}Cu_{17}$ aerogel [47]	Moderate O-modified Pd–Cu surface facilitates CO_2 chemical adsorption, synergistic Pd–Cu enables CH_3OH generation, nanochain structure enhances mass transfer	H-cell, 0.5 M H_2SO_4 with 25% [Bmim]BF_4	−2.1 V_{Ag/Ag^+}	80	32	24
$Cu_2O_{(OL\text{-}MH)}$/Ppy particles [51]	Polypyrrole shell concentrates CO_2, O vacancies, and high-index facets of Cu_2O enable CH_3OH production	0.5 M $KHCO_3$[a)]	−0.85 V_{RHE}	93	0.16	15
Partially oxidized (PO)-5-nm Co/single- layer N doped graphene (SL-NG) [49]	Partially oxidized 5-nm Co and single-layer N-doped graphene both enable methanol generation, their synergistic effect allows easier generation	H-cell, 0.1 M $NaHCO_3$	−0.9 V_{SCE}	71	4	10
CoPc/CNT [50]	CoPc supported by CNTs enables the domino pathway for further reaction of *CO	H-cell, 0.1 M $KHCO_3$	−0.94 V_{RHE}	44	10.6	<12
Co-corrole-carbon paper [52]	pH = 6 enables necessary protons for methanol production with the formic acid pathway	H-cell, 0.1 M $NaClO_4$ (pH = 6)	−0.515 V_{RHE}	59	~2.5	5

(continued)

Table 25.5 (Continued)

Catalyst	Structure–activity relationship	Reactor and catholyte	Applied potential	FE (%)	Activity ($mA\ cm^{-2}$)	Stability (h)
PD–Zn/Ag foam [53]	Ag shortens Zn–Zn distance for better *CO stabilization and 3D dendrite Zn sites enable $*CH_3O$-to-CH_3OH reaction	H-cell, 0.1 M $KHCO_3$	$-1.41\ V_{RHE}$	10.5	26	8[b)]
2D hierarchical Pd/SnO_2 [54]	Partially exposed Pd adsorbs *CO and Pd–O–Sn interface facilitates *CO reduction to CH_3OH	H-cell, 0.1 M $NaHCO_3$	$-0.24\ V_{RHE}$	55	~1.5	8
FeS_2/NiS [55]	Active sites at the interface between FeS_2 and NiS suppress HER and ladder structure facilitates CH_3OH production	H-cell, 0.5 M $KHCO_3$	$-0.6\ V_{RHE}$	64	3.1	4
FeP nanoarray [56]	Synergistic effect of two adjacent Fe atoms on FeP(211) surface favors CO_2 adsorption and *CO hydrogenation	H-cell, 0.5 M $KHCO_3$	$-0.2\ V_{RHE}$	80	~2	36
Boron phosphide nanoparticles [57]	P donates electrons to B, which absorbs and activates CO_2	H-cell, 0.1 M $NaHCO_3$	$-0.5\ V_{RHE}$	92	~ −0.04	18[c)]

a) Electrolysis cell type is not mentioned.
b) Stability test on $-1.38\ V_{RHE}$.
c) Stability test on $-0.6\ V_{RHE}$ in 0.1 M $KHCO_3$.

Table 25.6 Typical catalysts for ECDRR to CH_3CH_2OH.

Catalyst	Structure–activity relationship	Reactor and catholyte	Applied potential (V_{RHE})	FE (%)	Activity	Stability
3D dendritic copper–cuprous oxide composite [59]	The synergy between Cu^{I}/Cu^{0} improves CO_2 activation and CO dimerization, the latter of which is facilitated by the 3D structure	H-cell, 0.1 M KCl	−0.6	56	$21\ mA\ cm^{-2}$	24 h
$Ag_{0.14}$–$Cu_{0.86}$ catalyst [60]	Ag–Cu alloy phase contains multiple active sites involved in ethanol production	Flow cell, 1 M KOH	−0.67	41	$250\ mA\ cm^{-2}$	2 h
ZnO–CuO shell–core bimetal oxide [61]	CO binding on Cu sites was modified by Zn	H-cell, 0.1 M $KHCO_3$	−1.15	32	$32\ mA\ cm^{-2}$	NM
		Flow-cell, 1 M KOH	−0.68	~42	$200\ mA\ cm^{-2}$	10 h
Cu–N–C material [62]	Isolated sites transiently convert into metallic copper nanoparticles as active sites	H-cell, 0.1 M $CsHCO_3$	−1.2	55	$\sim 16\ mA\ cm^{-2}$	1 h
Ag-G-NCF [63]	*CO forms and desorbs from Ag dimerizes on triple-pyridine N of graphene	H-cell, 0.1 M $KHCO_3$	−0.6	85	$\sim 78\ mg\ l^{-1}\ h^{-1}$	2 h
RuPC/NPC [64]	Synergistic effect: Ru(II) catalyst for *CO production and N-doped C for C–C coupling	H-cell, 0.5 M $KHCO_3$	−0.97	27	$\sim 0.7\ mA\ cm^{-2}$	3 h
Co-corrole-carbon paper [52]	pH = 6 enables necessary protons for ethanol production with the formic acid pathway	H-cell, 0.1 M $NaClO_4$ (pH = 6)	−0.8	48	$\sim 3.2\ mA\ cm^{-2}$ [a)]	5 h
B, N co-doped nanodiamond [65]	B dopant enhances CO_2 capture and N dopant enhances *H transfer, and their synergetic effect is crucial for ethanol production	H-cell, 0.1 M $NaHCO_3$	−1.0	93.2	$\sim 72\ \mu g\ h^{-1}\ cm^{-1}$	48 h [b)]
N-doped ordered cylindrical mesoporous carbon [66]	Pyridinic N favors *CO and C–C formation, and electron-rich surface favored proton–electron transfer reactions	H-cell, 0.1 M $KHCO_3$	−0.56	77	$\sim 0.3\ mA\ cm^{-2}$	6 h

a) Current density at 5th hour.
b) Sixteen consecutive electrolysis with each run lasting for three hours.

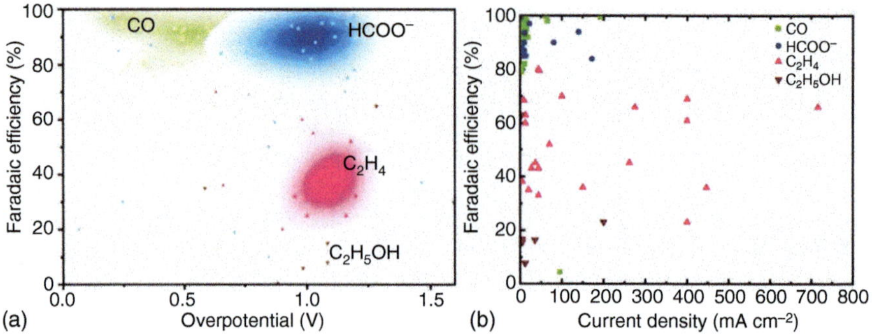

Figure 25.1 (a) Distribution of FE with corresponding overpotential. (b) Maximum FE with corresponding current density. Source: Ref. Kibria et al. [68]. Reproduced with permission of John Wiley & Sons, Inc.

on N sites in a carbon matrix, leading to final ethanol production. This design strategy could achieve high FE while relatively lower activity, typically 85% of FE and 78 mg l^{-1} h^{-1} on Ag-G-NCF. Though most metal-free catalysts contributed to two-electron-transfer products from ECDRR, e.g. CO and HCOOH, it showed a few designs that enabled *CO generation and, in the meantime, further reduction of *CO to ethanol, such as nanodiamond, the synergy between B dopant (for improved CO_2 adsorption) and N dopant (for facilitated *H transfer), moderately high N content, and high overpotential for HER on nanodiamond surface cooperated to a remarkable selective ethanol production of 93.2% of FE at −1 V, though the activity was limited (~72 g h^{-1} cm^{-1}) [65].

As presented above, advanced catalysts for CO, HCOOH, CH_3OH, and CH_3CH_2OH have been developed along with investigation on mechanisms and pathways. However, several challenges remain in the current ECDRR process for the large-scale application as a part of electrochemical carbon cycle (ECC) [67]. The most important one is that high current density, low overpotential, and high selectivity are still hard to be achieved on one catalyst except CO and HCOOH on noble metals, as shown in Figure 25.1. Gas diffuse electrode with directly supplying CO_2 gas rather than dissolved CO_2 in flow cell have recently demonstrated its potential to achieve the three indexes mentioned above for not only CO but also ethylene and acetate [69–71]. Also, catalysis stability is an important index that needs thousands of hours for large-scale applications, which is still far from current advances (as shown in Tables 25.2–25.5). A recent study found that the instability of Co complex catalyst was derived from the reduction of the phthalocyanine ligand, which can be suppressed by introducing electron-donating substituents into the ligand [50]. Another study that involved Co complex found that the stability of ECDRR catalysis was more related to the gradually decreased supply of proton at the reaction interface in the flow cell, which emphasized the effect of other components besides of catalyst itself when evaluating the stability of ECDRR at industrial current density [72].

References

1 Nitopi, S., Bertheussen, E., Scott, S.B. et al. (2019). *Chem. Rev.* 119: 7610–7672.
2 Ma, T., Fan, Q., Tao, H. et al. (2017). *Nanotechnology* 28: 472001.
3 Song, R.B., Zhu, W., Fu, J. et al. (2019). *Adv. Mater.* 32: 1903796.
4 Ooka, H., Figueiredo, M.C., and Koper, M.T. (2017). *Langmuir* 33: 9307–9313.
5 Cave, E.R., Shi, C., Kuhl, K.P. et al. (2018). *ACS Catal.* 8: 3035–3040.
6 Hori, Y.i. (2008). Heterogeneous catalysis of redox reaction. In: *Modern Aspects of Electrochemistry* (eds. J.O.'.M. Ockris, B.E. Conway and B.E. Conway), 89–189. Springer.
7 Gao, D., Arán-Ais, R.M., Jeon, H.S., and Cuenya, B.R. (2019). *Nat. Catal.* 2: 198–210.
8 Birdja, Y.Y., Pérez-Gallent, E., Figueiredo, M.C. et al. (2019). *Nat. Energy* 4: 732–745.
9 Wang, Y., Han, P., Lv, X. et al. (2018). *Joule* 2: 2551–2582.
10 Liu, M., Pang, Y., Zhang, B. et al. (2016). *Nature* 537: 382–386.
11 Jiang, H., Hou, Z., and Luo, Y. (2017). *Angew. Chem. Int. Ed.* 56: 15617–15621.
12 Gao, D., Zhou, H., Wang, J. et al. (2015). *J. Am. Chem. Soc.* 137: 4288–4291.
13 Sun, X., Chen, C., Liu, S. et al. (2019). *Angew. Chem. Int. Ed.* 58: 4669–4673.
14 Jiao, J., Lin, R., Liu, S. et al. (2019). *Nat. Chem.* 11: 222–228.
15 Luo, W., Zhang, J., Li, M., and Züttel, A. (2019). *ACS Catal.* 9: 3783–3791.
16 Yang, F., Song, P., Liu, X. et al. (2018). *Angew. Chem. Int. Ed.* 57: 12303–12307.
17 Kas, R., Hummadi, K.K., Kortlever, R. et al. (2016). *Nat. Commun.* 7: 1–7.
18 Rosen, J., Hutchings, G.S., Lu, Q. et al. (2015). *ACS Catal.* 5: 4586–4591.
19 Yan, C., Li, H., Ye, Y. et al. (2018). *Energy Environ. Sci.* 11: 1204–1210.
20 Han, J., An, P., Liu, S. et al. (2019). *Angew. Chem. Int. Ed.* 58: 12711–12716.
21 Zhang, J., Yin, R., Shao, Q. et al. (2019). *Angew. Chem. Int. Ed.* 58: 5609–5613.
22 Zheng, T., Jiang, K., Ta, N. et al. (2019). *Joule* 3: 265–278.
23 Huan, T.N., Ranjbar, N., Rousse, G. et al. (2017). *ACS Catal.* 7: 1520–1525.
24 Lin, S., Diercks, C.S., Zhang, Y.-B. et al. (2015). *Science* 349: 1208–1213.
25 Jeong, H.-Y., Balamurugan, M., Choutipalli, V.S.K. et al. (2019). *J. Mater. Chem. A* 7: 10651–10661.
26 Yang, H., Wu, Y., Lin, Q. et al. (2018). *Angew. Chem. Int. Ed.* 130: 15702–15706.
27 Liu, S., Tao, H., Zeng, L. et al. (2017). *J. Am. Chem. Soc.* 139: 2160–2163.
28 Lv, K., Teng, C., Shi, M. et al. (2018). *Adv. Funct. Mater.* 28: 1802339.
29 Han, N., Wang, Y., Yang, H. et al. (2018). *Nat. Commun.* 9: 1–8.
30 Gu, J., Héroguel, F., Luterbacher, J., and Hu, X. (2018). *Angew. Chem. Int. Ed.* 130: 2993–2997.
31 Lee, C.H. and Kanan, M.W. (2015). *ACS Catal.* 5: 465–469.
32 Gong, Q., Ding, P., Xu, M. et al. (2019). *Nat. Commun.* 10: 1–10.
33 Gao, S., Lin, Y., Jiao, X. et al. (2016). *Nature* 529: 68–71.
34 Ma, W., Xie, S., Zhang, X.-G. et al. (2019). *Nat. Commun.* 10: 1–10.
35 Gao, D., Zhou, H., Cai, F. et al. (2017). *Nano Res.* 10: 2181–2191.
36 Zheng, Y., Vasileff, A., Zhou, X. et al. (2019). *J. Am. Chem. Soc.* 141: 7646–7659.
37 Zu, X., Li, X., Liu, W. et al. (2019). *Adv. Mater.* 31: 1808135.

38 Ikemiya, N., Natsui, K., Nakata, K., and Einaga, Y. (2017). *RSC Adv.* 7: 22510–22514.

39 Chen, Z., Mou, K., Yao, S., and Liu, L. (2018). *J. Mater. Chem. A* 6: 11236–11243.

40 Manthiram, K., Beberwyck, B.J., and Alivisatos, A.P. (2014). *J. Am. Chem. Soc.* 136: 13319–13325.

41 Tang, C., Shi, J., Bai, X. et al. (2020). *ACS Catal.* 10: 2026–2032.

42 Liang, C., Kim, B., Yang, S. et al. (2018). *J. Mater. Chem. A* 6: 10313–10319.

43 Weng, Z., Wu, Y., Wang, M. et al. (2018). *Nat. Commun.* 9: 1–9.

44 Zhang, H., Chang, X., Chen, J.G. et al. (2019). *Nat. Commun.* 10: 1–9.

45 Piontek, S., Junge Puring, K., Siegmund, D. et al. (2019). *Chem. Sci.* 10: 1075–1081.

46 Wang, Y., Chen, J., Wang, G. et al. (2018). *Angew. Chem. Int. Ed.* 57: 13120–13124.

47 Lu, L., Sun, X., Ma, J. et al. (2018). *Angew. Chem. Int. Ed.* 130: 14345–14349.

48 Yang, D., Zhu, Q., Chen, C. et al. (2019). *Nat. Commun.* 10: 1–9.

49 Huang, J., Guo, X., Yue, G. et al. (2018). *ACS Appl. Mater. Interfaces* 10: 44403–44414.

50 Wu, Y., Jiang, Z., Lu, X. et al. (2019). *Nature* 575: 639–642.

51 Periasamy, A.P., Ravindranath, R., Kumar, S.M.S. et al. (2018). *Nanoscale* 10: 11869–11880.

52 Gonglach, S., Paul, S., Haas, M. et al. (2019). *Nat. Commun.* 10: 1–10.

53 Low, Q.H., Loo, N.W.X., Calle-Vallejo, F., and Yeo, B.S. (2019). *Angew. Chem. Int. Ed.* 58: 2256–2260.

54 Zhang, W., Qin, Q., Dai, L. et al. (2018). *Angew. Chem. Int. Ed.* 57: 9475–9479.

55 Zhao, S., Guo, S., Zhu, C. et al. (2017). *RSC Adv.* 7: 1376–1381.

56 Ji, L., Li, L., Ji, X. et al. (2020). *Angew. Chem. Int. Ed.* 132: 768–772.

57 Mou, S., Wu, T., Xie, J. et al. (2019). *Adv. Mater.* 31: 1903499.

58 Handoko, A.D., Wei, F., Yeo, B.S., and Seh, Z.W. (2018). *Nat. Catal.* 1: 922–934.

59 Zhu, Q., Sun, X., Yang, D. et al. (2019). *Nat. Commun.* 10: 1–11.

60 Li, Y.C., Wang, Z., Yuan, T. et al. (2019). *J. Am. Chem. Soc.* 141: 8584–8591.

61 Ren, D., Gao, J., Pan, L. et al. (2019). *Angew. Chem. Int. Ed.* 131: 15178–15182.

62 Karapinar, D., Huan, N.T., Ranjbar Sahraie, N. et al. (2019). *Angew. Chem. Int. Ed.* 58: 15098–15103.

63 Lv, K., Fan, Y., Zhu, Y. et al. (2018). *J. Mater. Chem. A* 6: 5025–5031.

64 Liu, Y., Fan, X., Nayak, A. et al. (2019). *Proc. Natl. Acad. Sci. U.S.A.* 116: 26353–26358.

65 Liu, Y., Zhang, Y., Cheng, K. et al. (2017). *Angew. Chem. Int. Ed.* 129: 15813–15817.

66 Song, Y., Chen, W., Zhao, C. et al. (2017). *Angew. Chem. Int. Ed.* 129: 10980–10984.

67 De Luna, P., Hahn, C., Higgins, D. et al. (2019). *Science* 364: eaav3506.

68 Kibria, M.G., Edwards, J.P., Gabardo, C.M. et al. (2019). *Adv. Mater.* 31: 1807166.

69 Yin, Z., Peng, H., Wei, X. et al. (2019). *Energy Environ. Sci.* 12: 2455–2462.

70 Dinh, C.-T., Burdyny, T., Kibria, M.G. et al. (2018). *Science* 360: 783–787.
71 Luc, W., Fu, X., Shi, J. et al. (2019). *Nat. Catal.* 2: 423–430.
72 Ren, S., Joulié, D., Salvatore, D. et al. (2019). *Science* 365: 367–369.

26

Electrochemical Fuel Oxidation

To enable the electrochemical carbon cycle (ECC), electrochemical oxidation of these fuels is expected to be complete processes toward the production of CO_2. However, with the increase of molecular weight of the fuel molecules from CO and HCOOH to CH_4, CH_3OH, and CH_3CH_2OH, difficulty in achieving such complete oxidation increases. Normally, the oxidation reaction of fuels involves multistep dehydrogenation and oxygenation processes [1]. Part of the elementary reaction steps during fuel oxidation will proceed along with the reverse pathways of electrochemical CO_2 reduction reaction (ECDRR), while most of them are different from ECDRR pathways due to the different catalysts used, different reaction media, and potentials. Typically, strongly adsorbed intermediates during fuel oxidation, such as CO*, COH*, CHOH*, and CH_2OH*, would be bonded closely with active sites on the catalyst surface, interdicting the subsequent fuel adsorption and oxidation toward CO_2 production [2–4]. This phenomenon is called as catalyst poisoning. To alleviate catalyst poisoning and promote fuel oxidation, we have previously summarized the promoting mechanisms as a bifunctional mechanism, electronic effect, and third body effect [5]. Most of the catalyst design has been focused on the improvement of these three effects.

In spite of some non-noble metals, electrocatalysts explored such as tungsten carbide [6] and Fe/Co/Ni-based materials [7–10] for the past 100 years of development, noble metals especially Pt- and Pd-based materials are still the most effective electrocatalysts for fuel oxidation up to date [11, 12]. To realize an optimal design, an ideal catalyst for fuel oxidation is suggested to have these features: (i) high oxygen insertion and/or dehydrogenation capacity to boost oxidation processes; (ii) high conductivity to endow fast electron transfer; (iii) high anti-poisoning properties to maintain the activity of active sites on catalyst surface; (iv) high stability for long-term applications. Overall, a catalyst with these superiorities provides the basis for fuel oxidation at the high activity (high peak current) and selectivity, ultimately leading to CO_2 production. With the well-developed research on the mechanism for the promotion of fuel oxidation, catalyst exploration with component/structure control, and performance evaluation, we now go to the summary on electrocatalysts for oxidation of CO, CH_4, HCOOH, CH_3OH, and CH_3CH_2OH fuels. Advances in the design principles of catalysts, the promising catalyst candidates, and the attractive performance

Electrocatalysis in Balancing the Natural Carbon Cycle, First Edition. Yaobing Wang.
© 2021 WILEY-VCH GmbH. Published 2021 by WILEY-VCH GmbH.

of catalysts and fuel cells will be focused on. A summary of this part is hoped to enhance the design of the link in fuel oxidation toward high-efficiency ECC.

CO is not only a simple inorganic fuel for ECC but also a poisoning intermediate during HCOOH, CH_3OH, and CH_3CH_2OH electrooxidation. Therefore, research on CO electrooxidation is usually combined with the study of other fuel electrooxidation. Individual CO electrooxidation research was reported early by N.P. Lebedeva et al.'s work in 2002 [13]. Pt-based catalysts were explored with crystal face control (Pt(111)) [13], alloying (PtRu) [14], support modulation with tin oxide [15], and carbon nanofibers (CNFs) [16] toward enhanced CO electrooxidation to CO_2. The CO electrooxidation performance on Pt catalysts can be influenced by multifactors, such as CO adsorption energy on different Pt-based catalyst surface, possible blocking effect by electrolyte, and geometric/defective/grain boundary effects [16, 17]. For example, CO electrooxidation on PtSn catalyst in NaOH medium showed a sharp peak compared to that in the H_2SO_4 electrolyte. This phenomenon was explained with the adsorbed sulfate anions on Pt active sites, which hindered CO mobility for oxidation. In addition to noble metal Pt, non-noble metals such as Sn [15, 18], Co [18], and Ni [18] were also reported for CO electrooxidation, with the maximum current of 630, 450, and 140 $\mu A\,mg^{-1}$ metal, respectively, obviously lower than that of Pt/C (13 200 $\mu A\,mg^{-1}$ metal). Noble metal Au was also used for CO electrooxidation [19]. However, among these reported metal materials, Pt-based catalysts exhibited the best activity for CO electrooxidation. The typical Pt-based alloy catalysts with Sn, Ru [14], Pb [5], and Cu [20] for CO electrooxidation to CO_2 are summarized in Table 26.1.

Besides CO_2, CH_4 is also a greenhouse gas with an effect more than 30 times stronger than CO_2 [24]. Therefore, its utilization for ECC is also significant to supplement the natural carbon cycle. However, the solubility of CH_4 in aqueous solution is extremely low, limiting their conversion efficiency to CO_2 [25]. The development of the gas diffusion electrode could increase the number of gas–liquid–solid interfaces and promote efficiency [26], as exemplified by the Ni-based hollow fiber electrode [27]. This strategy is also applicable to CO electrooxidation. On the other hand, CH_4 electrooxidation suffers from extremely low sluggish kinetics due to high dissociation energy (104 $kcal\,mol^{-1}$) of the C—H bonds in the CH_4 molecule [24]. In this case, high oxidation temperatures, more positive potentials, and additional oxidants such as O_2 and high-valent metals are usually needed [21, 28, 29]. Recently, J.H. Park and coworkers had used highly active Co_3O_4/ZrO_2 nanocomposite as the catalyst for CH_4 oxidation at ambient conditions, with the carbonate as the oxidizing agent [24]. Only the organic products (e.g. methanol, ethanol, formaldehyde, acetaldehyde, 1-/2-propanol, and acetone) were detected in the proton nuclear magnetic resonance spectroscopy. Among various possible products from CH_4 oxidation [23], CO_2 is the most likely product because of the low redox potential required [30]; thus, the CH_4–CO_2 cycle should be thermodynamically favorable. Up to date, Pt and Pd catalysts were studied systematically for complete CH_4 electrooxidation to CO_2 under ambient conditions [22, 31]. Pt could be the most active for CH_4 electrooxidation, during which *CO is the stable intermediate. The bifunctional mechanism

Table 26.1 Typical catalysts for CO/CH_4 electrooxidation to CO_2.

Catalyst	Reaction type	Noble metal content and loading on electrode	Structure–activity relationship	Electrolyte/scan rate	Peak potentials vs. RHE (Pt/C value)	Current density at peak potential except specified (Pt/C value) or fuel cell performance
Pt_3Sn_1/C [17]	CO oxidation	20 wt%/not mentioned	Sn provides –OH species to promote CO oxidation on Pt	0.1 M NaOH/20 mV s^{-1}	0.74 V	0.09 mA cm^{-2}
Pt_3Sn_1/CNF [17]	CO oxidation	20 wt%/not mentioned	CNF reduces CO adsorption and lowers the oxidation potential beyond Sn effects	0.1 M NaOH/20 mV s^{-1}	0.71 V	0.079 mA cm^{-2}
Pt_3Sn_1/C [16]	CO oxidation	20 wt%/not mentioned	Sn provides –OH species to promote CO oxidation on Pt	0.5 M H_2SO_4/20 mV s^{-1}	0.75 V	0.036 mA cm^{-2}
PtRu [14]	CO oxidation	20 wt%/ 0.0214 mg cm^{-2}	Ru provides active oxygen species from water molecule to promote CO oxidation	0.5 M H_2SO_4/20 mV s^{-1}	0.54 V (0.82 V)	1.18 mA cm^{-2} (0.42 mA cm^{-2})
PtPb [5]	CO oxidation	20.3 wt%/ 0.064 mg cm^{-2}	Pb provides active oxygen species and weakens Pt–CO binding	0.5 M H_2SO_4/50 mV s^{-1}	0.67 V (0.726 V)	4.1 mA cm^{-2} (2.24 mA cm^{-2}) at 0.62 V
MnA-PtCu/C [20]	CO oxidation	9.3 wt%/ 0.016 mg cm^{-2}	Cu weakens Pt–CO binding and core–shell structure increases Pt active sites	0.5 M H_2SO_4/50 mV s^{-1}	0.78 V (0.76 V)	6.08 mA cm^{-2} (3.2 mA cm^{-2})

(*continued*)

Table 26.1 (Continued)

Catalyst	Reaction type	Noble metal content and loading on electrode	Structure–activity relationship	Electrolyte/scan rate	Peak potentials vs. RHE (Pt/C value)	Current density at peak potential except specified (Pt/C value) or fuel cell performance
Pt/C-ATO [21] Pd/C-ATO [21]	CH_4 oxidation	1 mg cm^{-2}	Sb-doped SnO_2 (ATO) favors oxidation of *CO and *HCO intermediate	0.5 M H_2SO_4/10 mV s^{-1}	No obvious peak	0.275 and 0.45 mW cm^{-2} for Pt/C-ATO and Pd/C-ATO at 80 °C, respectively.
Pt [22]	CH_4 oxidation	100 wt%/6 cm^{-2}	Pt favors CH_4 dehydrogenation to form *CO and finally oxidation	0.5 M $HClO_4$/50 mV s^{-1}	0.75 V	0.02 mA $cm^{-2}{}_{ECSA}$
NiO-ZrO_2 [23]	CH_4 oxidation	0.03 mg $cm^{-2}{}_{catalyst}$	CO_3^{2-} adsorbs on ZrO_2 for CH_4 activation, and Ni^{3+} species oxidizes CH_4	0.1 M Na_2CO_3 + KOH/20 mV s^{-1}	0.66 V vs. SCE	0.127 μA cm^{-2}

All data were obtained at room temperature unless specified; otherwise, part of the data was obtained by reading from plots or involving a conversion in terms of metal loadings and/or reference electrode.

plays a crucial role in the complete oxidation of CH_4. Some of the typical catalysts for CH_4 oxidation were summarized in Table 26.1.

Formic acid is one of the most structurally simple organic fuels with a theoretical volumetric energy density of 1750 Wh l^{-1} [32]. Formic acid oxidation reaction (FAOR) normally undergoes two pathways, namely successive/concurrent dehydrogenation processes and dehydration processes. The dehydration process is undesired because it generates adsorbed CO, causing deactivation of noble metal catalysts [33]. The reactant type (either formic acid or formate) is dependent on the solution pH. A proper adjustment of pH to the pK_a of HCOOH would significantly promote the oxidation activity by increasing the content of $HCOO^-$ with higher reactivity [34] and reducing the surface-adsorbed oxydic species coverage [35]. Noble metals of Pd [36, 37], Pt [38, 39], Ir [40, 41] have activities for FAOR. Among them, Pd attracts more attention due to higher reaction activity even compared to Pt [34]. To further enhance the electrocatalytic activity and stability of Pd materials toward FAOR, strong interactions between Pd nanoparticles and tungsten oxide support were introduced to protect the alloy component from solution leaching, expand Pd lattice for enhancing formate adsorption, and reduce surface electron density to weaken CO adsorption [42]. Some supports or promoters such as metal phosphides can activate water to generate $-OH_{ad}$ for more effectively oxidative removal of CO via a so-called bifunctional mechanism, which cooperates with hydrogen spillover effect to achieve highly selective successive FAOR process to CO_2 [43]. Significantly, $CuPd/WO_{2.72}$ catalyst exhibited a mass activity of 2086 mA mg^{-1} Pd and maintained this activity for 12 hours at the potential of 0.40 V [44], showing its great potential for FAOR application. A. Manthiram and coworker reported a formate fuel cell using Pd catalyst in the anode and formate in the aqueous solution [34]. This fuel cell exported an open-circuit potential of 1.45 V and a maximum power density of c. 70 mW mg^{-1} Pd. In our previous work, we have also achieved a formic acid/formate oxidation current of 4.2 mA cm^{-2} at 0.3 V vs. reversible hydrogen electrode (RHE) on 3D porous Pd interconnected nanosheets catalyst and demonstrated its feasibility for CO_2-$HCOOH/HCOO^-$ cycle. Some typical Pd-based catalysts for FAOR to CO_2 were listed in Table 26.2.

Methanol, with the theoretical energy density of 4820 Wh l^{-1}, is a commonly-used fuel and has received the most advances in the fuel cell field [32]. The methanol oxidation reaction (MOR) is required to proceed via the elementary steps of methanol adsorption, dehydrogenation, and oxygen insertion. Compared to the only poisoning intermediate (*CO) from HCOOH oxidation, MOR may generate multiple strongly adsorbed intermediates such as *CO, *COH, and *HCHO, which can block the active sites from fresh methanol adsorption, thus inhibiting the overall oxidation efficiency [11]. Among various noble metals available such as Pd, Pt, and Rh catalysts [47–50], Pt is widely used as the best monometallic catalyst for methanol oxidation [51]. Many efforts have been made to overcome the essential drawbacks/limitations of Pt catalysts, namely slow kinetics and surface poisoning. The strategies include modifying Pt with other metals (oxides) and forming core–shell structures and other nanostructures [46, 48, 52–55]. These strategies could either improve charge transfer to surface Pt active sites or increase surface areas, modulate the electronic structure

Table 26.2 Typical catalysts for HCOOH oxidation to CO_2.

Catalyst	Noble metal content (wt%) and loading on electrode	Structure–activity relationship	Electrolyte/scan rate	Peak potentials vs. RHE except specified (Pd/C value)	Activity at peak potential (Pd/C value)	Stability/fuel cell performance
$CuPd/WO_{2.72}$ [44]	19.5 wt%/ 0.0195 mg cm^{-2}	CuPd (111) lattice expansion with stabilizing Cu and improving Pd activity	0.1 M $HClO_4$ + 0.1 M HCOOH/50 mV s^{-1}	0.75 V	2050 mA mg^{-1}$_{Pd}$	2086 mA mg^{-1}$_{Pd}$ after 12 h test at 0.4 V/not tested
$Pd/WO_{2.72}$ [42]	18.8 wt%/ 0.0188 mg cm^{-2}	Pd lattice expansion decreases electron density; $WO_{2.72}$ facilitates $-OH_{ads}$ generation, both promote CO oxidation	0.1 M $HClO_4$ + 0.1 M HCOOH/50 mV s^{-1}	0.79 V	1600 mA mg^{-1}$_{Pd}$	1600 mA mg^{-1}$_{Pd}$ after 12 h test at 0.4 V/not tested
Pd-CoP/C [43]	5 wt%/ 0.0095 mg cm^{-2}	CoP promotes hydrogen adsorption and transfer, and facilitate formation of $-OH_{ads}$ species	0.5 M H_2SO_4 + 0.5 M HCOOH/200 mV s^{-1}	0.495 V (0.445 V)	7000 mA mg^{-1}$_{Pd}$ (1050 mA mg^{-1}$_{Pd}$)	100% activity reservation after 1000 cycles/ 150 mW cm^{-1}

Pd@Graphene [36]	30 wt%/ 0.033 mg cm^{-2}	Small Pd particles, hierarchical structure, and strong affinity to HCOOH	0.5 M H_2SO_4 + 0.5 M HCOOH/50 mV s^{-1}	0.405 V (0.465 V)	300 mA mg$^{-1}{}_{Pd}$ (216 mA mg$^{-1}{}_{Pd}$)	Better than Pd–C/not tested
Pd/C [34]	40 wt%/ 0.64 mg cm^{-2}	Formate solution has higher reactivity compared to formic acid	1.0 M HCOOK + 1.0 M KOH/20 mV s^{-1}	0.1 V vs. SHE	156 mA mg$^{-1}{}_{Pd}$	c. 88% activity reservation after 8 h test at −0.25 V vs. SHE/ 150 mW mg$^{-1}{}_{Pd}$ at 60 °C
Pd nanowires [45]	10 wt%	Wavy, wire-like structure and numerous twin defects	0.1 M $HClO_4$ + 2 M HCOOH/50 mV s^{-1}	0.64 V vs. SHE (0.62 V vs. SHE)	2.42 mA cm^{-2} (0.65 mA cm^{-2})	Not tested
PdAg/$Ti_{0.5}Cr_{0.5}N$ [46]	10.2 wt%	Ag causes up-shift of Pd d-band center; $Ti_{0.5}Cr_{0.5}N$ has high conductivity	0.1 M $HClO_4$ + 1 M HCOOH/10 mV s^{-1}	0.65 V (0.66 V)	975 mA mg$^{-1}{}_{Pd}$ (330 mA mg$^{-1}{}_{Pd}$)	106 mA mg$^{-1}{}_{Pd}$ after 1 h test at 0.6 V/not tested

All data were obtained at room temperature unless specified otherwise, part of the data was obtained by reading from plots or involving a conversion in terms of metal loadings and/or reference electrode.

of Pt toward weaken the poisoning intermediates, and enhance chemical resistivity as well as mechanical strength. Recently, some non-noble metals of Ni, Co, and Cu-based materials were reported for MOR [42, 56–58] in alkaline media. Alkaline media could avoid the dissolution of these materials; however, it requires the development and application of highly chemically stable anion exchange membrane in practical applications. These non-noble metal materials catalyzed MOR based on their high oxidation state of metals [59, 60]. Most of them needed more positive potentials by 200–600 mV for MOR relative to noble metal Pt catalysts [61, 62], which limited their energy efficiency to supplement ECC. Recently, it was found that a few non-noble metal composites such as Cu_2O/polypyrrole-graphene oxide (PPy-GO) and SnO_2/mesoporous zeolite showed comparable peak potentials for MOR relative to noble metals [58, 63]. A dual-site pathway was revealed for SnO_2/mesoporous zeolite, in which zeolite was responsible for methanol molecule adsorption and activation with the frame acidic sites via formation of methyl groups, while SnO_2 could catalyze the dehydrogenation of generated methyl groups, and promote adsorbed CO reaction to CO_2, leading to complete methanol oxidation [58]. These discussions suggest Pt-based catalysts and some non-noble metal materials have a good prospect for MOR, with the typical ones summarizing in Table 26.3.

Ethanol oxidation reaction (EOR) was explored late relative to other fuel electrooxidation. However, it received a lot of research concerns due to many merits, including high energy density of ethanol (8030 Wh kg^{-1}) [32], abundant source, safety for storage, and use of ethanol [68]. EOR was rather complicated via a dual-path mechanism, involving multiple dehydrogenations and oxygen insertion steps. Incomplete electrooxidation of ethanol occurred in most cases, leading to the formation of acetaldehyde and acetic acid. Complete EOR to CO_2 is desired for ECC, whereas it remains difficult because it requires C—C bond cleavage and generates strongly adsorbed intermediates such as *CO, leading to the sluggish reaction kinetics. Ideal electrocatalysts should not only control the reaction mechanism of EOR toward C—C bond cleavage, but also possess surface anti-poisoning properties. Up to date, electrochemical oxidation of ethanol has been reported to perform mostly by noble metals Pt and Pd, and some Ni-/Co-based electrocatalysts. Pt-based electrocatalysts [68–70] were usually used in acid media, while Pd [71–73] and Ni-/Co-based [74, 75] electrocatalysts were normally used in alkaline media for maximization of their performance. Among these alternatives, only a few noble-metal catalysts could catalyze complete ethanol oxidation to CO_2. The first systematic research for EOR to CO_2 was reported by Nitopi et al. [76]. In situ infrared reflection-absorption spectroscopy (IRRAS) and in situ X-ray absorption spectroscopy (XAS), combined with density functional theory (DFT) calculations, have revealed that SnO_2 adsorbed water and produced –OH, Pt facilitated ethanol dehydrogenation, while Rh with modified electronic structure boosted C—C bond cleavage. The synergistic effects among these three constituents have promoted EOR to CO_2. PtRh-based electrocatalysts have been received further development recently: PtRhFe-PNS@MXene catalyst achieved current of 3407.7 mA mg^{-1} for EOR due to the synergistic effects among Pt, Rh, and Fe, as well as MXene with large surface termination groups and high conductivity [77]; Rh was revealed to provide

Table 26.3 Typical catalysts for CH_3OH oxidation.

Catalyst	Metal content (wt%) and loading on the electrode	Structure–activity relationship	Electrolyte/scan rate	Peak potentials vs. RHE (reference catalyst)	Activity at peak potential (reference catalyst)	Stability/fuel cell performance
Cu@Pt/C [62]	8.09 wt% Pt/ 0.14 mg cm^{-2} Pt	Electronic effect between Pt and underlying Cu; Pt-rich surface	0.5 M CH_3OH + 0.5 M H_2SO_4/50 mV s^{-1}	0.88 V (0.88 V, Pt/C)	620 mA cm^{-2} $mg^{-1}{}_{Pt}$ (170 mA cm^{-2} $mg^{-1}{}_{Pt}$)	36% activity reservation after 1200 s test at 0.645 V/not tested
Cross-double dumbbell-like Pt–Ni [61]	30 wt% Pt/ 0.0153 mg cm^{-2} Pt	High specific surface area; shifted d-band center to reduce intermediates adsorption	0.5 M CH_3OH + 0.5 M H_2SO_4/50 mV s^{-1}	0.88 V (0.88 V, Pt/C)	32 mA cm^{-2} (18 mA cm^{-2})	92% activity reservation after 1000 cycles/ not tested
Cu_2O/PPy–GO [58]	4 mg cm^{-2} for membrane electrode assemblies	Small and uniformly dispersed nanoparticles	2 M CH_3OH + 0.05 M H_2SO_4/50 mV s^{-1}	0.934 V (0.9 V PtRu/C)	7.8 mA g^{-1} (7 mA cm^{-2})	30% activity reservation after 900 s test at 0.66 V/ 31 mW cm^{-2} at 80 °C
SnO_2/m-ZSM-5 nanocomposite [63]	3.8 wt%	Frame acidic sites in ZSM-5 adsorb/activate methanol; SnO_2 catalyzes $-CH_3$ dehydrogenation	0.8 M CH_3OH + 0.5 M NaOH/50 mV s^{-1}	0.803 V (0.778 V, Pt/C)	14.7 A g^{-1} (14.7 A g^{-1}, Pt/C)	14.5 A g^{-1} after 1000 cycles/not tested

(continued)

Table 26.3 (Continued)

Catalyst	Metal content (wt%) and loading on the electrode	Structure–activity relationship	Electrolyte/scan rate	Peak potentials vs. RHE (reference catalyst)	Activity at peak potential (reference catalyst)	Stability/fuel cell performance
Au@CeO_2@Pt/C [64]	3.84 wt% Pt	Bifunctional effect for –OH radicals' production; electronic effect	1 M CH_3OH + 0.25 M H_2SO_4/50 mV s^{-1}	0.937 V (0.937 V, Pt/C)	1.36 A $mg^{-1}{}_{Pt}$ (0.67 A $mg^{-1}{}_{Pt}$)	70% activity reservation after 600 cycles/not tested
PtPdAg hollow nanodendrites [65]	48.9 wt% Pt/ 0.0254 mg cm^{-2} Pt	Rich atomic steps and grain boundaries with a high anti-poison capability	0.2 M CH_3OH + 0.1 M $HClO_4$/50 mV s^{-1}	1.06 V (0.87 V, Pt black)	1.56 A $mg^{-1}{}_{Pt}$ (0.36 A $mg^{-1}{}_{Pt}$)	58% activity reservation after 5000 cycles/ not tested
PtCoNiRh ultrathin nanowires [66]	90.3 wt% Pt	Rh stabilizes Pt atoms to form highly-active interatomic Pt–Rh sites	0.5 M CH_3OH + 0.1 M $HClO_4$/50 mV s^{-1}	0.87 V (0.87 V, Pt/C)	1.35 A $mg^{-1}{}_{Pt}$ (0.34 A $mg^{-1}{}_{Pt}$)	0.27 A $mg^{-1}{}_{Pt}$ after 10 000 s test at 0.8 V/not tested
PdCeO_2/N-doped mesoporous carbon sphere [67]	90.3 wt% Pd/ 0.024 mg cm^{-2} Pd	Uniform dispersion of Pd; electronic effect from CeO_2 to reduce CO adsorption energy	1.0 M CH_3OH + 1.0 M KOH/50 mV s^{-1}	0.866 V (0.865 V, Pd/C)	1.5 A $mg^{-1}{}_{Pt}$ (0.55 A $mg^{-1}{}_{Pd}$)	93% activity reservation after 1000 cycles/ 25 mW cm^{-2} at 20 °C

All data were obtained at room temperature unless specified; otherwise, part of the data was obtained by reading from plots or involving a conversion in terms of metal loadings and/or reference electrode.

Table 26.4 Typical catalysts for CH_3CH_2OH oxidation.

Catalyst	Metal content (wt%) and loading on electrode	Structure–activity relationship	Electrolyte/scan rate	Peak potentials vs. RHE (reference catalyst)	Activity at peak potential (reference catalyst)	Stability/fuel cell performance
$PtRhSnO_2/C$ [76]	37.3 wt% Pt in metals/0.029 mg cm^{-2} Pt	SnO_2 produces –OH, Pt facilitates dehydrogenation, Rh boosts C—C bond cleavage	0.1 M $HClO_4$ + 0.2 M CH_3CH_2OH/50 mV s^{-1}	0.87 V (0.975 V, $PtSnO_2/C$)	15.1 mA cm^{-2} (11.25 mA cm^{-2})	15.1 mA cm^{-2} after 2 h test at 0.45 V/not tested
PtRh nanowires [68]	20 wt% Pt/ 0.0127 mg cm^{-2} Pt	Rh provides –OH for alleviate poisoning effect, and tunes C_1 pathway selectivity	0.1 M $HClO_4$ + 0.5 M CH_3CH_2OH/50 mV s^{-1}	1.0 V (0.95 V, Pt/C)	19.1 mA cm^{-2} (6.6 mA cm^{-2})	51% activity reservation after 2000 cycles/not tested
PtRhFe nanospheres@ MXene [77]	86.4 wt% Pt in metals	Synergistic effects among Pt, Rh, and Fe; MXene with abundant surface termination groups and high conductivity	1 M KOH + 1 M CH_3CH_2OH/50 mV s^{-1}	0.72 V (0.7 V, Pt/C)	3407.7 mA mg^{-1} (800 mA mg^{-1})	72.8% activity reservation after 2000 cycles/not tested
Rh@Pt nanowires [70]	25 wt% of total metals/ 0.0153 mg cm^{-2} PtRh	Rh modifies Pt to promote C—C bond cleavage	0.1 M $HClO_4$ + 0.2 M CH_3CH_2OH/50 mV s^{-1}	0.9 V (0.9 V, Pt/C)	800 mA $mg^{-1}{}_{Pt}$ (220 mA $mg^{-1}{}_{Pt}$)	70 mA mg^{-1} Pt after 1 h test at 0.945 V /not tested

(*continued*)

Table 26.4 (Continued)

Catalyst	Metal content (wt%) and loading on electrode	Structure–activity relationship	Electrolyte/scan rate	Peak potentials vs. RHE (reference catalyst)	Activity at peak potential (reference catalyst)	Stability/fuel cell performance
Au@PtIr/C [78]	~0.02 mg cm^{-2} metals	Au induces lattice expansion; PtIr atomic steps enables full ethanol oxidation	1 M KOH + 1 M CH_3CH_2OH/20 mV s^{-1}	0.9 V	58 A mg$^{-1}{}_{PtIr}$	0.28 A mg^{-1} PtIr after 1 h test at 0.945 V/not tested
Pd/Ni(OH)$_2$/rGO [79]	51.3 wt% Pd/ 0.0145 mg cm^{-2} Pd	Ni(OH)$_2$ alleviates poisoning of Pd and shifts selectivity to C_1 pathway	1 M KOH + 1 M CH_3CH_2OH/20 mV s^{-1}	0.946 V (0.886 V, Pd/C)	1.5 A mg$^{-1}{}_{Pd}$ (1.03 A mg$^{-1}{}_{Pd}$)	0.4 A mg$^{-1}{}_{Pd}$ after 20 000 s test at 0.872 V/not tested
$Ni_{29}Pd_{34}Pt_{37}$/C [80]	40.5 wt% of metal/ 0.77 mg cm^{-2}	Ni provides –OH and triggers the C—C bond scission	0.5 M NaOH + 1 M CH_3CH_2OH/50 mV s^{-1}	1.26 V (1.16 V, Pt/C)	105 mA cm^{-2} (45 mA cm^{-2})	Not tested/ 41 mW cm^{-2} at 40 °C
PdAg single crystalline nanowires [71]	65.5 wt% Pd/ 0.0278 mg cm^{-2} Pd	Synergic electronic effect and bifunctional mechanism	1 M KOH + 1 M CH_3CH_2OH/50 mV s^{-1}	0.826 V (0.826 V, Pd/C)	2750 mA mg$^{-1}{}_{Pd}$ (900 mA mg$^{-1}{}_{Pd}$)	1230 mA mg$^{-1}{}_{Pd}$ after 2500 cycles/not tested

All data were obtained at room temperature unless specified; otherwise, part of the data was obtained by reading from plots or involving a conversion in terms of metal loadings and/or reference electrode.

abundantly -OH for enhancing anti-poisoning capacity and modify Pt electronic structure toward favorable C–C cleavage in PtRh nanowires [68, 70]. Beyond PtRh electrocatalysts, Au (core)–PtIr (shell) [78] and Pd catalysts coupled with $Ni(OH)_2$ [79, 80] could convert ethanol into CO_2 via a direct 12-electron pathway. A list of typical catalysts for EOR to CO_2 was summarized in Table 26.4.

Advanced catalysts for CO, CH_4, HCOOH, CH_3OH, and CH_3CH_2OH electrooxidation, along with investigation on promotion mechanism, have been discussed earlier. As listed in the mentioned Tables 26.1–26.4, the promising catalysts for favorable fuel electrooxidation could be concluded respectively: PtRu and PtSn-based catalysts benefit CO oxidation, Pd-based catalysts favor for FAOR, Pt-based catalysts make for CH_4 oxidation and MOR, and PtRh-based catalysts conduce to EOR. Despite these advances, several challenges remain in fuel electrooxidation, especially for large-scale applications to supplement ECC. (i) Limitation between activity and noble metal loading should be overcome. As aforesaid, noble metal catalysts are usually required to convert fuel into CO_2, while the activity is highly dependent on its loading on the electrode. Improvement in activity or decrease of noble metal content by over an order of magnitude is suggested toward its application. (ii) To supplement ECC, the fuel electrooxidation must lead to the generation of CO_2, which could be relatively easily realized in CO oxidation and FAOR. However, for other oxidation reactions, especially CH_4 oxidation, and EOR, electrocatalysts with high selectivity should be designed toward the complete oxidation reaction pathway to CO_2. (iii) Examination of advanced catalysts in devices with long-term durability. As shown in the Tables 26.1–26.4, many of the advanced catalysts were not examined by the device test. Only long-term operation of catalysts in devices such as fuel cells can practically supplement ECC. (iv) Other issues, such as fuel crossover (especially for methanol) and cathode flooding by generated water, have to be optimized in device operation for releasing electrocatalyst activity [32].

References

1 Martínez-Huerta, M. and Lázaro, M. (2017). *Catal. Today* 285: 3–12.
2 Skúlason, E. and Jonsson, H. (2017). *Adv. Phys.: X* 2: 481–495.
3 Muttaqien, F., Hamamoto, Y., Hamada, I. et al. (2017). *J. Chem. Phys.* 147: 094702.
4 Du, P., Gao, Y., Wu, P., and Cai, C. (2018). *Phys. Chem. Chem. Phys.* 20: 10132–10141.
5 Huang, Y., Zheng, S., Lin, X. et al. (2012). *Electrochim. Acta* 63: 346–353.
6 Akhairi, M. and Kamarudin, S. (2016). *Int. J. Hydrogen Energy* 41: 4214–4228.
7 Stathi, P., Deligiannakis, Y., Avgouropoulos, G., and Louloudi, M. (2015). *Appl. Catal., A* 498: 176–184.
8 Cuña, A., Plascencia, C.R., da Silva, E.L. et al. (2017). *Appl. Catal., B* 202: 95–103.
9 Yin, Z., Zheng, Y., Wang, H. et al. (2017). *ACS Nano* 11: 12365–12377.
10 Cao, S., Chen, Y., Wang, H. et al. (2018). *Joule* 2: 549–557.

11 Kakati, N., Maiti, J., Lee, S.H. et al. (2014). *Chem. Rev.* 114: 12397–12429.
12 Yin, Z., Lin, L., and Ma, D. (2014). *Catal. Sci. Technol.* 4: 4116–4128.
13 Lebedeva, N., Koper, M., Feliu, J., and Van Santen, R. (2002). *J. Electroanal. Chem.* 524: 242–251.
14 Ueda, A., Yamada, Y., Ioroi, T. et al. (2003). *Catal. Today* 84: 223–229.
15 Matsui, T., Fujiwara, K., Okanishi, T. et al. (2006). *J. Power Sources* 155: 152–156.
16 Asgardi, J., Calderón, J.C., Alcaide, F. et al. (2015). *Appl. Catal., B* 168: 33–41.
17 Rizo, R., Sebastián, D., Lázaro, M.J., and Pastor, E. (2017). *Appl. Catal., B* 200: 246–254.
18 Kheradmandinia, S., Khandan, N., and Eikani, M.H. (2016). *Int. J. Hydrogen Energy* 41: 19070–19080.
19 Diao, P., Zhang, D., Guo, M., and Zhang, Q. (2007). *J. Catal.* 250: 247–253.
20 Huang, Y., Zhao, T., Zhao, G. et al. (2016). *J. Power Sources* 304: 74–80.
21 Nandenha, J., Fontes, E., Piasentin, R. et al. (2018). *J. Fuel Chem. Technol.* 46: 1137–1145.
22 Boyd, M.J., Latimer, A.A., Dickens, C.F. et al. (2019). *ACS Catal.* 9: 7578–7587.
23 Spinner, N. and Mustain, W.E. (2013). *ECS Trans.* 53: 1.
24 Ma, M., Jin, B.J., Li, P. et al. (2017). *Adv. Sci.* 4: 1700379.
25 Meng, X., Cui, X., Rajan, N.P. et al. (2019). *Chem* 5: 2296–2325.
26 Rocha, R.S., Reis, R.M., Lanza, M.R., and Bertazzoli, R. (2013). *Electrochim. Acta* 87: 606–610.
27 Guo, Z., Chen, W., Song, Y. et al. (2020). *Chin. J. Catal.* 41: 1067–1072.
28 O'Reilly, M.E., Kim, R.S., Oh, S., and Surendranath, Y. (2017). *ACS Cent. Sci.* 3: 1174–1179.
29 Kim, R.S. and Surendranath, Y. (2019). *ACS Cent. Sci.* 5: 1179–1186.
30 Xie, S., Lin, S., Zhang, Q. et al. (2018). *J. Energy Chem.* 27: 1629–1636.
31 Tomita, A., Nakajima, J., and Hibino, T. (2008). *Angew. Chem. Int. Ed.* 47: 1462–1464.
32 Zhao, L., Dong, B., Li, S. et al. (2017). *ACS Nano* 11: 5800–5807.
33 Yu, X. and Pickup, P.G. (2009). *Electrochem. Commun.* 11: 2012–2014.
34 Yu, X. and Manthiram, A. (2015). *Appl. Catal., B* 165: 63–67.
35 Jeon, H., Jeong, B., Joo, J., and Lee, J. (2015). *Electrocatalysis* 6: 20–32.
36 Zhang, L.Y., Zhao, Z.L., and Li, C.M. (2015). *Nano Energy* 11: 71–77.
37 Klinkova, A., De Luna, P., Dinh, C.-T. et al. (2016). *ACS Catal.* 6: 8115–8120.
38 Huang, Y., Zhao, T., Zeng, L. et al. (2016). *Electrochim. Acta* 190: 956–963.
39 Huang, Y., Yang, R., Anandhababu, G. et al. (2018). *ACS Energy Lett.* 3: 1854–1860.
40 Kang, P., Zhang, S., Meyer, T.J., and Brookhart, M. (2014). *Angew. Chem. Int. Ed.* 53: 8709–8713.
41 Motoo, S. and Furuya, N. (1986). *J. Electroanal. Chem.* 197: 209–218.
42 Xi, Z., Erdosy, D.P., Mendoza-Garcia, A. et al. (2017). *Nano Lett.* 17: 2727–2731.
43 Feng, L., Chang, J., Jiang, K. et al. (2016). *Nano Energy* 30: 355–361.
44 Xi, Z., Li, J., Su, D. et al. (2017). *J. Am. Chem. Soc.* 139: 15191–15196.
45 Wang, Y., Choi, S.I., Zhao, X. et al. (2014). *Adv. Funct. Mater.* 24: 131–139.

46 Cui, Z., Yang, M., and DiSalvo, F.J. (2014). *ACS Nano* 8: 6106–6113.
47 Kang, Y.Q., Xue, Q., Zhao, Y. et al. (2018). *Small* 14: 1801239.
48 Wu, P., Huang, Y., Kang, L. et al. (2015). *Sci. Rep.* 5: 14173.
49 Jia, J., Seitz, L.C., Benck, J.D. et al. (2016). *Nat. Commun.* 7: 13237.
50 Zhu, J.-Y., Chen, S., Xue, Q. et al. (2020). *Appl. Catal., B* 264: 118520.
51 Ali, A. and Shen, P.K. (2019). *J. Mater. Chem. A* 7: 22189–22217.
52 Huang, Y., Cai, J., and Guo, Y. (2012). *Int. J. Hydrogen Energy* 37: 1263–1271.
53 Huang, Y., Cai, J., Zheng, S., and Guo, Y. (2012). *J. Power Sources* 210: 81–85.
54 Zhang, K., Wang, C., Bin, D. et al. (2016). *Catal. Sci. Technol.* 6: 6441–6447.
55 Arulmani, S., Krishnamoorthy, S., Wu, J.J., and Anandan, S. (2017). *Electroanalysis* 29: 433–440.
56 Saranya, D. and Selvaraj, V. (2018). *Int. J. Hydrogen Energy* 43: 13450–13461.
57 Liu, D., Lu, W., Wang, K. et al. (2016). *Nanotechnology* 27: 44LT02.
58 Pattanayak, P., Pramanik, N., Kumar, P., and Kundu, P.P. (2018). *Int. J. Hydrogen Energy* 43: 11505–11519.
59 Yang, W., Yang, X., Jia, J. et al. (2019). *Appl. Catal., B* 244: 1096–1102.
60 Dong, B., Li, W., Huang, X. et al. (2019). *Nano Energy* 55: 37–41.
61 Gong, W., Jiang, Z., Wu, R. et al. (2019). *Appl. Catal., B* 246: 277–283.
62 Long, X., Yin, P., Lei, T. et al. (2020). *Appl. Catal., B* 260: 118187.
63 Cui, X., Zhu, Y., Hua, Z. et al. (2015). *Energy Environ. Sci.* 8: 1261–1266.
64 Van Dao, D., Le, T.D., Adilbish, G. et al. (2019). *J. Mater. Chem. A* 7: 26996–27006.
65 Zhang, T., Sun, Y., Li, X. et al. (2020). *Small Methods* 4: 1900709.
66 Wang, W., Chen, X., Zhang, X. et al. (2020). *Nano Energy* 17: 104623.
67 Tan, Q., Shu, C., Abbott, J. et al. (2019). *ACS Catal.* 9: 6362–6371.
68 Yin, Z., Peng, H., Wei, X. et al. (2019). *Energy Environ. Sci.* 12: 2455–2462.
69 Wang, L., Wu, W., Lei, Z. et al. (2020). *J. Mater. Chem. A* 8: 592–598.
70 Liu, K., Wang, W., Guo, P. et al. (2019). *Adv. Funct. Mater.* 29: 1806300.
71 Lv, H., Wang, Y., Lopes, A. et al. (2019). *Appl. Catal., B* 249: 116–125.
72 Liu, J., Luo, Z., Li, J. et al. (2019). *Appl. Catal., B* 242: 258–266.
73 Guo, J., Chen, R., Zhu, F.-C. et al. (2018). *Appl. Catal., B* 224: 602–611.
74 Chen, D., Sun, P., Liu, H., and Yang, J. (2017). *J. Mater. Chem. A* 5: 4421–4429.
75 Sharma, P., Radhakrishnan, S., Khil, M.-S. et al. (2018). *J. Electroanal. Chem.* 808: 236–244.
76 Nitopi, S., Bertheussen, E., Scott, S.B. et al. (2019). *Chem. Rev.* 119 (12): 7610–7672.
77 Wang, H., Jiang, D., Huang, D. et al. (2019). *J. Mater. Chem. A* 7: 22848–22870.
78 Liang, Z., Song, L., Deng, S. et al. (2019). *J. Am. Chem. Soc.* 141: 9629–9636.
79 Zhao, S., Guo, S., Zhu, C. et al. (2017). *RSC Adv.* 7: 1376–1381.
80 Dutta, A., Adhikary, R., Broekmann, P., and Datta, J. (2019). *Appl. Catal., B* 257: 117847.

27

Evaluation and Management of ECC

27.1 Basic Performance Index

For evaluating the overall performance of electrochemical carbon cycle (ECC), here we are going to propose some basic metrics involving cycle current (I_{cycle}), voltage gap (ΔV), and cycle Faradaic/energy efficiency (FE_{cycle} and EE_{cycle}). Their physical significance and computational formula are introduced further.

Current (I) is an important parameter to determine the electrochemical reaction rate. The cycle current in ECC is defined as the smaller current between oxidation and reduction processes, as determined by $I_{cycle} = \left[I_{reduction/electrolyzer}, I_{oxidation/fuel\ cell}\right]_{min}$. As thus, the cycle current represents the rate bottleneck of ECC. Larger I_{cycle} represents a higher operation rate of ECC. On the other hand, it potentially provides a driving force for electrochemical reactions in the ECC. Additional overpotentials are usually required to overcome the reaction energy barrier and achieve desired redox reaction rates. ΔV is defined by $\Delta V = V_{electrolyzer} - V_{fuel\ cell}$ between oxidation and reduction processes at specific currents. Regarding the specific product, a smaller ΔV indicates less energy loss during ECC.

Faradaic efficiency (FE) signifies the selectivity toward a specific product during the reactions, and it can be evaluated from the charge ratio or mole number ratio of the specific product in all equivalent counterparts, with the calculated equation of $FE = \alpha nF/Q = mole_{product}/mole_{equivalent\ product\ of\ all\ electron}$ [1], in which α is the electron transferred number, n is the product mole number, and F is Faraday's constant of $96\,485\ C\ mol^{-1}$, Q is the total charge. Regarding a specific product in ECC, the FE_{cycle} can be represented by a lower FE between oxidation and reduction processes, as defined by $FE_{cycle} = \left[FE_{oxidation}, FE_{reduction}\right]_{min}$. Cycle energy efficiency defines the ratio between energy output (E_{output}) and total energy input (E_{input}) during ECC, as expressed by $EE_{cycle} = E_{output}/E_{input}$. The energy output involves electricity output from fuel cell, the chemical conversion, CO_2 capture, and fuel transport, while the latter three are still hard to evaluate. In addition, optimization of electrocatalysts in both fuel cell and electrolyzer could render low voltage gap, high cycle current, FE_{cycle}, and EE_{cycle}.

Electrocatalysis in Balancing the Natural Carbon Cycle, First Edition. Yaobing Wang.
© 2021 WILEY-VCH GmbH. Published 2021 by WILEY-VCH GmbH.

27.2 CO_2 Capture and Fuel Transport

CO_2 capture and fuel transport are necessary links that form a practical ECC together, besides electrochemical CO_2 reduction reaction (ECDRR) and fuel electrooxidation that implement key chemical conversion coupled with energy transfer. Capturing CO_2 from the air (e.g. ~400 ppm CO_2) or concentrated emission sources (e.g. 10–20% CO_2 in industrial exhaust gas) is an essential process prior to ECDRR, commonly followed by desorption and compression, which require high energy input [2]. Moreover, designing sorbents is the most challenging step, including cost, capacity, selectivity, stability, recyclability, and fast kinetics [3]. Toward this, different materials have been proposed as promising structures to capture CO_2, such as metal–organic frameworks (MOFs) [4], covalent triazine frameworks (CTFs) [5], and POPs [6]. However, large-scale CO_2 capturing technology is still under development [7]. On the other hand, there have been studies that directly reduced captured CO_2 but omitting the storage [8] and transport processes [9]. Once collected from the ECDRR electrolyzer, fuels are prepared to transport to widely spread destinations. Gaseous fuels, such as CO (or syngas) and CH_4, can be transported by three main methods: (i) directly transported by existed pipelines over land, which is efficient up to a few thousand kilometers. (ii) Compressed to a liquid state for oversea transport, which may up to 10 000 km. (iii) Chemically converted into liquid products for cheaper and safer transport [10]. Liquid fuels, including HCOOH, CH_3OH, and CH_3CH_2OH, can be mainly transported by existed pipelines for long-distance and tank car for short-distance overland. Metal-CO_2 batteries are also considered as a function cell for ECC, which can fix CO_2 into chemicals accompanied with electricity storage [11]. However, the relatively high weight of them results in a more possible application saturation without frequent movement and transport.

27.3 External Management

Beyond electrocatalyst optimizations, there are several strategies or apparatuses required in association with external management of ECC toward maximization of overall performance:

(a) Continuous operation mode of ECC is essential. To realize this mode, three aspects of work can be done: First, the CO_2 electrolyzer in ECC requires electricity input, which is converted from renewable energy such as wind and solar. These sources have the features of intermittent distribution in space and time. Therefore, mediation of different renewable energy sources or utilization of mature large-scale stored energy devices would guarantee electricity supply. Second, the fuel cell is supposed to be continuously operated to avoid the design of time-varying biases and achieve the capital utilization of 100% [12]. Finally, the efficiencies of CO_2 capture and fuel transportation are also required to be sufficient to ensure the reactant supplies in the ECC system.

(b) Flowing fuel cell and electrolyzer [13–15], which enable direct supplying of gaseous reactant and facilitate removal of gaseous/liquid products to increase the contact of reactant and catalyst active sites, are the ideal device type for ECC toward its highly efficiency operation. Thus, actuating devices such as peristaltic pumps must ensure the continuous supply of reactant and electrolyte flow.
(c) An external management apparatus and a controller for ECC are required. The external management apparatus is used to acquire information about the current/voltage of fuel cell and electrolyzer, flowing rate of gas and electrolyte, operation temperature, and so on. Control information is then generated and transmitted to the controller to tune the operation state of fuel cell and electrolyzer to maintain the supply–consumption balance of each link in ECC, with maximum electricity and chemical output.
(d) Product separation devices will also be required if a variety of products beyond the targeted product are generated from CO_2 electrolysis and fuel electrooxidation processes.

In summary, this chapter presented the overview of ECC to supplement natural carbon cycle (NCC) by discussing the basic principles of the cycle and introducing the overall performance index as well as the external management optimization of ECC. With focusing on the advances of CO_2 reduction and fuel oxidation in ECC, current requirements to form the efficient CO_2–CO/CH_4/$HCOOH$/CH_3OH/CH_3CH_2OH cycle were also proposed. In general, the ECC is still at a very early stage with great challenges remaining on the way to its practical applications. Herein we have summarized some primary challenges and the related tactics to adopt.

First, the limited intrinsic activity and selectivity of electrocatalysts for CO_2 reduction and fuel oxidation should be further improved. According to the advances summarized earlier, some promising electrocatalysts beneficial for the four kinds of cycle reaction can be concluded here: Au, Pd-based catalysts and N-doped carbons for CO_2 reduction to CO; Pb, Bi, In, and Pd-based catalysts for CO_2 reduction to HCOOH; Cu-based catalysts for CO_2 reduction to CH_3OH and CH_3CH_2OH; PtRu and PtSn-based catalysts for CO oxidation to CO_2; Pd-based catalysts for HCOOH oxidation to CO_2; Pt-based catalysts for methane and methanol oxidation to CO_2; and PtRh-based catalysts for ethanol oxidation to CO_2. Design of noble/nonnoble metal active sites for stabilizing *COOH and *OCHO intermediates would help the production of CO and HCOOH from CO_2 reduction, respectively. Tuning Cu active site to stabilize intermediate of *CHO formation and *CO for dimerization facilitates the production of CH_3OH and CH_3CH_2OH, respectively. Control noble metal active sites and the adjacent sites with high dehydrogenation capacity, weak CO adsorption strength, and easy water activation to generate $-OH_{ad}$ species are beneficial for CO, CH_4, HCOOH, and CH_3OH oxidation. Additional electronic structure modulation on Pt or Rh to facilitate C–C cleavage will promote CH_3CH_2OH oxidation to CO_2.

Second, the highly stable electrocatalysts/electrodes are indispensable for the long-term operation of ECC. Some strategies for enhancing the stability of electrocatalysts/electrodes are proposed: (i) selecting the relatively stable catalysts in an electrochemical environment. For supported catalysts, strong interactions between active phases (e.g. Pt) and supports would increase the stability; (ii) using the binder-free electrode to maximize surface active sites and avoid the binder decomposition and performance degradation; (iii) limiting the operation temperature and flowing rate of electrolyte in ECC, which would reduce the levels of surface reconstruction of catalysts, or peeling off from electrode; (iv) stabilizing the outer power input/output, especially avoiding the frequent on–off operation, to endow the electrocatalysts/electrodes a stable operation voltage.

Third, the overall performance of ECC has not been investigated yet by now, which remains a huge room for optimization. As suggested earlier, the flowing cells for CO_2 reduction and fuel oxidation are needed for increasing ECC performance. The CO_2 capture and fuel transport are not required to be continuous, whereas they should meet the demand of supplying reactant for flowing cells. A dynamic equilibrium among these four links in ECC is better to keep to minimize energy loss. To realize this equilibrium, the factors of current/voltage of fuel cell and electrolyzer, flowing rate of gas and electrolyte, operation temperature, and so on should be taken into full consideration. A professional management module for this is requisite.

Forth, formic acid and methanol have relatively high energy density and ease to the storage, and most importantly, give high feasibility for complete interconversion with CO_2 at high efficiency. Therefore, they exhibit the most promise to propel the ECC applications for supplementing NCC in the future.

Finally, a general outlook has been mentioned as a quick summary of the presented book, which can help the electrocatalytic carbon community for further developments and discoveries.

27.4 General Outlook

This book presented an overview of ECC and the related electrocatalysis reactions to supplement NCC. First, the natural carbon cycle and anthropogenic carbon cycle were introduced. The basic electrochemical catalysis process was discussed, following by a detailed discussion on each kind of electrocatalytic reaction and the electrochemical energy devices correspondingly. These electrocatalytic reactions include hydrogen oxidation reaction (HOR), hydrogen evolution reaction (HER), oxygen evolution reaction (OER), oxygen reduction reaction (ORR), small organic molecule oxidation, carbon dioxide reduction reaction (CDRR), which consisted of the water-splitting device, various fuel cells, and CO_2 electrolysis. The density functional theory (DFT) calculations and in situ characteristic techniques were also discussed to support these electrocatalysis toward understanding mechanism, active sites, intermediates interaction, and so on. Then, the advances of CO_2 reduction and fuel oxidation in ECC were focused, and the basic principles of the four links of ECC, the overall performance index, the external management optimization were discussed. Generally, although there are still some challenges on the way to supplement NCC by ECC, it also shows great promise. Herein we have summarized the primary challenges and propose the development direction supported by related tactics to adopt.

The biggest problem of the ECC and electrocatalysis reactions is it the poor performance of these processes, which has strongly related to the electrocatalysts, electrode design, electrolyte selection, and operation conditions such as temperature and pH. Among these alternative factors, the most crucial factor is to develop highly efficient, cost-effective electrocatalysts with long stability. The design of catalysts can be done by modifying crystalline, surface, and electronic structures, as well as the composition. The larger surface area and more exposed catalytic active sites and optimized pore structures are generally favorable for the diffusion of the reagents in the electrolyte. The understanding of the active sites of electrocatalysts, intermediates, and surface interactions by theoretical calculations and in situ characterization techniques would promote to survey the fundamental mechanism of these electrocatalysis processes (CDRR, small organic oxidation, HOR, OER, ORR, and HER), ultimately favoring the electrocatalyst design.

As for the hydrogen-related electrocatalytic reactions such as HER and HOR, previous metal catalysts, Pt-based catalysts are still the best choice. Meanwhile, a lot of the nonprevious metal catalysts were designed based on Mo, W, Co, Ni elements, and so on, by alloying with a second or more components. Although some HER/HOR mechanisms on these kinds of electrocatalysts were not fully understood, the general view is that the reaction kinetics is related to the H binding energy on the catalysts. In detail, the dissociative adsorption of H_2, electron transfer from H_2 to catalyst, and discharge of adsorbed H atom could be the rate-determining steps and should be emphatically considered. Besides, the adsorbed species of H_2 are different in acid (only hydrogen) and alkaline (hydrogen and hydroxide) media, which would influence the reaction mechanism of hydrogen-related reactions. The catalysts, which have a high affinity for H adsorption, the balance between H and OH adsorption, and high electron transferability, are recommended. To apply these electrocatalyst on a large scale, the ultralow content of precious metal catalysts and the nonprecious metal catalysts, together with their scaling-up synthesis methods, should be exploited. As for the oxygen-related electrocatalytic reactions (ORR and OER), the interactions between key intermediates of OH* and OOH* and active sites during the four-step reaction processes should be considered for the design of efficient electrocatalysts. Although previous metal Pt and Ir/Ru electrocatalytic materials have excellent activities for oxygen-related reactions, many nonprecious metal catalysts (e.g. Fe, Co, Ni-based catalysts and metal-N-containing materials) also developed, which also show outstanding oxygen-related performance comparable to the precious metals. However, it is noted that most of these nonprecious electrocatalysts worked, preferably in alkaline media. Another big challenge for the nonprecious metal catalysts is their stability. Water splitting and H_2–O_2 fuel cells are based on these hydrogen/oxygen-involved reactions, together with aqueous metal-air batteries. These techniques consisted of the next-generation energy conversion technologies on the basis of the high specific energy density. When considering their applications in electric vehicles, smart grids, and portable electronic devices, etc., the mechanical stability during long-term operation should be further improved. It needs a comprehensive optimization of the catalysts, membrane, electrodes, and device structure and operation conditions.

For the oxidation of small organic molecule, noble metal Pt and Pd-based materials are the best catalysts hard to replace by other nonprevious metals and nonmetal materials. Various poisoning intermediates, e.g. CO, are strongly adsorbed on the metal surface and are the main reason for the performance decay during C_1 organic molecule oxidation. For the multicarbon fuels, most catalysts cannot completely convert them into CO_2, which also reduces their electrocatalytic efficiency. Controlling the composition and nanometer/microstructure of the electrocatalyst is the key factor in improving its electrocatalytic activity and stability. In addition, combined with optimization, the hybrid catalyst shows great potential in fuel cell applications. However, there is a gap between the performance of the electrocatalyst characterized well in the three-electrode electrochemical test and the practical application. Besides, the total cost of fuel cells must be reduced, especially the cost of anodic catalysts and membranes. Despite these challenges, there is still much room for improvement in fuel cell performance by developing high-performance, low-cost electrode materials and membranes and optimizing device structure and operating parameters. To promote the applications of CDRR, many factors should be considered, such as electrolyte (type, concentration, and pH/cations/anions in the electrolyte) and electrochemical conditions (temperature and pressure). However, the electrocatalysts are still the main factor for control. A wide range of electrocatalysts could be utilized toward different CDRR products and requirements. For example, bulk metals (Au, Ag, Pd, Cu, and so on) were considered as an electrocatalyst where different products (C^1 and C^{2+}) have been reported for ECDRR but usually with high overpotentials and low yield. On the other hand, nanoscale materials (nanoparticles [NPs], nanocubes [NCs], nano diamonds [NDs], and so on) revealed superior catalytic activity/selectivity for ECDRR compare to bulk metal, while with the simple product of CO. Carbon-based electrocatalysts such as single-atom catalysts (SACs) and heteroatom-doped materials could also produce CO and at high FE, while their long-term durability of M–N–C catalysts in aqueous electrolyte should be further improved. As for the whole device optimization, it is necessary to promote the application from lab-used H-type cells to the large current-applied flow-type cell and overcome the intrinsic disadvantages such as "flooding" of gas diffusion electrodes (GDEs) when liquid electrolyte flows in the flow-type cells. Overall, the electrocatalysts, electrode, and device are the key factors in the CDRR electrolysis devices: (i) new electrocatalysts with high activity and stability, such as highly porous MOFs and covalent organic frameworks (COFs) materials, Boron-Nitrogen co-doped diamond (BND) and N-doped graphene quantum dots (NGQDs), a various 2D matrix such as Mxenes (carbides and nitrides), black phosphorus (BP), and Boron doped Nitrogen (BN)-decorated SACs are worth to further developed; (ii) engineering electrode design, especially long-term stable electrode design, especially to avoid leakage of GDEs; besides, selection of the relatively stable catalysts in the electrochemical environment, utilization of the binder-free electrode to maximize surface active sites and avoid the binder decomposition, and performance degradation; (iii) In various fuel-type cells such as suitable proton exchange membrane (PEM) and microfluidic structure, GDE, solid oxide electrolysis cell (SOEC), and differential electrochemical mass spectrometry (DEMS) cells need to be considered carefully for further lab-scale to industrial application. Additionally, the limitation of the operation temperature, flowing rate of electrolyte in the device, and stabilization of the outer power input/output are needed to achieve a high over performance.

Theoretic calculations and in situ characterizations are of considerable value for designing highly efficient electrocatalysts based on the mechanism explanation and activity prediction. The computing-related and simulating-related challenges of theoretic calculations and the complexity and low applicability of in situ techniques should be further overcome before their wide applications. With unifying these complex environment influence, and looking forward, it seems that creating an accessible database that indicates the material structure, reaction mechanism, interaction between active sites and intermediates, from the basic data of theoretic calculations and in situ characterization, and combined with the machine learning, would potentially provide effective catalyst design information to a new era. The combination of theoretic calculations and in situ characterization would create important opportunities to have a major impact in the future of electrocatalyst design.

To realize the ECC for efficiently supplement NCC, the processes of CO_2 capture and fuel transport are also important. These processes are not required to be continuous, whereas they should meet the demand of supplying reactant for flowing cells. A dynamic equilibrium among these four links in ECC is better to keep to minimize energy loss. For this equilibrium requirement, the factors of various operation conditions of fuel cell and electrolyzer, and other matched devices, should be taken full consideration. A professional management module for this is requisite. Among the proposed reactant/product enabling ECC, formic acid and methanol with relatively high energy density and ease to the storage, and most importantly, high feasibility for complete interconversion with CO_2 at high efficiency, exhibit the most promise to propel the ECC applications for supplementing NCC in the future.

References

1 Xie, J., Huang, Y., Wu, M., and Wang, Y. (2019). *ChemElectroChem* 6: 1587–1604.
2 Sanz-Perez, E.S., Murdock, C.R., Didas, S.A., and Jones, C.W. (2016). *Chem. Rev.* 116: 11840–11876.
3 Patel, H.A., Byun, J., and Yavuz, C.T. (2017). *ChemSusChem* 10: 1303–1317.
4 Lin, Y., Kong, C., Zhang, Q., and Chen, L. (2017). *Adv. Energy Mater.* 7: 1601296.
5 Wang, H., Jiang, D., Huang, D. et al. (2019). *J. Mater. Chem. A* 7: 22848–22870.
6 Wang, W., Zhou, M., and Yuan, D. (2017). *J. Mater. Chem. A* 5: 1334–1347.
7 Yuan, Z., Eden, M.R., and Gani, R. (2016). *Ind. Eng. Chem. Res.* 55: 3383–3419.
8 Aminu, M.D., Nabavi, S.A., Rochelle, C.A., and Manovic, V. (2017). *Appl. Energy* 208: 1389–1419.
9 Kar, S., Goeppert, A., and Prakash, G.S. (2019). *Acc. Chem. Res.* 52: 2892–2903.
10 Shih, C.F., Zhang, T., Li, J., and Bai, C. (2018). *Joule* 2: 1925–1949.

11 Xie, J. and Wang, Y. (2019). *Acc. Chem. Res.* 52: 1721–1729.
12 Zhao, L., Dong, B., Li, S. et al. (2017). *ACS Nano* 11: 5800–5807.
13 Lin, R., Guo, J., Li, X. et al. (2020). *Catalysts* 10: 473.
14 Brée, L.C., Wessling, M., and Mitsos, A. (2020). *Comput. Chem. Eng.* 139: 106890.
15 Noël, T., Cao, Y., and Laudadio, G. (2019). *Acc. Chem. Res.* 52: 2858–2869.

Index

Electrocatalysis in Balancing the Natural Carbon Cycle, First Edition. Yaobing Wang.
© 2021 WILEY-VCH GmbH. Published 2021 by WILEY-VCH GmbH.

f

g

Z